AF412792

Pathophysiology of Endotoxin

Handbook of Endotoxin

Series Editor: R.A. PROCTOR
University of Wisconsin Medical School, Madison, Wisconsin, U.S.A.

VOLUME 2

Pathophysiology of Endotoxin

Editor:

L.B. HINSHAW

University of Oklahoma Health Sciences Center,
Veterans Administration Medical Center,
Oklahoma City, U.S.A.

1985

ELSEVIER

Amsterdam – New York – Oxford

ISBN 0 444 90385 2

Published by:
Elsevier Science Publishers B.V.
P.O. Box 1126
1000 BC Amsterdam

Sole distributors for the USA and Canada:
Elsevier Science Publishing Co. Inc.
52 Vanderbilt Avenue
New York, NY 10017

Typeset in The Netherlands by Pecasse Intercontinental B.V., Maastricht
Printed in The Netherlands by Casparie, Amsterdam

Contributors

DR. H.R. ADAMS
Department of Veterinary Bio-
  medical Sciences
College of Veterinary Medicine
University of Missouri
Columbia, MO 65211
U.S.A.

DR. E. BALISH
Gnotobiotic Research Laboratory
University of Wisconsin Medical
  School
Medical Sciences Center, Room
  4638
1300 University Avenue
Madison, WI 53706
U.S.A.

DR. C.R. BAXTER
Department of Surgery
University of Texas Health Science
  Center
5323 Harry Hines Boulevard
Dallas, TX 75235
U.S.A.

DR. R.F. BOND
Department of Physiology
Oral Roberts University
7777 South Lewis Avenue
Tulsa, OK 74171
U.S.A.

DR. C.G. COCHRANE
Department of Immunology
Scripps Clinic and Research
  Foundation
10466 North Torrey Pines Road
La Jolla, CA 92037
U.S.A.

DR. D.B. COURSIN
Departments of Anesthesiology
  and Internal Medicine
University of Wisconsin Medical
  School
B6/387 Clinical Science Center
600 Highland Avenue
Madison, WI 53792
U.S.A.

DR. T.E. EMERSON Jr
Cutter Group of Miles
  Laboratories Inc.
Physiology Research Department
Fourth and Parker Streets
Berkeley, CA 94710
U.S.A.

DR. D.P. FINE
Infectious Diseases Section IIIc
Veterans Administration Medical
  Center
921 Northeast 13th Street
Oklahoma City, OK 73104
U.S.A.

DR. J.T. FLYNN
Department of Physiology
Jefferson Medical College
Thomas Jefferson University
Philadelphia, PA 19107
U.S.A.

DR. S.E. GREISMAN
Departments of Medicine and
 Physiology
University of Maryland School of
 Medicine
22 South Greene Street
Baltimore, MD 21201
U.S.A.

DR. N.J. GURLL
Department of Surgery
University of Iowa Hospitals and
 Clinics
Iowa City, IA 52242
U.S.A.

DR. L.B. HINSHAW
Department of Physiology and
 Biophysics
Department of Surgery
University of Oklahoma Health
 Sciences Center
Veterans Administration Medical
 Center
921 Northeast 13th Street
Oklahoma City, OK 73104
U.S.A.

DR. G.W. HUNNINGHAKE
Department of Internal Medicine
University of Iowa Hospitals and
 Clinics
Iowa City, IA 52242
U.S.A.

DR. S.D. IZENBERG
Department of Pharmacology
University of Texas Health Science
 Center
Basic Sciences Research Building
5323 Harry Hines Boulevard
Dallas, TX 75235
U.S.A.

DR. E.R. JACOBS
Department of Medicine
Rush-Presbyterian – St. Lukes
 Medical Center
1753 West Congress Parkway
Chicago, IL 60612
U.S.A.

DR. C.A. JOHNSTON
Department of Pathology
University of Maryland School of
 Medicine
22 South Greene Street
Baltimore, MD 21201
U.S.A.

DR. J.R. PARRATT
Department of Physiology and
 Pharmacology
University of Strathclyde
Royal College
204 George Street
Glasgow, G1 1XW
U.K.

DR. D.M. SHASBY
Department of Internal Medicine
University of Iowa Hospitals and
 Clinics
Iowa City, IA 52242
U.S.A.

DR. J.A. WILL
Department of Veterinary Sciences
College of Agricultural and Life
  Sciences
Department of Anesthesiology
University of Wisconsin Medical
  School
B6/387 Clinical Science Center
600 Highland Avenue
Madison, WI 53792
U.S.A.

Contents

Contents

Chapter 3
Peripheral circulatory responses to endotoxin
R.F. Bond

Chapter 4
Endotoxin and the lung
J.A. Will and D.B. Coursin

Chapter 5
Endotoxin-induced pulmonary leukostasis
D.M. Shasby and G.W. Hunninghake

Chapter 6
Role of complement in endotoxin shock
D.P. Fine

Chapter 7
Adrenergic aspects of endotoxin shock
H.R. Adams, S.D. Izenberg and C.R. Baxter

Chapter 8
Release and vascular effects of histamine, serotonin, angiotensin II and renin following endotoxin
T.E. Emerson Jr.

Chapter 9
The role of arachidonic acid metabolites in endotoxin shock I: lipoxygenase products
J.A. Parratt

Chapter 10
The role of arachidonic acid metabolites in endotoxin shock II: involvement of prostanoids and thromboxanes
J.T. Flynn

Chapter 11
The contact system in septic shock
C.G. Cochrane

Chapter 12
Endorphins in endotoxin shock
N. J. Gurll

Contents

Chapter 13
Endotoxin effects on germfree animals
E. Balish

Chapter 14
Mechanisms of endotoxin tolerance
C.A. Johnston and S.E. Greisman

Introduction

The key objective of this volume is to identify the multitude of pathophysiological mechanisms elicited by the introduction of endotoxin into the vascular systems of animals and humans. The myriad effects of endotoxin on the body became apparent when its administration to several species of animals gained importance in the years 1950-1965. Surgeons, internists and physiologists among others, working independently or in collaborative groups, steadily enlarged the scope of research in animals, stimulated in the first place by the seriousness of the clinical entity of 'septic shock' in man. The early workers quickly recognized parallels between their clinical observations of septic patients and the responses of dogs, rats, nonhuman primates and other species of animals when endotoxin was introduced into vascular systems of animals.

The study of endotoxin shock in animals has always been intriguing and challenging to the investigator. Virtually any study involving the use of endotoxin yields positive results, and the interplay between clinician and basic scientist grows steadily stronger, for the purpose of determining whether the results of 'animal shock models' are applicable to the response of the human patient with 'severe sepsis.' A strong consensus held currently is that the animal models have adequately served in providing an understanding of the many pathophysiological mechanisms at work in the human patient during early and late sepsis. Many investigators working in the more basic fields of research including microbiology, pharmacology, immunology and biochemistry, have pursued avenues of study designed to identify mechanisms of endotoxin action as important studies in their own right not necessarily having any obvious relevance to that which is occurring in the human septic patient.

It now seems to be more evident than ever that the most basic research imaginable in the area of endotoxin investigation has unavoidable clinical implications. It may be remembered that thirty years ago Professor Wesley W. Spink of the University of Minnesota commented to his colleagues that if one could only understand the many complex actions and reactions occurring during the first thirty seconds after endotoxin has been injected into the vascular bed of a dog, one might well have the answer to the pathogenesis of septic shock in man. At the same time and place, it was Professor Maurice B. Visscher who said that the problem of studying endotoxin shock is to get at the central core of actions following the administration of endotoxin into animals and to be able to sort out the primary factors and phenomena from the secondary and tertiary ones.

It was Professor Spink who set the stage for thirty years of subsequent research when he proposed that antibiotic administration to patients with large numbers of circulating gram-negative microorganisms might release endotoxin from the organisms, resulting in shock and death, and he also emphasized the view that septic shock in man was primarily due to release of endotoxin. Since that time many workers have pursued studies using live organisms or endotoxin, believing that either or both may perform vital roles in the pathogenesis of shock.

One of the purposes of this volume is to update the reader in traditional and new approaches which have been developed to better describe, define and evaluate the numerous mechanisms at work when endotoxin, its active chemical components, and live organisms enter the vascular system and set into play a myriad of responses which may function as separate entities or, more likely, in interaction with one another. What events occur and what ultimately happens to the animal or human patient seem to vary with the dosage, species and means of eliciting shock, but it is becoming increasingly apparent that a host of mechanisms can be identified which interlock in various ways, accounting for the complex pathophysiological condition we term 'endotoxin shock' or 'septic shock.' Investigators are also expected to follow courses of studies involving lesser degrees of challenge to animal models than those which elicit shock. The latter studies should yield valuable information about the animal's means of responding to endotoxin, its components and products.

It is hoped that the reader will find this volume of importance for its significance as a general review vehicle; as a reference text; and as a source of stimulation for currently conducted and proposed research. The volume may be enjoyed and appreciated in its own right as a source of wonderment that such strange and complex pathophysiological mechanisms even occur at all. An ultimate goal should be to distinguish factors which occur after endotoxin administration that are actually 'protective' from those that are 'detrimental' in order to better understand the problems that a living organism must deal with when confronted with this type of foreign invader.

Finally, it is believed that the most exciting research is yet to come: there is presently a multitude of young, capable investigators who will contribute greatly to our understanding of the pathophysiology of endotoxin in the years to come.

LERNER B. HINSHAW

Handbook of Endotoxin, Vol. 2: Pathophysiology of Endotoxin
L.B. Hinshaw, editor
© Elsevier Science Publishers B.V., 1985
ISBN 0 444 90385 2
$0.85 per article per page (transactional system)
$0.20 per article per page (licensing system)

CHAPTER 1

Overview of mediators affecting pulmonary and systemic vascular changes in endotoxemia

ELIZABETH ROGERS JACOBS

1. INTRODUCTION

Septic lung injury remains a major cause of respiratory failure and death among hospitalized patients despite early and meticulous application of appropriate antimicrobial drugs and sophisticated supportive therapy (Fulton and Jones 1975; Kaplan et al 1979; Zapol et al 1979). The respiratory injury in such cases is characterized by refractory hypoxemia with shunting, diminished pulmonary compliance, and the accumulation of extravascular lung water secondary to an increase in pulmonary microvascular permeability (Ashbaugh et al 1967; Staub 1974). Pulmonary hypertension frequently accompanies acute respiratory failure (Fairman and Glauser 1980, Sibbald et al 1978). The contributions of several blood elements, lipids and peptides towards the systemic hypotension and functional abnormalities of septic lung injury are recognized, but a more complete understanding of the pathophysiologic sequence leading to respiratory failure in this condition should suggest means of intervention with specific therapies which may block or inhibit pulmonary injury.

Infusion of *Escherichia coli* endotoxin in experimental animals reproduces many of the features of clinical respiratory failure including hypotension, alteration of pulmonary vascular tone, airway tone and microvascular injury. (Brigham et al 1979; Esbenshade et al 1982). Such a model of septic respiratory failure, although not identical in all respects to adult respiratory distress syndrome (ARDS), allows investigators to examine the sequence of hemodynamic and functional changes in endotoxemia. Pulmonary microvascular injury can be quantitated and followed temporally by measurements and characterization of lung lymph flow in sheep and by use of double-indicator techniques to evaluate extravascular lung water. Elevated levels of proposed mediators in serum and pulmonary lymph in animal models of endotoxemia, the known biologic activities of these compounds, and the absence or amelioration of the hemodynamic or functional abnormality in the presence of specific inhibitors, form the basis for presuming that several compounds mediate pulmonary vascular or microvascular alterations of endotoxemia. This chapter will attempt to review evidence implicating proposed mediators of pulmonary hypertension, pulmonary microvascular injury, airway abnormalities and systemic hypotension associated with endotoxemia.

2. PULMONARY HYPERTENSION

2.1. Arachidonic acid metabolites

2.1.1. Cyclooxygenase metabolites

A major nonrespiratory function of the lung is synthesis and metabolism of prostaglandin products (Ferreira and Vane 1967; Piper and Vane 1971). The pulmonary vascular endothelium is a rich source of prostacyclin (PGI_2) in the lung (Gryglewski 1978). Thromboxane A_2 is released by the pulmonary arteries and platelets (Moncada and Amezcua 1979; Salzman et al 1980). A marked increase in the serum or lymphatic levels of thromboxane A_2, PGI_2, PGF_2, and PGE_2 follows endotoxin infusion in nearly every species of experimental animal tested (Anderson et al 1975; Cook et al 1980; Demling et al 1981).

The effects of cyclooxygenase products on the pulmonary vascular tone vary from intense vasoconstriction to vasodilation. Prostaglandin endoperoxides are unstable precursors of PGE_2 and PGF_2 and are highly potent vasoconstrictors (Alpert et al 1973; Weir and Grover 1978). Infusion of a stable ether analog of PGH_2 in sheep causes an increase in pulmonary artery pressure and vascular resistance (Bowers et al 1979). Thromboxane A_2 is also a vasoconstrictor and platelet aggregator (Samuelsson et al 1978). PGF_2 causes pulmonary vasoconstriction when infused into calves (Lewis and Eyre 1972). In contrast, prostacyclin is a pulmonary and systemic vasodilator and an inhibitor of platelet aggregation and thromboxane synthesis (Dusting et al 1979; Kadowitz et al 1977). Infusion of arachidonate into unanesthetized sheep causes a dose-related increase in pulmonary artery pressure and pulmonary vascular resistance. Inhibition of prostaglandin synthesis with sodium meclofenamate in conscious dogs elicits a pulmonary pressor response suggesting that vasodilator prostaglandins contribute to the normally low vascular tone in the pulmonary system (Walker et al 1982). Therefore synthesis of many vasoactive products of the cyclooxygenase pathway greatly increases following endotoxin infusion. Pulmonary artery tone and resistance probably reflect the sum of the effects of all prostaglandin metabolites. This balance of vasoactive arachidonic acid products shifts in favor of vaso-constrictor metabolites in endotoxemia (Demling et al 1981; Frolich et al 1980).

Prevention of pulmonary hypertension by treatment with specific blockers or inhibitors of the cyclooxygenase pathway provides additional support for the role of these compounds in mediating hemodynamic changes of en-dotoxemia. Indometacin and sodium meclofenamate administration abolishes the early pulmonary hypertension of endotoxemia in the cat (Par-ratt and Sturgess 1975a, 1975b). Ibuprofen treatment in sheep protects against early pulmonary hypertension and decreases the protein-poor lymph flow in endotoxemia (Adams and Traber 1982). Similarly, imidazole di-minishes pulmonary hypertension in canine endotoxin shock (Smith et al 1980). In addition to amelioration of pulmonary hypertension, the prosta-glandin inhibitor flunixen meglumine appears to attenuate the pulmonary vas-cular endothelial damage of endotoxin in ponies (Turek et al 1983). Prelimi-nary data suggest diminished accumulation of extravascular lung water in septic goats treated with indometacin or dazoxiben (Price et al 1983). Im-proved survival with cyclooxygenase inhibitors in several models of endoto-xin shock has been reported (Emerson et al 1983; Fletcher and Ramwell 1980; Wise et al 1981).

2.1.2. *Lipoxygenase metabolites*

Leukotrienes, products of the lipoxygenase pathway of arachidonic acid metabolism, are potent constrictors of airway and lung parenchyma. In addi-tion leukotriene D_4 causes profound coronary artery vasoconstriction in sheep (Michclassi et al 1982). A marked transient increase in pulmonary artery pressure in the monkey has been described following infusion of leukotriene C_4 (Smedegård et al 1982). Increased lymphatic levels of 5-HETE are found in sheep following endotoxin infusion (Ogletree et al 1981). The bulk of experimental evidence, however, supports a more important role for these lipoxygenase products in mediating increased pulmonary mic-rovascular permeability and bronchoconstriction than pulmonary hyperten-sion (see Section 2).

2.2. Serotonin

Serotonin (5-hydroxytryptamine) has long been recognized as a potent pul-monary vasoconstrictor and as a potential mediator of pulmonary hyperten-sion and increased vascular permeability (Gilbert et al 1958; Kabins et al 1959). Free plasma serotonin is increased following endotoxin injection (Davis et al 1960), and blockade of endotoxin-induced alteration of lung mechanics with methysergide has been described (Stein and Thomas 1967). Brigham and coworkers concluded that serotonin infusion into sheep caused an increase in lung lymph flow which could be explained by an increase in transmural pressure gradient alone without an increase in vascular permeability (Brigham and Owens 1975a). Recently the contribution of serotonin to the pulmonary response of endotoxemia was questioned after endotoxin infusion into dogs failed to elicit platelet release of radiolabeled serotonin (Murphy et al 1981).

2.3. Vasoactive peptides

There are at least eighteen and possibly more peptides known to occur in the lungs (Said 1983). To date their pathophysiologic role has not been established, but two of these neuropeptides, substance P and vasoactive intestinal polypeptide (VIP), have been suggested as mediators of pulmonary vascular tone in some circumstances. Substance P, found in the nerves supplying airways and pulmonary vessels (Dey and Said 1981), constricts guinea pig pulmonary artery in vitro (Said 1982). VIP, in contrast, relaxes airways and pulmonary vascular tissue (Said 1980). Serum levels of VIP are elevated in canine endotoxin shock (Bone et al 1980), suggesting that VIP possibly functions as a modulator of the pulmonary hypertension and bronchoconstriction of endotoxemia. Evidence increases to suggest that VIP may in fact mediate nonadrenergic noncholinergic bronchodilation (Hamasaki et al 1983).

3. MICROVASCULAR PERMEABILITY

The primary mechanism of septic respiratory failure is an increase in capillary endothelial and perhaps alveolar epithelial permeability which results in accumulation of protein-rich fluid in the interstitial and alveolar spaces of the lung. This injury reaction is most graphically demonstrated and quantitated in sheep with chronic lung lymph fistulae. In this model endotoxin administration results in an early transient phase of severe pulmonary hypertension and is followed by a second phase of normal or moderately elevated pulmonary vascular pressures and high flow rates of protein-rich lung lymph (Brigham et al 1979). Although morphologic alterations at this early stage of endotoxin-induced lung injury are absent by light microscopy, microvascular permeability to urea is increased (Harris and Brigham 1982). This section will review the evidence for several proposed mediators of the microvascular permeability injury of endotoxemia.

3.1. Complement system

The proteins of the complement system mediate inflammation and opsonization, actions which enhance the clearance of microorganisms. Polypeptides C3a, C4a and C5a generated in the sequence of complement activation increase vascular permeability, cause histamine release from mast cells, and stimulate smooth muscle concentration (Hugli and Muller-Eberhard 1978). C5a is also a potent chemoattractant for neutrophils (Fernandez et al 1975). Despite the obvious importance of an intact complement system to host defense against microorganisms, not all consequences of complement activation may be beneficial to the host. Intravascular complement activation results in a profound neutropenia (McCall et al 1974) secondary to sequestration of the leukocytes in the peripheral vasculature of the lungs (Henson et al 1982). Dialysis-associated leukopenia is believed to be related to complement activation and C5a generation (Craddock et al 1977a), and it is accompanied by increased pulmonary lymph flow and pulmonary dysfunction in both laboratory animals and patients (Craddock et al 1977b). In a guinea pig model of pneumococcal sepsis, genetically C5-deficient or complement-depleted guinea pigs did not localize bacteria in the lungs while control animals did (Hosea et al 1980). Infusion of activated C5 increased alveolar capillary permeability; neutropenic animals similarly treated demonstrated no such increase in permeability. Also, C5-deficient mice are protected against the early pulmonary localization of *Pseudomonas aeruginosa* seen in their C5-sufficient counterparts (Larsen et al 1982). Recent evidence suggests that C3a and C5a promote disruption of endothelial cells of the pulmonary artery in vitro in dogs (Glovskyl et al 1983). All of these data suggest an important role for complement in the microcirculatory injury of the septic lung. Indeed, complement activation in patients with gram-negative sepsis is associated with the development of hypotension and high mortality (Fearon et al 1975).

While the contribution of complement to the pulmonary microvascular injury of endotoxemia is widely accepted, most investigators believe that the subsequent release of toxic oxygen radicals from neutrophils is probably the final mediator of injury. Lung injury in rats with cobra venom factor-induced complement activation is significantly attenuated by neutrophil depletion or treatment with superoxide dismutase or catalase (Till et al 1982). Henson suggests that neutrophil sequestration and complement activation alone do not result in increased microvascular permeability in rabbit lungs without an additional insult, such as hypoxemia (Webster et al 1982).

3.2. Toxic oxygen radicals

The association of prolonged oxygen therapy with pulmonary toxicity has long been recognized. The products of oxygen reduction, or free radicals, are believed to mediate the cellular destruction in such circumstances. Most experts maintain that the pulmonary toxicity caused by oxygen radicals represents the balance between the rate of free radical generation and the rate of removal by antioxidant protective enzyme systems (Jenkinson 1982). Microvascular injury attributable to toxic oxygen radicals has been implicated in circumstances other than oxygen toxicity including immune complex injury of the lung (Johnson and Ward 1981), pulmonary dysfunction in hemodialysis (Craddock et al 1977b), and cardiopulmonary bypass (Haslam et al 1980).

Polymorphonuclear cells are a rich source of superoxide and other free radicals which induce endothelial cell injury (Johnson and Ward 1981). Neutrophil sequestration in the lung and microcirculatory damage have been well described in *E. coli* septicemia and endotoxemia (Coalson et al 1978; Esbenshade et al 1982). Sublethal doses of endotoxin stimulate production of pulmonary superoxide dismutase, catalase and glutathione peroxidase in rats subjected to hyperoxic exposure, and give animals thus treated protection against pulmonary edema, and a survival advantage (Frank et al 1978, 1980). The broncho-alveolar lavage fluids of patients with ARDS contain an inactive antitrypsin and free elastase, which are believed to reflect activity of oxygen-derived free radicals (Lee et al 1981). Hence, toxic oxygen radicals are strongly implicated in causing pulmonary microcapillary injury of complement activation and oxygen toxicity and may contribute to the pulmonary edema of certain strains of endotoxin induced injury.

3.3. Leukotrienes

Leukotrienes are products of the lipoxygenase pathway of arachidonic acid metabolism which stimulate smooth muscle contraction and bronchoconstriction. Leukotrienes C and D increase microcapillary permeability in the peripheral vascular bed (Morley et al 1981). Infusion of arachidonate into an isolated perfused rabbit lung pretreated with indometacin to inhibit cyclooxygenase results in increased nonhydrostatic pulmonary edema (Albert 1982). Activated granulocytes produce lipoxygenase as a product of arachidonate (Ford-Hutchinson et al 1980), and endotoxin administration increases lymphatic levels of 5-HETE in sheep (Ogletree et al 1981). Preliminary reports of increased flow of protein-rich pulmonary lymph in sheep infused with leukotrienes C_4 and D_4 further support the role of these compounds in the production of microvascular permeability (Mojarad 1983).

3.4. Histamine

Application of histamine to the skin results in an immediate substantial leakage of protein, which is due to a separation of endothelial cells in venules (Byan and Manjo 1976). In the lung the venules of bronchial walls do leak after exposure to histamine, although the small vessels in pulmonary alveoli do not (Fishman and Pietra 1976). Prolonged infusion of histamine in sheep produces increased flow of low-protein pulmonary lymph, which is most consistent with altered microvascular permeability (Brigham and Owens 1975b). The action of histamine is highly evanescent, however, and there is no direct evidence that histamine release in endotoxin-induced injury is responsible for sustained increase in microvascular permeability.

4. AIRWAY TONE AND PULMONARY MECHANICS

Within several minutes of endotoxin infusion in sheep, dynamic lung compliance falls and lung resistance increases (Esbenshade et al 1982). The alveolar-arterial oxygen gradient increases early following endotoxin administration and correlates with diminished dynamic compliance, although not with extravascular lung water in this model. Similarly, an early increase in airway resistance and decrease in dynamic compliance are reported in dogs and baboons infused with live *E. coli* (McCaffree et al 1981). Some investigators believe that mechanical pulmonary dysfunction within the first 4 hours following endotoxin infusion is attributable to bronchoconstriction since gravimetric measurements of lung water are not consistently elevated during this period (McCaffree et al 1981).

4.1. Cyclooxygenase metabolites

Inhibition of cyclooxygenase with meclofenamate or indometacin attenuates early increase in airway resistance following endotoxin administration (Snapper et al 1981; Snapper 1982). In addition granulocyte depletion with hydroxyurea decreases early airway changes associated with endotoxemia (Heflin and Brigham 1979), possibly because of diminished generation of cyclooxygenase metabolites. Thromboxane is a potent smooth muscle constrictor, and may account for some portion of the airway response to endotoxin.

4.2. Lipoxygenase metabolites

As previously stated, leukotrienes C and D are constrictors of airways as well as pulmonary vasculature (Michelassi et al 1982; Smedegård et al 1982; Zapol 1982), and pulmonary lymphatic levels of 5-HETE, a lipoxygenase product, are increased following *E. coli* endotoxin infusion (Ogletree et al 1981). The availability of newly developed lipoxygenase inhibitors and blockers should further investigation of the contributions of these metabolites.

4.3. Serotonin and histamine

Attenuation of the early alterations in lung mechanics with methysergide, a serotonin antagonist, has been reported by some authors (Stein and Thomas 1967). More recently other workers have described conflicting results using the same serotonin antagonist in canine endotoxemia (Allison et al 1981). Histamine administered intravenously or by aerosol causes a reproducible decrease in dynamic lung compliance in sheep (Hutchinson et al 1982). Antihistamine pretreatment, however, does not abolish the acute pulmonary effects of endotoxin (Parratt and Sturgess 1977). Hence the contribution of these mediators to airway abnormalities of endotoxemia are uncertain.

5. SYSTEMIC HYPOTENSION

Refractory hypotension which responds poorly to intravenous fluid administration is characteristic of clinical sepsis and experimental endotoxemia (Fairman and Glauser 1980; Jacobs et al 1982a). This hypotension is clearly multifactoral in etiology and includes pre- and postcapillary pooling, alterations in regional vascular resistance, capillary leakage and myocardial dysfunction (Hinshaw et al 1982). Increased levels of 2 vasoactive mediators, however, seem to relate closely to endotoxin-induced hypotension. The evidence for involvement of β-endorphins and vasodilator prostaglandins in systemic hypotension of endotoxemia follows.

5.1. β-Endorphins

Endorphins, endogenous opioid compounds, are synthesized in and released from the anterior pituitary concomitantly with ACTH in response to stress (Guillemin et al 1977). These polypeptides bind competitively with opiates to stereospecific receptor sites where they affect hemodynamic and psychologic changes common to opioid drugs such as systemic hypotension and euphoria. Evidence for the contribution of endorphins comes from the following observations: (a) serum levels of β-endorphins are elevated in canine endotoxin shock (Bone et al 1981); (b) injection of β-endorphin in experimental animals or man causes hypotension and tachycardia (Lemaire et al 1978); (c) administration of naloxone, a specific opiate antagonist, reverses the hypotension and improves survival in canine endotoxin shock (Holaday and Faden 1978; Jacobs et al 1982b; Reynolds et al 1980). Preliminary data suggest a potential salutary effect of naloxone administration in clinical hypotension (Higgins et al 1983; Peters et al 1981).

Current evidence suggests that the protective effect of naloxone on endotoxin-induced hypotension is mediated through competitive binding to opioid receptors within the central nervous system rather than to peripheral receptors (Holaday et al 1981; Janssen and Lutherer 1980). Increases in serum levels of cyclic AMP after treatment with naloxone in dogs infused with live *E. coli* suggest that opiates may interfere at a tissue level with the adenylate cyclase cAMP system (Gahhos et al 1982). Some authors believe that naloxone has a protective effect on vital organs such as the myocardium, central nervous system, and gastrointestinal tract, all of which may not be accounted for by increases in cardiac output (Muraszko et al 1983).

5.2. Cyclooxygenase products

Infusion of prostacyclin (PGI_2) into unanesthetized sheep causes a marked systemic vasodilatory response (Gunther et al 1982; Kadowitz et al 1978). Elevated levels of the stable metabolite of prostacyclin, 6-keto-PGF_1, are found in the sera of endotoxin-treated dogs, rats and baboons (Cook et al 1980; Fletcher and Ramwell 1980; Wise et al 1981). Administration of cyclooxygenase inhibitors such as ibuprofen or indometacin improves systemic hypotension and survival in endotoxin models of shock as well (Fletcher and Ramwell 1980; Jacobs et al 1982a). On the basis of these data, it is postulated that the vasodilator cyclooxygenase products may contribute to the systemic hypotension of endotoxemia. The reader is cautioned, however, that most of the inhibitors used in these studies have a wide range of biochemical activities, such as diminished neutrophil adhesiveness and complement binding, which might also account for beneficial effects in the animal models studied.

6. SUMMARY

Endotoxin administration in experimental animals produces hemodynamic and functional alterations which appear to correspond to elaboration of various vasoactive mediators. Pulmonary hypertension is most closely correlated to thromboxane release, although the source of prostaglandin production is uncertain. In addition lipoxygenase products, serotonin and vasoactive peptides may contribute to increases in pulmonary vascular tone. Microvascular permeability increases in endotoxemia, and current evidence suggests a role for complement, toxic oxygen radicals, and perhaps leukotrienes in mediating this expression of pulmonary injury. Bronchoconstriction and altered lung mechanics are well described with endotoxemia and appear to be related to release of cyclooxygenase or lipoxygenase products of arachidonic acid metabolism. Systemic hypotension associated with endotoxin injection may be improved by cyclooxygenase inhibitors or opiate antagonists, implying involvement of prostaglandins and β-endorphins in mediating refractory hypotension of endotoxemia. Although our understanding of the mediators of vascular changes of endotoxemia and sepsis is still rudimentary, it is hoped that further knowledge will allow us to institute more specific therapy in patients with clinical sepsis and thereby improve the morbidity and mortality rates of this increasingly common disorder.

REFERENCES

Adams T Jr, Traber DL (1982) The effects of a prostaglandin synthetase inhibitor, ibuprofen, on the cardiopulmonary response to endotoxin in sheep. *Circ. Shock* 9, 481-489.

Albert RK (1982) Lipoxygenase products of arachidonic acid increase pulmonary vascular permeability in excised rabbit lungs. Paper presented at NIH Workshop on Arachidonic Acid Metabolites and the Pulmonary Circulation, September 30, 1982 (unpublished).

Allison RC, Murphy TL, Weisman IM, McCaffree DR, Gray BA (1981) Effect of methysergide on the acute lung mechanics response to endotoxin. *J. Appl. Physiol. 50*, 185-190.

Alpert JS, Haynes FW, Knutson PA, Dolen JE, Dexter L (1973) Prostaglandins and the pulmonary circulation. *Prostaglandins 3*, 759-765.

Anderson FL, Tsagaris TJ, Jubiz W, Kuida H (1975) Prostaglandin F and E levels during endotoxin-induced pulmonary hypertension in calves. *Am. J. Physiol. 228*, 1479-1482.

Ashbaugh DG, Bigelow DB, Petty TL (1967) Acute respiratory distress in adults. *Lancet 2*, 319-323.

Bone RC, Said SI, Hiller C, Wilson F (1980) Vasoactive intestinal polypeptide in endotoxin shock. *Am. Rev. Respir. Dis. 121*, 113.

Bone RC, Potter M, Wilson F, Hiller C (1981) Endorphins in endotoxin shock. *Am. Rev. Respir. Dis. 123*, 107.

Bowers RE, Ellis EF, Brigham KL, Oates JA (1979) Effects of prostaglandin cyclic endoperoxides on the lung circulation of unanesthetized sheep. *J. Clin. Invest. 63*, 131-137.

Brigham KL, Owens PJ (1975a) Mechanism of the serotonin effect on lung transvascular fluid and protein movement in awake sheep. *Circ. Res. 36*, 761-770.

Brigham KL, Owens PJ (1975b) Increased sheep lung vascular permeability caused by histamine. *Circ. Res. 37*, 647-657.

Brigham KL, Bowers R, Haynes J (1979) Increased sheep lung vascular permeability caused by *E. coli* endotoxin. *Circ. Res. 45*, 292-297.

Byan GB, Manjo G (1976) Acute inflammation. A review. *Am. J. Pathol. 86*, 185-276.

Coalson JJ, Benjamin B, Archer LT, Beller B, Gilliam CL, Taylor FB, Hinshaw LB (1978) Prolonged shock in the baboon subjected to infusion of *E. coli* endotoxin. *Circ. Shock 5*, 423-437.

Cook JA, Wise WC, Halushka P (1980) Elevated thromboxane levels in the rat during endotoxic shock: protective effects of imidazole, 13-azaprostanoic acid or essential fatty acid deficiency. *J. Clin. Invest. 65*, 227-230.

Craddock PR, Hammerschmidt D, White JG, Dalmasso AP, Jacob HS (1977a) Complement (C_{5a})-induced granulocyte aggregation in vitro. A possible mechanism of complement mediated leukostasis and leukopenia. *J. Clin. Invest. 60*, 260-264.

Craddock PR, Fehr J, Brigham KL, Kronenberg RS, Jacob HS (1977b) Complement and leukocyte-mediated pulmonary dysfunction in hemodialysis. *N. Engl. J. Med. 296*, 769-774.

Davis LB, Meeker WR, McQuarrio DG (1960) Immediate effects of intravenous endotoxin on serotonin concentrations and blood platelets. *Circ. Res. 8*, 234-239.

Demling R, Smith M, Gunther R, Flynn J, Gee M (1981) Pulmonary injury and prostaglandin production during endotoxemia in conscious sheep. *Am. J. Physiol. 240*, H348-353.

Dey RD, Said SI (1981) Localization of substance P-like immunofluorescence in nerves within dogs airways and pulmonary vessels. *Fed. Proc. 40*, 595.

Dusting GJ, Moncada S, Vane JR (1979) Prostaglandins, their intermediates and precursors: cardiovascular actions and regulatory roles in normal and abnormal circulatory systems. *Prog. Cardiovasc. Dis. 21*, 405.

Emerson TE Jr, McCarthy TF, Raymond RM (1983) Increased survival time with ibuprofen treatment during endotoxemia in the dog. *Fed. Proc. 42*, 323.

Esbenshade AM, Newman J, Lams P, Jolles H, Brigham K (1982) Respiratory failure after endotoxin infusion in sheep: lung mechanics and lung fluid balance. *J. Appl. Physiol. 53*, 967-976.

Fairman RP, Glauser FL (1980) Pathophysiologic alterations in endotoxemia. Similarities to an animal model. *Arch. Intern. Med. 110*, 788-790.

Fearon DT, Ruddy S, Schur PH, McCabe WR (1975) Activation of the properdin pathway of complement in patients with gram-negative bacteremia. *N. Engl. J. Med. 292*, 937-940.

Fernandez HN, Henson PM, Otani A, Hugli TE (1975) Chemotactic response to human C_3 and C_5 anaphylatoxins. I. Evaluation of C_3 and C_5 leukotaxis in vitro and under simulated in vivo conditions. *J. Immunol. 129*, 109-115.

Ferreira SH, Vane JR (1967) Prostaglandins: their disappearance from and release into the circulation. *Nature (London) 216*, 868-873.

Fishman AP, Pietra GG (1976) Permeability of pulmonary vascular endothelium. *Ciba Found. Symp. 38*, 29-48.

Fletcher JR, Ramwell PW (1980) Indomethacin treatment following baboon endotoxin shock improves survival. *Adv. Shock Res. 4*, 103-111.

Ford-Hutchinson A, Bray M, Doig M, Shipley M, Smith M (1980) Leukotriene B, a potent chemokinetic and aggregating substance released from polymorphonuclear leukocytes. *Nature (London) 286*, 264-265.

Frank L, Yam J, Roberts RJ (1978) The role of endotoxin in protection of adult rats from high oxygen induced lung toxicity. *J. Clin. Invest. 61*, 269-275.

Frank L, Summerville J, Massaro D (1980) Protection from oxygen toxicity with endotoxin. Role of the endogenous antioxidant enzymes of the lung. *J. Clin. Invest. 65*, 1104-1110.

Frolich JC, Ogletree M, Peskar BA, Brigham KL (1980) Pulmonary hypertension correlated to pulmonary thromoboxane synthesis. *Adv. Prostaglandin Thromboxane Res. 7*, 745-750.

Fulton RL, Jones CE (1975) The cause of post-traumatic pulmonary insufficiency in man. *Surg. Gynecol. Obstet. 140*, 179-186.

Gahhos F, Chiu R, Hinchey E, Richards G (1982) Endorphins in septic shock. *Arch. Surg. 117*, 1053-1057.

Gilbert R, Hinshaw I, Kuida H, Visscher M (1958) Effects of histamine, 5 hydroxytryptamine and epinephrine on pulmonary hemodynamics with particular reference to arterial and venous segment resistances. *Am. J. Physiol. 194*, 165-170.

Glovsky MM, Ryan U, Quiroga M, Hogg JC, Kunkel S, Ward PA (1983) Endothelial cell damage: a model for studying anaphylatoxin-induced cellular injury. *Immunobiology 164*, 247.

Gryglewski RJ, Korbut R, Ocetkiewicz A, Splawinski J, Wojtaszek B, Swies J (1978) Lungs as generators of prostacyclinhypothesis on physiological significance. *Naunyn Schmiedebergs Arch. Pharmakol. 304*, 45-50.

Guillemin R, Bloom F, Rossier J, Minick S, Henricksen S, Burgus R, Ling N (1977) Current physiological studies with the endorphins. In: MacIntyre I, Szelke M (Eds), *Molecular Endocrinology*, pp 251-267. North-Holland Biomedical Press, Amsterdam.

Gunther R, Zaiss C, Demling RH (1982) Pulmonary microvascular response to prostacyclin (PGI_2) infusion in unanesthetized sheep. *J. Appl. Physiol. 52*, 1338-1342.

Hamasaki Y, Mojarad M, Said SI (1983) Relaxant action of VIP on cat pulmonary artery: comparison with acetylcholine, isoproterenol, and PGE_1. *J. Appl. Physiol. 54*, 1607-1622.

Harris TR, Brigham KL (1982) The exchange of small molecules as a measure of normal and abnormal lung microvascular function. *Ann. N.Y. Acad. Sci. 384*, 417-434.

Haslam PL, Townsend PJ, Branthwaite MA (1980) Complement activation during cardiopulmonary bypass. *Anaesthesia 35*, 22-26.

Heflin AC, Brigham KL (1979) Granulocyte depletion prevents increased lung vascular permeability after endotoxemia in sheep. *Clin. Res. 27*, 399A.

Henson PM, Larson GL, Webster RO, Mitchell BC, Goins AJ, Henson JE (1982) Pulmonary microvascular alterations and injury induced by complement fragments: synergistic effect of complement activation, neutrophil sequestration and prostaglandins. *Ann. NY Acad. Sci. 384*, 287-300.

Higgins TL, Sivak ED, O'Neil D, Graves J, Foutch O (1983) Reversal of hypotension by continuous naloxone infusion in a ventilator-dependent patient. *Ann. Intern. Med. 98*, 47-48.

Hinshaw LB, Beller-Todd B, Archer L (1982) Review update: current management of the septic shock patient: experimental basis for treatment. *Circ. Shock 9*, 543-553.

Holaday JW, Faden AI (1978) Naloxone reversal of endotoxin hypotension suggests role of endorphins in shock. *Nature (London) 275*, 450-451.

Holaday JW, O'Hara M, Faden AI (1981) Hypophysectomy allows cardiorespiratory variables: central effects of pituitary endorphins in shock. *Am. J. Physiol. 241*,

12

H479-H485.

Hosea S, Brown E, Hammer C, Frank M (1980) Role of complement activation in a model of adult respiratory distress syndrome. *J. Clin. Invest. 66*, 375-382.

Hugli TE, Muller-Eberhard HJ (1978) Anaphylatoxins: C_{3a} and C_{5a} *Adv. Immunol. 26*, 1-53.

Hutchison A, Brigham KL, Snapper JR (1982) Effect of histamine on lung mechanics in sheep. *Am. Rev. Resp. Dis. 126*, 1025-1029.

Jacobs ER, Soulsby ME, Bone RC, Wilson F, Hiller F (1982a) Ibuprofen in canine endotoxin shock. *J. Clin. Invest. 70*, 536-541.

Jacobs ER, Bone RC, Wilson FJ, Hiller CH (1982b) Naloxone blockade of endorphins in canine endotoxin shock. *Microcirculation 2*, 19-27.

Janssen HF, Lutherer LO (1980) Ventriculocisternal administration of naloxone protects against severe hypotension during endotoxin shock. *Brain Res. 194*, 608-612.

Jenkinson S (1982) Oxygen toxicity. In: RC Bone (Ed), *Pulmonary Disease Review, Vol 3*, pp 79-112. Wiley Medical Publications, New York.

Johnson KJ, Ward PA (1981) Role of oxygen metabolites in immune complex injury of lung. *J. Immunol. 126*, 2365-2369.

Kabins S, Molina C, Katz L (1959) Pulmonary vascular effects of serotonin (5-OH-tryptamine) in dogs: its role in causing pulmonary edema. *Am. J. Physiol. 197*, 955-958.

Kadowitz PJ, Spannhake EW, Knight DS, Hyman AL (1977) Vasoactive hormones in the pulmonary vascular bed. *Chest 71 (Suppl 2)*, 257-261.

Kadowitz PJ, Chapnick B, Feigen L, Hyman P, Nelson P, Spannhake E (1978) Pulmonary and systemic vasodilator effects of the newly discovered prostaglandin, PGI_2. *J. Appl. Physiol. 45*, 408-413.

Kaplan RL, Sahn SA, Petty TL (1979) Incidence and outcome of the respiratory distress syndrome in gram-negative sepsis. *Arch. Intern. Med. 139*, 867-869.

Larsen GL, Mitchell B, Harper T, Hensen P (1982) The pulmonary response of C_5 sufficient and deficient mice to *Pseudomonas aeruginosa*. *Am. Rev. Respir. Dis. 126*, 306-311.

Lee CT, Fein AM, Lippmann M, Holtzman H, Kimbel P, Weinbaum G (1981) Elastolytic activity in pulmonary lavage fluid from patients with adult respiratory-distress syndrome. *N. Engl. J. Med. 304*, 192-196.

Lemaire I, Tseng R, Lemaire S (1978) Systemic administration of beta-endorphin: potent hypotensive effect involving a serotonergic pathway. *Proc. Natl Acad. Sci. USA 75*, 6240-6242.

Lewis AJ, Eyre P (1972) Some cardiovascular and respiratory effects of prostaglandins E_1, E_2, and F_2 alpha in the calf. *Prostaglandins 2*, 55-64.

McCaffree DR, Gray BA, Pennock BE, Coalson J, Bridges C, Taylor FB, Rogers RM (1981) Role of pulmonary edema in the acute pulmonary response to sepsis. *J. Appl. Physiol. 50*, 1198-1205.

McCall CE, Dechatelet LR, Brown D, Lachmann P (1974) New biological activity following intravascular activation of the complement cascade. *Nature (London) 249*, 841-843.

Michelassi F, Landa L, Hill R, Lowenstein E, Watkin W, Petkau A, Zapol WM (1982) Leukotriene D_4: a potent coronary artery vasoconstrictor associated with impaired ventricular contraction. *Science 217*, 841-843.

Mojarad M (1983) Research data on file, Pulmonary Disease Section, University of Oklahoma Health Sciences Center, Oklahoma City, Oklahoma.

Moncada S, Amezcua JL (1979) Prostacyclin, thromboxane A_2, interactions in haemostasis and thrombosis. *Haemostasis 8*, 252-265.

Morley J, Page C, Paul W (1981) Leukotrienes, SRS-A and the vascular manifestations of PCA. *Agents Actions 11*, 585-587.

Muraszko KM, Chen R, Carlin R, Antunes J, Chien S (1983) Effects of naloxone on regional blood flow in hypovolemia. *Fed. Proc. 42*, 981.

Murphy TL, Allison RC, Weisman IM, McCaffree DR, Gray BA (1981) Role of platelet serotonin in the canine pulmonary response to endotoxin. *J. Appl. Physiol. 50*, 178-184.

Ogletree ML, Brigham KL, Oates JA, Hubbard WC (1981) Increased flux of 5-HETE in sheep lung lymph during pulmonary leukostasis after endotoxin. *Fed. Proc. 40*, 767.

Parratt JR, Sturgess RM (1975a) *E. coli* endotoxin shock in the cat–treatment with indomethacin. *Br. J. Pharmacol. 53*, 485-488.

Parratt JR, Sturgess RM (1975b) The effects of the repeated administration of sodium meclofenamate, an inhibitor of prostaglandin synthetase, in feline endotoxin shock. *Circ. Shock 2*, 301-310.

Parratt JR, Sturgess RM (1977) The possible roles of histamine, 5-hydroxytryptamine and prostaglandin F_2 alpha as mediators of the acute pulmonary effects of endotoxin. *Br. J. Pharmacol. 60*, 209-219.

Peters WP, Johnson MW, Friedman PA, Mitch WE (1981) Pressor effect of naloxone in septic shock. *Lancet I*, 529-532.

Piper P, Vane J (1971) The release of prostaglandins from lung and other tissues. *Ann. NY Acad. Sci. 180*, 363-385.

Price S, Harlan J, Winn R, Carrico CJ, Hildebrandt J (1983) Indomethacin and dazoxiben prevent increased lung water in septic goats. *Fed. Proc. 42*, 1274.

Reynolds DG, Gurll NJ, Vargish T, Lechner RB, Faden AI, Holaday JW (1980) Blockade of opiate receptors with naloxone improves survival and cardiac performance in canine endotoxin shock. *Circ. Shock 7*, 39-48.

Said SI (1980) Vasoactive intestinal peptide (VIP): isolation, distribution, biological actions, structure-function relations and possible functions. in: Jerzy Glass GB (Ed), *Gastrointestinal Hormones*, pp 245-273. Raven Press, New York, NY.

Said SI (1982) Vasoactive peptides and the pulmonary circulation. *Ann. NY Acad. Sci. 384*, 207-212.

Said SI (1983) Vasoactive peptides, state-of-the-art review. *Hypertension 5, Suppl 1*, I17-26.

Salzman PM, Salmon JA, Moncada S (1980) Prostacyclin and thromboxane A_2 synthesis by rabbit pulmonary artery. *J. Pharmacol. Exp. Ther. 215*, 240-247.

Samuelsson B, Goldyne M, Granstrom E, Hamberg M, Hammarstrom S, Malmsten C (1978) Prostaglandins and thromboxanes. *Ann. Rev. Biochem. 47*, 997-1029.

Sibbald WJ, Paterson NAM, Holliday RL, Anderson RA, Lobb TR, Duff JH (1978) Pulmonary hypertension in sepsis: measurement by the pulmonary arterial-diastolic pulmonary wedge pressure gradient and the influence of passive and active factors. *Chest 73*, 583-591.

Smedegård G, Hedqvist P, Dahlen S, Revenas B, Hammarström S, Samuelsson B (1982) Leukotriene C_4 affects pulmonary and cardiovascular dynamics in monkey. *Nature (London) 295*, 327-329.

Smith EF, Tabas JH, Lefer AM (1980) Beneficial actions of imidazole in endotoxin shock. *Prostaglandins Leukotrienes Med. 4*, 215-225.

Snapper JR (1982) Airway and hemodynamic responses to endotoxin. *Ann. N.Y. Acad. Sci. 384*, 435-450.

Snapper JR, Ogletree ML, Hutchison A, Brigham KL (1981) Meclofenamate prevents increased resistance of the lung (R_L) following endotoxemia in unanesthetized sheep. *Am. Rev. Respir. Dis. 123*, 200.

Staub NC (1974) State of the art: pathogenesis of pulmonary edema. *Am. Rev. Resp. Dis. 109*, 358-372.

Stein M, Thomas DP (1967) Role of platelets in the acute pulmonary responses to endotoxin. *J. Appl. Physiol. 23*, 47-52.

Till G, Johnson K, Kunkel R, Ward P (1982) Intravascular activation of complement and acute lung injury. *J. Clin. Invest. 69*, 1126-1135.

Turek JJ, Templeton CB, Bottoms GD, Fessler JF (1983) Changes in the pulmonary vascular endothelium of the pony in a model for endotoxin shock. *Fed. Proc. 42*, 980.

Walker BR, Voelkel NF, Reeves JT (1982) Pulmonary pressor response after prostaglandin synthesis inhibition in conscious dogs. *J. Appl. Physiol. 52*, 705-709.

Webster RO, Larsen GL, Mitchell BC, Goins AJ, Henson PM (1982) Absence of inflammatory lung injury in rabbits challenged intravascularly with complement-derived chemotactic factors. *Am. Rev. Respir. Dis. 125*, 335-340.

Weir EK, Grover RF (1978) The role of endogenous prostaglandins in the pulmonary circulation. *Anesthesiology 48*, 201-212.

Wise WC, Cook JA, Haluska PV (1981) Implications for thromboxane A_2 in the pathogenesis of endotoxic shock. *Adv. Shock Res. 6*, 83-91.

Zapol WM, Snider MT, Hill JD, Fallat RJ, Bartlett RH, Edmunds LH, Morns AH, Pierce EC, Thomas AN, Proctor HJ, Drinker PA, Pratt PC, Bagniewski A, Miller RG (1979) Extracorporeal membrane oxygenation in severe acute respiratory failure. A randomized prospective study. *J. Am. Med. Assoc. 242*, 2193-2196.

Zapol WM (1982) Leukotriene D_4: a potent coronary artery vasoconstrictor associated with impaired ventricular contraction. *Science 217*, 841-843.

Handbook of Endotoxin, Vol. 2: Pathophysiology of Endotoxin
L.B. Hinshaw, editor
© Elsevier Science Publishers B.V., 1985
ISBN 0 444 90385 2
$0.85 per article per page (transactional system)
$0.20 per article per page (licensing system)

CHAPTER 2

Cardiodepressant effects of endotoxin

LERNER B. HINSHAW

1. INTRODUCTION: HISTORY OF ENDOTOXIN RESEARCH

The insidious development of septic shock in man involves mechanisms that are not clearly understood. A multitude of potential mechanisms has been proposed, each of which could account for a substantial degree of the pathophysiological manifestations. Among the many possibilities, cardiovascular changes have been reported to occur within seconds following an intravenous injection of endotoxin in dogs and at variable times after injection of endotoxin or live *Escherichia coli* organisms in the canine, primate, and other species. Although the animal shock model has been studied extensively in recent years, the precise mechanisms involved in the early development of shock are still not clearly identified.

Recently, attention has been focused on the causes of inadequate tissue perfusion in this form of shock. Abnormal hemodynamic responses comprise a primary hallmark of endotoxic or septic shock. They are of great complexity and variability and differ between the species, depending on the severity, sequential time periods, and duration of shock. In animal models and in humans, at least four sequential phases of shock have been recognized: at first, a 'preshock' period has been identified which persists from minutes to a few hours, characterized by a moderate decrease in mean systemic pressure and the development of tachycardia, tachypnea, fever, restlessness, and irritation. Cardiac output may be significantly elevated. Secondly, slowly or rapidly developing systemic hypotension, decreased venous return and depressed cardiac output have been reported in many species. Thirdly, a compensatory period precipitated by systemic hypotension, involving neurohumoral and autonomic reflex activation, regularly follows; lastly, a period of improvement is observed if survival is to be attained; alternatively, progressive circulatory deterioration with sequentially developing multiple organ failure terminating in irreversible shock and death may also follow.

The role of the heart in the pathogenesis of septic (endotoxin) shock has become a controversial issue, particularly its role during the intermediate phases of shock after augmentation of compensatory mechanisms has occurred. The contribution of the heart to the pathogenesis of shock is not easy to determine because of peripheral vascular events which influence cardiac

16

performance (Hinshaw 1974). This makes it difficult to decide whether the heart is failing in its pumping function after endotoxin as a result of some intrinsic myocardial defect, or whether cardiac output falls simply because venous return to the right ventricle is decreasing due to peripheral pooling (Weil et al 1956). In the latter instance, myocardial function could be entirely normal. The earlier historical view was that the heart is not adversely affected in septic shock or endotoxin-induced shock. In an extensive review of the hemodynamics of endotoxin shock, Gilbert in 1960 stated that there was no evidence that endotoxin altered myocardial work capacity and therefore the effects of endotoxin must be primarily vascular. This view was substantiated earlier by the observations of Weil and others in 1956, who found no evidence for myocardial dysfunction in the early phase of endotoxin shock, and by the subsequent report of Alican et al (1962) that the canine heart only fails terminally if unfavorable circulatory conditions persist. Furthermore, it was subsequently discovered that large doses of endotoxin in dogs produced no adverse effects on myocardial extraction of oxygen or coronary hemodynamics (Londe et al 1967).

The earliest evidence for myocardial dysfunction in endotoxin shock was provided by the studies of Solis and Downing in 1966 when cardiac depression was demonstrated early after endotoxin administration. Ventricular contractile force was diminished and was found to be depressed even when arterial pressure was maintained at the preshock level. Brunson and others (1966) pointed out that in a large series of dogs given endotoxin during the Oklahoma Shock Tour of 1964, the earliest detectable lesions were observed in the heart, liver and gall bladder. Myocardial hemorrhage was a common finding and increased in severity during the subsequent 5–12 hour period post endotoxin (Brunson et al 1966).

A turning point has now been reached, however, in our views on the role of the heart in clinical septic shock or shock produced by the administration of endotoxin or live organism: the preponderance of evidence now supports the position that the heart may fail in animal shock models during any period of shock, with the probable exception of the early phase (1–3 hours). There is a possibility that adverse myocardial effects may be precipitated earlier than expected (Janssen et al 1983) since decreases in cardiac taurine levels were found very early (0–90 minutes) after endotoxin.

Clinical findings in severe sepsis/septic shock in man have documented the presence of myocardial dysfunction which is clearly precipitated by the state of shock as opposed to the presence and influence of pre-shock, preexisting myocardial dysfunction (Cerra et al 1978; Clowes et al 1970; MacLean et al 1965, 1967; Siegel et al 1967; Weisel et al 1975, among others).

The purpose of the review of this important issue is to explore past and recent findings regarding direct and indirect factors influencing myocardial performance in endotoxin shock. It will be apparent that recent research is revealing numerous pathophysiological influences each of which may account for some degree of myocardial dysfunction during shock and could significantly contribute to lethality.

1.1. Myocardial function in endotoxin shock

Earlier reports failed to document the development of myocardial dysfunction during the first hour of endotoxin shock (Hinshaw et al 1958; Weil et al 1956). Several studies provided evidence that the heart is depressed during later stages of shock in animals given endotoxin (Alican et al 1962; Solis and Downing 1966), while clinical reports documented cardiac dysfunction in human septic shock (MacLean et al 1965, 1967; Siegel et al 1967). Using the following criteria for the verification of myocardial dysfunction in endotoxin shock (Hinshaw 1979): increase in left ventricular end diastolic pressure (LVEDP) and decreases in positive dP/dt_{max}, negative dP/dt_{max}, cardiac power and efficiency, we can summarize the results obtained in canine studies as follows:
(a) within 1–3 hours after endotoxin, there is no adverse cardiac effect and only the typical decrease in venous return causes cardiac output to fall;
(b) 3–5 hours after endotoxin there are significant indications of myocardial dysfunction;
(c) from 7 to 9 hours severe myocardial mitochondrial edema and disruption are evident (Hinshaw 1979). Recent studies of live *E. coli*-induced shock in dogs have also demonstrated that the majority of hearts fail within 3–5 hours (Archer et al 1982; Benjamin 1980).

2. INDIRECT EFFECTS OF ENDOTOXIN ON THE MYOCARDIUM

2.1. Peripheral alterations: decreased venous return

Initial studies carried out in canine endotoxin shock suggested that cardiac function is unimpaired during the early phase of shock (Weil et al 1956). The early decrease in cardiac output after infusion of endotoxin was found to be due to hepatosplanchnic pooling of blood and the resultant decrease in venous return to the right ventricle. Notable decreases in venous return during the first hour after endotoxin have been reported in both canine and nonhuman primate species (Hinshaw et al 1958, 1970; Weil et al 1956). Hemodynamic measurements in human septic shock have clearly demonstrated a correlation between cardiac output and survival (Nishijima et al 1973), diminished cardiac output being a harbinger of impending irreversibility. Decreases in venous return in the early phase of shock in canine and nonhuman primates were shown to clearly account for the drop in cardiac output and mean arterial pressure and suggested that peripheral pooling mechanisms predominated in the early period of endotoxin shock. Intravascular pooling is a relatively early event in the dog, and the site of pooling is apparently not confined to the hepatosplanchnic region (Guntheroth and Kawabori 1977). It is also important to distinguish between transitory intravascular pooling in the dog and the subsequent progressive loss of perfusate into extravascular regions which has a more prolonged devastating effect (Brunson et al 1966; Coalson et al 1972; Guntheroth and Kawabori 1977). On the other hand, intravascular pooling appears to be the predominating feature in the nonhuman primate (Hinshaw et al 1970), with only a small fraction of the blood volume lost to the extravascular compartment. Regardless of the site of pooling there is reason to believe that all animal species challenged with endotoxin suffer a decrease in venous return which strongly diminishes cardiac output.

2.1.1. Cardiac responses to coronary hypoperfusion and endotoxin

Findings have clearly emphasized the importance of coronary perfusion in endotoxin shock (Elkins et al 1973; Hinshaw et al 1974b). The separate roles of coronary hypotension and endotoxin have been evaluated in the pathogenesis of heart failure. Experiments have been performed on isolated canine hearts supported by blood from intact dogs (Hinshaw et al 1974b). In these studies, the first series of hearts was subjected to 4 hours of coronary hypotension combined with endotoxin administration, while the second series was exposed to the same degree of low coronary pressure but in the absence of endotoxin. Ninety percent of hearts subjected to both endotoxin and decreased coronary perfusion pressure demonstrated dysfunction at 4 hours after endotoxin, as evidenced by abnormally increased LVEDP, decreased negative dP/dt_{max} and depressed efficiency. In contrast, 40% of the hearts subjected to low pressure alone demonstrated dysfunction at 4 hours (Hinshaw et al 1974b). These findings suggest that during endotoxin shock inadequate coronary blood flow alone may have a significant role in the precipitation of heart dysfunction.

Experiments demonstrating an important role of coronary perfusion after endotoxin were reported in the retrograde-perfused canine heart with supported coronary circulation (Elkins et al 1973). It was found that perfusion of the coronary vascular system by pump perfusion improved heart function during 5 hours of endotoxin exposure by improving length–tension relationships (Elkins et al 1973). In support of these studies, marked cardiac dysfunction was observed in dogs following LD_{100} injections of endotoxin, in which severe systemic hypotension occurred and early deaths were recorded (Hinshaw et al 1972a, 1973a). In similar experiments which elicited marked systemic hypotension, extensive cardiac mitochondrial edema and disruption were observed at times of heart failure, 5–7 hours after endotoxin injection (Coalson et al 1972). Experiments strongly suggesting the protective effect of increased coronary flow on the myocardium after endotoxin were those using methylprednisolone sodium succinate, sodium nitroprusside, and insulin (Hinshaw 1979; Hinshaw et al 1976). In those experiments increased coronary blood flow was commonly associated with improved myocardial performance, including depressed LVEDP and increased positive and negative dP/dt_{max}.

2.1.2. Effects of intracardial redistribution of coronary blood flow

Nonuniform perfusion of the myocardium after endotoxin in dogs has been reported (Adiseshiah and Baird 1978; Hess and Krause 1981; Kleinman et al 1980). Nonuniform perfusion in the presence of increased coronary vascular resistance after endotoxin, appearing to disfavor perfusion of the subendocardial region, is considered to be a primary pathophysiological event causing severe heart pathology (Kleinman et al 1980) and leading to decreased myocardial contractility resulting from a defect at the level of the contractile proteins (Hess and Krause 1981). An explanation might be that the myocardium is additionally unresponsive to the usual calcium stimulus, as revealed in the studies by McCaig and Parratt (1980). Unfavorable perfusion gradients due to opposing extravascular compression forces are responsible for the diversion of the blood flow away from the subendocardial region, and this action is exacerbated by concomitant tachycardia (Adiseshiah and Baird 1978).

2.1.3. Cardiac responses to increased preload

Greenfield and others (1974) reported that when nonhuman primates were administered endotoxin and subsequently 'fluid loaded' (infusion of colloid), there was a failure of the heart to perform increased work at the same LVEDP (12–15 mmHg) as control nonshock primates receiving the same volume of infusate. This observation was confirmed by Falk (1981) and Falk and others (1982) in the feline species given *E. coli*: stroke volume, positive dP/dt_{max} and cardiac work increased less than in control animals after dextran infusion despite large increases in LVEDP. Weisel et al (1977) documented myocardial depression in human sepsis and also found that both surviving and nonsurviving septic patients responded subnormally to fluid loading in terms of myocardial function. On the other hand, Vincent et al (1981) found

20

that the nonsurviving group of patients in septic shock (with peritonitis) were those who failed to respond to fluid administration and had increases in both right and left ventricular filling pressures with a failure to perform adequate ventricular work. These studies implicated biventricular failure as a causative factor in the fatal progression of septic shock. Right heart failure appears early and is related to augmented pulmonary vascular resistance while left heart failure associated with a reduction in work capability emerges later as the lethal factor (Vincent et al 1981). These observations are important in that prior to fluid loading, as a result of a functional hypovolemia derived from peripheral pooling leading to a depressed cardiac output, myocardial dysfunction is masked because of the presence of both decreased preload and afterload (mean aortic pressure) (Guntheroth et al 1982). Fluid challenge unmasks this deficiency and myocardial dysfunction is thereby revealed.

2.2. Abnormal neurohumoral influences and depressed responsiveness of the heart to inotropic and chronotropic stimuli after endotoxin

Animal and human clinical studies have yielded evidence that ventricular performance in the initial stage of endotoxin (septic) shock is supported by a sympathoadrenal stimulus initiated subsequently to the development of systemic hypotension (Hinshaw 1979). Increased neurohumoral activity associated with the eary compensatory period appears to give substantial cardiovascular support in the face of diminished venous return. Cann et al (1972) pointed out that as the heart is eventually depressed in function, it is dependent on intensive sympathetic stimulation to maintain cardiac output.

During the intermediate phase of shock following the early compensatory period when myocardial performance begins to falter or fail altogether, depressed responses of the myocardium to circulating catecholamines follow the period of increased sympathetic activity (Hinshaw et al 1979). Results from canine and feline studies have shown decreased responsiveness of the left ventricle to the infusion of epinephrine and norepinephrine on the basis of depressed changes in cardiac output, coronary blood flow, heart rate, systolic and diastolic pressures, positive dP/dt_{max}, negative dP/dt_{max}, cardiac power and efficiency, in contrast to myocardial responses to which endotoxin was not administered (Archer et al 1975; Parratt 1973). It was also noted in isolated working heart studies that within 5–6 hours after endotoxin, the heart could not perform at elevated afterloads unless driven by inotropic agents (Hinshaw et al 1972b). It is now generally agreed that myocardial contractile and relaxation characteristics and coronary vascular responses to catecholamines are depressed in endotoxin shock during the period of depressed function. Goldfarb (1982) and Goldfarb et al (1983) concluded that survival after endotoxin in dogs is associated with the maintenance of normal cardiac contractility, and that reduced inotropicity precipitated by endotoxin leads to significantly depressed myocardial function.

2.3. Intracardiac ionic and fluid disturbances

Reports demonstrating myocardial dysfunction in endotoxin shock have shown close associations between the concomitant appearance of heart edema, decrease in peak negative dP/dt_{max}, increase in left ventricular end diastolic pressure, and abnormal concentrations of K^+, and perhaps also Ca^{2+}, at the time when cardiac dysfunction is first revealed (Archer et al 1978; Coalson et al 1972; Hinshaw 1974; Hinshaw et al 1972).

The mechanism of the development of heart edema after endotoxin is unknown but it appears to be an early critical stimulus for the onset of myocardial dysfunction (Mela et al 1974). There is a possibility that coronary capillary permeability is increased after endotoxin. Resultant elevations in extravascular pressure in myocardial tissue due to the progressive fluid accumulation (Coalson et al 1972) most probably would increase coronary postcapillary resistance and thereby provide a further stimulus for edema formation. This defect would be further intensified by elevations in right atrial pressures as cardiac function is progressively worsening (Clowes et al 1970; Hinshaw et al 1979; Vincent et al 1981). Intramyocardial tissue pressure differences found to be influential in diverting blood away from subendocardial regions may lead to subnormal regional perfusion (Adiseshiah and Beard 1978). Kleinman et al (1980) documented severe heart pathology in canine endotoxin shock including diffuse intramyocardial hemorrhage and swelling and distortion of the subendocardial sarcoplasmic reticulum. Severe mitochondrial edema coincident with myocardial dysfunction after live organism infusion in dogs was reported by Postel and Schloerb (1977). These findings occurred independently of metabolic derangements and rates of lethality. Bardakhchian et al (1978) observed ultrastructural lesions in myocardial cells early after endotoxin in dogs (30 minutes) which included intravascular coagulation, increased permeability, reduction in numbers of myofibrils and decreased myocardial contractility. A constellation of cardiac morphologic effects observed at autopsy following septicemia in human patients has been reported by Bohm (1982); this included myocarditis accompanied by endothelial damage, interstitial edema, hemorrhages, foci of necrotic heart muscle fibers, fibrinous transudation and granulocytic infiltration. These findings suggest generalized damage of endothelial cells, vascular walls and adjacent interstitial and parenchymal tissues. These observations in animals and humans underscore a probable significant role of edema in the pathogenesis of myocardial dysfunction in endotoxin- or live organism-induced shock and human septic shock.

Ionic imbalances may have resulted from increases in coronary capillary and myocardial cell membrane permeabilities, exacerbated by the accumulation of edema fluid between and within contractile elements and in the mitochondria.

Peak negative dP/dt is regularly reduced in the failing heart following administration of endotoxin or *E. coli* and therefore becomes a reliable 'marker' for myocardial dysfunction. It has been found to be the most reliable measure of the rate of ventricular relaxation and is reported to be influenced by or related to a number of factors: intrinsic contractile properties of the myocardium, volume of the heart, catecholamine activity, and the level of inotropic stimuli, end-systolic fiber length, end-diastolic fiber length, amount of calcium uptake by the sarcoplasmic reticulum, and the development of myocardial edema (Hinshaw et al 1979).

The actions of pharmacological agents in improving heart function and morphology after endotoxin suggest the underlying causes of dysfunction: digoxin is very effective in preventing and reversing cardiac dysfunction in endotoxin-shocked animals. Both the severe mitochondrial edema and disruption of mitochondria are prevented by early digoxin treatment (Coalson et al 1972). Intracardiac infusions of insulin in an isolated working heart preparation are effective in restoring heart function to normal after endotoxin. Infusions of insulin alone reverse cardiac failure and maintain normal myocardial performance in spite of wide ranges in blood glucose concentrations (5–120 mg/ml). Insulin infusions restore depressed negative dP/dt_{max} values to normal in endotoxin treated hearts and also increase negative dP/dt_{max} values significantly even in normal, nonshocked hearts. A consistent fall in arterial blood concentrations of potassium occurs following infusions of insulin after endotoxin, suggesting the restoration of cell membrane potential leading to improvement of the myocardial contractile state. Negative dP/dt_{max} increases after insulin, which may occur following restoration of normal fluid balance (removal of edema fluid) with the return of ionic distributions to normal (Archer et al 1978; Hinshaw et al 1976).

2.4. Blood-borne myocardial depressant substances

Attempts by our laboratory to implicate a myocardial depressant factor released during shock capable of exerting a significant adverse influence on myocardial performance in vivo have been fruitless (Hinshaw et al 1973b, 1974a, 1979). We have concluded that a myocardial depressant factor does not play a significant role in endotoxin shock. The basis for this conclusion is that (a) blood from dogs in all stages of endotoxin shock (6–21 hours) does not depress myocardial function; (b) β-adrenergic blockade fails to unmask a depressant effect on function of the heart; (c) blood from dogs in lethal

23

splanchnic arterial occlusion shock does not depress myocardial function of a normal heart or even one previously placed on the brink of failure by prior coronary hypoperfusion; and (d) acute pancreatectomy does not prevent myocardial dysfunction after endotoxin but, on the other hand, without endotoxin it does not result in depressed cardiac function either (Hinshaw et al 1972a, 1973b, 1974a, 1979). From these observations it can be concluded that if a myocardial depressant agent is released into the blood during endotoxin shock at any time period, even under conditions in which splanchnic perfusion is drastically reduced, there is no detrimental effect on myocardial work performance. It is conceivable that other forces opposing a depressant factor, such as increased myocardial perfusion, elevated sympathoadrenal myocardial support or other humoral stimulating influences, exactly counterbalance the myocardial effect of a circulating depressant factor (Lefer 1970). That this possibility is a viable option is derived from the experimental observations of Maksad et al (1979) and McConn et al (1979), who have reported the presence of a plasma factor in the blood of shocked animals which depresses myocardial function. Parker and Adams (1981) found that atrial muscle isolated from guinea pigs subjected to endotoxin for 16–18 hours showed decreases in contractile tension and dP/dt in vitro.

The confusion of results in this area of research may be due to the possibility that there are many humoral factors put into play during shock which influence myocardial function both negatively and positively while the net effect of their combined participations in vivo is a preserved myocardial functional status.

2.5. Elevated plasma vasopressin concentrations

Markedly elevated plasma vasopressin concentrations after endotoxin or *E. coli* administration in dogs and nonhuman primates may contribute to myocardial dysfunction (Wilson et al 1980, 1981). It was shown that infusing vasopressin into the left atrium in concentrations found during shock can cause myocardial dysfunction which is not associated with generalized ischemia of the heart.

2.6. Alterations of myocardial cellular energy mechanisms and substrate utilization

Endotoxin administration to dogs stimulates phospholipase A activities in myocardial mitochondria, which results in an adverse effect on cellular energy metabolism ending in depressed myocardial contractility (Liu and Takeda 1982). Spitzer and Hinshaw (1975) observed alterations of myocardial substrate utilization which occurred soon after the administration of endotoxin and several hours before the appearance of cardiac dysfunction. The major change was that the contribution of free fatty acid oxidation to myocardial oxidative metabolism decreased and the contribution of lactate increased after the administration of endotoxin.

3. DIRECT EFFECTS OF ENDOTOXIN IN THE MYOCARDIUM

Attempts to assay the direct effects of endotoxin and thus elucidate the mechanisms contributing to the failure of the heart have led to several in vitro models which include papillary strips, atrial strips, myofibrils, myocardial homogenates and isolated heart cell tissue cultures. Results from these preparations are summarized in the following paragraphs.

Using isolated cat papillary muscles, Lefer and Rovetto (1970) demonstrated no change in inotropic responses when endotoxin (10 µg/ml, final concentration) was added directly to the muscle chamber. Raffa and Trunkey (1978) found that endotoxin or *E. coli* added to an isolated rabbit interventricular septum preparation in vitro produced no direct effect on function or morphology. Kutner and Cohen (1966) observed that endotoxin added to a left ventricular papillary muscle preparation in vitro did not change the force of muscle contraction or rate of force development. McCaig et al (1979) reported that endotoxin has no effect on cardiac muscle in vitro after 60 minutes of exposure.

Carli et al (1981) found that endotoxin produced a depression of rat heart cells in vitro to isoproterenol and ascribed the mechanism of depression to release of a humoral lipid-soluble cardiodepressant factor. Macnicol et al (1973) reported that endotoxin administered to a rat left ventricular trabecular corneal preparation in vitro caused a reduction in peak developed tension. Cho (1972) found that endotoxin exerted a direct vasoconstrictor effect on canine coronary vessels. Parker and Sardesai (1973) incubated heart myofibrils with *E. coli* endotoxin (0.3 mg/ml) in vitro and found a 60% depression of ATPase activity.

4. CLINICAL CONSIDERATIONS

4.1. Hyperdynamic and hypodynamic states

Early reports in the clinical literature seemed to indicate that cardiac output was always reduced in septic shock, but later studies began to describe a hyperdynamic state with elevated cardiac output in sepsis (Levison 1982). Recent clinical observations have confirmed both low and high cardiac out-

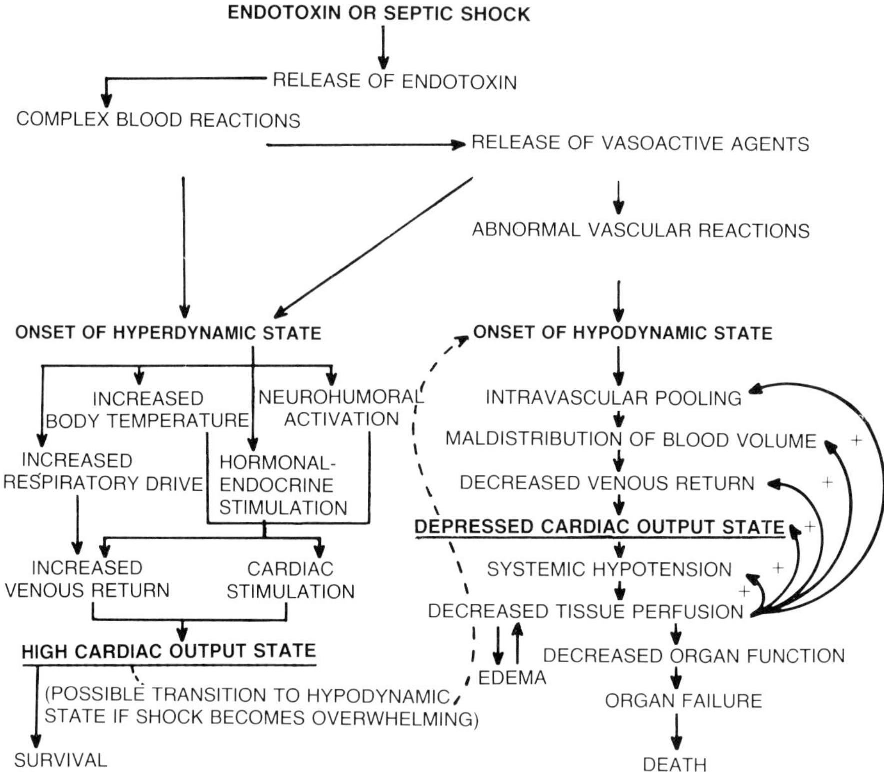

Fig. 1   + = *positive feedback*
Reproduced from Hinshaw et al (1982), with permission.

put shock with impaired myocardial performance in man and animals (Gahhos 1981; Heng-Yi and Ren-Yu 1982; Hinshaw 1982, 1983; Levison 1982).

The hyperdynamic (high flow) and hypodynamic (low flow) states represent most probably the difference between survival (the former state) and nonsurvival (the latter state) since increased mortality and decreased cardiac output are clearly related. Nishijima et al (1973) summarized the results of seven clinical reports and showed that as cardiac output decreased, mortality rate increased.

Suggested mechanisms accounting for the hyperdynamic and hypodynamic states are shown in Figure 1. This diagram portrays the concept that the state of elevated cardiac output is consistent with survival, but that there may be a progression of this early condition to the hypodynamic state which possesses ominous lethal overtones. It should be pointed out that as diminished

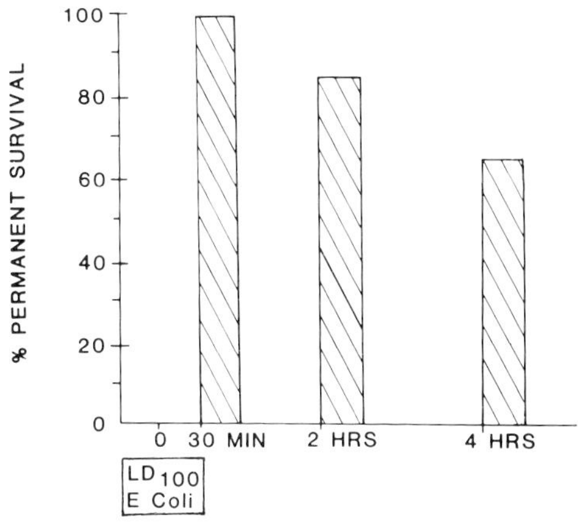

Fig. 2 *Effect on percent survival (baboons) of delaying initiation of therapy with methylprednisolone sodium succinate/gentamicin sulfate after administration of LD$_{100}$ E. coli at time = 0. Reproduced from Wilson et al (1981), with permission.*

tissue perfusion in shock becomes a sustained feature, positive feedback pathways set up 'vicious circles' resulting in an enhanced rate of progressive deterioration.

A 'hyperdynamic' phase in a 2-hour LD$_{100}$ live *E. coli* infusion baboon model has been reported (Hinshaw et al 1983). It is particularly significant that this lethal baboon septic model exhibits a hyperdynamic phase similar to that in the human patient and that morphologic findings in nonsurviving baboons closely parallel those found in patients who die in septic shock (Hinshaw 1982).

The results of this review suggest that the onset of myocardial dysfunction may occur relatively early during severe sepsis/septic shock. Recent findings in the baboon (Hinshaw et al 1983) suggest that the hyperdynamic phase occurring early in this particular shock model (1–2 hours) heralds the time when treatment procedures should be instituted. Figure 2 indicates the critical importance of early therapy with infusions of methylprednisolone sodium succinate and gentamicin sulfate. Since the onset of the hypodynamic phase of the baboon in these experiments begins by 2 hours after the onset of LD$_{100}$ *E. coli* infusion, the time for the beginning of the development of myocardial dysfunction may be approximately 2 hours. The implication of these observations is that exceptionally early treatment of septic shock may be a crucial necessity in order to prevent progressive myocardial deterioration. The precise role of a dysfunctioning myocardium in contributing to death in endotoxin (septic) shock is not clearly defined at the present time, but the emerging concept of progressively developing multiple organ failure in this form of shock places a high priority on protecting myocardial integrity.

5. SUMMARY

Recently, the emphasis in septic shock therapy has been on the causes of inadequate tissue perfusion. The myocardium was previously not considered to be an important link in this process inasmuch as its magnificent reserve capabilities were thought to protect it particularly well from the ravages of shock.

Much earlier information pointed to a peripheral problem in perfusion and return of blood back to the heart during shock. Moreover, many experimental studies were of short duration during which time the heart could be driven effectively by sympathoadrenal influences. In addition, the very nature of shock itself, especially the 'cold shock' phase with diminished cardiac output and depressed mean arterial pressure, permitted the heart to function in an understressed fashion: the heart would appear to be performing adequately even though the actual work it produced was very much diminished. Similar to an underworking human-devised machine, there was nothing to indicate that its function was impaired. This condition probably led many investigators to discard any consideration of the heart performing a crucial role in survival in terms of its quality of performance.

A turning point came when results of investigative and clinical studies and experiences showed that cardiac function was impaired in endotoxic and/or septic shock: 'loading experiments' in which perfusates of various types were infused in shocked animals or patients revealed that the heart had difficulty ejecting the blood: the high price it exacted was a larger end diastolic volume and pressure for a stroke volume comparable to that of a normal heart. This observation pointed to some sort of defect in the ejecting machinery, which was further verified by studies in isolated working left ventricles of dogs. It was discovered that hearts removed from dogs receiving lethal injections of endotoxin could not eject the same volume of blood as a control (non-shocked) heart against an elevated afterload (mean aortic pressure) at the same end diastolic pressure. It soon became apparent that this effect was time-dependent: hearts removed from endotoxin-shocked dogs performed better against higher afterloads if they were transferred earlier to the perfusion circuit. Hearts stressed by afterloads of 150 mmHg could do so successfully 2 hours after endotoxin but not at 4 hours. In the latter instance, LVEDP continued to rise in the face of the maintained elevated afterload till complete and massive irreversible failure of the heart. This feat, however, was no problem for the heart removed from a normal, nonshocked animal.

Initially, myocardial dysfunction in man during severe sepsis/septic shock was considered to be only a terminal event, in which the heart somehow escaped the onslaught of shock to the very edge of death. Much evidence is now amassed to show that myocardial dysfunction in man and animals subjected to this form of shock clearly demonstrates the central role of a dysfunctioning heart in the pathogenesis of shock.

The main question at present has evolved from 'Is the heart affected by endotoxin?' and 'Is myocardial function in clinical septic shock an important link in its pathogenesis and lethal characteristics?' to 'How is the heart adversely affected?' and 'What are the mechanisms resulting in its impaired performance?'

This overview of factors pertaining to the cardiotoxic effects of endotoxin began by showing how investigators and clinicians became aware of the existence of myocardial dysfunction in endotoxic (septic) shock. It was found at an early stage in research (1950-1960) that there were at least two basic ways in which cardiac output could be depressed: (a) by a decrease in venous return to the right ventricle because of peripheral trapping of blood and/or ultrafiltrate, and (b) by failure of the ventricles to properly eject the blood entering their chambers during diastole. This chapter has attempted to distinguish peripheral (pooling) from central (cardiac function) mechanisms in accounting for the reduction in cardiac output during endotoxin shock. At the same time a new phenomenon arrived on the scene: the 'hyperdynamic state', in which cardiac output is elevated above control values. At first this phenomenon appeared to occur only in the human patient, and it was thought that it was triggered by large volumes of exogenous perfusate given as treatment to patients. This, if true, would only be part of the explanation since the hyperdynamic state has been found in animal shock models, including pigs, dogs and nonhuman primates, and does not depend on fluid administration for the significant increase in cardiac output. It appears that cardiac output is elevated in this form of shock in response to the invasive stimulus of endotoxin or live organisms and represents a massive marshalling of the animal's (or human) defense against the foreign invasion. Research on baboons in our laboratory suggests that the peak of the hyperdynamic phase represents the latest time that the most effective treatment should be commenced in order to maximally increase the probability of survival.

Mechanisms may well be at work during the early compensatory state of shock when cardiac output is normal or elevated, which are setting the stage for eventual myocardial dysfunction and the subsequent hypodynamic state leading to increased mortality.

Abnormal neurohumoral influences and depressed responsiveness of the heart to inotropic and chronotropic stimuli, intracardiac ionic and fluid disturbances and, possibly, blood-borne myocardial depressant substances together with diminished coronary blood flow are the factors most commonly considered to elicit impaired myocardial performance in endotoxic (septic) shock.

Figure 3 illustrates the probable mechanisms accounting for cardiac dysfunction in both *E. coli-* and endotoxin-induced shock. The basic factors that influence myocardial performance adversely include coronary hypoperfusion, depressed ventricular responses to catecholamines, increased coronary capillary permeability, myocardial edema formation, intramyocardial ionic imbalances and myocardial mitochondrial damage. The combination of these pathophysiological factors results in depressed ventricular relaxation, decreased ventricular compliance, and depressed ventricular contractility leading to significantly severe myocardial dysfunction and diminished cardiac output.

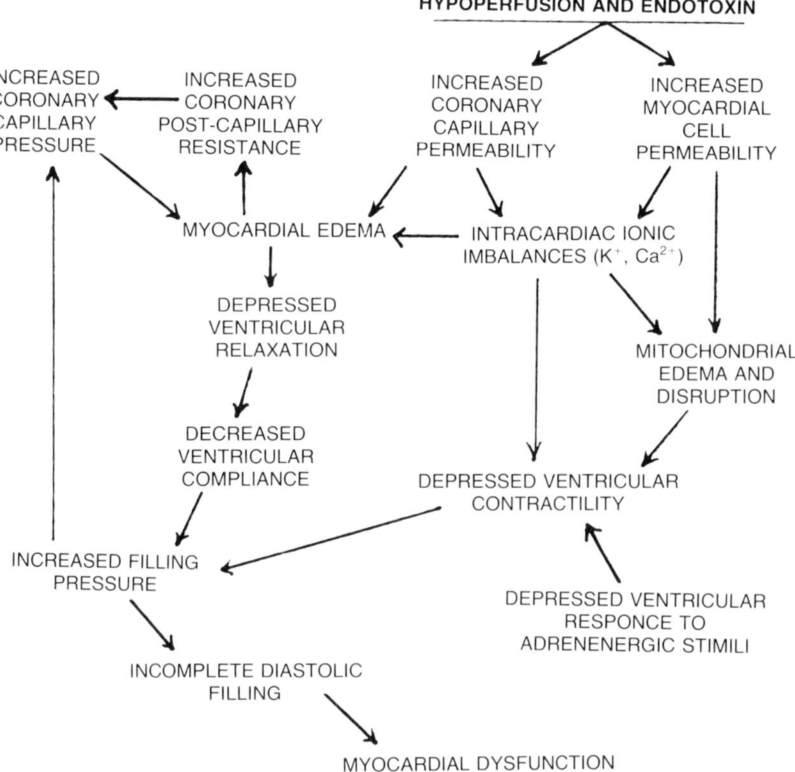

Fig 3 *Diagrammatic representation of probable mechanisms accounting for cardiac dysfunction in both live E. coli- and endotoxin-induced shock. Reproduced from Hinshaw et al (1979), with permission.*

6. PROBLEMS TO BE RESOLVED BY FUTURE RESEARCH FOR CLINICAL PRACTICE

A fundamental question requiring further clarification is the role of the heart with its depressed function in severe sepsis/septic shock as related to ultimate survival of the patient. At what level of depressed cardiac performance will requirements for optimal blood circulation of the septic patient have to be met to achieve an acceptable recovery rate?

A second question concerns the means of recognizing the onset of myocardial dysfunction. Can adequate biochemical or physiological 'markers' be developed for diagnostic purposes?

A third question pertains to the development of a schema describing the probable multiple mechanisms accounting for myocardial dysfunction. Do these occur together, does one precipitate another, and are time relationships, severity of shock and differences between species important in the precipitation of the dysfunction?

A final question concerns the development of effective therapies in treating shock and specifically therapies that also benefit the myocardium during sepsis/septic shock. Should the therapy be directed toward specifically protecting the heart or toward a general protection of the patient?

REFERENCES

Adiseshiah M, Baird RJ (1978) Correlation of the changes in diastolic myocardial tissue pressure and regional blood flow in hemorrhagic and endotoxic shock. *J. Surg. Res. 24*, 20-25.

Alican F, Dalton ML, Hardy JD (1962) Experimental endotoxin shock. *Am. J. Surg. 103*, 702-708.

Archer LT, Black MR, Hinshaw LB (1975) Myocardial failure with altered response to adrenaline in endotoxin shock. *Br. J. Pharmacol. 54*, 145-155.

Archer LT, Beller BK, Drake JK, Whitsett TL, Hinshaw LB (1978) Reversal of myocardial dysfunction in endotoxin shock with insulin. *Can. J. Physiol. Pharmacol. 56*, 132-138.

Archer LT, Benjamin BA, Beller-Todd BK, Brackett DJ, Wilson MF, Hinshaw LB (1982) Does LD_{100} *E. coli* shock cause myocardial failure? *Circ. Shock 9*, 7-16.

Bardakhchian EA, Gordeeva-Gavrikova TV, Cherepanov YP (1978) Ultrastructural changes in dog myocardium in the early phase of endotoxin shock. *Cor Vasa (USSR) 20*, 324-327.

Benjamin B (1980) *Myocardial Dynamics in Experimental Septic Shock*. Thesis, University of Oklahoma, Oklahoma City, Oklahoma.

Bohm N (1982) Adrenal, cutaneous and myocardial lesions in fulminating endotoxinemia. *Pathol. Res. Pract. 174*, 92-105.

Brunson JG, Schulz DM, Angevine DM, Imrie ST, Douglas BH (1966) Evaluation of therapeutic agents: morphologic changes and survival data in dogs subjected to endotoxin during the shock tour. *J. Oklah. State Med. Assoc. 59*, 479-484.

Cann M, Stevenson T, Fiallos E, Thal AP (1972) Depressed cardiac performance in sepsis. *Surg. Gynecol. Obstet. 134*, 759-763.

Carli A, Auclair MC, Benassayag C, Nunez E (1981) Evidence for an early lipid soluble cardiodepressant factor in rat serum after a sublethal dose of endotoxin. *Circ. Shock 8*, 301-312.

Cerra FB, Hassett J, Siegel JH (1978) Vasodilator therapy in clinical sepsis with low output syndrome. *J. Surg. Res. 25*, 180-183.

Cho YW (1972) Direct cardiac action of *E. coli* endotoxin. *Proc. Soc. Exp. Biol. Med. 141*, 705-707.

Clowes GHA, Farrington GH, Zuschneid W, Cossette GR, Saravis C (1970) Circulating factors in the etiology of pulmonary insufficiency and right heart failure accompanying severe sepsis (peritonitis). *Ann. Surg. 171*, 663-678.

Coalson JJ, Woodruff HK, Greenfield LJ, Guenter CA, Hinshaw LB (1972) Effects of digoxin on myocardial ultrastructure in endotoxin shock. *Surg. Gynecol. Obstet. 135*, 908-912.

Elkins RC, McCurdy JR, Brown PP, Greenfield LJ (1973) Effects of coronary perfusion on myocardial performance during endotoxin shock. *Surg. Gynecol. Obstet. 137*, 991-996.

Falk A (1981) *Septic Shock: an Experimental Study on Cardiovascular Mechanisms.* Thesis, University of Göteborg, Göteborg, Sweden.

Falk A, Myrvold HE, Haglund U (1982) Cardiopulmonary function as related to intestinal mucosal lesions in experimental septic shock. *Circ. Shock 9*, 419-432.

Gahhos FN, Chiu RCJ, Bethune D, Dion Y, Hinchey EJ, Richards GK (1981) Hemodynamic responses to sepsis: hypodynamic versus hyperdynamic states. *J. Surg. Res. 31*, 475-481.

Gilbert RP (1960) Mechanisms of the hemodynamic effects of endotoxin. *Physiol. Rev. 40*, 245-279.

Goldfarb RD (1982) Cardiac mechanical performance in circulatory shock: a critical review of methods and results. *Circ. Shock 9*, 633-653.

Goldfarb RD, Tambolini W, Wiener SM, Weber PB (1983) Canine left ventricular performance during LD_{50} endotoxemia. *Am. J. Physiol. 244*, H370-H377.

Greenfield LJ, Jackson RH, Elkins RC, Coalson JJ, Hinshaw LB (1974) Cardiopulmonary effects of volume loading of primates in endotoxin shock. *Surgery 76*, 560-572.

Guntheroth WG, Kawabori I (1977) The contribution of splanchnic pooling to endotoxin shock in the dog. *Circ. Res. 41*, 467-472.

Guntheroth WG, Jacky JP, Kawabori I, Stevenson JG, Moreno AH (1982) Left ventricular performance in endotoxin shock in dogs. *Am. J. Physiol. 242*, H172-H176.

Heng-yi G, Ren-yu S (1982) Experimental investigations on canine endotoxin shock. *Chin. Med. J. 95*, 613-621.

Hess ML, Krause SM (1981) Contractile protein dysfunction as a determinant of depressed cardiac contractility during endotoxin shock. *J. Mol. Cell. Cardiol. 13*, 715-723.

Hinshaw LB (1974) Role of the heart in the pathogenesis of endotoxin shock. *J. Surg. Res. 17*, 134-145.

Hinshaw L (1979) Myocardial function in endotoxin shock. *Circ. Shock Suppl 1*, 43-51.

Hinshaw L (1982) Overview of endotoxin shock. In: Cowley RA, Trump BF (Eds), *Pathophysiology of Shock, Anoxia and Ischemia*, pp 219-234. Williams and Wilkins, Baltimore, Maryland.

Hinshaw LB, Gilbert RP, Kuida H, Visscher MB (1958) Peripheral resistance changes and blood pooling after endotoxin in eviscerated dogs. *Am. J. Physiol. 195*, 631-634.

Hinshaw LB, Shanbour LL, Greenfield LJ, Coalson JJ (1970) Mechanisms of decreased venous return: subhuman primate-administered endotoxin. *Arch. Surg. 100*, 600-606.

Hinshaw LB, Greenfield LJ, Owen SE, Archer LT, Guenter CA (1972a) Cardiac response to circulating factors in endotoxin shock. *Am. J. Physiol. 222*, 1047-1053.

Hinshaw LB, Greenfield LJ, Owen SE, Black MR, Guenter CA (1972b) Precipitation of cardiac failure in endotoxin shock. *Surg. Gynecol. Obstet. 135*, 39-48.

Hinshaw LB, Archer LT, Black MR, Greenfield LJ, Guenter CA (1973a) Prevention and reversal of myocardial failure in endotoxin shock. *Surg. Gynecol. Obstet. 136*, 1-11.

Hinshaw LB, Archer LT, Black MR, Greenfield LJ (1973b) Myocardial performance in splanchnic arterial occlusion shock. *J. Surg. Res. 15*, 417-428.

Hinshaw LB, Archer LT, Black MR, Elkins RC, Brown PP, Greenfield LJ (1974a) Myocardial function in shock. *Am. J. Physiol. 226*, 357-366.

Hinshaw LB, Archer LT, Spitzer JJ, Black MR, Peyton MD, Greenfield LJ (1974b) Effects of coronary hypotension and endotoxin on myocardial performance. *Am. J. Physiol. 227*, 1051-1057.

Hinshaw LB, Archer LT, Benjamin B, Bridges C (1976) Effects of glucose or insulin on myocardial performance in endotoxin shock. *Proc. Soc. Exp. Biol. Med. 152*, 529-534.

Hinshaw LB, Benjamin B, Archer LT, Peyton MD (1979) The heart and endotoxin shock. *Texas Rep. Biol. Med. 39*, 173-191.

Hinshaw LB, Brackett DJ, Archer LT, Beller BK, Wilson MF (1983) Detection of the hyperdynamic state of sepsis in the baboon during lethal *E. coli* infusion. *J. Trauma 23*, 361-365.

Janssen HF, Lombardini JB, Lust RM (1983) Cardiac taurine levels during endotoxemia. *Proc. Soc. Exp. Biol. Med. 172*, 407-411.

Kleinman WM, Krause SM, Hess ML (1980) Differential subendocardial perfusion and injury during the course of gram-negative endotoxemia. *Adv. Shock Res. 4*, 139-152.

Kutner FR, Cohen J (1966) Effect of endotoxin on isolated cat papillary muscle. *J. Surg. Res. 6*, 83-86.

Lefer AM (1970) Role of a myocardial depressant factor in the pathogenesis of circulatory shock. *Fed. Proc. 29*, 1836-1847.

Lefer AM, Rovetto MJ (1970) Influence of myocardial depressant factor on physiologic properties of cardiac muscle. *Proc. Soc. Exp. Biol. Med. 134*, 269-273.

Levison M (1982) Myocardial failure. *Surg. Clin. N. Am. 62*, 149-156.

Liu M-S, Takeda H (1982) Endotoxin-induced stimulation on phospholipase A activities in dog hearts. *Biochem. Med. 28*, 62-69.

Londe SP, Massie H, Monafro WW Jr, Bernard HR (1967) Resistance of the isolated canine heart to endotoxin. *Surgery 61*, 466-470.

MacLean LD, Duff JH, Scott HM, Peretz DI (1965) Treatment of shock in man based on hemodynamic diagnosis. *Surg. Gynecol. Obstet. 120*, 1-16.

MacLean LD, Mulligan WG, McLean APH, Duff JH (1967) Patterns of septic shock in man – a detailed study of 56 patients. *Ann. Surg. 166*, 543-562.

Macnicol MF, Goldberg AH, Clowes GHA Jr (1973) Depression of isolated heart muscle by bacterial endotoxin. *J. Trauma 13*, 554-558.

Maksad AK, Chung-Ja C, Stuart RC, Brosco FA, Clowes GHA Jr (1979) Myocardial depression in septic shock: physiologic and metabolic effects of a plasma factor on an isolated heart. *Circ. Shock Suppl 1*, 35-42.

McCaig DJ, Parratt JR (1980) Reduced myocardial response to calcium during endotoxin shock in the cat. *Circ. Shock 7*, 23-30.

McCaig DJ, Kane KA, Bailey G, Millington PF, Parratt JR (1979) Myocardial function in feline endotoxin shock: a correlation between myocardial contractility, electrophysiology, and ultrastructure. *Circ. Shock 6*, 201-211.

McConn R, Greineder JK, Wasserman R, Clowes GHA Jr (1979) Is there a humoral factor that depresses ventricular function in sepsis? *Circ. Shock Suppl 1*, 9-22.

Mela L, Hinshaw LB, Coalson JJ (1974) Correlation of cardiac performance, ultrastructural morphology, and mitochondrial function in endotoxemia in the dog. *Circ. Shock 1*, 265-272.

Nishijima H, Weil MH, Shubin H, Cavanilles J (1973) Hemodynamic and metabolic studies on shock associated with gram-negative bacteremia. *Medicine 52*, 287-294.

Parker JL, Adams HR (1981) Contractile dysfunction of atrial myocardium from endotoxin-shocked guinea pigs. *Am. J. Physiol. 240*, H954-H962.

Parker CJ Jr, Sardesai VM (1973) The effect of *E. coli* endotoxin on cardiac and skeletal muscle myofibrillar ATPase. *Biochem. Med. 8*, 11-17.

Parratt JR (1973) Myocardial and circulatory effects of *E. coli* endotoxin: modification of responses to catecholamines. *Br. J. Pharmacol. 47*, 12-25.

Postel J, Schloerb PR (1977) Cardiac depression in bacteremia. *Ann. Surg. 186*, 74-82.

Raffa J, Trunkey DD (1978) Myocardial depression in sepsis. *J. Trauma 18*, 617-622.

Siegel JH, Greenspan M, Del Guercio LRM (1967) Abnormal vascular tone, defective oxygen transport and myocardial failure in human septic shock. *Ann. Surg. 165*, 504-517.

Solis RT, Downing SE (1966) Effects of *E. coli* endotoxemia on ventricular performance. *Am. J. Physiol. 211*, 307-313.

Spitzer JJ, Hinshaw LB (1975) Myocardial substrate utilization in experimental shock. *Circ. Shock 2*, 137-142.

Vincent J-L, Weil MH, Puri V, Carlson RW (1981) Circulatory shock associated with purulent peritonitis. *Am. J. Surg. 142*, 262-270.

Weil MH, MacLean LD, Visscher MB, Spink WW (1956) Studies on the circulatory changes in the dog produced by endotoxin from gram-negative microorganisms.

J. Clin. Invest. 35, 1191-1198.
Weisel JP, O'Donnell TF Jr, Stone MA, Clowes GHA Jr (1975) Myocardial performance in clinical septic shock: effects of isoproterenol and glucose potassium insulin. *J. Surg. Res. 18*, 357-363.
Weisel RD, Vito L, Dennis RC, Valeri CR, Hechtman HB (1977) Myocardial depression during sepsis. *Am. J. Surg. 133*, 512-521.
Wilson MF, Brackett DJ, Archer LT, Hinshaw LB (1980) Mechanisms of impaired cardiac function by vasopressin. *Ann. Surg. 191*, 494-500.
Wilson MF, Brackett DJ, Tompkins P, Benjamin B, Archer LT, Hinshaw LB (1981) Elevated plasma vasopressin concentrations during endotoxin and *E. coli* shock. In: Schumer W, Spitzer JJ, Marshall BE (Eds), *Advances in Shock Research, Vol 6*, pp 15-26. Alan R. Liss, New York.

Handbook of Endotoxin, Vol. 2: Pathophysiology of Endotoxin
L.B. Hinshaw, editor
© Elsevier Science Publishers B.V., 1985
ISBN 0 444 90385 2
$0.85 per article per page (transactional system)
$0.20 per article per page (licensing system)

CHAPTER 3

Peripheral circulatory responses to endotoxin

ROBERT F. BOND

1. INTRODUCTION

It has been recognized for many years that a profound state of systemic hypotension and shock will accompany the existence of a significant concentration of gram-negative *Escherichia coli* endotoxin in the blood of mammals (Gilbert 1960; Weil et al 1956). The precise mechanism(s) responsible for this hypotensive episode are not well understood but must be related to uncompensated decreases in peripheral vascular tone (Bond 1983; Gilbert 1960) and/or cardiac output (Bohs et al 1979; Lefer 1970). This chapter will specifically examine the known and hypothesized mechanisms concerned with the failure of the peripheral vascular system to respond appropriately to the hypotension which occurs during endotoxemia. The cardiac response to endotoxin has been covered in considerable detail in Chapter 2.

Before one can properly analyze the peripheral vascular consequences of this pathological situation, one must first have a good understanding of the normal response of the vasculature to a fall in systemic pressure. In this normal response a delicate balance exists between intrinsic (i.e., autoregulatory) and extrinsic (i.e., neural and/or humoral) mechanisms (Bond 1982). The primary function of the intrinsic mechanism is to protect the metabolic integrity of the peripheral tissue through vasodilation as perfusion pressure falls, while the baroreceptor-induced extrinsic mechanism(s) serve to support systemic pressure by causing vascular resistance to rise. In the normal compensatory response to a fall in systemic pressure, the extrinsic mechanisms dominate, which results in a selective vascular vasoconstriction through the baroreceptor mechanism (see Fig. 2, below) in an effort to return pressure back toward normal (Bond and Green 1969). Since each tissue and organ has its own unique function, it is not surprising that each should also have its own individual set of vascular controls. Therefore, in order to evaluate the role of these controlling mechanisms, one must be able to distinguish between the influences of intrinsic and of extrinsic mechanisms in all major

36

vascular beds which, when taken together, add up to the total peripheral vascular response. This separation can be accomplished by analysis of the pressure/vascular conductance relationships in each vascular bed during progressive decreases in perfusion pressure (Bond 1983, Green et al 1980) induced by: (a) stepwise occlusion of the cognate artery to determine the autoregulatory or intrinsic ability of the vascular bed, (b) systemic hypotension induced by hemorrhage, and (c) systemic hypotension induced by endotoxin administration. Since vascular conductance (flow/pressure) is the reciprocal of vascular resistance (pressure/flow), any rise above control conductance as pressure falls represents a decrease in vascular tone (i.e., an intrinsic autoregulatory or vasodilatory response) while a decrease below control represents an increased vascular tone (i.e., an extrinsic baroreceptor-induced vasoconstriction). The differences between the pressure/conductance curves obtained during local hypotension and those obtained during systemic hypotension will provide insight into the relative strengths of the intrinsic and extrinsic mechanisms operating in each vascular bed during hemorrhagic and endotoxin shock.

The first major division of this chapter will include a general discussion of the cardiovascular response to systemic endotoxin. Most of the discussion will be centered around data derived from dog experiments, but where species differences are known to exist, they will be pointed out and discussed. This section will be followed by an in-depth comparison of the hemodynamic characteristics of six selected vascular beds which when added together have been reported (Green 1950) to make up approximately 99% of the total body vascular conductance. These vascular beds are: cerebral, myocardial, mesenteric (visceral), renal, skeletal muscle, and skin. The next section will examine the summated responses of these six vascular beds in an effort to establish the relative roles played by each in the overall systemic response. A discussion of the known microvascular responses to endotoxin, including the pre- and postcapillary resistance and capillary permeability changes which occur during endotoxemia, will then be presented. Finally, the concluding section will attempt to integrate the entire picture of the influence that endotoxemia has on the normal mechanism for the control of blood pressure. This section will also include a discussion of areas where future studies are needed.

2. GENERAL CARDIOVASCULAR RESPONSES TO ENDOTOXIN

Considerable confusion exists in the literature about the role of the heart and total peripheral vascular conductance in response to endotoxin. All agree, however, that significant hypotension does follow the toxic insult even though the responsible mechanisms are not well understood. The controversy is made even more difficult to interpret because of the total lack of consistency in experimental designs between investigators (i.e., endotoxin dose, species, anesthetics, etc.). For example, in human sepsis, Clowes et al (1966), Cohn et al (1968), and Siegel et al (1979) all report a decreased cardiac output and an increased total peripheral conductance. Conversely, however, though Weil and Nishijima (1978) agree that cardiac output falls, they also provide evidence of a decreased conductance. Perhaps more realistically, Mac Lean et al (1967) report that either an increased or decreased conductance can be seen in patients depending upon a number of predisposing factors. Studies conducted in unanesthetized (Elsberry et al 1969; Rutherford et al 1976; Wyler et al 1969) and anesthetized (Hinshaw 1972) nonhuman primates suggest a reduction in cardiac output and rise in the conductance. Though the use of the dog as an animal model in shock has been criticized because of its reported sensitivity to visceral pooling (Hinshaw 1972; Lillehei and Mac Lean 1959; Vaughn et al 1967), there are numerous reports which indicate that canine cardiac output and vascular tone both respond in a manner very similar to those of the unanesthetized monkey (Bond 1983; Brockman et al 1967; Seyfer et al 1977). In addition, Hinshaw et al (1961a, 1961b) reported a fall in both cardiac output and total body conductance followed by a return to control conductance in response to endotoxin.

Therefore, while bearing the above discussion in mind, we can conclude that the systemic pressure, cardiac output and total peripheral vascular responses of unanesthetized and anesthetized nonhuman primates and dogs are probably quite similar. For this reason the following more in-depth description of the systemic responses to endotoxin from the anesthetized dog will be presented as representative of the typical mammalian response (Bond 1983). Figure 1 depicts the cardiovascular response of the pentobarbitalized dog to a slow intravenous infusion of 2 mg/kg endotoxin. Vascular conductances which are the reciprocals of resistances were calculated by subtracting the right atrial pressure from the mean systemic arterial pressure and dividing the result into the flow. Note that the aortic pressure began to fall immediately after the start of the infusion from a control of 128 mmHg to a minimum of 55 mmHg at 105 minutes, after which there was a gradual recovery to a maximum of 80 mmHg at 240 minutes. Heart rate fell transiently from 139 to 120 beats per minute at 30 minutes, then returned to control values. Modest bradycardia has also been described by Vick (1965), who attributes this to a systemic endotoxin release of a cholinergic-like substance which presumably acts directly on the cardiac pacemaker. The total body flow was monitored by the use of a Swanz-Ganz thermodilation catheter placed into the pulmonary artery. The injection port in the right atrium

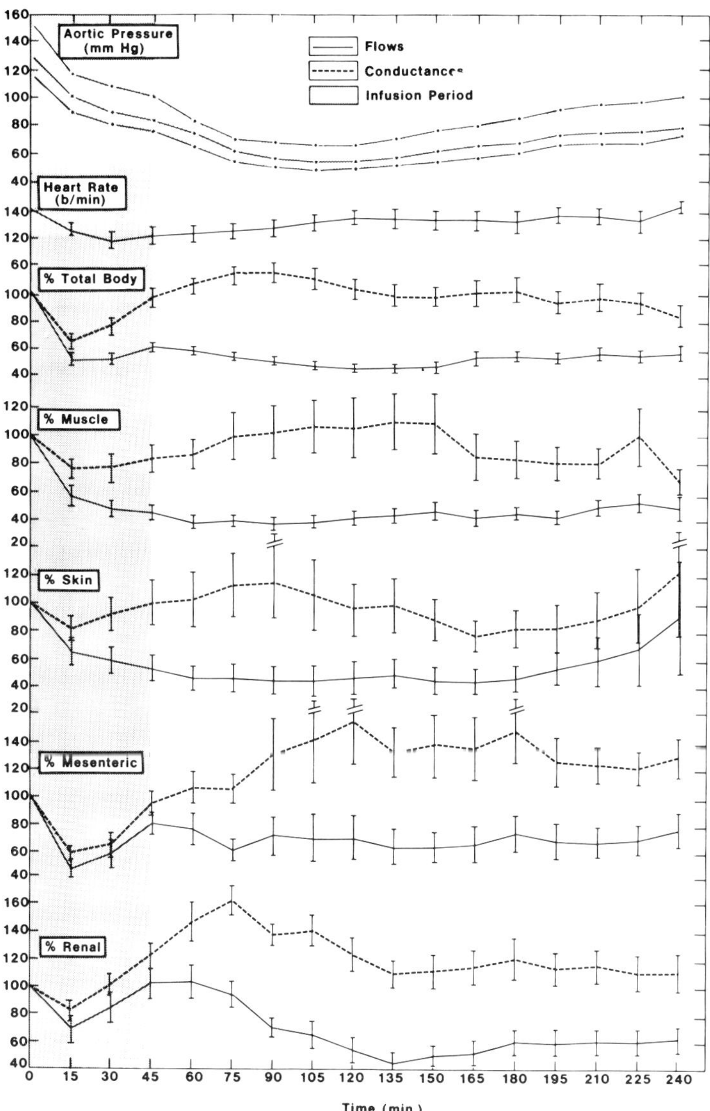

Fig. 1 *Composite of hemodynamic data ($\bar{x} \pm 1$ SEM) obtained during intravenous infusion of 2 mg/kg Escherichia coli endotoxin. Twenty dogs were observed for 4 hours. The vertical shaded area denotes the period of infusion (0-45 min). All data are presented as percentages of the pre-endotoxin control except for aortic pressure (mmHg) and heart rate (beats/min). The aortic pressure is divided into systolic, mean and diastolic pressures. To avoid confusion, SEM for pressure was not provided. Reproduced from Bond (1983), with permission.*

allowed constant monitoring of central venous pressures (CVP). Although not included in Figure 1, CVP never exceeded 2 mmHg, suggesting the lack of right ventricular decompensation during endotoxemia. Panel 3 in Figure 1 shows a fall in total body flow (cardiac output) to approximately 50% of control within 15 minutes where it remained for the duration of the 4-hour study. Total body conductance, however, fell transiently to 70% of control by 15 minutes after which it returned to control or above by 45 minutes. A similar vasoconstriction followed by arteriolar dilation was reported by Lillehei et al (1965).

In summary, this representative study demonstrates an initial decrease in total body conductance suggestive of an initial total peripheral vasoconstrictor response followed by a return to control or above, indicating a secondary loss of total vascular tone. The CVP data do not support myocardial depression, an observation which is consistent with the work of Hinshaw et al (1971), Hinshaw (1979), and Weil et al (1956).

3. REGIONAL VASCULAR RESPONSES TO LOCAL AND SYSTEMIC HYPOTENSION

When systemic vascular responses are to be critically evaluated in an attempt to determine discrete differences in controlling factors, one must consider several factors. First, how does the vascular bed respond to a fall in local perfusion pressure while systemic pressure remains unchanged? That is, how well does the vascular bed autoregulate its own blood flow, without the extrinsic influence of baroreceptor activation? Second, what level of control does the baroreceptor activation of the sympathetic nervous system have on the vascular bed? Third, when systemic blood pressure and local perfusion pressures are both reduced simultaneously by a stepwise hemorrhage or *Escherichia coli* endotoxin administration, which influence dominates: that of the local intrinsic vasodilator or of the extrinsic baroreceptor-induced vasoconstrictor? Answers to all these questions can be found through the careful analysis of pressure/vascular conductance relationships obtained from isolated vascular beds during (a) local hypotension induced by progressive occlusion of the arterial supply, (b) baroreceptor-induced hypotension during bilateral common carotid artery occlusion, (c) systemic hypotension induced by stepwise hemorrhage, and (d) systemic hypotension produced by intravenous administration of *E. coli* endotoxin. This approach will be used in the following subsections to evaluate vascular control in cerebral, myocardial, mesenteric (visceral), renal, skeletal muscle, and skin vascular beds.

3.1. Cerebral vasculature

3.1.1. Local hypotension

Of all the vascular beds discussed in this chapter, the cerebral vasculature exhibits the most efficient autoregulation over the widest pressure range. Rapela and Green have reported (1964) that the cerebral vessels are capable of intrinsically regulating their flow between perfusion pressures of 30 and 200 mmHg. This effective autoregulatory response is graphically presented in Figure 3 (below) in which vascular conductance increases from 100% to 225% as pressure is reduced from 100 to 40 mmHg (Green et al 1980). The primary factors responsible for this intrinsic regulatory response are the tissue levels of CO_2, pH and O_2, with elevated Pco_2 being the most dominant factor (Slater et al 1975). Thus, as perfusion pressure falls, flow also falls initially which reduces both the rate of CO_2 and H^+ removal and decreases O_2 availability. The elevated CO_2 and H^+ feed back upon the cerebral arterioles and cause them to relax allowing flow to return toward control.

3.1.2. Carotid sinus hypotension

Work published by Rapela et al (1967) indicates that carotid sinus denervation results in an elevated systemic pressure equivalent to the rise seen by Bond and Green (1969) during bilateral carotid sinus occlusion, without causing a significant change in cerebral venous outflow. Conversely, when systemic pressure was lowered by carotid sinus nerve stimulation, cerebral blood flow again remained unchanged. These authors conclude that the baroreceptor reflex has little or no control over the cerebral vasculature and that this vascular bed responds only to changes in perfusion pressure through the autoregulatory mechanism discussed above and not to any known extrinsic influence. The data in Figure 2, therefore, show that cerebral vascular conductance falls to 70% of control when the carotid sinus pressure in vagotomized dogs is reduced from 101 to 75 mmHg causing an increase in perfusing systemic pressure from 114 to 166 mmHg (see Table 1). Vagotomized animals were used in this study to minimize the conflicting influence of the aortic baroreceptors.

3.1.3. Hemorrhagic hypotension

The response of the cerebral vasculature to a stepwise reduction in systemic arterial pressure induced by hemorrhage is illustrated in Figure 3 (Green et al 1980). Note that over the pressure range of 100 to 40 mmHg, local hypotension caused an increase of conductance to 225%; yet over the same

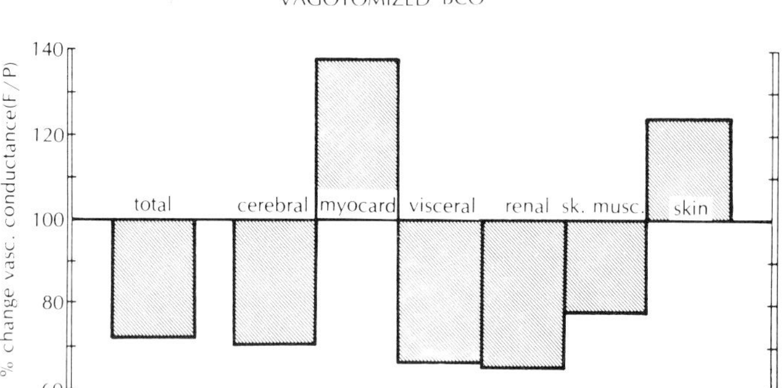

Fig. 2   *Histogram illustrating the changes in total and regional vascular conductances when both common carotid arteries of an aortic arch baroreceptor denervated dog (vagotomized) are occluded (BCO). The data in Table 1 show the carotid sinus and systemic pressure responses to this occlusion. Note that all vascular beds except myocardial and skin show some degree of vasoconstriction.*

TABLE 1   *Effect of bilateral common carotid artery occlusion (BCO) on mean aortic pressure (MAP) and mean carotid sinus pressure (MCSP) in normal and vagotomized dogs*

	Control (mmHg)		During BCO (mmHg)	
	MAP	MCSP	MAP	MCSP
Normal	98	88	117	73
Vagotomized	114	101	162	75

Data taken from Bond and Green (1969).

pressure range, conductance only increased to 200% of control during acute hemorrhage. This modest decrease from 225% to 200% (shaded area) is caused by some as yet undetermined mechanism. One possible explanation could be related to the decreased arterial Pco_2 which occurs as a result of the hyperventilation known to be present during hemorrhagic hypotension (Bond et al 1977). Because of the cerebral autoregulation described above this fall in Pco_2 would increase vascular tone. When pressure is reduced much below 40 mmHg by either hemorrhage or local hypotension, the cerebral autoregulatory capacity is exceeded, and conductance will fall proportionally faster than pressure (Green et al 1980).

In another study using radioactive microspheres, Slater et al (1975) discovered that even though total cerebral flow did decrease somewhat during

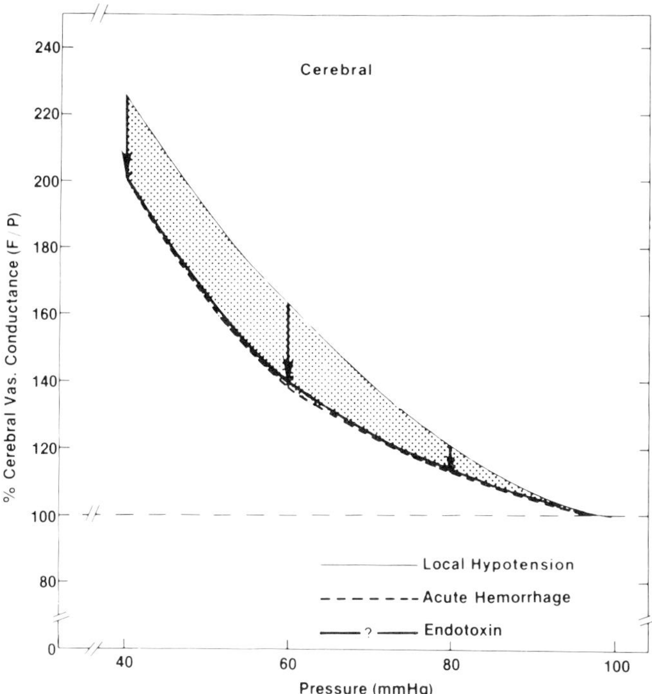

Fig. 3 *Pressure/cerebral vascular conductance responses to local hypotension, acute hemorrhage and endotoxin hypotension. The local hypotension and acute hemorrhage lines were drawn from data presented by Green et al (1980) and suggest strong intrinsic vascular control in this bed. The endotoxin response is designated with a question mark since there are no definitive pressure/conductance data available in the literature. The arrows which transverse the shaded area illustrate a small fall in conductance during systemic hypotension.*

hemorrhage, the flow distribution pattern to the gray matter, cerebellum, hypothalamus, white matter and medulla were equally reduced during both early and late oligemia. These authors suggest the likelihood of a loss of central nervous system function during both of these stages. In 1972 Kovách reported that hypothalamic and frontal cortical flow fell more than the flow in other areas of the brain. He reported that even though cerebral O_2 utilization was increased during shock, it was not metabolically coupled, and this resulted in localized areas of cerebral hypoxia.

3.1.4. Endotoxin hypotension

There is a surprising level of agreement in the literature concerning the cerebral vascular hemodynamic behavior during endotoxic hypotension. It would appear that even though there is a modest fall in total cerebral blood flow similar to that seen during hemorrhagic hypotension, the fall in pressure is even greater, suggesting significant arterial vasodilation (or increased vascular conductance (Fig. 3)) in an attempt to protect cerebral function. Since a specific pressure/conductance study during endotoxin has not yet been performed, the data plotted in Figure 3 have been labeled with a question mark. The present author is assuming the line to be similar to the hemorrhage line. These conclusions were also reached by Wyler et al (1969) and Rutherford et al (1976) in unanesthetized Rhesus monkeys, and by Raymond and Emerson (1978) and Ekström-Jodal et al (1982) in anesthetized dogs.

In summary, even though the cerebral vasculature has been shown to exhibit excellent autoregulatory ability during systemic hypotension induced by both acute hemorrhage and endotoxin, there is little doubt that sustained hypotension to levels less than 40 mmHg does result in cerebral dysfunction. In patients with preexisting cerebral vascular disease, the extent of cerebral dysfunction would be even greater. For a more comprehensive review of this subject, the reader is referred to reviews by Kovách and Sándor (1976) and Kuschinsky and Wahl (1978).

3.2. Myocardial vasculature

3.2.1. Local hypotension

Because of the constant motion of the beating heart, the autoregulatory ability of the myocardial vasculature is difficult to evaluate, but studies suggest that this vascular bed exhibits good autoregulatory ability over a wide pressure range (Rubio and Berne 1975). Berne (1963) reported that a fall in myocardial perfusion pressure from 100 mmHg to 75 mmHg resulted in a transient fall in flow followed by a return to near control levels. The 'local hypotension' line in Figure 4 illustrates this as a rise in conductance from 100% at 100 mmHg, to over 150% at 60 mmHg, after which it falls to control by 40 mmHg. Berne suggested that since the heart extracts nearly all of the available oxygen from the perfusing blood, even a slight flow reduction would cause a relative ischemia and the accumulation of vasodepressor nucleotides. These vasoactive metabolites feed back on the arterioles causing them to dilate, allowing conductance to increase and flow to return toward the control level. However, once the perfusion pressure falls between 60 and

40 mmHg, the downstream arterioles would have attained maximum dilation and any further pressure fall would result in passive relaxation within the elastic elements of the vasculature.

The pressure/conductance relationships illustrated in Figure 4 summarize the myocardial vascular responses to local hypotension, acute hemorrhagic hypotension and endotoxic hypotension. Note that as the local perfusion pressure is reduced, the conductance increases to a maximum of 150% of 60 mmHg in an effort to maintain flow almost constant (i.e., autoregulation).

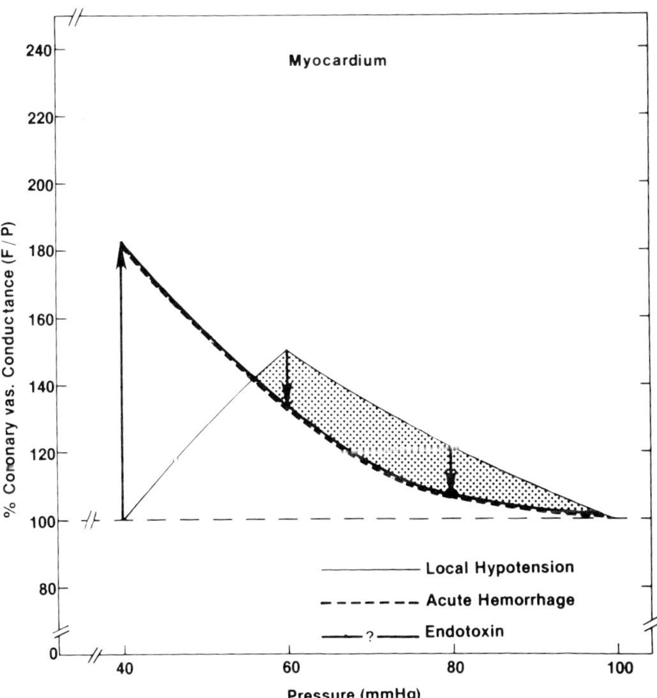

Fig. 4 *Pressure/myocardial vascular conductance responses to local hypotension, acute hemorrhage and endotoxin hypotension. The rise in conductance during local hypotension above the 100% line as pressure falls suggests good intrinsic vascular control down to pressures of about 60 mmHg. The downward shift in the acute hemorrhage and endotoxin curves suggests a slightly reduced conductance probably due to a lowering in cardiac work because of the systemic hypotension. Data published by Bond et al (1973) verify that the coronary vascular conductance continues to rise as systemic pressure falls to 40 mmHg during hemorrhagic hypotension. Available data on the coronary vasculature during endotoxin are lacking at pressures below 50 mmHg; therefore a question mark has been applied to the endotoxin curve.*

When the systemic pressure is reduced, however, by either stepwise hemorrhage or endotoxin, there is a small reduction in flow at each step between 100 and 40 mmHg resulting in a 180% conductance. This could be explained by either a slightly increased vascular tone induced by baroreceptor activation of the sympathetic nervous system controlling the myocardial vasculature at the lower pressures, or, more likely, it can be the result of the reduced work of the heart which occurs during systemic hypotension and the decreased cardiac output. The continued rise in conductance during systemic hypotension between 60 and 40 mmHg cannot be explained.

3.2.2. *Carotid sinus hypotension*

DeGeest et al (1964) have reported that carotid sinus hypotension reflexly increases in sympathetic activity to the heart and causes increased contractility and tachycardia. According to Berne (1963), and Rubio and Berne (1975) the increased metabolic demand would result in a relative hypoperfusion and hypoxia and an autoregulatory vasodilation. Since the cardiac output did not change but systemic pressure rose from 114 to 162 mmHg during carotid sinus hypotension from 101 to 75 mmHg (Bond and Green 1969; see also Table 1), the external work of the heart increased. Consequently, myocardial blood flow increases proportionally more than perfusion pressure, which causes a rise in myocardial vascular conductance to 138% of control (Fig. 2). These observations confirm the hypothesis that baroreceptor hypotension results in a metabolically increased myocardial vascular conductance.

3.2.3. *Hemorrhagic hypotension*

Wiggers and Werle (1942) were the first to suggest the possibility of a relationship between reduced coronary blood flow and irreversible shock during hemorrhage. In 1968 Gregg, in collaboration with Granata, published a series of studies on shock which showed a rise in myocardial vascular conductance in spite of a moderate fall in coronary flow. This increased conductance occurred in both compensatory hypovolemic and decompensatory normovolemic hypotension. This tendency to maintain flow is consistent with the previously described coronary vascular response to a local hypotension. In addition to this autoregulatory response, a declining diastolic tissue perfusion pressure gradient has been shown to occur as a result of the decreased work of the heart (Adiseshiak and Baird 1978). The results of a study by Bond et al (1973), in which pentobarbitalized dogs were used, have been supported by Archie and Mertz (1978), Lee and Downing (1976), and Rutherford et al (1976) in unanesthetized Rhesus monkeys and suggest that there is only a modest reduction in myocardial blood flow and oxygen consumption during compensatory hypovolemia and decompensatory normovolemia. Archie and Mertz (1978) reported that when arterial oxygen content is reduced prior to the hemorrhage, evidence of subendocardial ischemia followed by cardiac failure is often seen. It would therefore appear that when arterial blood gases are normal, the animal models described above are capable of maintaining a metabolically healthy myocardium during both local and hemorrhagic hypotension down to pressures of 40 mmHg (Fig. 4).

3.2.4. Endotoxin shock

A careful search of the literature failed to reveal any studies in which the human myocardial vascular dynamics were evaluated during septic shock. However, Wyler et al (1969), Hinshaw et al (1971), Spitzer et al (1974), and Rutherford et al (1976) all reported that myocardial blood flow in both anesthetized and unanesthetized nonhuman primates and anesthetized dogs tended to be maintained as the systemic blood pressure fell during endotoxic shock (i.e., a response of the autoregulatory type). More recent studies by Bruni et al (1978), and Kause et al (1980) report that myocardial flow varied during different stages of endotoxic shock depending upon the work load of the heart. Since no specific data are currently available for the pressure/conductance relationships during endotoxin, the curve in Figure 4 is indicated with a question mark.

In summary, it would appear that the myocardial vascular response to endotoxic shock is quite similar to that of hemorrhagic shock and that the final determinants of blood flow and vascular resistance in the heart depend upon the myocardial metabolic need rather than being directly under the extrinsic influence of the sympathetic nervous system or circulating vasoactive agents. Therefore, whenever the systemic pressure and cardiac output are reduced as in either hemorrhagic or endotoxic hypotension, the cardiac power (power = mean aortic pressure × cardiac output (MAP × CO)) is also reduced, which decreases myocardial metabolic demand. The result is a shift in the pressure/conductance curve in Figure 4 down and to the left. The clinician is reminded that since the myocardial blood flow is a function of tissue perfusion pressure measured distal to an obstruction, patients having coronary artery disease will show signs of myocardial ischemia and failure, even though systemic arterial pressure remains within the autoregulatory range.

3.3. Mesenteric (visceral) vasculature

3.3.1. Local hypotension

Autoregulation of blood flow to the intestine has been described by Johnson (1964) as a 'labile phenomenon which is less pronounced than that seen in either brain or kidney.' Note in Figure 5 that mesenteric vascular conductance increases to 118% of control when pressure is reduced from 100 to 60 mmHg, compared to a rise in cerebral vascular conductance to 162% of control over this same pressure range (Fig. 3). As pressure is reduced from 60 to 40 mmHg, conductance falls to control. Since Johnson was able to demonstrate intestinal autoregulation by either increasing venous pressure or decreasing arterial pressure, the suggestion was made that autoregulation in the intestine is not flow-dependent. Other experiments by him in which he exposed the tissue to decreased oxygen tension indicated no change in vascular resistance. From these studies, it would appear that the intestine regulates its flow in response to a myogenic mechanism directed at maintaining a constant capillary pressure rather than a constant flow.

3.3.2. Carotid sinus hypotension

Reports by Polosa and Rossi (1961), and Ernsting and Parry (1957) suggest that the baroreceptors are capable of initiating reflex control over the visceral vasculature. This concept was supported by Bond and Green (1969), whose data are summarized in Figure 2 and Table 1. These data indicate a fall in visceral vascular conductance to 66% of control when carotid sinus pressure is reduced from 101 to 75 mmHg in the vagotomized dog (aortic baroreceptor denervation). This confirms the hypothesis that the carotid baroreceptor reflex does exercise considerable extrinsic control over the visceral vascular tone.

3.3.3. Hemorrhagic hypotension

The broken line in Figure 5 presents the pressure/conductance relationship obtained when the perfusion pressure to the viscera is reduced by stepwise hemorrhage. Note that as the pressure is reduced from 100 to 40 mmHg, the vascular conductance falls to 55% of control. The difference between the local hypotension curve and the hemorrhage curve (shaded area) symbolizes the intensity of the vasoconstriction induced by the extrinsic factors such as circulatory neurohumors and augmented sympathetic neural activity (Bond 1983; Green et al 1980). This fall in conductance causes a significant reduction in oxygen utilization by both the liver and intestines (Selkurt and

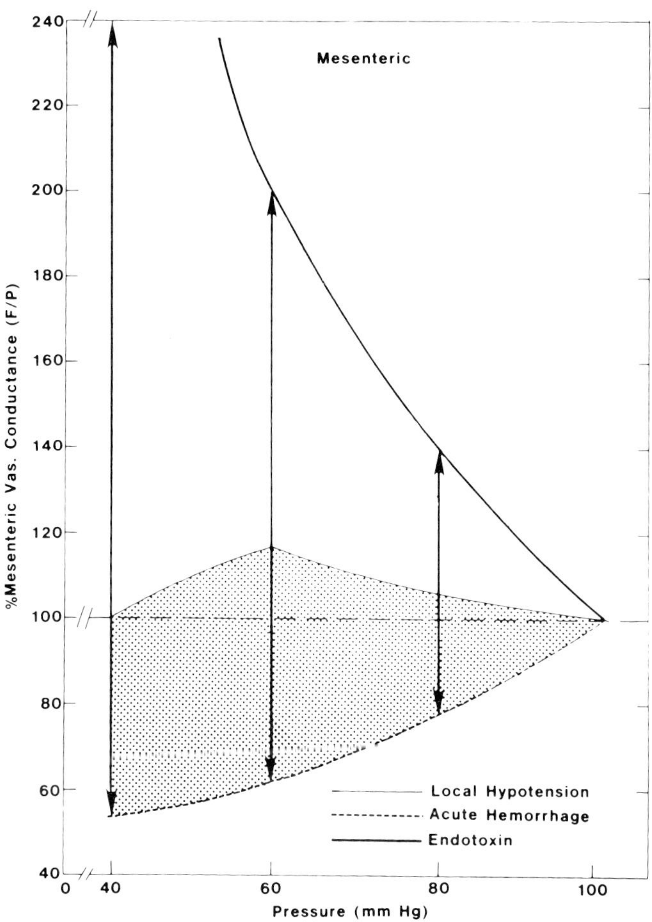

Fig. 5 *Pressure/mesenteric vascular conductance responses to local hypotension, acute hemorrhage and endotoxin. The mesenteric circulation shows only a modest degree of intrinsic control during local hypotension as indicated by the small rise in conductance to 118% at 60 mmHg. The conductance falls well below this intrinsic control line during hemorrhagic hypotension providing evidence of a strong extrinsic override of intrinsic mechanisms (shaded area). The great increase in conductance as pressure falls during endotoxin suggests that not only are the extrinsic controlling factors inhibited but there must also be some active vascular smooth muscle inhibition. Modified from Bond (1983), with permission.*

Brecher 1956). In addition, the dog often exhibits a dramatic increase in hepatic vascular resistance giving rise to a condition favoring intestinal blood pooling. Selkurt and Brecher have suggested that the impaired oxygen utilization by the viscera may be an important factor in the inability of the splanchnic circulation to compensate for hemorrhagic shock. In support of these reports, Bohlen et al (1975) have documented vasoconstriction in the rat intestinal villi during hemorrhage. When these experiments were repeated in denervated preparations, however, the vasoconstrictor response was converted to a vasodilation suggestive of autoregulation. These data would suggest that the intestinal vasoconstriction was predominantly under the control of the sympathetic nervous system rather than circulatory vasoactive agents and that it was activated by baroreceptor hypotension. In addition, these authors were able to show that intestinal denervation serves to protect the integrity of the intestinal mucosa during shock.

3.3.4. Endotoxin shock

It has long been recognized that considerable visceral damage occurs in dogs subjected to endotoxin (Lillehei and Mac Lean 1958). This damage has been reported to occur as a result of intense visceral vasoconstriction causing portal hypertension and visceral pooling (Brobmann et al 1970; Brungardt et al 1972; Hinshaw and Nelson 1962). Studies by Wyler et al (1969), Swan et al (1971), and Rutherford et al (1976) all agree that the response of the intestinal vasculature of both anesthetized and unanesthetized nonhuman primates to endotoxin is vasodilation rather than the expected vasoconstrictor baroreceptor response to systemic hypotension seen in dogs (Fig. 2). This apparent species difference between dogs and nonhuman primates may have some anatomical basis in that the vascular architecture of these species has been shown to be different (Reynolds et al 1971). This has been used as an argument against the use of dogs in shock studies. In a recent report on anesthesized dogs by Bond (1983), which is presented in Figure 1, endotoxin infusion resulted in an initial decrease in mesenteric blood flow to 55% of control at 15 minutes. The calculated conductance fell to 60% of control indicating significant vasoconstriction, which is in agreement with the above mentioned studies. However, between 15 and 120 minutes this intense vasoconstriction gave way to a powerful loss of vascular tone such that the mesenteric conductance increased from 60% to over 150% of control by 150 minutes post endotoxin (Fig. 1). This biphasic response was consistent with an earlier report by Vaughn et al (1967), who also used dogs.

The pressure/conductance data for the mesenteric vascular bed taken from Figure 1 between the minimum conductance at 15 minutes and maximum conductance at 150 minutes are plotted in Figure 5. Note that if one normalizes the minimum conductance to 100% (since that point also coincides with 100 mmHg systemic pressure), the conductance increases to over 200% of control by 60 mmHg. This tremendous loss of vascular tone far exceeds the vasodilator response of 118% that occurs as a result of intrinsic autoregulatory mechanisms and further indicates a marked inhibition of the normal vascular tone.

In summary, the visceral vascular response to hemorrhagic hypotension seems to be dominated by extrinsic factors leading to intensive vasoconstriction in both canine and nonhuman primate animal models. Though it is generally accepted that a species difference does exist between dogs and nonhuman primates during hypotension induced by endotoxin, data are presented which suggest that the difference may not be as great as previously thought.

3.4. Renal vasculature

3.4.1. Local hypotension

The local hypotension curve in Figure 6 illustrates that the renal vascular conductance increases to about 140% of control as pressure drops from 100 to 70 mmHg after which there is a fall to 100% at 50 mmHg (Navar et al 1982). Most investigators agree that this autoregulatory vasodilator response between 100 and 70 mmHg primarily occurs in the afferent arterioles, even though considerable disagreement concerning the mechanism exists. In a recent review of the intrinsic control of renal hemodynamics, Navar et al (1982) provide evidence of both myogenic and local humoral control. The humoral aspect seems to include interaction between the renin-angiotensin system and tubuloglomerular feedback as well as a complex interaction with prostaglandins.

3.4.2. Carotid sinus hypotension

The response of the renal vasculature to carotid sinus hypotension is qualitatively similar to the response seen in the viscera suggesting active vasoconstriction. The report by Bond and Green (1969) summarized in Figure 2 describes a fall in conductance to 65% of control as carotid sinus pressure falls to 75 mmHg, which indicates strong extrinsic control. Since the renal vascular conductance actually falls below control during the systemic hypotension which occurs during carotid sinus hypotension, the response may involve more than just vascular autoregulation. It may, in fact, include a strong extrinsic neurohumoral mechanism such as reviewed by Navar et al (1982). McGiff and Fasy (1965) were the first to report that the baroreceptor

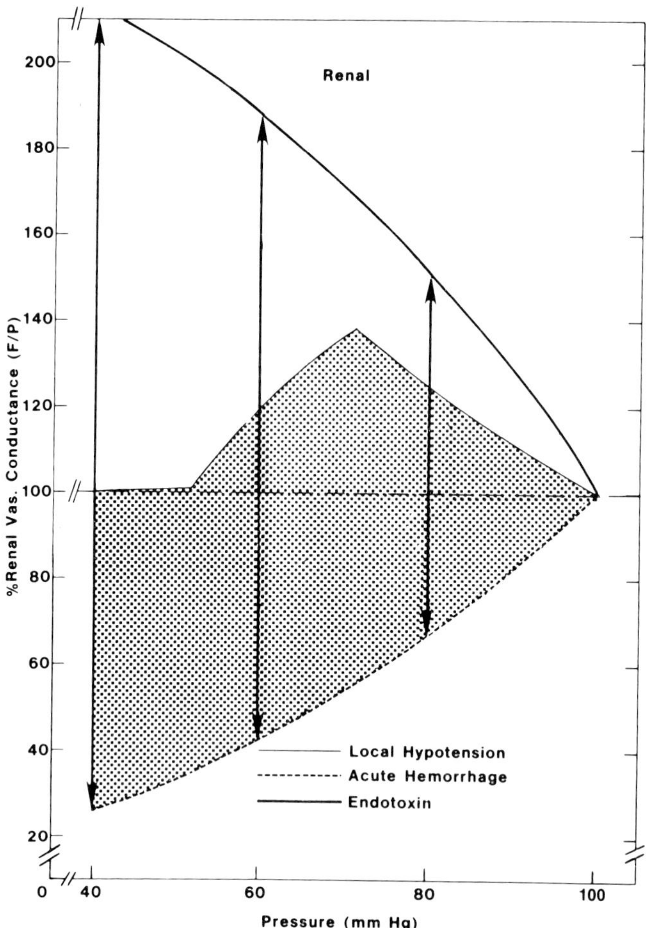

Fig. 6 *Pressure/renal vascular conductance responses to local hypotension, acute hemorrhage and endotoxin. This vascular bed shows a strong intrinsic response to local hypotension between 100 and 70 mmHg below which the conductance falls back to the 100% control line. The sharp fall in conductance seen during hemorrhagic hypotension below the local hypotension line (shaded area) illustrates the strong extrinsic control present in this vascular bed. The rise in conductance above the local hypotension line in response to hypotension induced by endotoxin again suggests an inhibition of smooth muscle tone. Modified from Bond (1983), with permission.*

reflex can initiate the renal release of angiotensin II, which in concert with an intact autonomic nervous system may act synergistically to cause a more powerful vasoconstriction than that seen in the viscera.

3.4.3. Hemorrhagic hypotension and shock

As can be seen in Figure 6, the renal vasculature responds to hemorrhagic hypotension by an intense vasoconstriction (i.e., conductance falls to 25% of control as pressure is reduced from 100 to 40 mmHg). Even following the reestablishment of normal blood volume, both systemic pressure and renal flow have been shown to return toward control but are not well maintained as the animals enter terminal shock. Thus, the renal vascular conductance remains depressed below control throughout the entire shock protocol (Beaty et al 1976; Green et al 1980). This persistent renal vasoconstriction can be accounted for by both neural and humoral mechanisms. First, the baroreceptor response to systemic hypotension includes both an increased activation of renal α-adrenergic activity, and second, an elevation in plasma catecholamine concentration (Watts and Westfall 1964). Since the kidney has been reported to respond to adrenergic stimuli with great intensity (Green and Kepchar 1959; Green et al 1965), it is not surprising that such a powerful vasoconstriction should occur during hemorrhage. In addition, increased plasma levels of the potent renal vasoconstrictor renin have been reported to occur in response to hypovolemic hypotension (Beaty et al 1976; Navar et al 1982).

3.4.4. Endotoxin shock

The renal hemodynamic pattern seen in anesthetized dogs during and after 2 mg/kg endotoxin administration is presented in Figure 1 (Bond 1983). Note that the initial fall in conductance to approximately 85% of control suggests moderate vasoconstriction. This is followed at 15 minutes by the onset of a progressive loss of vascular tone until 75 minutes. At this point a maximum of 160% conductance is reached. This loss of tone is then followed by a return toward control over the succeeding 60 minutes. When the data between 15 and 75 minutes are plotted on the pressure/conductance graph in Figure 6, one can readily appreciate the magnitude of the endotoxin-induced secondary vasodilation, which greatly exceeds the normal autoregulatory (local hypotension) curve, suggesting that some factors other than those described above to explain autoregulation in the kidney are involved. Similar data have been reported by Wyler et al (1969) and Rutherford et al (1976) in unanesthetized nonhuman primates, and Vaughn et al (1967) and Emerson et al (1966) in anesthetized dogs. Wyler et al (1969) suggest that the massive renal vasodilation is due to endotoxin acting either directly or indirectly upon some unidentified plasma substrate which causes the release of an agent such as kinins which are strong enough to overcome the baroreceptor-induced vasoconstriction.

In summary, the renal vasculature does exhibit excellent intrinsic control of blood flow when local perfusion is reduced to 70 mmHg. However, when perfusion pressure is reduced by hemorrhage, this vascular bed responds by intense vasoconstriction brought about by both adrenergic and renin angiotension stimulation. An initial modest vasoconstriction is seen with endotoxin which is followed by vasodilation suggesting an inhibition of the normal extrinsic response to systemic hypotension.

3.5. Skeletal muscle vasculature

3.5.1. Local hypotension

The data which describe the vascular response of skeletal muscle to local hypotension are presented in Figure 7 and were taken from studies by Bond et al (1969). These data show a rise in conductance between 100 and 60 mmHg to 133% with a passive fall below 60 mmHg. In 1964, Stainsby reported a positive relationship between the skeletal muscle autoregulatory capability and increased metabolic activity. Also in 1964, Guyton et al reviewed several studies which indicated vasodilation in both a hindlimb preparation and small isolated arteries in response to reduced oxygen. The studies by Stainsby (1964) and Guyton et al (1964) were later confirmed by those of Hutchins et al (1974), who noted significant small artery dilation when inspiratory oxygen was reduced. Thus, there is little doubt that the skeletal muscle vasculature does dilate in response to tissue hypoxia and that a metabolic mechanism is an important part of the autoregulatory response in this tissue. In 1969, however, Bond et al provided evidence that tissue hypoxia is not the only determinant of skeletal muscle autoregulation. Their data indicate that the vasculature of skeletal muscle is capable of exhibiting good autoregulatory control even under hyperoxic conditions, which suggests that both myogenic and metabolic autoregulatory mechanism(s) reside in the skeletal muscle vasculature.

3.5.2. Carotid sinus hypotension

The vascular response of skeletal muscle to a decreased carotid sinus pressure is diagramatically depicted in Figure 2. This histogram illustrates a fall in vascular conductance of skeletal muscle to 78% of control when carotid sinus perfusion pressure falls from 101 to 75 mmHg in the vagotomized dog, caus-

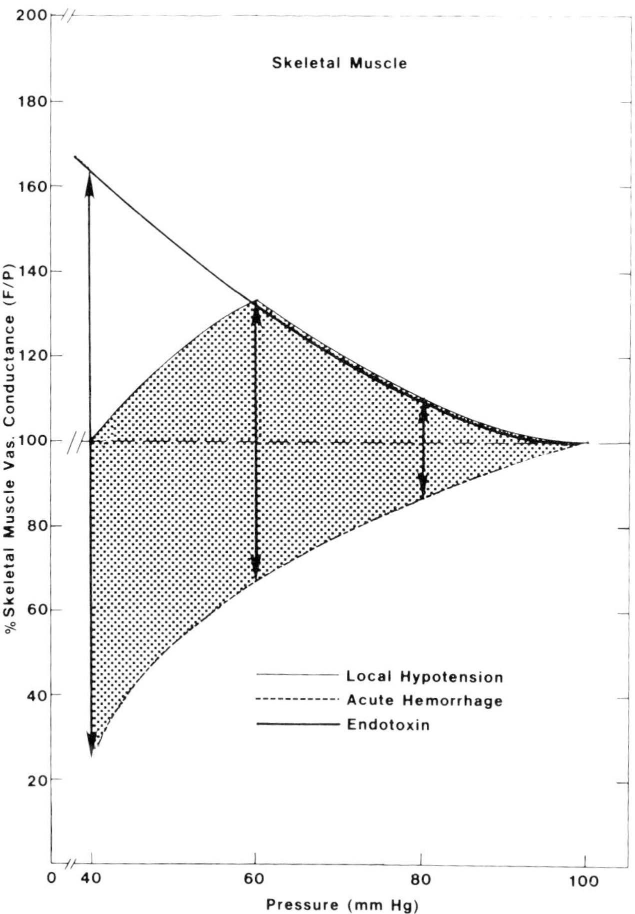

Fig. 7   *Pressure/skeletal muscle vascular responses to local hypotension, acute hemorrhagic hypotension and endotoxic hypotension. Note that the intrinsic regulation of this vascular bed is good between 100 and 60 mmHg after which it falls to control. The response to hemorrhage suggests that extrinsic mechanisms are capable of overcoming the local regulatory mechanisms (shaded area). The endotoxin curve follows the local hypotension curve very closely between 100 and 60 mmHg after which it continues to rise as the local hypotension curve falls, suggesting that endotoxin can also inhibit the normal intrinsic responses in this vascular bed. Modified from Bond (1983), with permission.*

ing systemic pressure to increase from 114 to 162 mmHg (Table 1). This level of baroreceptor control is not as strong as that seen in either visceral or renal vasculature but it has been shown to initiate the rise in systemic pressure coincident with the onset of carotid sinus hypotension (Bond and Green 1969). Confirmation by data on the microvascular response was provided by Hutchins et al (1974), who reported that vasoconstriction in the small (40 µm) distributing arterioles supplying the rat cremaster occurs during bilateral carotid artery occlusion. Skeletal muscle has been shown to autoregulate up to pressures of 90-100 mmHg, but above this pressure this bed behaves as a passive vascular bed, that is, as the skin (Green et al 1965).

3.5.3. *Hemorrhagic hypotension and shock*

The typical vascular response of skeletal muscle to a stepwise hemorrhage to 40 mmHg is seen in Figure 7 (Bond et al 1967). Note that the skeletal muscle vascular bed responds to a controlled graded local hypotension in an autoregulatory manner (conductance increases) between 100 and 60 mmHg, but responds to graded hemorrhagic hypotension by reducing the vascular conductance to less than 30% of control at 40 mmHg. The difference between these two curves (shaded area) illustrates the dominance of the extrinsic over the intrinsic mechanisms. In a series of studies designed to evaluate the roles of the autonomic nerves and circulating adrenergic transmitters during acute hemorrhage, it has been determined that approximately half of the extrinsic response can be eliminated by denervation and the remaining half by pretreatment with adrenergic blockers (Bond et al 1967, 1973, 1977, 1979; Gonzales and Bond 1974).

3.5.4. *Endotoxin shock*

Rutherford et al (1976) and Wyler et al (1969) report a reduced skeletal muscle flow and conductance in nonhuman primates subjected to endotoxin. In canine endotoxin shock, Hinshaw and Owen (1971), and Hopkins and Damewood (1974) agree that blood flow and conductance in skeletal muscle are both severely reduced. These canine studies are in some conflict with the reports of Wright et al (1971), who report that the blood flow and conductance in skeletal muscle both increase during septic and endotoxin shock. Wright et al (1971) state that the increased flow is accompanied by a reduced arteriovenous oxygen extraction suggesting the opening of nonnutritive flow channels.

The data presented in Figure 1 (Bond 1983) exhibit an immediate fall in skeletal muscle flow to 40-50% of control. In addition, the conductance shows an initial fall to about 78% of control followed by return to near control over the succeeding two hour period. The subsequent loss of vascular tone (conductance rise) is plotted against the mean aortic pressure in Figure 7. Note that as pressure falls from 100 to 40 mmHg, the vascular conductance increases in a manner equal and parallel to the local hypotension curve between 100 and 60 mmHg, but continues to rise below 60 mmHg while the autoregulation curve falls to control.

In summary, these data suggest that the extrinsic influence which dominates the skeletal muscle vasculature during hemorrhagic hypotension has been totally eliminated during endotoxic hypotension. In fact, when systemic pressure falls below 60 mmHg during endotoxin, the loss of vascular tone in skeletal muscle even exceeds the autoregulatory ability to dilate, which suggests that endotoxin causes an inhibition of vascular smooth muscle tone.

3.6. Cutaneous vasculature

3.6.1. *Local hypotension*

The thin solid line in Figure 8 represents the pressure/conductance relationship in the canine cutaneous vasculature when the local perfusion pressure is reduced in a stepwise manner (Bond 1983; Bond et al 1967; Green et al 1980). Note that, in contrast to all other vascular beds presented, this line is curvilinear with the convexity directed toward the pressure axis. This fall in conductance to nearly 60% of control at 40 mmHg indicates a more rapid decrease in flow than pressure and is characteristic of a non-autoregulation or passive vascular bed (Green et al 1972). The term passive vascular bed implies that the vascular walls behave in a predictably elastic manner in accordance with changes in intraluminal pressure. The apparent vasoconstriction is caused by a passive release of tension on the vessel walls with the fall in intravascular pressure and, conversely, passive dilation takes place when pressure rises.

3.6.2. *Carotid sinus hypotension*

Cutaneous vascular conductance increases to 124% of control when carotid sinus pressure falls to 75 mmHg (Fig. 2), and this causes a reflex increase in the pressure perfusing the skin from 114 to 162 mmHg (Table 1). Note that the myocardial vascular bed is the only other bed exhibiting an increased conductance during carotid sinus hypotension. This vascular response of the skin is consistent with the characteristic passive behavior of the vascular bed (discussed above) in the case of a rise in systemic perfusion pressure which accompanies the carotid sinus hypotension (Bond and Green 1969). Even

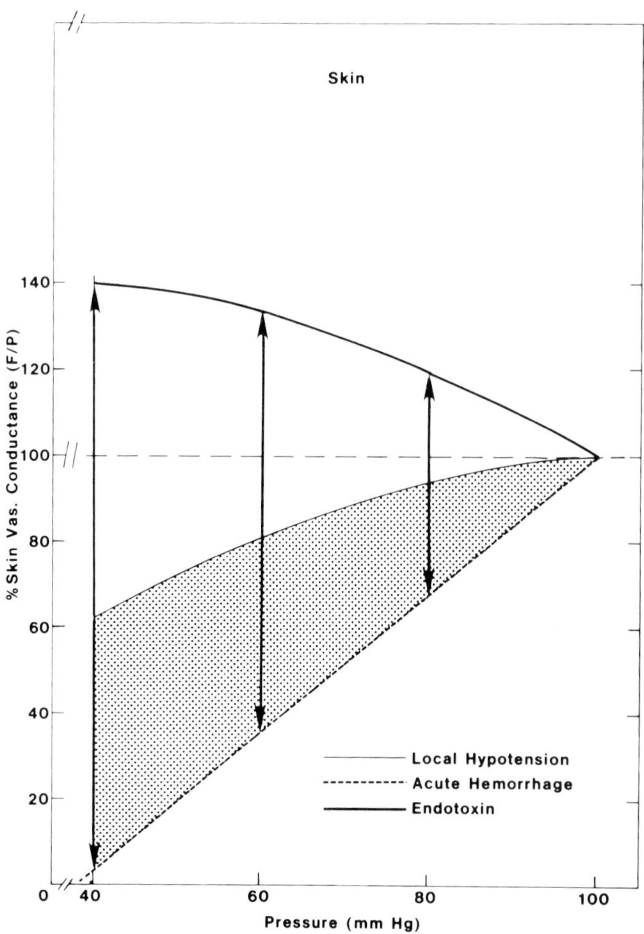

Fig. 8 *Pressure/skin vascular conductance responses to local hypotension, acute hemorrhagic hypotension and hypotension induced by endotoxin. The gradual fall in conductance as local hypotension progresses verifies that this vascular bed does not intrinsically control its vasculature but rather behaves in a passive manner as pressure falls. The curve falls even more rapidly during hemorrhage which suggests that the extrinsic control does overpower the passive response (shaded area). The increase in conductance above control again verifies that endotoxin can act either directly or indirectly to inhibit the normal extrinsic control. Modified from Bond (1983), with permission.*

58

though the skin vasculature is known to contain α-adrenergic vasoconstrictor fibers (Green and Kepchar 1959), these fibers apparently do not participate in the regulation of blood pressure through the baroreceptor reflex (Bond and Green 1969).

3.6.3. Hemorrhagic hypotension

The configuration of the curve (Fig. 8) shows that there is a passive response during acute hemorrhage (systemic hypotension), but the downward shift of the curve indicates a more pronounced vasoconstriction at each pressure level. Unlike the skeletal muscle vasculature in which denervation caused a shift of the pressure/conductance relationship back toward the autoregulatory curve, denervation of the cutaneous vasculature has virtually no effect on the pressure/conductance relationships during hemorrhage (i.e., the curves would be coincident). This observation suggests the total absence of an active neural vasoconstriction component in cutaneous vasculature during hemorrhagic hypotension. Therefore, the downward shift of the hemorrhage curve can be explained on the basis of elevated plasma vasoconstrictor agents, the influence of which is sustained throughout the irreversible shock stage (Bond et al 1967). In another study which confirms this hypothesis (Bond et al 1970), the cutaneous vasculature was isolated and pump-perfused at a constant flow with blood derived from the aorta of the shocked animal. As the mean arterial pressure was reduced by hemorrhage, there was a progressive increase in the mean pressure perfusing the cutaneous vascular bed, indicating an increasing vascular resistance (i.e., vasoconstriction). Since denervation failed to alter the perfusion pressure, additional evidence favoring the lack of a neural vasoconstriction was provided. The possibility that the blood-borne vasoactive agent might be renin (Beaty et al 1976; McGiff and Fasy 1965; Navar et al 1982), which is known to be released from the kidney during hemorrhage, was rejected when repetition of these studies in nephrectomized dogs (Bond et al 1970) failed to eliminate the vascular response. The response was eliminated, however, by the prehemorrhage administration of the adrenergic inhibitor phenoxybenzamine. It was therefore concluded that the primary circulating agent(s) responsible for the powerful vasoconstriction during hemorrhagic hypotension and shock are the catecholamines released into the circulation by both the adrenal medulla and all active peripheral adrenergic nerve terminals, and that little or no effect comes from the baroreceptor activation of the adrenergic nerves directly innervating the cutaneous vasculature (Bond and Green 1969; Watts and Westfall 1964; see also Fig. 2).

3.6.4. Endotoxin shock

The first study that specifically investigated the cutaneous vascular response to endotoxin shock was published by Wyler et al (1969). Using the microsphere method of blood flow measurement, these authors determined that the cutaneous vascular resistance in the unanesthetized monkey fell to about 60% of control 24 hours following the infusion of 10 mg/kg endotoxin. The data presented by Bond (1983) can be seen in panel 5 of Figure 1. Note that both skin flow and conductance fall over the first 15 minutes of endotoxin infusion. After this period, conductance increases from a minimum of 85% to a maximum of 115% even though flow continues to fall moderately. When the pressure/conductance data from panels 1 and 5 are plotted with 85% being normalized to 100% in Figure 8, the skin vascular tone shows an increased conductance to 140% at 40 mmHg.

In summary, the large shaded area between the intrinsically derived curve (i.e., local hypotension) and acute hemorrhage curve (Fig. 8) suggests that there is a strong extrinsic cutaneous vasoconstriction during hemorrhage. Endotoxin causes a significantly increased conductance at each pressure step compared with the local hypotension data suggesting that not only is this vascular bed made insensitive to the extrinsic control seen in systemic hypotension induced by hemorrhage, but endotoxin also inhibits the vascular smooth muscle elasticity or contractile mechanism. In any event, these data, which are consistent with reports on all other vascular beds, suggest that endotoxin causes a secondary loss of vascular tone which the extrinsic mechanisms cannot overcome.

4. SUMMED REGIONAL VASCULAR RESPONSES

Since systemic blood pressure is equal to the product of the cardiac output and total peripheral vascular resistance, the role played by any individual bed in pressure regulation must be directly proportional to the fraction of the cardiac output perfusing that bed. The data in column 1 of Table 2 provide the percentages of distribution of the cardiac output in dogs that perfuses the six regional vascular beds presented in this chapter (Green 1950). Note that the sum of these percentages is nearly 100%. The second column in Table 2 depicts the calculated expected flow to each vascular bed assuming a cardiac output of 3000 ml/min which is the value published by Bond (1983) and used in Figure 1. The third column illustrates the calculated vascular conductance within each vascular bed based on a perfusion pressure of 100 mmHg. These individual conductances represent the control of 100% values used to construct Figures 3-8. Since the regional beds are in a parallel

TABLE 2   *Control canine regional vascular flows and conductance distribution*

	% of cardiac output	Flow (ml/min)	$F/P \left(\dfrac{ml \cdot mmHg}{min} \right)$
Cerebral	3.6	108	1.08
Myocardial	4.7	141	1.41
Mesenteric	25.1	753	7.53
Renal	25.4	762	7.62
Skeletal muscle	28.2	846	8.46
Skin	12.9	387	3.87
Total	99.9	2997	29.97

Data based on a cardiac output of 3 l/min (Bond and Green 1969) and a mean systemic pressure of 100 mmHg; the percentages of distribution were taken from Green (1950).

rather than a series arrangement, these values can be summed to arrive at total body conductance. The curves presented in Figure 9 were determined by summation of the regional vascular curves with appropriate normalizations.

4.1. Local hypotension

The local hypotension line in Figure 9 was determined by summation of the individual vascular beds as described above using 29.97 as the 100% control conductance (see Table 2). As pressure was reduced, the total body conductance increased to about 132% at 60 mmHg after which it fell back to 100% at 40 mmHg. This indicates that the vascular beds which do exhibit autoregulation within this pressure range dominate the response because of their greater contribution to the total cardiac output.

4.2. Carotid sinus hypotension

When carotid sinus pressure in the anesthetized and vagotomized dog is reduced from 101 to 75 mmHg, which causes a reflex increase in systemic pressure from 114 to 162 mmHg (Table 1), the cardiac output does not change significantly (Bond and Green 1969). The total body conductance, however, falls to 72% of control (Fig. 2), indicating a marked total body vasoconstriction.

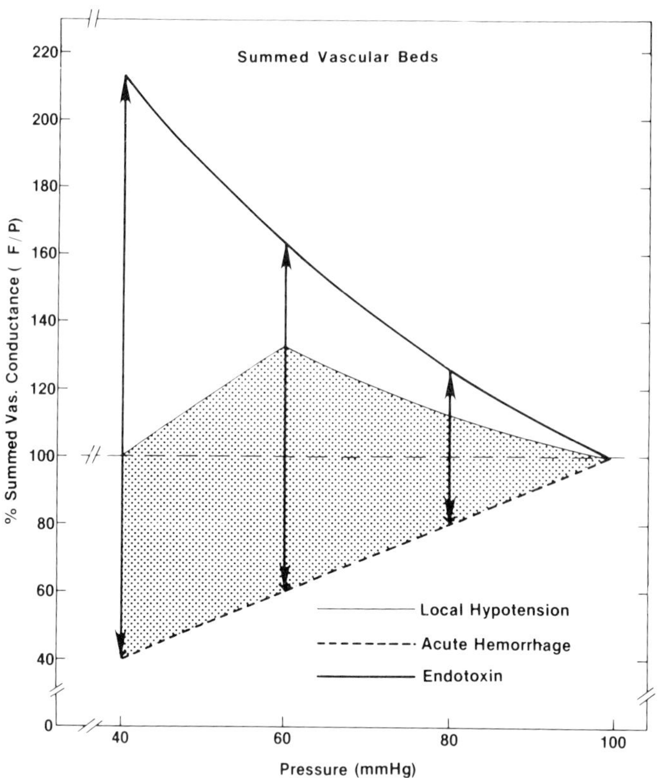

Fig. 9 *Pressure/summed vascular conductance responses to local hypotension, acute hemorrhagic hypotension and hypotension induced by endotoxin. These curves were derived using the cardiac output distribution to the regional vascular beds published by Green (1950) which are presented in Table 2. These summed responses show that the total body vasculature autoregulates down to pressures of 60 mmHg and that hemorrhagic hypotension causes a significant extrinsic lowering of total conductance as pressure falls to 40 mmHg (shaded area). When pressure falls during endotoxemia, the total vascular conductance increases to over 200% of control, thus indicating that endotoxin does cause a profound loss of total vascular tone which would explain the hypotension accompanying this pathological condition.*

4.3. Hemorrhagic hypotension

When systemic pressure is reduced from 100 to 40 mmHg, the conductance falls to 40% of control; this indicates strong extrinsic control. The shaded area in Figure 9 illustrates the magnitude of this extrinsic dominance over the intrinsic autoregulatory response.

4.4. Endotoxin hypotension

The sum of the responses in vascular conductance which accompany endotoxin hypotension is seen in Figure 9 as a significant increase in conductance to 217% of control at a hypotensive level of 40 mmHg. The unshaded area between the local hypotension and endotoxin curves illustrates the extent of total body vasodilation above the intrinsic regulatory ability of the system. This response is even more significant when one considers that the anticipated extrinsic response to this level of systemic hypotension would be on the acute hemorrhage curve. The summated response to endotoxin in Figure 9 is no different from the response of the total body conductance curve seen in Figure 1 between 15 and 75 minutes which was calculated by the thermodilation method for cardiac output.

In summary, the increase in total body conductance following endotoxin to levels that are even higher than the normal intrinsic response to hypotension points to a massive total peripheral vascular loss of tone which contributes significantly to the overall hypotensive episode. Possible mechanisms for this profound circulatory collapse will be discussed in the following sections of this chapter.

5. PATHOGENESIS OF ENDOTOXIN SHOCK

5.1. Macrovascular responses

The data presented in Figure 1 show that the initial response of the cardiac output to endotoxin is a fall to approximately 50% of control. Also note that the systemic pressure falls from 128 mmHg to 100 mmHg during the same 15-minute period. Since flow is falling more rapidly than pressure, the calculated total vascular conductance falls to 65% of control which is indicative of vasoconstriction. The regional circulations represented in Figure 1 indicate that the initial vasoconstriction is most marked in the mesenteric vascular bed in which the conductance falls to 60% of control at 15 minutes, compared with 80–90% for muscle, skin and kidney. All four vascular beds show a progressive loss of tone between 15 and 120 minutes after endotoxin. Those demonstrating the greatest loss of tone during this phase are the mesenteric and renal beds which increase to 160% of control conductance compared with increases in the skin and muscles of only 110%. The loss of vascular tone in each of these beds far exceeds the expected intrinsic response to hypotension, and this suggests that not only are the extrinsic constriction factors inhibited, but some active vasodepression must also occur. These regional responses support the concept of peripheral vascular pooling as reported by Hinshaw and Nelson (1962), which seems to be more prominent in the dog than other species (Hinshaw 1972; Lillehei et al 1965).

5.2. Microvascular responses

A comprehensive search of the literature suggests the existence of two conflicting schools of thought concerning the microvascular responses to endotoxin. One school reports arterial constriction (Hauman and Rosenberg 1966), while the other provides evidence favoring arterial dilation (Freed and Proctor 1967; Rai et al 1974). The interpretation of these studies is confused even further by the report of Baker and Davis (1980), who report that the terminal mesenteric arteries of the cat constrict, but the larger arteries dilate in response to 3 mg/kg endotoxin. Perhaps the most detailed description of the microcirculatory responses to endotoxin was published by Branemark and Urbaschek (1967). Applying vital microscopic techniques to the hamster cheek pouch and the mesentery of both guinea pigs and rabbits, these authors report arteriole constriction and venular dilation 10 minutes after endotoxin administration. Intravascular lumina were reduced even further by pericyte and endothelial cell edema. Microemboli as well as structural damage to the endothelial lining with extravascular fluid accumulation were also reported. Balis et al (1978) also report evidence of capillary damage including capillary engorgement, endothelial swelling and rouleau formation in primates exposed to endotoxin. Litton (1969) demonstrated that topical application of endotoxin resulted in dilation of the arteries and constriction of the veins, while systemic administration caused uniform constriction of the arterioles with decreased blood flow to the capillary bed. These observations suggest that endotoxin may possess both local and systemic actions which may help explain the biphasic response seen in Figure 1. The following two sections will concentrate first on the initial arterial vasoconstrictor phase, then the secondary vasodilator phase.

5.2.1. Initial phase

Arterial constriction and venous dilation   Zweifach and Thomas (1957) were the first to describe an initial hyperactivity of the arterioles and precapillary sphincters in rat mesentery when epinephrine was topically applied to endotoxic rats. The vascular sensitivity was diminished in animals made tolerant to endotoxin. Lillehei et al (1965) and Hauman and Rosenberg (1966) provide evidence favoring that the initial vasoconstriction was due to a sympathomimetic action of endotoxin which causes an increase in the peripheral catecholamine level. Using microangiographic techniques, Rai et al (1974) presented data from canine brain, lung, heart, liver, kidney, and viscera which suggest that the primary cause of endotoxin shock is the excessive production and release of neurohumors by a direct stimulation of adrenergic nerve endings which causes the severe reduction in blood flow. In 1980, Baker used microvascular flow velocity and indicator dispersion techniques to verify that endotoxin does cause vasoconstriction in arterioles and vasodilation in venules; this initial response was, however, followed by arteriole dilation and venous constriction. These reports are all consistent with the recent data presented by Pomerantz et al (1982), who reported an initial norepinephrine hyperactivity followed by a hypoactivity in rat aortic strips subjected to 6 mg/kg endotoxin. In addition, Hall and Hodge (1971) showed an immediate rise in plasma angiotensin followed by an elevation in plasma catecholamines in response to endotoxin. For additional information on the adrenergic and angiotensin interaction in endotoxin shock, the reader is referred to Chapters 7 and 8.

In summary, the literature suggests that the primary or initial decreased vascular conductance (i.e., vasoconstriction) which lasts approximately 15 minutes is the result of a direct activation of the adrenergic nervous system together with an increased arterial sensitivity to norepinephrine and elevated plasma angiotensin levels.

5.2.2. Secondary phase

Arterial dilation and venous constriction   In 1961, Hinshaw et al (1961a) provided data which indicated that endotoxin stimulates histidine decarboxylase to increase histamine synthesis. Three years later Kobol et al (1964) reported that endotoxin not only causes release of histamine from tissue homogenates but catecholamines and serotonin are released as well. Previously Haddy et al (1959) demonstrated that serotonin has a vasodilator effect on the small resistance vessels. The kinins are another family of biological amines which have been shown to occur in greater amounts in the blood of animals treated with endotoxin and which also have strong vasodilator properties (Nies et al 1968; Robinson et al 1975; Shah et al 1970). Branermark and Urbaschek (1967) and Urbaschek et al (1969) reported an initial vasoconstriction followed in 10–15 minutes by a degranulation of the mast cells and subsequent loss of arterial tone together with venous constriction. These authors state that this secondary response could be duplicated by topical administration of either histamine or serotonin with the important difference that these microcirculatory changes occurred immediately rather than after a 10-minute latent period. The reader is referred to Chapters 8 and 11 for a more detailed discussion of the roles of histamine, serotonin and kinins in endotoxin shock.

In summary, it would appear that the secondary arterial vasodilation (i.e., loss of muscle tone) and venous constrictor responses come as a result of the tissue release of the several biogenic amines which include histamine, serotonin and bradykinin.

5.3. Conclusions

The flow chart presented in Figure 10 illustrates a rather simplistic analysis of the pathogenesis of circulatory shock induced by gram-negative endotoxemia. This diagram is consistent with the previous discussion in this chapter as well as with two outstanding reviews published by Christy (1971) and Kováts (1971).

The initial effect of systemic endotoxin appears to be a massive stimulation

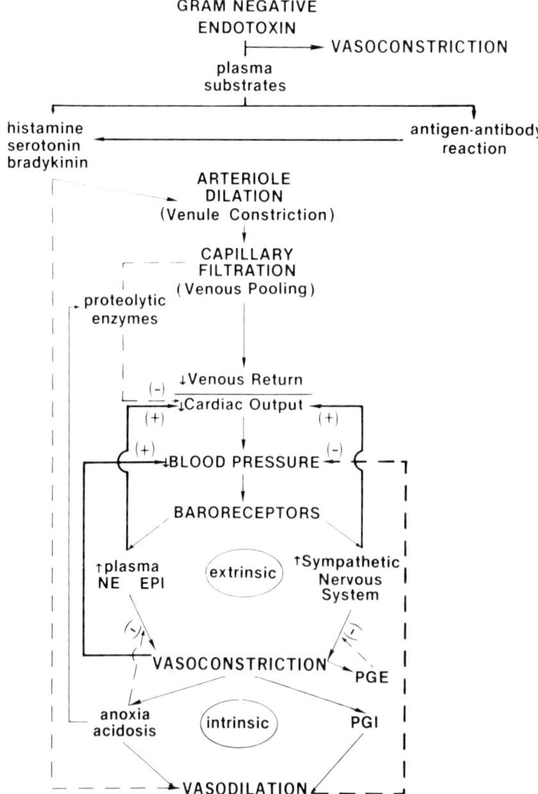

Fig. 10 *Summary of the proposed sequence of cardiovascular events which accompany endotoxemia.*

of all adrenergic structures causing the immediate release of catecholamines and sympathetic nerve stimulation which culminates in more or less uniform arterial vasospasm. Prager et al (1975) and more recently Ekström-Jodal et al (1982) have presented data which confirm a notable increase in the plasma levels of both epinephrine and norepinephrine within a few minutes of the endotoxin administration. These agents reach peaks of 358% of control for epinephrine and 2000% of control for norepinephrine at 60 minutes after the administration of endotoxin. The secondary effect of endotoxin is exerted on certain cellular targets (e.g. mast cells) by means of the immunological system. These interactions, which involve complex antigen–antibody reactions including the role of complement, are discussed in detail in Chapter 6. Briefly, however, these actions initiate the synthesis and release of histamine, serotonin and bradykinin, which all act to inhibit the primary adrenergically induced elevated arterial vascular tone. The postcapillary resistance vessels (e.g. venules) seem to be more resistant to this inhibition than the precapillary arterioles; consequently, capillary hydrostatic pressure increases. This rise in capillary pressure together with the damage inflicted on capillaries by endotoxin (Baris et al 1978, 1980) act synergistically to promote both capillary pooling and interstitial edema (Rayner et al 1960). The venous congestion and hypoperfused state of the tissues which occurs most prominently in the viscera (Hinshaw and Nelson 1962) provides an optimal situation for the cellular release of proteolytic enzymes from damaged cells. Fragments of these catabolic enzymes have been implicated in a subsequent myocardial dysfunction (Lefer 1970) which, together with the fall in venous return (i.e., venous pooling) to the heart, potentiates the fall in cardiac output and systemic pressure. This systemic hypotension is sensed by the baroreceptors which reflexly excite the adrenergic nervous system leading to a further increase in plasma levels of catecholamines (Prager et al 1975). The increased tension placed on the vascular endothelium during this compensatory response has been shown by Bond et al (1981) to activate the synthesis and release of prostaglandins of the E series (PGE), which sets off a chain of events terminating with the inhibition of adrenergic neurotransmitter release. Recent studies by Tempel et al (1982) support this concept; they found that the PGE precursor arachidonic acid improves the circulatory hemodynamics during endotoxemia. In addition, tissues that are hypoperfused become hypoxic and acidotic, thereby causing the vascular tissue to become less sensitive to the high levels of plasma catecholamines (Pomerantz et al 1982). The adrenergic inhibition is further augmented by the reported elevated concentrations of the powerful vasodilator prostacyclin (PGI) (Fletcher et al 1981; Flynn and Demling 1982). For more details on the role of prostaglandins in endotoxin shock the reader is referred to Chapters 9 and 10. All of these inhibitor influences act synergistically to counteract the extrinsically induced peripheral vascular constriction resulting in a profound peripheral vascular failure and irreversible hypotension.

In summary, the initial response of the peripheral vasculature to endotoxin is a precapillary vasoconstriction induced by a hyperactive sympathetic nervous system through some as yet undetermined cellular mechanism. This high vascular tone, which seems to be most pronounced in the visceral and renal vascular beds, is subsequently inhibited by several interacting factors which include: (a) the synthesis and release of endogenous vasodepressor peptides from the mast cells (e.g., histamine, serotonin and bradykinin); (b) tissue hypoperfusion leading to acidosis which inhibits the effectiveness of the circulating catecholamines to act on the vascular smooth muscle; and (c) the synthesis and release of prostanoids which depress adrenergic tone by either direct inhibition of vascular smooth muscle tone or by inhibiting the release of neurotransmitters. The precapillary dilation, which far exceeds the pre-directed intrinsic response in all regional vascular beds, is also accompanied by venous constriction with capillary congestion and edema. The profound loss of peripheral vascular tone to all regional vascular beds causes a fall in venous return and cardiac output leading to the vicious cycle of circulatory hypotension and shock.

6.   AREAS IN NEED OF FURTHER STUDY

6.1.   Normal response

The literature review presented in this chapter suggests that there is much to be learned about the fundamental compensatory and decompensatory adjustments to hypotension and shock. Since shock is basically a hypotensive disorder resulting in tissue hypoperfusion and cellular death, the understanding of the mechanisms responsible for tissue hypoperfusion is essential. Future studies should include the correlation of neural, endocrine and metabolic dysfunction with the basic hemodynamic responses during shock. Since all forms of circulatory shock culminate in progressive hypotensive episodes as a direct result of cardiac and/or peripheral vascular failure, future research needs to be directed toward clarifying our understanding of the normal physiological response to hypotension and toward establishing the point(s) in the shock cycle at which these mechanisms might fail. Therefore, before the physician can provide a rational approach to shock prevention or management, carefully controlled studies focusing upon the elucidation of the normal compensatory responses to uncomplicated systemic hypotension must be forthcoming. This understanding in turn will provide a basis for examining the physiological dysfunctions which occur as a result of the prolonged tissue hypoperfusion and shock.

6.2. Venous return and myocardial function

The data presented in Figure 1 show an initial fall in both cardiac output and blood pressure with a calculated fall in total peripheral vascular conductance (i.e., vasoconstriction). This suggests that the cardiac output is falling more rapidly than arterial pressure, which means that the initial fall in pressure must be due to a pump problem in the face of an increased sympathetic outflow (Weil et al 1956). This could mean inadequate venous return, but since CVP remains unchanged, this seems unlikely. The work reported by Hinshaw in this volume (Chapter 2) also argues against myocardial failure; yet the fact remains that the cardiac output is the first hemodynamic variable to show depression following endotoxin. More well-controlled studies are needed to examine the effect of endotoxin on venous return and myocardial function.

6.3. Management of regional circulations

The data presented in this chapter further verify that each vascular bed has its own unique set of vascular control mechanisms. For example, the cerebral and myocardial circulations are controlled by intrinsic mechanisms, while renal and visceral vasculation are strongly influenced by extrinsic factors. The vascular bed in skeletal muscle has a strong element of external control which can be easily overriden by increasing metabolic demand. The cutaneous circulation does not participate in the baroreceptor reflex response to hypotension or exhibit autoregulation, yet it is very sensitive to circulating catecholamines. Therefore, since the basic controlling factors are unique to each vascular bed, work directed toward individually manipulating each vascular bed as a means of shock therapy should be developed. For example, the present chapter provides evidence that though all vascular beds are dilated by endotoxin, the visceral bed seems to be most strongly affected. A shotgun therapeutic approach which might affect all beds uniformly may have a devastating effect on blood pressure, which in turn may have an undesirable effect on the work load of the heart. Therefore, future work should be directed toward therapeutic management of individual vascular beds during hypotension and shock.

6.4. Structural correlation with hemodynamics

Some investigators believe that the initiating factor in irreversible shock induced by hemorrhage is a paradoxical loss of vascular tone of skeletal muscle which causes a further fall in systemic pressure to levels below the autoregulatory capabilities of myocardial and cerebral vasculatures. Although this vascular response of skeletal muscle is beneficial to the local tissue, the added fall in pressure elicits a reflex stimulation of the myocardium which increases myocardial metabolic demand aggravating the myocardial dysfunction. One suggested mechanism to explain this loss of vascular tone is that the shock-induced prostaglandins released from the vascular endothelium feed back on the adrenergic receptors in the vasculature, blocking the peripheral adrenergic compensatory response (Bond et al 1981; Bond 1982). Light and electron microscopic studies aimed at correlating the structural integrity of the vascular endothelium with both prostaglandin release and basic hemodynamics may well answer this question.

6.5. Common denominators

Another area needing careful study is the comparison of the responses of regional circulations to all forms of circulatory shock (i.e., hemorrhagic, burn, splanchnic artery occlusion (SAO), endotoxin, and cardiogenic). If similar techniques are used in conjunction with a protocol allowing comparison of data during equivalent stages of shock, it might be possible to establish one or more common denominators by which the physician could better predict the course of his patient and thus institute a more beneficial treatment regime.

REFERENCES

Adiseshiak M, Baird RJ (1978) Correlation of the changes in diastolic tissue pressure and regional coronary blood flow in hemorrhagic and endotoxin shock. *J. Surg. Res. 24*, 20-25.

Archie JP Jr, Mertz WR (1978) Myocardial oxygen delivery after experimental hemorrhagic shock. *Ann. Surg. 187*, 205-210.

Baker CH, Davis DL (1980) Endotoxin effects on capillary transit times of RBC and plasma as measured by indicator dilution. *Microvasc. Res. 20*, 242-252.

Balis JU, Rappaport ES, Gerber LJ, Fareer J, Buddingh F, Messmore HL (1978) A primate model for prolonged endotoxin shock. Blood vascular reactions and effects of glucocorticoid treatment. *Lav. Invest. 38*, 511-523.

Baris C, Guest MM, Frazer ME (1980) Direct effects of endotoxin on the microcirculation. *Adv. Shock Res. 4*, 153-160.

Beaty O III, Sloop CH, Schmid HE Jr, Buckalow VM Jr (1976) Renin response and angiotensinogen control during graded hemorrhage and shock in the dog. *Am. J. Physiol. 231*, 1300-1307.

Berne RM (1963) Cardiac nucleotides in hypoxia: possible role in regulation of coronary blood flow. *Am. J. Physiol. 204*, 318-322.

Bohlen HG, Hutchins PM, Rapela CE, Green H (1975) Microvascular control in testinal mucosa of normal and hemorrhaged rats. *Am. J. Physiol. 299*, 1159-1164.

Bohs CT, Fish JC, Miller TH, Traber DL (1979) Pulmonary vascular response to endotoxins in normal and lymphocyte depleted sheep. *Circ. Shock 6*, 13-21.

Bond RF (1982) A review of the skin and muscle hemodynamics during hemorrhagic hypotension and shock. *Adv. Shock Res. 8*, 53-70.

Bond RF (1983) Peripheral vascular adrenergic depression during hypotension induced by *E. coli* endotoxin. *Adv. Shock Res. 9*, 157-169.

Bond RF, Green HD (1969) Cardiac output redistribution during bilateral common carotid occlusion. *Am. J. Physiol. 216*, 393-403.

Bond RF, Manley ES Jr, Green HD (1967) Cutaneous and skeletal muscle vascular response to hemorrhage and irreversible shock. *Am. J. Physiol. 212*, 488-497.

Bond RF, Blackard RF, Taxis JA (1969) Evidence against oxygen being the primary factor governing autoregulation. *Am. J. Physiol. 216*, 788.

Bond RF, Lackey GF, Taxis JA, Green HD (1970) Factors governing cutaneous vasoconstriction during hemorrhage. *Am. J. Physiol. 219*, 1210-1215.

Bond RF, Manning ES, Gonzales NM, Gonzalez RR Jr, Becker VE (1973) Myocardial and skeletal muscle responses to hemorrhage and shock during α-adrenergic blockade. *Am. J. Physiol. 225*, 247-257.

Bond RF, Peissner LC, Manning ES (1977) Skeletal muscle vascular decompensation in dogs subjected to prolonged hypovolemia: neural versus humoral mechanisms. *Circ. Shock 4*, 327-344

Bond RF, Manning ES, Peissner LC (1979) Correlation between skeletal muscle vascular decompensation and survival: roles of tissue ischemia and innervation. *Circ. Shock 6*, 43-54.

Bond RF, Bond CH, Peissner LC, Manning ES (1981) Prostaglandin modulation of adrenergic vascular control during hemorrhagic shock. *Am. J. Physiol. 241* (*Heart Circ. Physiol. 10*), H85-H90.

Branemark PI, Urbaschek B (1967) Endotoxins in tissue injury. Vital microscopic studies on the effect of endotoxin from *E. coli* on the microcirculation. *Angiology 18*, 667-671.

Brobmann GF, Ulano HB, Hinshaw LB, Jacobson ED (1970) Mesenteric vascular responses to endotoxin in the monkey and dog. *Am. J. Physiol. 219*, 1464-1467.

Brockman SK, Thomas CS Jr, Vasko JS (1967) The effect of *Escherichia coli* endotoxin on the circulation. *Surg. Gynecol. Obstet. 125*, 763-774.

Brungardt JM, Reynolds DG, Swan KG (1972) Route of endotoxin delivery; effects on canine mesenteric hemodynamics. *Am. J. Physiol. 223*, 565-568.

Bruni FD, Komwatana P, Soulsby ME, Hess ML (1978) Endotoxin and myocardial failure: role of the myofibril and venous return. *Am. J. Physiol. 235*, H150-H156.

Christy JH (1971) Pathophysiology of gram-negative shock. *Am. Heart J. 81*, 694-701.

Clowes JG, Vucinic N, Weidner N (1966) Circulatory and metabolic alterations associated with survival or death in peritonitis. *Ann. Surg. 163*, 866-885.

Cohn JD, Greenspan M, Goldstein CR, Gudwin AL, Siegel JH, Del-Guercio LRM (1968) Arteriovenous shunting in high cardiac output shock syndromes. *Surg. Gynecol. Obstet. 127*, 282-288.

DeGeest H, Levy N, Zieske H Jr (1964) Carotid sinus baroreceptor reflex effects upon myocardial contractility. *Circ. Res. 15*, 327-342.

Ekström-Jodal B, Häggendal J, Larsson LE, Westerlind A (1982) Cerebral hemodynamics, oxygen uptake and cerebral arteriovenous differences of catecholamines following *E. coli* endotoxin in dogs. *Acta Anaesthesiol. Scand. 26*, 446-452.

Elsberry DD, Rhoda DA, Beisel WR (1969) Hemodynamics of staphylococcal B enterotoxemia and other types of shock in monkeys. *J. Appl. Physiol. 27*, 164-169.

Emerson TE Jr, Wagner TD, Gill CC (1966) In situ kidney pressure-flow-resistance relationship during endotoxin shock in dogs. *Proc. Soc. Exp. Biol. Med. 122*, 366-368.

Ernsting J, Parry DJ (1957) Some observations on the effects of stimulating the stretch receptors in the carotid artery of man. *J. Physiol. 137*, 45P.

Fletcher JR, Ramwell PW, Harris RH (1981) Thromboxane, prostacyclin, and hemodynamic events in primate endotoxin shock. *Adv. Shock Res. 5*, 143-148.

Flynn JT, Demling RH (1982) Inhibition of endogenous thromboxane synthesis by exogenous prostacyclin during endotoxemia in conscious sheep. *Adv. Shock Res. 7*, 199-207.

Freed SC, Proctor JH (1967) Arterial tone in endotoxin hypotension. *Angiology 18*, 62-68.

Gilbert RP (1960) Mechanisms of the hemodynamic effects of endotoxin. *Physiol. Rev. 40*, 245-279.

Gonzalez RR Jr, Bond RF (1974) Sympathetic nerve and skeletal muscle vascular responses to hemorrhage and shock with and without cholinergic and beta adrenergic blockade in the dog. *Circ. Shock 1*, 39-50.

Green HD (1950) Circulatory system: physical principles. In: Glasser O (Ed), *Medical Physics, Vol 2*, pp 228-251. The Year Book Publishers, Chicago.

Green HD, Kepchar JH (1959) Control of peripheral resistance in major systemic vascular beds. *Physiol. Rev. 39*, 617-686.

Green HD, Schmid HE, Rapela CE (1965) Autoregulation in vascular beds. Possible role of autoregulation in the increased resistance to flow in hypertension. In: Johnson PC (Ed), *Hypertension XIII, Proceedings of the Council for High Blood Pressure*, pp 144-159. American Heart Association, New York.

Green HD, Rapela CE, Bond RF (1972) Cerebral vs. cutaneous and skeletal muscle vascular responses to localized hypotension and to systemic hypotension induced by hemorrhage and shock. In: Malinin TI, Zeppa R, Gollan F, Callahan AB (Eds), *Reversibility of Cellular Injury Due to Inadequate Perfusion*, pp 135-152. Charles C Thomas, Springfield, Ill.

Green HD, Bond RF, Rapela CE, Schmid HE, Manley E, Farrar DJ (1980) Competition between intrinsic and extrinsic controls of resistance vessels of major vascular beds during hemorrhagic hypotension and shock. *Adv. Shock Res. 3*, 77-104.

Gregg DE (1968) The microcirculation of the heart in reduced flow states. In: Shepro D, Fulton GP (Eds), *Microcirculation as Related to Shock*, pp 51-67. Academic

Press, New York-London.

Guyton AC, Ross JM, Carrier O Jr, Walker JR (1964) Evidence for tissue oxygen demand as the major factor causing autoregulation. *Circ. Res. 15 Suppl. 1*, I-60-I-68.

Haddy FJ, Fleishman M, Emanuel DA (1959) The influence of tone upon responses of small and large vessels to serotoxin. *Circ. Res. 7*, 123-130.

Hall RC, Hodge RL (1971) Vasoactive hormones in endotoxin shock: a comparative study in cats and dogs. *J. Physiol. 213*, 69-84.

Hauman RL, Rosenberg JC (1966) Effect of celiac blockade on intestinal arteriolar resistance and critical closing pressures during endotoxin shock. *Surg. Forum 17*, 7-8.

Hinshaw LB (1972) Comparison of responses of canine and primate species to bacteria and bacterial endotoxin. In: Forscher B, Lillehei RC, Stubbs SS (Eds), *Shock in Low- and High-Flow States*, pp 245-249. Excerpta Medica, Amsterdam.

Hinshaw LB (1979) Myocardial function in endotoxin shock. *Circ. Shock Suppl 1*, 43-51.

Hinshaw LB, Nelson DL (1962) Venous response of intestine to endotoxin. *Am. J. Physiol. 203*, 870-872.

Hinshaw LB, Owen SE (1971) Correlation of pooling and resistance changes in the canine forelimb in septic shock. *J. Appl. Physiol. 30*, 331-337.

Hinshaw LB, Jordan MM, Vick JA (1961a) Histamine release and endotoxin shock in the primate. *J. Clin. Invest. 40*, 1631-1637.

Hinshaw LB, Vick JA, Wittmers LE, Worthen DM, Nelson DL, Swenson OP (1961b) Changes in total peripheral resistance in endotoxin shock. *Proc. Soc. Exp. Biol. Med. 108*, 24-27.

Hinshaw LB, Greenfield LJ, Archer LT, Guenter CA (1971) Effects of endotoxin on myocardial hemodynamics, performance, and metabolism during beta adrenergic blockade. *Proc. Soc. Exp. Biol. Med. 137*, 1217-1224.

Hopkins RW, Damewood CA (1974) Septic shock: hemodynamics of endotoxin and inflammation. *Am. J. Surg. 127*, 476-483.

Hutchins PM, Bond RF, Green HD (1974) Participation of oxygen in the local control of skeletal muscle microvasculature. *Circ. Res. 34*, 85-93.

Johnson PC (1964) Origin, localization and homeostatic significance of autoregulation in the intestine. *Circ. Res. 15 Suppl 1*, I-225-I-232.

Kause SM, Kleinman W, Hess ML (1980) Cardiogenic endotoxin shock: coronary flow and contractile protein dysfunction as determinants of depressed cardiac contractility. *Adv. Shock Res. 3*, 105-116.

Kobol EE, Lovell R, Katz W, Thal AP (1964) Chemical mediators released by endotoxin. *Surg. Gynecol. Obstet. 118*, 807-813.

Kováts TG (1971) The role of targets and mediators in endotoxin shock. *Adv. Exp. Med. Biol. 23*, 347-363.

Kovách AGB (1972) Blood flow and metabolism in the brain and in adipose tissue during hemorrhagic shock. In: Forscher BK, Lillehei RC, Stubbs SS (Eds), *Shock in Low- and High-Flow States*, pp 65-76. Excerpta Medica, Amsterdam.

Kovách AGB, Sándor P (1976) Cerebral blood flow and brain function during hypotension and shock. *Annu. Rev. Physiol. 38*, 571-596.

Kuschinsky W, Wahl M (1978) Local chemical and neurogenic regulation of cerebral vascular resistance. *Physiol. Rev. 58*, 656-689.

Lee JC, Downing SE (1976) Myocardial oxygen availability and cardiac failure in hemorrhagic shock. *Am. Heart J. 92*, 201-209.

Lefer AM (1970) Role of myocardial depressant factor in the pathogenesis of circulatory shock. *Fed. Proc. 28*, 1836-1847.

Lillehei RC, Mac Lean LD (1958) Intestinal factor in irreversible endotoxin shock. *Ann. Surg. 148*, 513-525.

Lillehei RC, Longerbeam JK, Block JH, Manax WG (1965) Hemodynamic changes in endotoxin shock. In: Mills LC, Moyer JH (Eds), *Shock and Hypotension: Pathogenesis and Treatment*, pp 442-462. Grune and Stratton, New York-London.

Litton A (1969) Microcirculatory effects of endotoxin. *Bibl. Anat. 10*, 334-339.

Mac Lean LD, Mulligan WG, Mc Lean APN, Duff JH (1967) Patterns of septic shock in man – a detailed study of 56 patients. *Ann. Surg. 166*, 543-562.

McGiff JC, Fasy TM (1965) The relationship of the renal vascular activity of angiotensin II to the autonomic nervous system. *J. Clin. Invest. 44*, 1911-1923.

Navar GL, Marsh DJ, Blantz RC, Hall J, Ploth DW, Nasjlett A (1982) Intrinsic control of renal hemodynamics. *Fed. Proc. 41*, 3022-3030.

Nies AS, Forsyth RP, Williams HE, Melmon KL (1968) Contribution of kinins to endotoxin shock in unanesthetized rhesus monkeys. *Circ. Res. 22*, 155-164.

Polosa C, Rossi G (1961) Cardiac output and peripheral blood flow during occlusion of carotid arteries. *Am. J. Physiol. 200*, 1185-1190.

Pomerantz K, Casey L, Fletcher JR, Ramwell PW (1982) Vascular reactivity in endotoxin shock: effect of lidocaine or indomethacin pretreatment. *Adv. Shock Res. 7*, 191-198.

Prager RL, Dunn EL, Seaton JF (1975) Increased adrenal secretion of norepinephrine and epinephrine after endotoxin and its reversal with corticosteroids. *J. Surg. Res. 19*, 371-375.

Rai DK, Gupta LP, Singh RH, Udupa KN (1974) A study of microcirculation in endotoxin shock. *Surg. Gynecol. Obstet. 139*, 11-16.

Rapela CE, Green HD (1964) Autoregulation of canine cerebral blood flow. *Circ. Res. 15 Suppl 1*, I-205 – I-211.

Rapela CE, Green HD, Denison AB Jr (1967) Baroreceptor reflexes and autoregulation of cerebral blood flow in the dog. *Circ. Res. 21*, 559-568.

Raymond RM, Emerson TE Jr (1978) Cerebral metabolism during endotoxin shock in the dog. *Circ. Shock 5*, 407-414.

Rayner R, Mac Lean LD, Grim E (1960) Intestinal tissue capillary flow in shock due to endotoxin. *Circ. Res. 8*, 1212-1217.

Reynolds DG, Brungardt JM, Swan KG (1971) Effects of endotoxin on the vascular architecture of intestinal mucosa. *Adv. Exp. Med. Biol. 23*, 113-126.

Robinson JA, Klondnycky ML, Loeb HS, Racic MR, Gunnar RM (1975) Endotoxin, prekallikrein, complement and systemic vascular resistance. Sequential measurements in man. *Am. J. Med. 59*, 61-67.

Rubio R, Berne RM (1975) Regulation of coronary blood flow. *Prog. Cardiovasc. Dis. 18*, 105-122.

Rutherford RB, Balis JV, Trow RS, Graves GM (1976) Comparison of hemodynamic and regional blood flow changes at equivalent stages of endotoxin and hemorrhagic shock. *J. Trauma 16*, 886-897.

Selkurt EE, Brecher IA (1956) Splanchnic hemodynamic and oxygen utilization during hemorrhagic shock in the dog. *Circ. Res. 4*, 693-704.

Seyfer AE, Zajtchuk R, Hazlett DR, Mologne LA (1977) Systemic vascular performance in endotoxic shock. *Surg. Gynecol. Obstet. 145*, 401-407.

Shah JP, Shah US, Appert HE, Howard JM (1970) Studies on the release of bradykinin by the splanchnic circulation during endotoxic shock. *J. Trauma 10*, 255-259.

Siegel JH, Cerra FB, Coleman B, Giovannini I, Sheyte M, Border JR, McMenamy RH (1979) Physiological and metabolic correlations in human sepsis. *Surgery 86*, 163-193.

Slater G, Vladeck BC, Bassin R, Brown RS, Shoemaker WC (1975) Sequential changes in cerebral blood flow and distribution of flow within the brain during hemorrhagic shock. *Ann. Surg. 181*, 1-4.

Spitzer JJ, Bechtel AA, Archer LT, Black MR, Hinshaw LB (1974) Myocardial substrate utilization in dogs following endotoxin administration. *Am. J. Physiol. 277*, 132-136.

Stainsby WN (1964) Autoregulation in skeletal muscle. *Circ. Res. 15 Suppl 1*, I-39 – I-45.

Swan KG, Barton RW, Reynolds DG (1971) Mesenteric hemodynamics during endotoxemia in the baboon. *Gastroenterology 61*, 872-876.

Tempel GE, Cook PA, Wise WC, Halushka PV (1982) The improvement in endotoxin induced redistribution of oxygen blood flow by inhibition of thromboxane and prostoglandin synthesis. *Adv. Shock Res. 7*, 209-218.

Urbaschek B, Huth K, Krech HJ (1969) Endotoxin-induced microcirculatory disturbances and interaction to various metabolic parameters including the coagulation system. *Bibl. Anat. 10*, 442-448.

Vaughn DT, Kirschbaum TH, Bersentes T, Assali NS (1967) Effects of corticosteroid hormones on regional circulation in endotoxin shock. *Proc. Soc. Exp. Biol. Med. 124*, 760-764.

Vick JA (1965) Bioassay of the prominent humoral agents involved in endotoxin shock. *Am. J. Physiol. 209*, 75-78.

Watts DT, Westfall V (1964) Studies on peripheral catecholamine levels during hemorrhagic shock in dogs. *Proc. Soc. Exp. Biol. Med. 115*, 601-604.

Weil MH, Nishijima II (1978) Cardiac output in bacterial shock. *Am. J. Med. 64*, 920-922.

Weil MH, Mac Lean LD, Visscher MB, Spink WW (1956) Studies on the circulatory changes in the dog produced by endotoxin from gram-negative microorganisms. *J. Clin. Invest. 35*, 1191-1198.

Wiggers CJ, Werle JM (1942) Cardiac and peripheral resistance as determinants of circulatory failure in hemorrhagic shock. *Am. J. Physiol. 136*, 421-432.

Wright CJ, Duff JH, Mc Lean APH, Mac Lean LD (1971) Regional capillary blood flow and oxygen uptake in severe sepsis. *Surg. Gynecol. Obstet. 132*, 637-644.

Wyler F, Forsyth RP, Nies AS, Neutze JM, Melmon KL (1969) Endotoxin-induced regional circulatory changes in the unanesthetized monkey. *Circ. Res. 24*, 777-786.

Zweifach BW, Thomas L (1957) The relationship between the vascular manifestations of shock produced by endotoxin, trauma and hemorrhage. I. Certain similarities between the reactions in normal and endotoxin-tolerant rats. *J. Exp. Med. 106*, 385-401.

Handbook of Endotoxin, Vol. 2: Pathophysiology of Endotoxin
L.B. Hinshaw, editor
© Elsevier Science Publishers B.V., 1985
ISBN 0 444 90385 2
$0.85 per article per page (transactional system)
$0.20 per article per page (licensing system)

CHAPTER 4

Endotoxin and the lung

JAMES A. WILL AND DOUGLAS B. COURSIN

1. INTRODUCTION

The complex interaction of bacterial endotoxins with the lung is challenging to review. During the past twenty to thirty years, studies on the effects of endotoxin have been confounded by the evolving recognition that the lung acts not merely for gas exchange, but is also highly active as a metabolic organ. Within this same period, appreciation has developed that control of the pulmonary circulation is not passively mediated by alteration of peripheral vascular reactivity and myocardial function. Instead, it is regulated dynamically by pharmacologic receptors including adrenergic, cholinergic, and peptidergic. Furthermore, this circulatory bed is important in the uptake and removal of various vasoactive and metabolic substances.

Bacterial endotoxins, and in particular the lipopolysaccharide portion (LPS), have a broad spectrum of target organs, the lung being prominent among these. The sequelae of pulmonary effects involve the whole organism and result, in turn, in microscopic and macroscopic structural changes in lung tissue. These changes cause major alterations of gas exchange and metabolic functions of the lung. Ultimately, death of the animal or patient may ensue.

One of the earlier references to the pulmonary vascular effects of endotoxin was made in 1958 by Kuida et al, who demonstrated changes in pulmonary vascular resistance. A summary of the early studies of vascular change due to endotoxin is provided by Bessa et al in 1974, who used the isolated perfused dog lung as a model. In essence, increases in pulmonary arterial pressure, resistance, and alveolar-capillary permeability have been reported.

Pulmonary function changes have also been investigated. Ouellette et al (1967) described enhanced methacholine responses in patients given 20 µg of *Salmonella enteritidis* intravenously and concluded that endotoxin caused release of epinephrine from the adrenal medulla (Ouellette et al 1967).

Two reviews, published in 1976 and 1977, effectively cover the spectrum of endotoxin effects in the animal as a whole (Berry 1977; Elin and Wolff 1976). This chapter will focus more specifically on the lung.

76

2. HISTORICAL PERSPECTIVES

'Shock lung' initiated by endotoxemia occurs, not uncommonly, in man. Endotoxemia is too often overlooked in the early stages of disease and can rapidly result in a near-fatal or fatal condition in both septic and nonseptic disorders. Caridis et al presented 13 case reports which indicated that the Limulus lysate test demonstrated endotoxemia in patients, some of whom would not ordinarily be suspect (Caridis et al 1972). The value and reliability of the Limulus lysate test is discussed in Volume 4 (see chapter by Jorgensen).

The presence of endotoxemia in experimental shock was also demonstrated by Cuevas et al in 1972. They demonstrated positive Limulus lysate tests in four different shock models using rabbits. The four methods of shock induction were: (a) injection of *Escherichia coli* organisms intraperitoneally; (b) occlusion of the superior mesenteric artery; (c) bradykinin given intravenously for 3 hours; and (d) *E. coli* endotoxin given intravenously. Lung lesions occurred only in those animals where endotoxemia persisted until death.

Despite the many studies which have preceded and followed this publication by Caridis, the mechanistic questions posed by the interaction between endotoxin and the lung have remained largely unanswered. In trying to establish the evolution of knowledge of the effects of endotoxin on the lung, several factors must be considered. The problem of purity and multiplicity of endotoxin sources is probably related to the extreme variation in doses reported to cause effects. Furthermore, many studies report doses as a percentage of LD_{50}. This was an acceptable approach at the time many of the earlier studies were performed, but it is impossible to duplicate those results or to extrapolate from one report to another without redoing the LD_{50} studies.

Finally, animal models present marked inter- as wel as intraspecies differences in susceptibility and responses to endotoxin. A statement from Hinshaw et al (1977) characterizes the problem: 'Significant differences in responses between models should elicit caution in the application of findings in non-human primates to the human patient.' If this is true for the baboon, much more caution need be taken in extending findings from phylogenetically lower animals to man.

3. ENDOTOXIN AND PULMONARY FUNCTION

The pulmonary functional characteristics of endotoxin insult have not been as extensively studied as the pulmonary vascular aspects. Esbenshade et al 1982, have, however, published a fairly complete study of respiratory failure in sheep given infusions of a high dose of (2.0–5.0 μg/kg) *E. coli* endotoxin. Eight of 11 animals developed respiratory failure and 7 died in 6.3 ± 1.1 hours.

The critical measurement was the $P(A-a)_{O_2}$ gradient. When the gradient exceeded 42 mmHg (actually 58 ± 5), animals died. Pulmonary resistance increased sixfold and dynamic compliance decreased fourfold. These changes have usually been considered secondary to the development of pulmonary edema. In this study, the changes occurred within 15–30 minutes after endotoxin infusion. Edema usually does not develop for several hours, and therefore, the authors concluded that the changes were not secondary to edema formation. This model has many features which resemble respiratory failure in gram-negative septicemic human patients.

In a study by the same group, dynamic compliance decreased to $41 \pm 10\%$ of baseline, total lung resistance increased to $283 \pm 71\%$, and the alveolar-to-arterial oxygen gradient was 47 ± 5 mmHg (Hinson et al 1983). A dose of 1.0 μg/kg was used. If granulocytes were depleted, the changes were about half those cited. The pulmonary vascular response in this early stage was unaffected by granulocyte depletion. The authors concluded that granulocytes are important in altered respiratory function but not in the early pulmonary vascular reactivity generated by endotoxin.

In a 1977 study, Ogawa et al found a significant decrease in alveolar surface activity with sublethal doses of endotoxin given to dogs. As in the study by Esbenshade et al, the $P(A-a)_{O_2}$ was widened. When a lethal dose of LPS was given to the dogs, less effect on surface activity was found as determined by stability, expansion, and percent lung volume indices.

Other effects have been reported to be: bronchoconstriction in the guinea pig (Kovács and Görög 1968), development of circulating immune complexes (Berdischewsky et al 1980; Rylander et al 1980), metabolic acidosis and hyperpnea (Gemer et al 1974). Manifestations of these changes and their implications will be discussed more fully in Section 6 and in Chapters 1, 5 and 10 in relation to leukocytes, platelets, and other cellular mechanisms.

4. THE EFFECTS OF LIPOPOLYSACCHARIDE ON THE PULMONARY CIRCULATION

LPS has a biphasic effect on the pulmonary circulation. The phases are described as Phase I, the 'hypertension phase' and Phase II, the 'permeability' phase (Brigham et al 1979a; Demling et al 1981a). Figure 1 illustrates many of the effects described in this chapter.

4.1 Phase I

In Phase I, mean pulmonary arterial pressure ($\bar{P}$pa) and vascular resistance are elevated and systemic vascular pressure and cardiac output are decreased, at least transiently for the latter two variables. Systemic pressure is increased after the transient hypotensive periods in some species, but in others it remains abnormally low. The hypertensive changes in the pulmonary circulation occur in 5–30 minutes and last one hour or more. Lung-lymph flow ($\dot{Q}_L$) is elevated, but the total protein lymph-to-plasma ratio

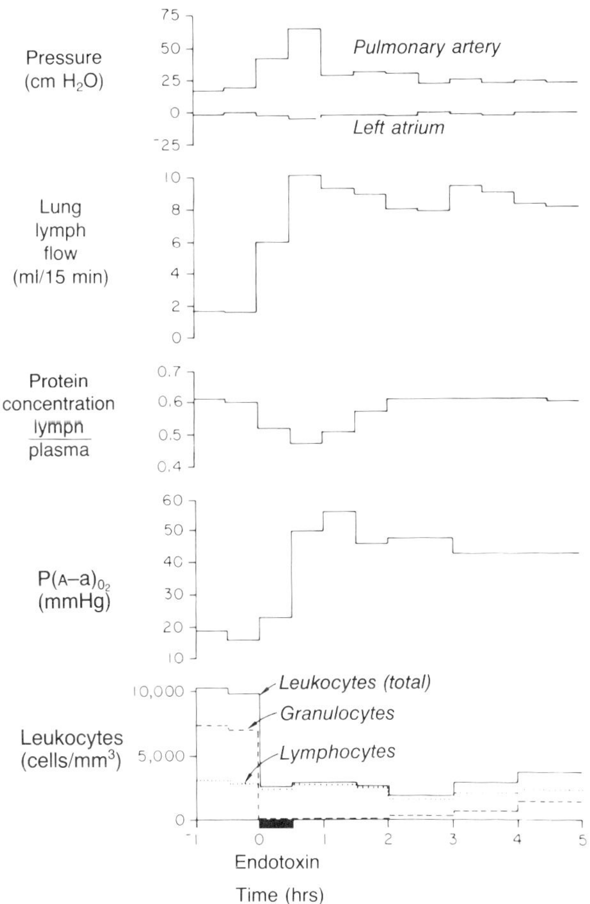

Fig. 1 *Effects of endotoxin on pulmonary circulation. Reproduced from Snapper et al (1983), with permission.*

(L/P) is decreased which results in little, if any, increase in protein transport from the plasma to the lymph.

Because endotoxin-induced pulmonary hypertension could be blocked by acetylsalicylic acid or indometacin, the role of arachidonic acid derivatives in this response was investigated more extensively (Anderson et al 1973; Reeves et al 1972). Vane had suggested that aspirin-like drugs probably inhibited prostaglandin synthesis (Vane 1971). The arachidonic acid pathways are extensively reviewed in Chapters 9 and 10 and a schema is shown on page 204.

These investigations culminated in the report of Frölich et al (1980), who reported that a derivative of the cyclooxygenase pathway, thromboxane A_2 (TXA_2), was the most likely derivative of arachidonic acid to cause endotoxin-induced pulmonary hypertension. This conclusion has support from Peterson et al 1982, who showed that TXA_2 is capable of causing pulmonary vasoconstriction and that the lung is the primary site of synthesis (Peterson et al 1982). The temporal relationship between thromboxane and pulmonary hypertension is confirmed by Fletcher et al (1981).

Many studies by multiple groups led to these conclusions. A review by Rietschel et al (1980) presents an overview of prostaglandins in endotoxicosis. Parratt and Sturgess (1977) reported that pulmonary responses to endotoxin were prevented by pretreatment with the prostaglandin antagonist polyphoretin phosphate. Previous studies by Parratt and Sturgess (1974, 1975, 1976) indicated that indometacin, sodium meclofenamate, and flurbiprofen all antagonized the effects of endotoxin on pulmonary vascular responses in the cat model. These agents all act as prostaglandin synthetase inhibitors. Fletcher and Ramwell (1978a, 1978b) found that indometacin protected the baboon and dog from endotoxin-induced mortality. Pulmonary arterial pressures after endotoxin did not change in either of the models these authors tested. The failure of pulmonary arterial pressure increase in these is different from the studies reported by Kuida et al (1958) and Pennington et al (1973). Many subsequent reports have also demonstrated endotoxin-induced pulmonary hypertension in dogs, but Starzecki et al reported a resistant population to the general effects of endotoxin (Starzecki et al 1967). The mechanism for the resistance to endotoxin in dogs has never been identified.

Prostaglandin synthesis is also blocked at the phospholipid-to-arachidonic acid site by steroids (Langham et al 1983). Using methylprednisolone, Demling et al (1981c) attempted to block the pulmonary hypertensive response in sheep. They found that methylprednisolone ameliorated, but did not prevent, the hypertensive response or the increased lung-lymph flow in Phase I. Decreases in cardiac output were abolished by methylprednisolone treatment. The authors did not try to explain these Phase I results.

Other derivatives of the cyclooxygenase pathway, PGE_1 and prostacyclin (PGI_2), are potent pulmonary vasodilators and have been studied. PGE_1 in the presence of endotoxin, decreased the Phase I increment in pulmonary perfusion pressure of dog lung lobes. When the PGE_1 infusion was stopped, perfusion went up again (Sorrells et al 1972). Smith et al in 1982 had similar results when they studied awake sheep. They infused either PGE_1 or PGI_2. During Phase I, PGI_2, but not PGE_1, significantly attenuated the pulmonary hypertension. Both were effective in reducing lung-lymph flow. Cardiac outputs were not significantly different during this phase, but a rise in aortic pressure was prevented in all cases. Immediately after the discontinuation of a 5-hour infusion, pulmonary arterial pressures, systemic pressures, and lung-lymph flows rebounded to almost match the untreated endotoxin controls. Although both of these cyclooxygenase derivatives are potent local vasodilators, they probably are not normally circulating hormones (Christ-Hazelhof and Nugteren 1981).

In a similar but earlier study, Demling et al (1981b) infused PGI_2 to see whether lung injury from endotoxin could be prevented. In this study, like the other, PGI_2 significantly attenuated pulmonary arterial hypertension at one hour and from then on. However, at one-half hour, the $\bar{P}pa$ was significantly elevated; therefore, the hypertension phase was attenuated in duration and magnitude but not abolished. Lung-lymph flow, while lower, was still significantly elevated during Phase I. $PGF_{2\alpha}$ and PGI_2 levels were significantly elevated in the hypertensive phase with endotoxin alone. Levels of PGF_2 (a potent pulmonary vasoconstrictor) were not affected by PGI_2 infusion.

In summary, PGE_1 and PGI_2 may play protective roles in lung injury. While it has been demonstrated that PGI_2 levels rise after LPS infusion, the balance would appear to be shifted in the direction of vasoconstricting arachidonic acid derivatives. The findings of Hales et al (1981) led the authors to conclude that a complex interaction exists between thromboxane and PGI_2 which controls the vasomotor responses to low-dose endotoxin. Since $PGF_{2\alpha}$ did not increase significantly, some other moiety must be the mediator of the elevated $\bar{P}pa$ (Demling et al 1981b; Frölich et al 1980).

The studies by Parratt and his collaborators reinforce this suggestion. Like Frölich et al (1980), Ball et al (1982) measured TXB_2, the stable breakdown product of TXA_2. They found its level to be markedly increased at 2 and 7 minutes after LPS administration. Pretreatment with a thromboxane-synthetase inhibitor, dazoxiben (UK-37248-01), significantly reduced the pulmonary hypertensive response and TXB_2 levels. $PGF_{2\alpha}$ was not affected by pretreatment with dazoxiben, which reinforces the importance of the TXA_2 in Phase I.

These results were published in a more complete form in 1983 (Ball et al 1983). To further validate the importance of TXA_2, Ball and Parratt (1983) also studied a thromboxane receptor blocker (AH-23848) and measured PGI_2 levels in the presence of this agent, and dazoxiben with and without endotoxin. They wanted to test whether blocking the thromboxane synthesis might shift endoperoxide activity toward PGI_2 synthesis. Their results indicate that although dazoxiben increased PGI_2 levels, the main effect was on thromboxane synthesis. AH-23848 was effective in blocking the pulmonary hypertensive effect.

Another study was more conclusive in demonstrating the shift in pathways. An imidazole derivative was effective in reducing the pulmonary hypertension and in decreasing the levels of TXB_2 while 6-keto-PGF_1 (the measurable breakdown product of PGI_2) and PGF_2 levels were much higher (Watkins et al 1982). These findings demonstrate the shifting towards prostacyclin and prostaglandins and away from thromboxanes in the presence of a thromboxane synthetase inhibitor.

The conclusions of Harlan et al (1983) were that dazoxiben may reduce the pulmonary hypertensive consequences of endotoxemia, but that the changes in permeability are not affected and therefore are not attributable to TXA_1.

The cyclooxygenase pathway is thereby demonstrated to be important in the genesis of pulmonary hypertension. Although PGI_2 levels are increased with endotoxin, the mediator of endotoxin-induced pulmonary hypertension seems to be mainly TXA_2. The effect of TXA_2 on lung-lymph flow, which is thought to be mainly the result of increased capillary hydrostatic pressure in Phase I, is also inhibited by indometacin (Frölich et al 1980). Other studies do not show complete attenuation during Phase I (Demling et al 1981b).

Spannhake et al (1983) suggest that the current data are most consistent with the hypothesis that endotoxemia facilitates synthesis of prostaglandins by increasing the availability of PGH_2 to the terminal enzymes by enhancing cyclooxygenase activity.

The role of the lipoxygenase pathway is less clear. The hypothesis has been that leukotrienes, a product of the 5-lipoxygenase pathway, may contribute to the damage initiated during Phase I which becomes evident during Phase II. Smith et al (1981b) infused lipoxygenase to stimulate the production of HPETE (hydroperoxy fatty acids) and HETE (hydroxy fatty acids) and the leukotrienes. The Phase I pulmonary hemodynamic response to lipoxygenase infusion was very much like that to endotoxin infusion, but delayed. Systemic pressure and cardiac output remained relatively constant. No discussion of the effects of this infusion on the actual generation of lipoxygenase or cyclooxygenase products was included. Ogletree et al in 1982 measured levels of 5-HETE (5-hydroxy-6,8,11,14 eicosatetraenoic acid) during the $2\frac{1}{2}$ hours after endotoxin infusion (Phase I). Peak lymph levels increased significantly. Since TXA_2 appears to be predominant in mediating the pulmonary hypertensive response in Phase I, the importance of 5-HETE during Phase I may not be recognized until Phase II; however, 5-HETE levels returned to baseline during this period. Although there were no apparent correlations between the peak levels of 5-HETE and other variables indicative of lung injury, this does not exclude the possibility that 5-lipoxygenase products participate in the mechanism of lung injury.

4.2. Phase II

Demling et al (1981b) described Phase II as the 'permeability' phase. In this phase, there is a marked increase in pulmonary endothelial permeability and 'capillary leak' which results in pulmonary edema. The lymph-to-plasma protein ratio is elevated. During Phase I, there is some influence of the cellular components of blood, suggested by the release of arachidonic acid products, but in the lung-injury phase (II) the influence of leukocytes, platelets, and free radicals seems to be greater. Hüttemeier et al (1982) believe, however, that leukocytes contribute substantially to the peak plasma concentrations of thromboxane following endotoxin infusion. Endothelial dysfunction with cellular edema occurs early on. Normal cell integrity is lost and endothelial cells are destroyed. Phase II changes progress to the death of the animal if the amount of LPS is large enough.

Only one recent publication argues against the occurrence of increased capillary permeability. The study of Gabel et al (1978) concludes that *E. coli* infusion at a very high dose does not damage the pulmonary capillary membrane, but the model on which this hypothesis was tested may not have been appropriate. In their study in dogs, they used a LD_{100} of 4 mg/kg. The time course of action of this dose and the intestinal hemorrhage it promotes may cause changes not directly related to the pulmonary effect. In fact, substances released into the circulation following the gastrointestinal injury may significantly influence the pulmonary responses to endotoxin, hence the difference seen between studies on the isolated perfused lung and the intact dog might be accounted for by this mechanism. In Gabel's study, animals were anesthetized with halothane which, in itself, could prevent the pulmonary arterial vasoconstriction and account for the decreased pulmonary arterial pressure found after 1 hour. The differences between the results of their study and those found generally in the literature raise questions as to the validity of the techniques used in that study. It emphasizes the questionable appropriateness of the intact dog as a model for the pulmonary effects of endotoxin, particularly because of the predominance of massive hemorrhagic necrosis of the gastrointestinal tract coupled with minimal pulmonary edema. These gastrointestinal influences may also account for the differences in response profiles between sheep and dogs reported by Connelly et al (1982).

The preponderance of evidence drawn from numerous data indicates that alterations of alveolar-capillary permeability are, in fact, the most important lesions in endotoxin-induced lung injury (Bessa et al 1974; Cintora et al 1972, 1973; Harrison et al 1969). At the end of Phase I, the pulmonary arterial pressure has returned toward baseline, the lymph-to-plasma total protein ratio is frequently at the same level as control, and lung-lymph flow varies between baseline and some elevation. These observations are true for the awake, unanesthetized sheep and vary in magnitude and duration between animals, but are easily reproducible (Brigham et al 1979a; Traber et al 1983a).

Phase II begins 2–5 hours after endotoxin infusion. At low doses, cardiac output has returned nearly to baseline levels and systemic pressure is usually at control or slightly elevated. For the sheep model, low doses are those under or equal to 1–3 µg/kg *E. coli* LPS. At higher doses in the mg/kg range, the depressed cardiac output and systemic hypotension persist. Pulmonary artérial pressures may remain slightly elevated. Lung-lymph flow usually increases up to 5 to 6 times the control value. Lymph-to-plasma total protein may be at baseline or only slightly above, but with the markedly increased $\dot{Q}_L$, the transport of protein from the vascular compartment to the interstitium-lymph compartment is tremendous. While changes in hydrostatic pressure always accompany pulmonary edema to some extent, in this case it is of little consequence. As the lung-lymph clearance (lymph flow × lymph-to-plasma ratio) increases, protein molecular radius of the lost proteins also increases, implying increased permeability. The concepts of lung vascular permeability and the inferences from the measurement of plasma to lung-lymph protein transport are fully described by Brigham et al (1979b).

Identification of the mechanisms and mediators of Phase II has not been confirmed. Free radical metabolites such as O_2^-, O_1^-, hydrogen peroxide (H_2O_2) and OH^- released from the sequestered leukocytes and platelets are implicated in the mechanism of damage and change in permeability.

Lysosomal enzyme release after endotoxin administration was reported by Warren et al (1973) and Demling et al (1981a). Warren et al used a dog model and 1 mg/kg *E. coli* endotoxin to show plasma elevations of cathepsin D and β-glucuronidase. In the sheep model, Demling et al (1981a), reported an increase of 200-800% in lysosomal enzymes occurring in the lymph early in the hypertensive phase. Lymph β-glucuronidase levels correlated well with lung-lymph flow. Plasma levels of enzymes increased but were poorly correlated with lung injury as assessed by lung-lymph flow. Enzyme levels of over 200% in the lymph were indicative of impeding systemic shock and death. The early appearance of the increases in lysosomal enzymes would imply that lung injury is initiated during Phase I. The manifestations of this early damage are seen in Phase II. Demling et al (1981c) used methylprednisolone to ameliorate the effects of endotoxin (see Section 4.1.). They found that methylprednisolone depressed the rise in β-glucuronidase and seemed to protect against platelet aggregation. Warren et al found that methylprednisolone did not prevent the increment in enzyme levels or improve survival rates. The differences in the results of these two studies are primarily due to different doses and models. Demling used a dose which typically causes pulmonary hypertension in sheep, while the doses Warren used were high and atypical for the dog. Demling measured enzymes in lymph and plasma; Warren could only measure levels in the plasma.

Studies by Brigham's group help to eliminate some possible mechanisms and direct our attention to others. As reported by Demling et al (1981c), Brigham et al (1981) also found that methylprednisolone decreased the $\dot{Q}_L$ in the permeability phase when administered beforehand, but unlike Demling et al, Brigham and his colleagues found no effect on lung injury when methylprednisolone was given after endotoxin infusion. These investigators did not attribute the effect of the steroid therapy to prevention of prostaglandin synthesis. They cited a study by Ogletree and Brigham (1979), in which the permeability phase was actually enhanced by indometacin treatment and speculated that the effect might instead be on inhibition of the lipoxygenase pathway. In a study of the effects of histamine on the lung vascular response, Brigham et al (1976) found that histamine produced edema and permeability changes that are not unlike those produced by endotoxin. Pretreatment with diphenhydramine predictably abolished these effects. Brigham et al (1980) demonstrated that diphenhydramine reduced the endotoxin-induced permeability in Phase II.

Lysosomal enzyme release, stimulation of the complement cascade, sequestration of platelets or leukocytes, and granulocyte recruitment in endotoxin-induced lung injury continue to be subjects of controversy.

5. METABOLIC PATHWAYS

The effects of endotoxin on metabolic pathways of major importance to the lung are numerous.

5.1. Angiotensin-converting enzyme

Endotoxin has been shown to decrease angiotensin-converting enzyme (ACE). In a mouse model, Hollinger (1983) found that with *Serratia marcescens* and *E. coli* endotoxin, the decrease was not dose-dependent in a range from 0.5 to 8.0 mg/kg. The decrease occurs in 2 hours and returns to normal within 24 hours. The time course is similar to the changes in permeability seen in the sheep model and may reflect reversible injury to endothelial cells. However, the use of serum ACE levels as a diagnostic aid is still fraught with the problem of deciding whether the decrease is due to a change in metabolism of endothelial cells from injury or to altered perfusion (in this case the decreased cardiac output) and therefore changes in the time and surface area of contact with endothelial cells.

5.2. Ion transport

Sayeed and his coworkers have paid particular attention to nucleotide pathways (Klein and Sayeed 1980). In 1982, Sayeed reviewed the actions of endotoxin on pulmonary vascular pressure, permeability and metabolic and ion transport alterations. He reported that in the anesthetized guinea pig model, glucose oxidation to lactate and CO_2 is significantly increased with a decrease in lung ATP after 10 mg/kg of *Salmonella enteritidis* LPS intravenously. In vitro Na^+-K^+ transport and oxidation of glucose and lactate production were measured in lung slices. Radionuclide studies demonstrated an increase in intracellular volume after endotoxin. Extravascular-extracellular volume was also increased. It could not be determined whether this increase was due to interstitial or alveolar fluid volume. This change in fluid volumes could be attributed to an impairment of Na^+-K^+ transport coupling.

5.3. Oxygen interactions

Endotoxin interacts with both extremes of oxygen-induced responses of the pulmonary circulation. In 1974, Reeves and Grover reported that 15 µg/kg of *E. coli* endotoxin partly or completely blocked the response to hypoxia in dogs. This loss of the hypoxic vasopressor response occurred within 30–90 minutes and lasted up to 4 hours. Responses to serotonin were not affected, but the response to prostaglandin $F_{2\alpha}$ disappeared. These data suggest that endotoxin pretreatment causes the elaboration of an antagonist to $PGF_{2\alpha}$ vasoconstriction but not to serotonin. Serotonin has been considered one of the possible mediators of the hypoxic pulmonary vasopressor response, but differences between species exist. Since inhibition of arachidonic acid metabolism with acetylsalicylic acid did not prevent the blockade of hypoxic pulmonary hypertension, the authors concluded that PGE_1 was not the vasodilator mediator in this case.

Further evidence from this laboratory, however, does not eliminate the possibility that prostaglandins are involved (Mlczoch et al 1978; Weir et al 1976). Weir et al established that sodium meclofenamate and indometacin would prevent the loss of the hypoxic pulmonary hypertensive response with

LPS. These reports tested the role of leukocytes and platelets in this mechanism and found that neither platelets nor leukocytes had an apparent role. They proposed that endotoxin caused the stimulation of a vasodilator prostaglandin release or production from the systemic circulation or pulmonary vessels themselves.

Either the mechanism or mediator of the pulmonary hypertensive response induced by hypoxia differs from species to species; or some other factor was involved in the study of the interaction between endotoxin and hypoxia in the pig (Orr et al 1977). A dose of 4 µg/kg *E. coli* endotoxin failed to alter the hypoxic vasopressor response in pigs 35 minutes after the LPS bolus was given. Several possibilities for the differences exist: (a) different mediators of the vasopressor response between species, (b) the awake state versus anesthesia and, (c) dosage differences. The alternative hypothesis of 'normoxic pulmonary vasodilatation', proposed by Weir in 1978, allows for the differences in species. He stated that in the normoxic state, pulmonary vasodilatation is present in all species. In the hypoxic state, active vasodilatation is depressed, allowing whichever vasoconstrictors are present to cause the hypertension. The relative importance of each constrictor varies with the species.

The role of endotoxin in hyperoxia appears to be much different. Alteration or induction of lung enzymes secondary to sublethal injection of endotoxin and prior to hyperoxic exposure in experimental animals is an exciting area of recent and future investigation. In 1978, Frank et al rather serendipitously showed that animals given plasmanate inadvertently contaminated with endotoxin, survived hyperoxic exposures that were normally lethal. In an additional study in rats, Frank and Roberts (1979) used sublethal LPS dosages, either 250 µg/kg per day at 0, 24, and 48 hours; or 500 µg/kg once just prior to exposure to more than 95% oxygen. These dosages are calculated to be 1/100 to 1/50 of LD_{50}. In adult rats, exposure to 95% oxygen is usually lethal within 60 to 72 hours. Twenty-seven percent of control animals survived at the end of 72 hours versus a 95% survival in LPS-pretreated animals. At necropsy, treated rats had significantly decreased pulmonary edema and pleural fluid, as well as less perivascular interstitial, and alveolar edema and hemorrhage when examined by light microscopy. The treated animals appeared to have a protected pulmonary capillary endothelium when examined by electron microscopy. Neither indometacin nor methylprednisolone altered the results.

The most widely accepted mechanism for lethal hyperoxic damage to the lung is the 'free radical' hypothesis of oxygen toxicity which was first proposed by Gerschman in 1964. Clark and Lambertson's classic review of pulmonary oxygen toxicity (1971), and more recent reviews by McCord and Fridovich (1978), and Deneke and Fanburg (1980) have examined the generation of free-radical metabolic products of cellular oxidative metabolism with production of O_2^-, OH^-, O_1^-, via the Haber-Weiss reaction. These substances are also released by inflammatory cells such as granulocytes, macrophages, and platelets.

Aerobic organisms have developed potent antioxidant enzyme systems, including superoxide dismutase (SOD), catalase, selenium-dependent glutathione peroxidase (GP), glucose-6-phosphodiesterase (G-6-PD), vitamins C and E, and possibly indoleamine-2,3-dioxygenase (IDO), to metabolize these potentially toxic substances. With this in mind, Frank et al (1980) studied several antioxidant enzymes and competitive enzyme inhibition with controls and animals pretreated with endotoxin, which were subsequently exposed to lethal hyperoxia, as previously described. The animals pretreated with endotoxin had significant increases in the four enzymes studied (SOD, G-6-PD, catalase, and GP) compared with controls. When the authors blocked SOD induction using diethyldithiocarbamate (DDC), the mortality of pretreated animals was not different from that of controls. They concluded that antioxidant enzyme induction, most importantly SOD, protects animals from subsequently lethal hyperoxia.

Frank (1981a) then studied the effects of low-dose endotoxin pretreatment prior to exposure to progressively increasing oxygen concentrations, that is, 40% for 2 days, 60% for 2 days, 85% for 2 days, followed by an exposure of more than 95%, or, alternatively, 40% oxygen for 5 days and then 95% oxygen exposure. Normally, rats preexposed to a sublethal concentration of 85% oxygen appeared to have antioxidant enzyme induction, again most importantly SOD, which allowed survival on subsequent exposure to more than 95% oxygen. The rats in this study, which were preexposed to 40% oxygen without LPS, or 40-60-85% and then 95% oxygen without LPS, died earlier than saline controls in 95% oxygen. The endotoxin-treated rats survived statistically longer periods of time than either the saline controls in 95% oxygen or the progressively exposed animals. Frank theorized that the effect of a low dose of LPS made the 'rat lung much more resistant biochemically to hyperoxic challenge in the presence of increased oxygen free radicals.' He believed this occurred because of activation of protein synthesis which was not induced by hyperoxia alone. Another study demonstrated prolonged survival of rats with endotoxin prior to paraquat and hyperoxic exposure (Frank 1981b). Paraquat is a known pulmonary toxin, thought to cause lung injury by free radical generation.

Block (1983) showed that pretreatment with endotoxin preserved the metabolic function of endothelial cells as measured by lung-amine uptake in animals prior to 100% oxygen exposure in one atmosphere for one week. The treated animals had normal uptake of ^{14}C-labeled serotonin and ^3H-labeled norepinephrine at 24 hours and 7 days. Control animals had significantly decreased amine uptake after 24 hours oxygen exposure, although they appeared normal and no different from air-exposed saline-treated rats. Endotoxin also prolonged survival in rats exposed to 100% oxygen.

A morphometric study by Thet et al (1983) provided insights which could not be attained with the indirect observations of Block and Frank. Endotoxin reduced the interstitial edema (34% less volume of noncellular interstitium) and the cellular portion of the interstitium. Endothelial surface area was decreased in a way similar to that found in 100% oxygen controls without endotoxin. While the decreased edema and cellularity confirm the indirect observations of better survival rates, the morphometric interpretation of the endothelial surface area findings are different from Block's interpretation (1983). His study implies that endothelial function is not significantly altered with endotoxin protection; the study of Thet et al, however, may mean that the decreased endothelial surface area actually has increased metabolic activity of the remaining cells so that function appears to be in the normal range. The authors observed that, even though the visible damage from oxygen as assessed by electron microscopy might be similar with or without endotoxin, the function of endothelium might be preserved with endotoxin, as reflected by a reduction of interstitial edema and cellularity.

5.4. Endorphins

Another pathway recently investigated is one of the most interesting. Some manifestations of endotoxin effects seem to be through the endorphins. Faden and Holaday (1980) have suggested that pituitary endorphins are pathophysiologic factors in shock and that opiate antagonists may be of therapeutic benefit in the treatment of shock. Their work suggests the relationship shown in Figure 2. While they have collected data relating to the systemic circulation and the cardiodepressant effect which they believe is mediated through opiate receptors in the central nervous system, they did not study the effects on the pulmonary circulation. Faden and Holaday proposed that the effects of naloxone may be mediated by parasympathetic pathways. Earlier Hildebrand and Seys (1965) suggested that the pulmonary vasoconstrictor responses from endotoxin might be partly the result of a depolarizing action of endotoxin on the vascular smooth muscle with increased reactivity to acetylcholine.

Faden and Holaday, however, found that regardless of the duration of naloxone treatment, no differences were found in 24-hour survival in rats but they were found in dogs. This led the authors to the conclusion that death from endotoxin may be species-related and that, in the case of the rat, noncardiovascular factors may be important contributors to death.

Traber et al (1983b) found that naloxone reduced Phase II elevation of

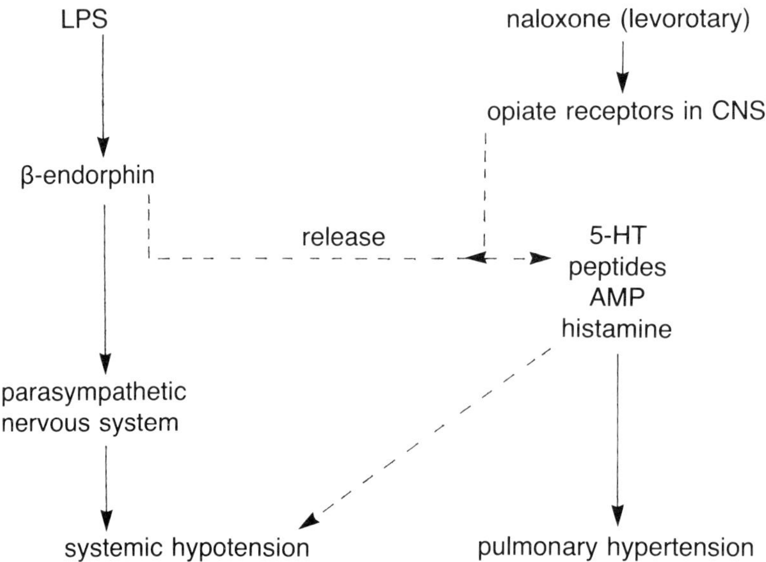

Fig. 2 *Relationship between opiate antagonists and endorphins in endotoxin shock.*

lung-lymph flow but had little effect on cardiac output. Opiate receptor binding activity was also elevated in sheep given 0.75 μg/kg of endotoxin. They concluded that endorphins or endorphin-like substances may play an important role in the pulmonary response to endotoxin.

Some patients appear to respond positively to naloxone treatment in sepsis. Groeger et al (1983) found a 50% response to 0.3 mg/kg naloxone. Systemic blood pressure increased significantly; pulmonary arterial mean pressure increased from 26±3 to 31±5 mmHg, but no difference in survival resulted.

5.5. Central nervous system effects

The suspicion that a CNS-mediated pathway might be the key to endotoxin injury was investigated earlier (Simmons et al 1969). The increased sensitivity of man and animals to endotoxin entering via the CNS indicated to these authors that endotoxin acted primarily on this organ system. They compared the effect of lethal doses (LD_{80}) of endotoxin administered intracisternally and intravenously to three species, the dog, rabbit, and monkey, and compared the gross lesions with those previously reported.

This study revealed that the lesions by intracisternal administration were similar between species; all were characterized by the absence of intracortical or intramedullary lesions and the presence of massive hemorrhagic pulmonary edema. Of interest was the lack of gastrointestinal lesions typical of intravenously administered endotoxin in the dog in particular. All animals in this study showed only hyperemic mucosa with a few ulcerations. The conclusions of Simmons et al were that since the spectrum of lesions in the intracisternally administered endotoxin were markedly different from those with the intravenous route of administration, it is unlikely that the primary effect of endotoxin is mediated by the CNS.

6. PATHOPHYSIOLOGY OF ENDOTOXIN-INDUCED LUNG INJURY

The histopathology of endotoxic shock is dependent upon the producing organism, the dose, and the animal species. There are common features in all reports and the lung is always affected. Rhoades et al (1967) describe endotoxin-induced lesions as 'hemorrhagic septicemia'. This report is cited because it contrasts the effects of live organisms with those of purified endotoxin and characterizes the differences in pathology seen when different routes of administration are used. In this study, *Pasteurella multocida* organisms and endotoxin were administered to calves and pigs. Dosages and routes of administration were varied, but the end result was death in 40–72 hours after administration of organisms and death in 4.5 hours for the one calf given endotoxin extracted from the animals infected with the whole organisms.

The septicemic lesions were characterized by widespread petechial hemorrhages. In the lung, the lesions for the lower respiratory tract and related structures varied with the type of exposure. Hyperemia was observed in the lungs of all calves. Those exposed by aerosol had well-developed interlobular edema accompanied by markedly increased fluid and fibrin in the pleural cavity. Other organs were also affected with characteristic edema, hemorrhages, and hyperemia. In another study by other authors, dogs given a culture of intact or disintegrated *Pseudomonas* bacteria had quite different pathological findings (Myrvold and Svalander 1977). Those animals given intact bacteria had no gross or light microscopic lesions.

In contrast, in the study by Rhoades et al, the calf that died following endotoxin administration had an almost pure respiratory manifestation. The trachea and bronchi were nearly completely filled with yellowish stable froth. Both the parenchymatous tissues and the interlobular septa showed extensive pulmonary edema. Increased pleural fluid and subpleural hemorrhages were present.

Microscopically, lesions differed with route of exposure. Aerosol exposure resulted in multifocal suppurative pneumonia while intranasal and intramuscular exposure resulted in interstitial pneumonia. The aerosol-route lesions were characterized by large numbers of neutrophils, but macrophages and other cell types were also seen. With the intranasal and intramuscular routes, lymphocytes and macrophages were much more prominent than neutrophils. The endotoxemic calf died much more rapidly and one would not expect the same lesions. Hyperemia, hemorrhage and edema were extensive in both the interlobular tissue and the parenchyma.

In pigs, these authors found that extensive lung lesions were a constant feature. In addition to the expected edema, congestion, and increased pleural cavity fluid and fibrin, the anteroventral portions of the lung were consolidated. Since the pigs all received the organism by aerosol and intranasal administration, the lesions would be expected to be similar to those of the calf. Neutrophils were predominant; macrophages also occurred in large numbers in the perivascular, interlobular and subpleural edematous areas, as well as in the pleural exudate.

The authors concluded that the lesions of the endotoxemic infection were compatible with shock and that the aerosol exposure was consistent with a primary respiratory infection while the interstitial pneumonia was probably a part of the general reaction to septicemia.

The awake sheep has become an important model to study endotoxin in the lung, despite the fact that there is some contamination that results from a nonlung (diaphragmatic) contribution to the caudal mediastinal lymph node efferent flow (Drake et al 1981). This contamination is considered to be insignificant by many groups (Gunther and Demling 1981; Staub et al 1975). Meyrick and Brigham (1983) have described the temporal relationship between structure and function in anesthetized sheep in Table 1.

The route of administration and the injection site of LPS can be critical. Freudenberg et al (1982) injected *Salmonella* endotoxin via the tail vein in rats. Whereas other investigators have shown that in larger animals the first lesions appear in the lung, presumably by the use of ear, jugular or cephalic vein injections, Freudenberg et al demonstrated the LPS in the liver first. They localized the presence of the LPS by immunoperoxidase methods, and the appearance of lesions was correlated with the appearance of immunoreactivity at 7 hours.

Important concepts emerged from the study by Freudenberg in rats. The distribution of the LPS in the lung was not confined to a single cell type. Immunoreactivity was present in interstitial cells thought to be macrophages, in capillaries of the alveolar walls and in some mononuclear cells migrating through the walls of vessels. Neither endothelial cells nor epithelial type II cells appeared to contain immunoreactivity, but granulocytes in alveolar walls (perhaps in capillaries), and alveolar and bronchial macrophages

92

TABLE 1 *Temporal relationship between structural and physiologic changes after infusion of E. coli endotoxin to anesthetized sheep*

Time after start of endotoxin infusion	Structural changes	Physiologic changes
15 min	Accumulation, margination, degranulation, and fragmentation of granulocytes and accumulation of activated lymphocytes in pulmonary microcirculation.	Increase in pulmonary artery pressure and in protein-poor lung-lymph flow; decreased Pa_{O_2} and leukopenia.
30 min	As at 15 min but more severe; migration of leukocytes into interstitium; some interstitial edema.	Peak pulmonary artery pressure.
60 min	Interstitial leukocytes, endothelial and vessel wall damage; damaged type I and interstitial cells; perivascular edema.	Pulmonary artery pressure decreasing toward baseline.
2 hr and onward	Endothelial layer disruption.	Pulmonary artery pressure stable; increased lung vascular permeability.

Reproduced from Meyrick and Brigham (1983), with permission.

showed strong immunoreactivity. The significance of the localization of LPS to these cell types is that the granulocytes and macrophages would appear to have a phagocytic role. The authors imply that these cells provide a means to remove LPS or its antigenic components from the lung. The origin of the macrophages and granulocytes could, in fact, be the liver (Easton 1952). The fact that LPS was not localized in endothelial cells or epithelial cells themselves may imply that even though these cells are primary targets for the LPS action in the lung, the effects may be secondary to the direct action of LPS on other cells or metabolic pathways. Platelets and leukocytes are considered important candidates mediated, perhaps, through complement (Evenson et al 1975).

At higher doses of endotoxin (5 mg/kg), the immunoreactivity in this Freudenberg study was present for up to 3 days, but the lung was almost clear at 5 days. Whether this persistence is dictated by cell turnover, or whether the labeled complexes were still active, could not be determined.

Other species and other endotoxins produce similar results. In rabbits, Goodman et al (1979) observed pulmonary congestion, swollen endothelial cells, local fibrin thrombi and margination of white blood cells using a dose of 20–100 μg/kg of intravenous *E. coli* endotoxin. If animals lived 36 hours or more, there were no microthromboemboli present.

In the disintegrated organism model, platelet aggregates resulting in microthromboemboli formation, polymorphonuclear leukocyte (PMN) disintegration, and ultrastructural pulmonary endothelial damage were found 2–5 minutes after toxin administration. Since the experiments only lasted 2 hours, it is not unreasonable to suppose that more severe changes did not become apparent (Myrvold and Svalander 1977). Ogawa et al (1977) believe that the formation of microthromboemboli and disseminated intravascular coagulation (DIC) played an important role in the dog *E. coli* model of Adult Respiratory Distress Syndrome (ARDS). The importance of DIC is supported by the results of ultrastructural studies of monkeys by McKay et al (1967). In one group, killed 5 hours after 3 mg/kg of the endotoxin, Ogawa et al found that the lung lesions consisted of congestion, atelectasis, thrombosis, and smooth muscle swelling. Other animals treated with 1 mg/kg and killed 24 hours later had lesions usually seen in other species. Dose undoubtedly plays a role, in that the lower dose can be lethal but only after sufficient time to cause the broad spectrum of lesions typical of patients with sepsis-producing endotoxin. The lesions of the 1 mg/kg group included the acute changes described above, plus interstitial and intra-alveolar edema and platelet megakaryocytic emboli. These authors concluded that the increase in pulmonary vascular resistance may be due to pulmonary capillary wall thickening and DIC, which in turn leads to the changes of permeability.

The importance of the contribution of complement, leukocytes, and platelets to the pathophysiology of endotoxin is discussed at length in Chapters 5 and 6. The following are brief descriptions of some of the work particularly focused on the lung.

Complement almost certainly plays a role. Traber et al (1983b) speculated on some of the interactions that might be important. Stimler et al (1980) demonstrated that complement anaphylatoxin $C5_a$ stimulates the release of slow reacting substance of anaphylaxis-like activity from the lung, and Rampart et al (1982) showed that this activation of complement contributed to the rise in blood levels of 6-oxo-prostaglandin $F_{1\alpha}$ during endotoxemia. Most studies indicate that the properdin or alternate complement pathway is the biologically active one (Fearon et al 1975).

The studies of Traber et al (1983a, 1983b) raise a number of interesting points. This group, like those of Demling, Brigham and others, investigated the physiological effects of endotoxin at low doses (0.75–3 μg/kg) which, in the case of the sheep model, are fatal in a percentage of animals. These doses are in contrast to the apparently very high doses used by some investigators which could not be achieved under any naturally occurring condition (Ochoa and Kern 1980). In this report by Traber et al (1983a), 3 of 16 original animals or approximately 19% died as a result of this dose and were excluded from the study. Traber studied the effects of two different endotoxins, previously in dogs and in the present study in sheep. He compared the effects of *Pseudomonas aeruginosa* and *E. coli* in dogs and of *E.coli* and *Salmonella typhimurium* in sheep. In both studies, he found no significant differences between measured variables using either endotoxin randomly in the first study and following this with the other endotoxin in a week. The authors cite evidence which supports the lack of cross-antigenicity between endotoxins, pointing out that the immunologic effects probably depend on the polysaccharide portion of the molecule, whereas the responses important in lung injury are probably related to the lipid portion of the molecule. Our own studies with lipid X reinforce these findings. We have demonstrated that all of the cardiopulmonary pathophysiology normally attributed to LPS can be initiated by this 711 molecular weight lipid fraction of lipid A (Burhop et al 1983).

The Traber study also documents the depressed white blood cell count during the 0–6 hour segment of the study which, they state, results from an early neutropenia and a later fall in lymphocytes.

The leukopenia has been more extensively studied by the group of Brigham and presented by Snapper et al (1983). While it has been clearly demonstrated that arachidonate products mediate the early phase of pulmonary hypertension, the increased permeability phase has only been pharmacologically attenuated by an H_1 antagonist. Brigham's group, in other studies, identified a different course of events relating to leukocyte depletion. They found that both granulocytes and lymphocytes migrated into the interstitium and that granulocyte depletion markedly attenuated the permeability phase. In the present study, they tested the hypothesis that leukopenia had no effect on the pulmonary hypertensive phase but did, in fact, find that leukopenia correlated well with the change in $P(A-a)_{O_2}$ gradient at 4 hours, and concluded that this decrease in circulating leukocytes probably contributed to hypoxemia and to increased pulmonary vascular permeability.

A more recent publication by the same group (Hinson et al 1983) investigated granulocyte depletion and effects on lung mechanics. They concluded that the action of granulocytes within the lung contributes to the change in lung mechanics observed following endotoxemia. They further suggest that the process causing early pulmonary hypertension and alterations in lung mechanics are different. Their conclusions are based on previous studies in their laboratory, which showed that granulocyte depletion had no effect on early pulmonary hypertension, and on this study, which showed that granulocyte depletion attenuated the decrease in dynamic compliance, the increase in lung resistance and the alveolar-to-arterial oxygen gradient (Heflin and Brigham 1981). Endotoxin also caused lymph thromboxane B_2 and 6-keto-PG_1 concentrations to increase at 1 hour post administration whether or not the sheep were granulocyte-depleted. These findings support a nongranulocyte source of thromboxane B_2 and 6-keto-PGF_1.

In contrast, Bohs et al (1979) found that T-lymphocyte depletion, in particular, moderated the increases in pulmonary arterial pressure. T-lymphocyte depletion had a specific effect on myocardial contractility which was markedly augmented in the absence of these leukocytes. The difference between these studies is not clear.

Hudson et al (1977) exposed guinea pigs and hamsters to an aerosol containing *Salmonella typhosa* endotoxin. Leukocyte numbers by bronchopulmonary lavage increased by more than 500% within 4 hours of exposure and were also significantly increased in cardiac blood. Most of the leukocytes were PMN. Interestingly, neither the PMN nor the airway cells showed damage despite a rather high dosage of LPS.

There has been some controversy as to whether leukocytes are the key to damage or may actually be protective. Smith and his colleagues (1981a) believe, as most of the literature suggests, that leukocytes are initiators of damage. Walker et al (1982) concluded that the early presence and disappearance of PMN during post-damage resolution implied a protective mechanism. They looked at differences between microcirculatory changes in C3Heb/FeJ endotoxin-sensitive mice, and C34/HeJ endotoxin-resistant mice, using 1 mg *Salmonella typhi* endotoxin which was lethal only to the sensitive mice within 24 hours. This study tested the hypothesis that resistant animals should possess mechanisms to dissipate and resolve initial inflammatory responses. At 1 hour after endotoxin, both strains of mice had similar numbers of PMN in pulmonary capillaries; by 3 hours, the number of PMN was decreasing in resistant mice, and by 6 hours the capillaries had returned to normal in this strain. The inflammatory response persisted until death in the C3Heb/FeJ mice. Because these authors ascribe the damage to the pulmonary capillaries by endotoxin to platelets and interactions with humoral factors, it might be reasonable to conclude that the PMN are not part of the damage-initiating factors, but that instead their early presence and disappearance at later stages in the resistant mice indicates a protective mechanism that is suggested in the resistant mice but fails in the sensitive animals.

In an effort to relate ultrastructural changes in the lung to platelet and leukocyte responses during the initial phase of endotoxin-induced shock, Myrvold and Svalander (1977) injected disintegrated *Pseudomonas* bacteria intravenously in dogs. Although no light microscopic lesions were seen, ultrastructural damage to endothelial cells was found 2–5 minutes post infection. No signs of fibrin precipitate appeared in 2 hours. When dogs were made thrombocytopenic or defibrinogenated, the changes were no different. Thrombocytopenia did, however, attenuate the pulmonary vasoconstriction. These data seem to imply that in Phase I in the dog, platelets are a major source, but not the only source, of the arachidonic acid metabolites which cause the pulmonary hypertension. Smith et al (1981a) suggest that platelets stick transiently to already damaged endothelium and that the degree of sequestration of leukocytes in the lung is indicative of injury.

This is not to say that platelets do not play an important role in the pathogenesis of endotoxin-induced lung injury. Bredenberg et al (1980) showed that in thrombocytopenic dogs, the pulmonary hypertensive phase did not occur. There is a strong possibility that platelet effects are the result of arachidonic acid metabolism and are not the only possible source for the activation of this pathway.

7. BYSSINOSIS AND AEROSOL EXPOSURE TO ENDOTOXIN

One of the manifestations of endotoxin and its effects on the lung is byssinosis. This phenomenon of endotoxin entry with cotton dust inhaled during the milling process can be extended to other dust inhalation as well. While endotoxin does not appear to be responsible for all of the problems associated with cotton dust inhalation, it is present in substantial amounts in unwashed cotton trash dust. Some estimates put it at 0.2–2.5 $\mu g/m^3$ of air, which also contains 10^2–10^5 gram-negative bacteria (Cinkotai et al 1977). While other compounds in the cotton bracts are being studied, endotoxin effects appear to play a major role in the development of occupational lung disease characteristic of cotton mill workers. In other organic, dust-initiated syndromes such as those connected with the production or processing of agricultural products or sewage sludge disease, endotoxin appears to have a role, but some investigators believe this may be indirect through the alternate complement pathway, creation of a hypersensitivity state, or adjuvant activity.

Rylander et al do not believe complement is an important factor in airway inflammation but that leukocytes are the important factor (Rylander et al 1979). The data presented by them indicate that not the fluctuation in LPS antibody level but rather a change in the size of the leukocyte population was the decisive factor in inducing the subjective symptoms.

Macrophages might be important in byssinosis, other dust-related diseases, and gram-negative pulmonary airway pneumonias. Using human alveolar macrophages collected from nonsmoking subjects, Davis et al (1980) incubated the collected cells with *E. coli* endotoxin for 1 and 24 hours at concentrations of 1, 10, and 100 µg/ml. The two highest concentrations proved cytotoxic to half of the macrophages and reduced phagocytic activities significantly in the viable cells at 24 hours. Indometacin had no protective effect on the cytotoxicity or deleterious effect on the phagocytic activities.

The cytotoxicity and impaired phagocytic activity may affect the pathogenesis of these conditions. The dose used, however, may be out of the naturally occurring range, and therefore during acquired infection it may take longer than this study indicates to have the same effect.

8. SUMMARY

In this chapter we have attempted to review the dynamic areas of investigation of endotoxin and its derivatives on the lung. Great strides in the understanding of basic humoral, cellular, and metabolic function within the lung have recently been made, but many issues remain controversial and unresolved. New data must be scrutinized to avoid continuation of old biases especially with regard to animal models, the dose of endotoxin, the type of endotoxin administered, and the route of administration. Intraspecies and interspecies conclusions must be tempered by phylogenic differences.

Future areas that appear to be most fruitful center around:

a. arachidonic acid metabolism with induction of protective pathways and inhibition of destructive pathways;

b. central nervous system–pulmonary interaction via peptidergic, cholinergic, and adrenergic pathways;

c. cellular mediated responses, most importantly those produced by leukocytes, platelets, and plasma proteins including kinins, complement, and clotting factors;

d. the potential of low-dose LPS or nontoxic LPS derivatives as pharmacological agents, which could potentially prevent free radically induced pulmonary injury secondary to hyperoxia, drugs, and toxins;

e. further understanding of the role of sepsis in the development of the adult respiratory distress syndrome in man and its prevention.

REFERENCES

Anderson FL, Kralios AC, Tsagaris TJ, Kuida H (1973) Hemodynamic effects of small amounts of endotoxin in the bovine. *Proc. Soc. Exp. Biol. Med. 143*, 1172-1175.
Ball HA, Parratt JR (1983) Thromboxane and the pulmonary response to endotoxin in the cat. *Circ. Shock 10*, 263.
Ball HA, Parratt JR, Zeitlin IJ (1982) Thromboxane and the pulmonary hypertensive

response to endotoxin. *Br. J. Pharmacol. 76*, 285P.

Ball HA, Parratt JR, Zeitlin IJ (1983) Effect of dazoxiben, a specific inhibitor of thromboxane synthetase, on acute pulmonary responses to *E. coli* endotoxin in anaesthetized cats. *Br. J. Clin. Pharmacol. 15*, 127S-131S.

Berdischewsky M, Pollack M, Young LS, Chia D, Osher AB, Barnett E (1980) Circulating immune complexes in cystic fibrosis. *Pediatr. Res. 14*, 830-833.

Berry LJ (1977) Bacterial toxins. *CRC Crit. Rev. Toxicol. 5*, 239-318.

Bessa SM, Dalmasso AP, Goodale RL Jr (1974) Studies on the mechanism of endotoxin-induced increase of alveolocapillary permeability. *Proc. Soc. Exp. Biol. Med. 147*, 701-705.

Block ER (1983) Endotoxin protects against hyperoxic alterations in lung endothelial cell metabolism. *J. Appl. Physiol. 54*, 24-30.

Bohs CT, Fish JC, Miller TH, Traber DL (1979) Pulmonary vascular response to endotoxin in normal and lymphocyte depleted sheep. *Circ. Shock 6*, 13-21.

Bredenberg CE, Taylor GA, Webb WR (1980) The effect of thrombocytopenia on the pulmonary and systemic hemodynamics of canine endotoxin shock. *Surgery 87*, 59-68.

Brigham KL, Bowers RE, Owen PJ (1976) Effects of antihistamines on the lung vascular response to histamine in unanesthetized sheep. *J. Clin. Invest. 58*, 391-398.

Brigham KL, Bowers RE, Haynes J (1979a) Increased sheep lung vascular permeability caused by Escherichia coli endotoxin. *Circ. Res. 45*, 292-297.

Brigham KL, Harris TR, Bowers RE, Roselli RJ (1979b) Lung vascular permeability: inferences from measures of plasma to lung lymph protein transport. *Lymphology 12*, 177-190.

Brigham KL, Padore SJ, Bryant D, McKeen CR, Bowers RE (1980) Diphenhydramine reduces endotoxin effects on lung vascular permeability in sheep. *J. Appl. Physiol. 49*, 516-520.

Brigham KL, Bowers RE, McKeen CR (1981) Methylprednisolone prevention of increased lung vascular permeability following endotoxemia in sheep. *J. Clin. Invest. 67*, 1103-1110.

Burhop KE, Helgerson RB, Proctor RA, Raetz CRH, Will JA (1983) Lipid X(LX), a monosaccharide subunit of lipid A reproduces the pathophysiological effects of endotoxin (LPS). *Fed. Proc. 42*, 1108.

Caridis DT, Reinhold RB, Woodruff PW, Fine J (1972) Endotoxemia in man. *Lancet 1*, 1381-1385.

Christ-Hazelhof E, Nugteren DH (1981) Prostacyclin is not a circulating hormone. *Prostaglandins 22*, 739-748.

Cinkotai FF, Lockwood MG, Rylander R (1977) Airborne micro-organisms and prevalence of byssinotic symptoms in cotton mills. *Am. Ind. Hyg. Assoc. J. 38*, 554-559.

Cintora I, Goodale RL, Yamoor A (1972) The effect of endotoxin of the alveolocapillary permeability coefficient in the dog. *J. Surg. Res. 13*, 59-62.

Cintora I, Bessa S, Goodale RL, Motsay GW, Borner JW (1973) Further studies of endotoxin and alveolocapillary permeability: effect of steroid pre-treatment and complement depletion. *Ann. Surg. 179*, 372-375.

Clark JM, Lambertson CJ (1971) Pulmonary oxygen toxicity: a review. *Pharmacol. Rev. 23*, 37-133.

Connelly AS, Feeley TW, Rosenthal MH, Mihm FG, Watkins DW, Ashton JPA (1982) Mechanisms of endotoxin-induced pulmonary edema in the dog. *Anesthesiology 47*, A135.

Cuevas P, De la Maza LM, Gilbert J, Fine J (1972) The lung lesion in four different types of shock in rabbits. *Arch. Surg. 104*, 319-322.

Davis WB, Barsoum IS, Ramwell PW, Yeager H Jr (1980) Human alveolar macrophages: effects of endotoxin in vitro. *Infect. Immun. 30*, 753-758.

Demling RH, Proctor R, Grossman J, Duy N, Starling J (1981a) Lung injury and lung lysosomal enzyme release during endotoxemia. *J. Surg. Res. 30*, 135-141.

Demling RH, Smith M, Gunther R, Gee M, Flynn J (1981b) The effect of prostacyclin infusion on endotoxin-induced lung injury. *Surgery 89*, 257-263.

Demling RH, Smith M, Gunther R, Wandzilak T (1981c) Endotoxin-induced lung injury in unanesthetized sheep: effect of methylprednisolone. *Circ. Shock 8*, 351-360.

Deneke SM, Fanburg BL (1980) Normobaric oxygen toxicity of the lung. *N. Eng. J. Med. 303*, 76-86.

Drake R, Adair T, Traber D, Gabel J (1981) Contamination of caudal mediastinal lymph node efferent. *Am. J. Physiol. 241 (Heart Circ. Physiol. 10)*, H354-H357.

Easton TW (1952) The role of macrophage movements in the transport and elimination of intravenous thorium dioxyde in mice. *Am. J. Anat. 90*, 1-33.

Elin RJ, Wolff SM (1976) Biology of endotoxin. *Annu. Rev. Med. 27*, 127-141.

Esbenshade AM, Newman JH, Lams PM, Jolles H, Brigham KL (1982) Respiratory failure after endotoxin infusion in sheep: lung mechanics and lung fluid balance. *J. Appl. Physiol. 53*, 967-976.

Evensen SA, Pickering RJ, Batbout J, Shepro D (1975) Endothelial injury induced by bacterial endotoxin: effect of complement depletion. *Eur. J. Clin. Invest. 75*, 463-469.

Faden AI, Holaday JW (1980) Experimental endotoxin shock: the pathophysiologic function of endorphins and treatment with opiate antagonists. *J. Infect. Dis. 142*, 229-238.

Fearon DT, Ruddy S, Schur PD, McCabe WR (1975) Activation of the properdin pathway of complement in patients with gram-negative bacteremia. *N. Engl. J. Med. 292*, 937-940.

Fletcher JR, Ramwell PW (1978a) *E. coli* endotoxin shock in the dog; treatment with lidocaine or indomethacin. *Br. J. Pharmacol. 64*, 185-191.

Fletcher JR, Ramwell PW (1978b) Lidocaine or indomethacin improves survival in baboon endotoxin shock. *J. Surg. Res. 24*, 154-160.

Fletcher JR, Ramwell PW, Harris RH (1981) Thromboxane, prostacyclin, and hemodynamic events in primate endotoxin shock. *Adv. Shock Res. 5*, 143-148.

Frank L, Yam J, Roberts RJ (1978) The role of endotoxin in protection of adult rats from oxygen-induced lung injury. *J. Clin. Invest. 61*, 264-275.

Frank L, Roberts RJ (1979) Oxygen toxicity: protection of the lung by bacterial lipopolysaccharide (endotoxin). *Toxicol. Appl. Pharmacol. 50*, 371-380.

Frank L, Summerville J, Massaro D (1980) Protection from oxygen toxicity with endotoxin: role of endogenic antioxidant enzymes of the lung. *J. Clin. Invest. 65*, 1104-1110.

Frank L (1981a) Endotoxin reverses the decreased tolerance of rats to greater than 95% O_2 after preexposure to lower O_2. *J. Appl. Physiol. 51*, 577-583.

Frank L (1981b) Prolonged survival after paraquat: role of the lung antioxidant enzyme system. *Biochem. Pharmacol. 30*, 2319-2323.

Freudenberg MS, Freudenberg N, Galanos C (1982) Time course of cellular distribution on endotoxin in liver, lungs and kidneys of rats. *Br. J. Exp. Pathol. 63*, 56-65.

Frölich JC, Ogletree M, Peskar BA, Brigham KL (1980) Pulmonary hypertension correlated to pulmonary thromboxane synthesis. *Adv. Prostaglandin Thromboxane Res. 7*, 745-750.

Gabel JC, Drake RE, Arens JF, Taylor AE (1978) Unchanged pulmonary capillary filtration coefficients after *Escherichia coli* endotoxin infusion. *J. Surg. Res. 25*, 97-104.

Gemer A, Hirsch EF, Hayes JA, Cuevas W, Ishikawa S (1974) Controlled endotoxemia and the lung. *Chest 65*, 47S-50S.

Gerschman R (1964) Biological effects of oxygen. In: Dickens F, Neil E (Eds), *Oxygen in the Animal Organism*, pp 474-494. Macmillan, New York.

Goodman ML, Way BA, Irwin JW (1979) The inflammatory response to endotoxin. *J. Pathol. 128*, 7-14.

Groeger JS, Carlon GC, Howland WS (1983) Naloxone in septic shock. *Crit. Care Med. 11*, 650-654.

Gunther R, Demling R (1981) Effect of diaphragmatic lymph contamination on caudal mediastinal node. *Physiologist 24*, 53.

Hales CA, Sonne L, Peterson M, Kong D, Miller M, Watkins WD (1981) Role of thromboxane and prostacyclin in pulmonary vasomotor changes after endotoxin in dogs. *J. Clin. Invest. 68*, 497-505.

Harlan J, Winn R, Hildebrandt J, Harker L (1983) Selective inhibition of thromboxane synthesis during experimental endotoxemia in the goat: effects on pulmonary haemodynamics and lung lymph flow. *Br. J. Clin. Pharmacol. 15*, 123S-126S.

Harrison LH, Beller JJ, Hinshaw LB, Coalson JJ, Greenfield LJ (1969) Effects of endotoxin on pulmonary capillary permeability, ultrastructure, and surfactant. *Surg. Gynecol. Obstet. 129*, 723-733.

Heflin AC Jr, Brigham KL (1981) Prevention by granulocyte depletion of increased vascular permeability of sheep lung following endotoxemia. *J. Immunol. 127*, 2245-2251.

Hildebrand GJ, Seys Y (1965) Depolarizing actions of endotoxin. *Nature (London) 206*, 406-407.

Hinshaw LB, Benjamin B, Holmes DD, Beller B, Archer LT, Coalson JJ, Whitsett T (1977) Responses of the baboon to live Escherichia coli organism and endotoxin. *Surg. Gynecol. Obstet. 145*, 1-11.

Hinson JM Jr, Hitchison AA, Ogletree ML, Brigham KL, Snapper JR (1983) Effect of granulocyte depletion on altered lung mechanics after endotoxemia in sheep. *J. Appl. Physiol. 55*, 92-99.

Hollinger MA (1983) Effect of endotoxin on mouse serum angiotensin-converting enzyme. *Am. Rev. Respir. Dis. 127*, 756-757.

Hudson AR, Kilburn KH, Halprin GM, McKenzie WN (1977) Granulocyte recruitment to airways exposed to endotoxin aerosols. *Am. Rev. Respir. Dis. 115*, 89-95.

Hüttemeier PC, Watkins WD, Peterson MB, Zapol WM (1982) Acute pulmonary hypertension and lung thromboxane release after endotoxin infusion in normal and leukopenic sheep. *Circ. Res. 50*, 688-694.

Klein DM, Sayeed MM (1980) Lung cyclic 3',5'-adenosine monophosphate and adenylate cyclase in endotoxin shock. *Exp. Lung Res. 1*, 191-199.

Kovács IB, Görög P (1968) Bronchoconstrictor effect of endotoxin in guinea-pig; the protective effect on antihistamine and anti-inflammatory compounds. *Eur. J. Pharmacol. 4*, 91-95.

Kuida H, Hinshaw LB, Gilbert RP, Visscher MB (1958) Effect of gram-negative endotoxin on pulmonary circulation. *Am. J. Physiol. 192*, 335-344.

Langham CS, Parente L, Persico P, Wood J (1983) Suppression of arachidonate oxidation by glucocorticoid-induced antiphospholipase peptides. *Adv. Prostaglandin Thromboxane Leukotriene Res. 11*, 65-71.

McCord JM, Fridovich I (1978) The biology and pathology of oxygen radicals. *Ann. Intern. Med. 89*, 122-127.

McKay DG, Margaretten W, Csavossy I (1967) An electron microscope study of endotoxin shock in rhesus monkeys. *Surg. Gynecol. Obstet. 125*, 825-832.

Meyrick B, Brigham KL (1983) Acute effects of *Escherichia coli* endotoxin on the pulmonary microcirculation of anesthetized sheep. *Lab. Invest. 48*, 458-470.

Mlczoch J, Weir E, Grover RF, Reeves JT (1978) Pulmonary vascular effects of endotoxin in leukopenic dogs. *Am. Rev. Respir. Dis. 118*, 1097-1099.

Myrvold HE, Svalander C (1977) Pulmonary microembolism in early experimental septic shock: a morphological study in dogs. *J. Surg. Res. 23*, 65-74.

Ochoa R, Kern SR (1980) The effects of Clostridium perfringens type A enterotoxin in shetland ponies – clinical, morphologic and clinicopathologic changes. *Vet. Pathol. 17*, 738-747.

Ogawa R, Takano Y, Fujita T (1977) Disseminated intravascular coagulation in the pathogenesis of adult respiratory distress syndrome: 2. experimental study. *Jpn. J. Surg. 7*, 223-229.

Ogletree ML, Brigham KL (1979) Indomethacin augments endotoxin induced increased lung vascular permeability in sheep. (Abstract). *Am. Rev. Respir. Dis. 119*, 383.

Ogletree ML, Oates JA, Brigham KL, Hubbard WC (1982) Evidence for pulmonary release of 5-hydroxyeicosatetraenoic acid (5-HETE) during endotoxemia in unanesthetized sheep. *Prostaglandins 23*, 459-468.

Orr JA, Ungerer T, Bisgard GE, Will JA (1977) Hemodynamic effects of acute hypoxia and minute amounts of endotoxin in awake swine. *Am. J. Vet. Res. 38*, 1753-1756.

Ouellette JJ, Chosy JJ, Reed CE (1967) Catecholamine excretion and bronchial response to methacholine after Salmonella enteritidis endotoxin in a normal and an asthmatic subject. *J. Allergy 39*, 234-237.

Parratt JR, Sturgess RM (1974) *E. coli* endotoxin shock in the cat; treatment with indomethacin. *Br. J. Pharmacol. 53*, 485-488.

Parratt JR, Sturgess RM (1975) The protective effect of sodium meclofenamate in experimental endotoxin shock. *Br. J. Pharmacol. 53*, 466P.

Parratt JR, Sturgess RM (1976) The effect of a new anti-inflammatory drug, flurbiprofen, on the respiratory, haemodynamic and metabolic responses to *E. coli* endotoxin in the cat. *Br. J. Pharmacol. 58*, 547-551.

Parratt JR, Sturgess RM (1977) The possible roles of histamine, 5-hydroxytryptamine and prostaglandin F2a as mediators of the acute pulmonary effects of endotoxin. *Br. J. Pharmacol. 60*, 209-219.

Pennington DG, Hyman AL, Jaques WE (1973) Pulmonary vascular response to endotoxin in intact dogs. *Surgery 73*, 246-255.

Peterson MB, Hüttemeier PC, Zapol WM, Martin EG, Watkins WD (1982) Thromboxane mediates acute pulmonary hypertension in sheep extracorporeal perfusion. *Am. J. Physiol. 243*, H471-H479.

Rampart M, Bult H, Herman AG (1982) Contribution of complement activation to the rise in blood levels of 6-oxo-prostaglandin F_1 during endotoxin-induced hypotension in rabbits. *Eur. J. Pharmacol. 78*, 91-99.

Reeves JT, Grover RF (1974) Blockade of acute hypoxic pulmonary hypertension by endotoxin. *J. Appl. Physiol. 36*, 328-332.

Reeves JT, Daoud FS, Estridge M (1972) Pulmonary hypertension caused by minute amounts of endotoxin in calves. *J. Appl. Physiol. 33*, 739-743.

Rhoades KR, Heddleston KL, Rebers PA (1967) Experimental hemorrhagic septicemia: gross and microscopic lesions resulting from acute infections and from endotoxin administration. *Can. J. Comp. Med. Vet. Sci. 31*, 226-233.

Rylander R, Mattsby I, Snella MC (1979) Airway cellular and immune response after exposure to inhaled endotoxins. *Chest 75*, 278-279.

Rylander R, Mattsby I, Snella MC (1980) Airway immune response after exposure to inhaled endotoxin. *Bull. Eur. Physiopathol. Respir, 16*, 501-509

Sayeed MM (1982) Pulmonary cellular dysfunction in endotoxin shock: metabolic and transport derangements. *Circ. Shock 9*, 335-355.

Schade U, Rietschel ET (1982) The role of prostaglandins in endotoxic activities. *Klin. Wochenschr. 60*, 743-745.

Simmons RL, Ducker TB, Martin AM Jr (1969) Comparative pathology after intrathecal endotoxin in the rabbit, dog and monkey. *Specialia 15*, 622-623.

Smith ME, Gunther R, Gee M, Flynn J, Demling RH (1981a) Leukocytes, platelets and thromboxane in endotoxin-induced lung injury. *Surgery 90*, 102-107.

Smith ME, Gunther R, Zaiss C, Flynn J, Demling RH (1981b) Pulmonary microvascular injury from lipoxygenase infusion; comparison with endotoxemia. *Circ. Shock 8*, 647-656.

Smith ME, Gunther R, Zaiss C, Demling RH (1982) Prostaglandin infusion and endotoxin-induced injury. *Arch. Surg. 117*, 175-180.

Snapper JR, Bernard GR, Hinson JM Jr, Hitchison AA, Loyd JE, Ogletree ML, Brigham KL (1983) Endotoxemia-induced leukopenia in sheep. *Am. Rev. Respir. Dis. 127*, 306-309.

Sorrells K, Erdos EG, Massion WH (1972) Effect of prostaglandin E_1 on the pulmonary vascular response to endotoxin. *Proc. Soc. Exp. Biol. Med. 140*, 310-313.

Spannhake EW, Colombo JL, Craigo PA, McNamara DB, Hyman AL, Kadowitz PJ (1983) Evidence for modification of pulmonary cyclooxygenase activity by endotoxin in the dog. *J. Appl. Physiol. 54*, 191-198.

Starzecki B, Reddin JL, Gran A, Spink WW (1967) Distribution of endotoxin-^{51}Cr in normal and endotoxin-resistant dogs. *Am. J. Physiol. 213*, 1065-1071.

Staub N, Bland R, Brigham K, Demling R (1975) Preparation of chronic lung lymph fistulas in sheep. *J. Surg. Res. 19*, 315-320.

Stimler NP, Brocklehurst WE, Bloor CM, Hugli TE (1980) Complement anaphyla-toxin C5a stimulates release of SRS-A-like activity from guinea-pig lung frag-ments. *J. Pharm. Pharmacol. 32*, 804.

Thet LA, Wrobel DJ, Crapo JD, Shelburne JD (1983) Morphologic aspects of the protection by endotoxin against acute and chronic oxygen-induced lung injury in adult rats. *Lab. Invest. 48*, 448-457.

Traber DL, Adams T Jr, Henriksen N, Traber LD, Thomson PD (1983a) Reproduci-bility of cardiopulmonary effects of different endotoxin in the same sheep. *J. Appl. Physiol. 54*, 1167-1171.

Traber DL, Thromson PD, Blalock JE, Smith EM, Adams T Sr, Sziebert LA, Traber LD (1983b) Action of an opiate receptor blocker on bovine cardiopulmonary responses to endotoxin. *Am. J. Physiol. 245 (Heart Circ. Physiol. 14)*, H189-H193.

Vane JR (1971) Inhibition of prostaglandin synthesis as mechanism of action for aspirin-like drugs. *Nature New Biol. 231*, 232-235.

Walker RI, Casey LC, Ramwell PW, Fletcher JR (1982) Association of prostacyclin production with resistance of C3H/HeJ mice to endotoxin shock. *Adv. Shock Res. 7*, 125-132.

Watkins WD, Hüttemeier PC, Kong D, Peterson MB (1982) Thromboxane and pul-monary hypertension following E. coli endotoxin infusion in sheep: effect of an imidazole derivative. *Prostaglandins 23*, 273-285.

Warren HS, Ferguson WW, Wangensteen SL (1973) Lysosomal enzyme release in canine endotoxin shock treated with methylprednisolone. *Am. Surg. 39*, 608-618.

Weir EK, Mlczoch J, Reeves JT, Grover RF (1976) Endotoxin and prevention of hypoxic pulmonary vasoconstriction. *J. Lab. Clin. Med. 88*, 975-983.

Weir EK (1978) Does normoxic pulmonary vasodilatation rather than hypoxic vaso-constriction account for the pulmonary pressor response to hypoxia? *Lancet 4*, 476-478.

Handbook of Endotoxin, Vol. 2: Pathophysiology of Endotoxin
L.B. Hinshaw, editor
© Elsevier Science Publishers B.V., 1985
ISBN 0 444 90385 2
$0.85 per article per page (transactional system)
$0.20 per article per page (licensing system)

CHAPTER 5

Endotoxin-induced pulmonary leukostasis

D. MICHAEL SHASBY AND GARY W. HUNNINGHAKE

1. INTRODUCTION

Acute edematous respiratory failure, or the adult respiratory distress syndrome, is one of the most lethal consequences of gram-negative sepsis in man (Braude et al 1953; Fowler et al 1983). A similar type of respiratory failure occurs in animal models following intravenous administration of endotoxin derived from gram-negative organisms (Esbenshade et al 1982). Although the exact pathogenesis of the respiratory failure remains undetermined, an increasingly large body of evidence suggests that activated blood granulocytes localized in the pulmonary microvasculature play an important role in the pathogenetic sequence. In this chapter we will review the evidence for this important role of granulocytes. We will also review what processes may be responsible for localization of granulocytes in the lung microvasculature and discuss the mechanisms that may contribute to the acute respiratory failure. Since platelets are found as well in the pulmonary microcirculation following endotoxemia, we will also review the potential processes responsible for platelet sequestration in the lungs and its potential consequences. We will close with some suggestions for areas of future investigation.

2. ROLE OF POLYMORPHONUCLEAR LEUKOCYTES

The role of polymorphonuclear leukocytes (PMN) in mediating the microvascular consequences of endotoxemia has been a subject of investigation for many years. Many of the earlier studies used the localized Shwartzman reaction, in which an area of rabbit skin is preinjected with endotoxin; 24 hours later the rabbit is given an intravenous injection of the same endotoxin. Following the intravenous injection, the preinjected site demonstrates hemorrhagic necrosis, while this does not occur in the rest of the skin. Stetson (1951) observed that prior to the intravenous injection, the preinjected site showed extravascular cuffing of the microvasculature with PMN, and that following the intravenous injection the prepared site showed hemorrhagic necrosis with PMN and platelet plugs in the microvasculature. The microvascular cuffing seen before the injection and the PMN plugs and hemorrhagic necrosis seen afterwards were not present at sites not preinjected (Stetson 1951).

To determine whether the PMN microvascular cuffing before the injection and the postinjection PMN plugs were important for the development of hemorrhagic necrosis, Stetson and Good (1951) depleted rabbits of PMN by pretreating them with nitrogen mustard. If the rabbits were then given a dermal injection and a subsequent intravenous injection, hemorrhagic necrosis did not develop at the preinjection site. It was important that they were granulocytopenic at the time of both the preparative dermal injection and the intravenous injection. If the abdominal aorta of the rabbits was clamped during the treatment with nitrogen mustard to preserve the femoral marrow, – and hence a supply of PMN –, a typical localized Shwartzman reaction with hemorrhagic necrosis occurred. These results confirmed Stetson's impressions that PMN were important in the development of the microvascular injury.

Subsequent observations were made with the use of the generalized Shwartzman phenomenon, in which an initial intravenous injection of endotoxin is followed with a second intravenous injection 24 hours later. The most consistent pathologic findings with the generalized Shwartzman reaction are renal glomerular thrombosis and cortical necrosis, although pulmonary arterial thrombosis also frequently occurs. Thomas et al (1952), using a similar protocol of PMN depletion with nitrogen mustard, observed that granulocytopenia prevented the development of the characteristic lesions of the generalized Shwartzman reaction. If the femoral marrow had been preserved at the time of nitrogen mustard treatment by clamping the aorta, pretreatment with nitrogen mustard did not prevent the generalized reaction. Hence, an important role for leukocytes in the generalized as well as the localized reaction was established.

A marked increase in circulating PMN following the first intravenous injection of endotoxin was a consistent observation of investigators studying the generalized Shwartzman reaction. Horn and Collins (1968) hypothesized that the effect of the first intravenous injection was to mobilize the marrow PMN and that the second injection activated the mobilized leukocytes. To test this they used rabbits pretreated with nitrogen mustard, in which the aorta had been clamped during nitrogen mustard injection to preserve the femoral marrow. They then gave the first intravenous injection at a time when the marrow should have been depleted of PMN. Since the femoral marrow was preserved, this resulted in a systemic leukocytosis. They then occluded the aorta for the second injection of endotoxin, thereby limiting the participation of the femoral marrow to the leukocytosis following the first injection, and observed the typical renal lesions of the generalized Shwartzman reaction. Hence, it seemed the marrow did not have to participate during the second injection phase. To confirm these results they also gave intravenous infusions of PMN prepared from peritoneal exudates to rabbits pretreated with nitrogen mustard without preservation of femoral marrow. Without the leukocyte infusion, these rabbits did not respond to one or two intravenous injections of endotoxin, but following leukocyte infusion they developed a generalized Shwartzman reaction after a single endotoxin injection. Injection of the PMN alone did not produce similar pathology. The authors felt these experiments supported their hypothesis that the first injection functioned only to mobilize the PMN. The peritoneal exudate PMN, however, are recruited and activated cells, and a similar activation of the PMN by the first injection which makes them more responsive to the second injection is an important alternative interpretation of the experiments.

Another alternative function of the first injection was suggested by Gaynor (1973). She gave a single injection of endotoxin to control and PMN-depleted rabbits and then examined blood for desquamated endothelial cells and sections of aorta for evidence of endothelial injury. She found evidence for endothelial cell desquamation and vacuolation and disruption of aortic endothelial cells whether or not PMN were present. These studies suggest that the initial injection may alter the endothelium as well as affect the PMN.

More recently Williams and coworkers (Wedmore and Williams 1981) have investigated the role of PMN in the local inflammatory response to intradermal injection of activated complement, leukotriene B_4 (LTB_4) and N-formyl-methionyl-leucyl-phenylalanine (fMLP). They used the nitrogen mustard protocol of Stetson and Good and demonstrated that local skin edema occurred after local injection of activated complement, LTB_4 or fMLP when PMN were present, but not in rabbits depleted of PMN. Preservation of femoral marrow preserved the edematous response, and transfusion of whole blood containing PMN to rabbits depleted of PMN with nitrogen mustard also restored the edematous response. Hence, PMN also participate in the acute inflammatory response to stimuli other than endotoxin.

Many histologic studies have documented the presence of PMN in the microvasculature following endotoxemia. For example, McKay and coworkers used the electron microscope to observe sequential changes in the microvasculature following endotoxemia in rats (McKay et al 1966) and rhesus monkeys (McKay et al 1967). In the rats, platelet aggregates were present in lung capillaries by 15 minutes, and by 1 hour PMN and platelet aggregates were present with evidence of PMN degranulation. While PMN were also in the liver and kidneys by 1 hour, there was evidence for PMN degranulation only in the lung. In addition, there was no evidence for endothelial cell injury in any organ. In the monkeys, PMN aggregates in the pulmonary capillaries were more prominent at 15 minutes, and there was less participation of platelets than in rats. By 2 hours, the liver was the only organ with occlusive masses of PMN, platelets, and fibrin, and there was apparent ischemic injury distal to the occluding masses. Again frank evidence of endothelial cell injury was absent. Coalson and coworkers (Coalson et al 1970) made similar observations in Rhesus monkeys given either endotoxin or live *E. coli* intravenously. They observed increased numbers of PMN in the lung microvasculature and apparent degranulation of the PMN was also noted.

Pietra and coworkers (1974) examined the lung microvasculature of dogs given an intravenous injection of endotoxin. Using colloidal carbon to mark the sites of vascular leak, they observed initial leakage of carbon from bronchial microvessels and subsequent (1 hour) leakage from pulmonary capillaries which contained large numbers of PMN.

Using the rabbit ear chamber, Goodman also observed an early margination of leukocytes to the endothelium following an intravenous endotoxin injection. Edema and extravasation of blood cells occurred subsequently, and the visceral organs showed swelling of endothelium and margination of PMN (Goodman et al 1979). This study suggests that PMN and endotoxin affect microvascular beds outside the lung as well.

Bachofen and Weibel (1977) examined the ultrastructure of human lungs obtained shortly after the onset of respiratory distress associated with sepsis. They also noted the association between PMN in the microvasculature and areas of edema, and the apparent lack of endothelial cell injury.

More recently Meyrick and Brigham (1983) described the sequential changes of the lung microvasculature in sheep following endotoxemia. The first change noted at 15 minutes was the accumulation of PMN in the microcirculation, which continued to increase over the next 4 hours. By 60 minutes, the endothelium of the small vessels was more electron dense than normal and showed increased numbers of pinocytotic vesicles. There was also damage to the type I pneumocytes. By 2–4 hours, the appearance of the endothelium of the small arteries and veins reverted to normal, but the capillary endothelium showed disruption of the continuous endothelial cell layer with gaps between adjacent cells. Most of the pathologic changes occurred in areas where PMN had accumulated.

Walker and Parker (1980) also offered histologic support for an important role of PMN in the response to endotoxin. They injected lethal doses of endotoxin into mice genetically sensitive or resistant to endotoxin. While both groups of mice demonstrated sequestration of PMN in the lung microvasculature at 1 hour, the PMN began to clear from the lungs of resistant mice by 3 hours and the lungs of these mice were normal by 6 hours. In contrast, PMN continued to accumulate in increasing numbers in the lungs of sensitive mice and persisted even at 12 hours.

Hence, early studies with the Shwartzman reaction and more recent reports with other stimuli which cause acute inflammation have suggested an important role of blood PMN in the microvascular response to endotoxin. These studies demonstrate PMN in the affected microvasculature prior to development of pathologic changes. While these studies support an important role for PMN in the microvascular consequences of endotoxemia, the report of Gaynor (1973) also suggests that endotoxin may have a direct adverse effect on the endothelial cell membrane.

To focus more on the lung microvascular consequences of endotoxemia Hinshaw and coworkers (Hinshaw et al 1957) perfused isolated lobes of dog lung with various perfusates and endotoxin. While perfusion with whole heparinized blood caused vasoconstriction of lung arteries and veins with subsequent lung edema, similar vasoconstriction and edema were not observed when the formed elements of the blood were not in the perfusate. In a similar study, Kux et al (1972) were not able to confirm this observation. However, they used a filter to deplete the perfusate of PMN and platelets and they acknowledged that the filtration procedure itself may have activated inflammation, a contention supported by the observation of increased vascular resistance in lungs perfused with filtered versus control perfusate. Moreover, even the filtered perfusate had leukocyte counts of approximately $500/mm^3$, and no estimate of the numbers of PMN in the lungs was given. Bessa et al (1974) also used isolated perfused canine lungs and concluded that endotoxin increased vascular permeability in the absence of blood cells, but they did not observe an increase in pulmonary vascular resistance with endotoxin, a consistent finding of the other studies. Hence, these studies do not agree on whether PMN are essential participants in the consequences of endotoxemia on the lung microvasculature.

Also focusing on the lung, Brigham and coworkers (Brigham et al 1974) used the sheep lung lymph fistula model to measure the effects of gram-negative bacteremia on transvascular fluid and protein flux in the lung. They observed an increased rate of lymph flow and enhanced transvascular flux of protein following a slow intravenous infusion of bacteria, even when the data were standardized for changes in vascular pressures. This was a hallmark study demonstrating the usefulness of the model to study acute alterations in pulmonary vascular permeability. A subsequent study (Brigham et al 1979) also demonstrated that *E. coli* endotoxin infusion reproduced the effects of the *Pseudomonas* infusion. Because the endotoxin infusion is shortly followed by severe leukopenia with sequestration of PMN in the lung microvasculature, as demonstrated by the histologic studies cited above, Heflin and Brigham (1981) depleted sheep of granulocytes with hydroxyurea prior to the infusion of endotoxin. Granulocyte depletion of the sheep prior to injecting endotoxin reduced the increases in lymph flow and transvascular protein flux observed in control sheep. These studies were therefore consistent with the early observations of the Shwartzman reaction and suggested an important role for PMN in the pulmonary microvascular response to endotoxemia. Hüttemeier and coworkers more recently confirmed these effects of granulocyte depletion on endotoxin-induced changes in vascular permeability in sheep (Hüttemeier et al 1982). Whereas Heflin and Brigham noted no differences in the pulmonary artery pressor responses to endotoxin after granulocyte depletion, Hüttemeier et al noted a blunted pressor response that correlated with decreased release of thromboxane across the pulmonary vasculature. These authors concluded that the PMN contributes significantly to the release of thromboxane and the pressor response as well as to the increased vascular permeability. Hinson and coworkers (1983) have recently confirmed Heflin's earlier results and failed to demonstrate that depletion of PMN reduced thromboxane generation.

It is important to note that PMN depletion did not completely block the increase in microvascular permeability of the lung following endotoxin. Whether this represents the effect of residual PMN or an effect of endotoxin independent of PMN remains uncertain.

Loyd and coworkers more recently reported that intravascular injection of phorbol myristate acetate (PMA) closely mimicked the endotoxin response in awake sheep with an initial increase in vascular resistance and a subsequent increase in lymph flow and protein clearance consistent with increased vascular permeability (Loyd et al 1983). Also similar to the endotoxin response, the number of lung granulocytes increased dramatically after PMA injection. Shasby and coworkers (1982) had previously shown that granulocyte depletion in rabbits prevented the development of the severe hemorrhagic lung edema observed in control rabbits given PMA. Hence, PMN contribute to the microvascular consequences of PMA and endotoxin infusions, the Shwartzman reaction, and the local acute inflammatory reactions studied by Williams and coworkers. While these studies have identified PMN as important participants, it is not clear whether they act directly and/or alone to alter vascular permeability or whether other blood cells and plasma constituents are also necessary coparticipants.

Shasby and coworkers (1982) extended these observations by adding PMN from purified blood to isolated rabbit lungs perfused with a physiological salt solution. Plasma constituents and blood cellular elements were washed from the lungs so that only very small concentrations and numbers remained. When PMN and PMA were added to the perfusates, lungs became edematous while lungs exposed to PMN or PMA alone did not. These experiments strongly suggest that PMN can act directly and alone to adversely effect the microvasculature of the lung, but they do not rule out a complementary role of other blood elements in the in vivo response.

More recently, Shasby and coworkers (1983) demonstrated that PMN can interact directly with endothelium to alter its permeability. In these experiments they used monolayers of cultured endothelial cells of the pulmonary artery as a model of the endothelium. Again purified PMN and PMA together increased the permeability of endothelial cells while neither PMN or PMA alone had this effect. These findings were consistent with previous in vitro reports from Sacks (Sacks et al 1978) and Weiss (Weiss et al 1981) that activated PMN could lyse endothelial cells, and with a report from Harlan (Harlan et al 1981) that activated PMN caused detachment of endothelial cells from tissue culture plates. The time course of the increased permeability was, however, much shorter than that for lysis and detachment, and Shasby suggested that changes in the shapes of endothelial cells, which result in gaps beween adjacent endothelial cells, might contribute to the increases in permeability.

Several other groups of investigators have identified PMN as important participants in the lung microvascular consequences of acute inflammation initiated by stimuli other than endotoxin and PMA. Craddock and coworkers (1977) had observed leukopenia and hypoxemia during hemodialysis, and they linked these two observations by demonstrating activation of complement by the dialysis membrane which caused PMN to localize in the lung microvasculature. They then demonstrated that complement activation caused hypoxemia and leukopenia in rabbits and that PMN depletion blocked the hypoxemia. Sacks and coworkers from the same laboratory then demonstrated that complement-activated PMN caused a small but detectable increase in ^{51}Cr release from cultured endothelial cells, which implies injury of endothelial cells (Sacks et al 1978). Hosea and coworkers also confirmed an important role for PMN in the complement-initiated processes by depleting guinea pigs of PMN and demonstrating that this prevented the increase in microvascular permeability in the lung caused by complement infusion in control guinea pigs (Hosea et al 1980). Till and coworkers recently confirmed these observations, demonstrating that granulocyte depletion prevented an increase in radiolabeled albumin in the lungs of rats infused with cobra venom factor (Till et al 1982). Since endotoxemia results in an activation of complement, these studies may have direct relevance to endotoxin-initiated pulmonary leukostasis.

Investigations using other models of acute lung edema have also identified PMN as important participants. Flick et al (1981a) demonstrated that PMN depletion protected sheep against increased alveolar capillary permeability following microembolization with glass beads or air (Flick et al 1979), and Johnson and Malik (1980a) made similar observations in dogs. Tahamont and Malik (1983) found that PMN depletion prevented the increases in lymph flow and lymph protein clearance in sheep after thrombin embolism. These investigators also selectively depleted the sheep of platelets and found that the platelet depletion did not prevent the increased lymph protein clearance. Interestingly, the platelet-depleted sheep had PMN counts of only 500–700 per mm^2, suggesting that only small numbers of circulating PMN are sufficient to alter the microvascular permeability.

The studies demonstrating that PMN are essential participants in the consequences for the lung microvasculature of endotoxemia and other stimuli of acute inflammation have focused on changes in microvascular permeability. However, as noted above, while PMN depletion did not reduce the increased pulmonary vascular resistance in the report from Heflin and Brigham (1981), it did in the investigation of Hüttemeier et al (1982) where there was an associated reduction in thromboxane levels. Endotoxemia also causes an acute fall in arterial oxygen tension in sheep. Hüttemeier's report suggested that this was partly prevented with PMN depletion and totally prevented by cyclooxygenase inhibition. Hinson and coworkers (1983) recently confirmed this effect of PMN on the decrease in arterial oxygen tension following endotoxin. These authors also demonstrated that PMN depletion partly prevented the acute rise in lung resistance and the acute fall in dynamic compliance that occurs after endotoxin, and that both changes were totally prevented by cyclooxygenase inhibition.

In summary, endotoxemia results in the accumulation of PMN in the microvasculature. In the skin, PMN depletion prevents the edema of the acute inflammatory response to endotoxin and several other mediators of acute inflammation. In the lung, the best characterized consequence of endotoxemia is an increase in microvascular permeability. Depletion of PMN reduces this increase in microvascular permeability, and this important role of PMN is supported by their important contribution in several other models of increased permeability of the lung microvasculature. PMN are able to increase alveolar capillary permeability of isolated perfused lungs in the absence of normal numbers of other blood cells or normal concentrations of plasma constituents, and PMN can interact directly with cultured endothelial monolayers to increase their permeability to albumin. PMN also seem to contribute to endotoxin-initiated changes in lung mechanics, independent of those caused by edema. Hence, the localization of PMN to the lung microvasculature is an important step in the pathogenesis of the consequences of endotoxemia on the lung.

3. ROLE OF PLATELETS

Some of the histologic studies demonstrating an accumulation of PMN in the microvasculature of laboratory animals after endotoxemia showed platelets deposited with the PMN. This was species-dependent, with platelets localizing to the lungs of rats (McKay et al 1966) but not monkeys (Coalson et al 1970; McKay et al 1967). Similarly, Meyrick and Brigham (1983) commented on the lack of the accumulation of platelets in sheep lung microvasculature after endotoxemia. Human lungs which became edematous in the clinical setting of sepsis demonstrated platelets aggregated in areas of injury. It is not known wether this represents a response to injury or a role for platelets in causing the injury (Bachofen and Weibel 1977; Tomashefski et al 1983). Similarly, Schneider and coworkers (1980) observed increased platelet consumption and sequestration of platelets in the lungs of patients suffering from acute respiratory failure, but they were not able to determine whether this was a response to injury or a step in the disease process.

There have been several studies in which platelets were depleted from laboratory animals to determine whether they contributed to the pulmonary consequences of endotoxemia or other inflammatory stimuli. Snapper and coworkers (1983) recently reported that platelet depletion of sheep prior to infusing endotoxin did not alter the increases in airway and vascular resistance and vascular permeability caused by endotoxin. Binder and coworkers (Binder et al 1980) also depleted sheep of platelets and observed the subsequent response to glass bead microemboli. Their results suggested that platelets contributed to vasoconstriction observed after microemboli, perhaps to decreases in oxygenation of arterial blood, but not to changes in microvascular permeability. Tahamont and Malik (1983) infused thrombin into sheep depleted of platelets and observed no effect of thrombocytopenia on the increase in pulmonary vascular resistance or in microvascular permeability seen with control sheep. Minnear et al (1982) had previously caused platelet aggregation in sheep with infusion of adenosine diphosphate (ADP) and observed hypoxemia and increased lymph flow with no significant change in pulmonary vascular resistance. The increased lymph flow seemed to be due to enhanced surface area for filtration, and the authors did not observe a rise in vascular permeability. Bredenberg et al (1980) observed that platelet depletion blunted the increase in pulmonary vascular resistance in dogs which received endotoxin. Busch and coworkers (1974) had previously infused thrombin into dogs and observed increases in lung weight and edema. This response was not blocked with depletion of platelets or PMN. However, this study evaluated only lung weight and histology which may not be sensitive or accurate indices of vascular permeability.

In addition to these depletion experiments, selective aggregation of platelets has also been used to determine the effects of platelet aggregation in the lung. The effects of ADP infusion described by Minnear and Malik were noted above. Bo and Hognestad (1972) infused collagen into cats and noted platelet aggregation with associated pulmonary vasoconstriction and a decrease in lung compliance. Radegran and McAslan (1972) noted similar effects of platelet aggregation in dogs induced with protamine sulfate. They observed marked increases in pulmonary vascular resistance, airway resistance and arterial oxygen tension, and all of these effects were blunted by pretreatment with acetylsalicylic acid, which did not block the platelet aggregation.

Platelets and platelet stimuli have also been added to blood-free perfusates of isolated lungs to observe the effects of stimulated platelets on the lung. Vaage and coworkers (Vaage et al 1976) added platelets and collagen to isolated perfused rabbit lungs and observed an increased rate of fluid filtration by the lungs. This increase was similar to that noted by Minnear (Minnear et al 1982) in sheep, which appeared to be due to changes in vascular surface area in the sheep. While the protocol for the experiments with the isolated lungs makes an increase in surface area for fluid filtration unlikely, there is no evidence for an increase in vascular permeability to serum proteins. Heffner (Heffner et al 1983) added acetyl glyceryl ether phosphoryl choline (AGECP, platelet activating factor) and platelets to isolated rabbit lungs and observed a marked increase in pulmonary vascular resistance that appeared to be due to thromboxane synthesis by the platelets, but he did not observe an increase in alveolar capillary permeability.

While there is no doubt platelets are a potent source of thromboxane, McDonald and coworkers (1983) have suggested that cells in the lung itself may also be important sources during reactions involving platelet and PMN aggregation in the lung. They found that platelet depletion did not block the increase in pulmonary vascular resistance or the release of thromboxane observed after infusion of activated complement. Smith and coworkers (1981) also observed a dissociation between platelet aggregation and the release of thromboxane when endotoxin was infused into sheep. Since several of the studies cited above also noted a dissociation between thrombocytopenia and vascular and airway effects of endotoxin and other stimuli, platelets are probably not the only sources of thromboxane and other humoral mediators in these models.

In addition to these direct effects of platelets, there are potentially important interactions of platelets with PMN in the endotoxin response. Boogaerts et al (1982) demonstrated that while complement and platelets did not directly affect cultured endothelial cells, platelets did enhance the cytotoxic effects of PMN and activated complement. This enhancement seemed to be due to increased PMN adherence to the endothelial cells and the enhanced adherence could be reproduced with serotonin, suggesting that platelet release of serotonin was responsible for the effect. In addition to this interaction, the vasoconstriction attributable to platelets following endotoxin infusion in dogs causes a marked slowing of blood flow in the pulmonary microvasculature (Bredenberg et al 1980). Since PMN adherence to the pulmonary endothelium is enhanced with decreases in pulmonary blood flow, this interaction could also potentiate effects of PMN on the lung microvasculature (Martin et al 1982).

In summary, endotoxin causes platelet aggregation in the lung microvasculature. While there is some variability with different species and different inflammatory stimuli, most experiments have suggested that the platelet aggregation has some effects on pulmonary vascular resistance, lung mechanics and/or ventilation perfusion imbalance. There is very little evidence to support the contention that platelets contribute directly to increased vascular permeability. Interactions of platelets with PMN and endothelium may potentiate effects of PMN during endotoxemia.

4. MECHANISM OF PULMONARY LOCALIZATION OF POLYMORPHONUCLEAR LEUKOCYTES

Many of the histologic studies previously described showed that PMN accumulated in the lung microvasculature of animals after endotoxemia. PMN also accumulated in the liver and kidneys, and studies with the rabbit ear chamber demonstrated PMN margination in the ear microcirculation with the subsequent development of PMN emboli and increased microvascular permeability. Similar microvascular changes occurred in the rabbit's visceral organs as well (Goodman et al 1979; McKay et al 1966, 1967). Northrup and Humphrey (1976) cannulated both the right and left thoracic ducts in dogs to characterize the changes in microvascular permeability occurring in the lungs (right thoracic duct) and the visceral and systemic circulation (left thoracic duct) following endotoxin infusion. After endotoxin, lymph flow in both thoracic ducts increased acutely by a factor of 6–8 and then stabilized at a level of 2–3 times control. The concentration in lymph relative to that in plasma for both albumin and globulin increased in both ducts with a sustained clearance of both proteins. From these experiments it appears that the lungs are not the only sites of leukoagglutination and microvascular injury following endotoxemia.

While the studies mentioned above suggest that the lung is not the only microvascular bed involved in leukostasis after endotoxemia, the unique function of the lung as a reservoir for PMN has been suggested for many years. Both Bierman et al (1951) and Ambrus et al (1954) put forward the hypothesis that the lung functions as a reservoir of PMN that could respond to produce leukocytosis, and that PMN do not simply enter and exit the lung microvasculature, but rather enter to be sequestered for a time and then to be released after other PMN are sequestered (Martin et al 1982). Staub and coworkers recently quantitated the PMN in the lungs of normal sheep and estimated approximately 60×10^9 PMN per lung, more than 3 times the number of circulating PMN (Staub et al 1982). The reasons for this lung localization of PMN even in control settings is not understood. Staub suggested that it might be related to the extremely pulsatile nature of pulmonary blood flow where diastolic flow is almost zero. This is consistent with the findings of Martin (1982) that PMN sequestration in the lung was inversely related to the rate of blood flow. Whether there is also a peculiarity of pulmonary endothelium that attracts PMN is not known.

PMN adherence is increased in the setting of acute inflammation, and since the lung receives all of the cardiac output, many PMN would be expected to accumulate even if there was no particular characteristic of lung endothelium to attract these cells (Lentnek et al 1976). Craddock and coworkers (1977) identified activated complement, and especially C5, as an important mediator of pulmonary leukostasis. Their early reports focused on PMN aggregation by C5 as a mechanism of localization, but more recently they have emphasized direct adherence to endothelium (Boogaerts et al 1982). Endotoxin activates complement with generation of C5 (Snyderman et al 1969), and several investigators have confirmed the ability of C5 to augment PMN adherence (O'Flaherty et al 1978). Hosea and coworkers (1980) demonstrated that complement was important for the lung localization of intravenously injected pneumococci, which localized more to the liver and spleen in complement-deficient guinea pigs and mice. However, several studies have demonstrated that even in complement-depleted animals or animals genetically deficient in some complement components the fall in PMN or tissue injury following intravenous endotoxin is not prevented (Garner et al 1974; Ulevitch et al 1975, 1978). Similarly, Freudenberg and Galanos (1978) demonstrated a dissociation of complement activation and the toxicity of endotoxin.

While endotoxin does activate complement, several reports demonstrate that endotoxin also increases PMN adherence in the absence of complement. Spagnuolo and Ellner (1980) showed that endotoxin increased PMN adherence in the absence of plasma or other sources of complement and that lipopolysaccharide (LPS) increased release of thromboxane B_2 from PMN. Inhibitors of thromboxane synthesis and antibodies to thromboxane also blocked the increase in adherence caused by LPS (Spagnuolo et al 1981). The authors interpreted these studies as evidence for the hypothesis that thromboxane mediates augmented PMN adherence in the presence of endotoxin. Dahinden et al (1983) also found that endotoxin stimulated PMN adherence in the absence of plasma and complement, but they were not able to confirm an important role of cyclooxygenase products such as thromboxane in the process. Of particular interest was their finding that activation of PMN production of superoxide anions and release of lysosomal enzymes induced by endotoxin were dependent on adherence of PMN to a surface. Hence, while endotoxin activates complement and complement increases PMN adherence to endothelium, endotoxin is able to produce this effect without complement as well. Whether active complement fragments might be important for preferential lung localization of PMN, as suggested by the studies of Hosea et al (1980), remains uncertain.

Other mechanisms for increased PMN adherence have also been described. Oseas and coworkers (1981) have demonstrated that release of the specific granule product lactoferrin from PMN enhances PMN adherence. Hoover and colleagues (1980) confirmed that a culture filtrate from *E. coli* increased PMN adherence to endothelium, and also showed that pretreatment of endothelium with the same filtrate without exposing PMN to the filtrate, increased PMN adherence to endothelium as well. These results suggest that endothelium as well as PMN can respond to endotoxin to augment PMN adherence.

Endotoxemia is associated with increased synthesis of metabolites of arachidonate (Demling et al 1981a). In analyzing the differences between mice resistant to and mice sensitive to endotoxin, Walker and Parker (1980) noted that in sensitive mice PMN were not cleared from the lung microvasculature while in resistant mice they were. In a subsequent report, Walker and coworkers (1982) reported that after endotoxin injection, prostaglandin release from lungs of resistant mice had a much higher ratio of prostacyclin to thromboxane than lungs of sensitive mice. Since prostacyclin inhibits PMN adherence and thromboxane augments it, these differences might account for the inability of sensitive mice to clear PMN from the lung microcirculation. The deposition of fibrin fragments is a frequent finding in the microvasculature of animals after endotoxemia. Malik et al (1982) have suggested that the fibrin network may entrap PMN and thereby enhance the accumulation of PMN in the small vessels. While defibrination and heparin did not prevent the increase in microvascular permeability in sheep lung (Binder et al 1979), these modalities did protect against lung edema in dogs (Johnson and Malik 1980b; Malik and Vander Zee 1978). Since endotoxin causes release of procoagulant activity directly from endothelial cells, these observations with models initiating intravascular coagulation – models in which PMN also play an important role in the microvascular pathology – are probably relevant to the lung consequences of endotoxemia (Colucci et al 1983).

5. MECHANISM OF PLATELET LOCALIZATION TO THE PULMONARY VASCULATURE

The direct effects of endotoxin on platelets and the important participation of complement in this response are discussed in Chapter 6. As noted previously, localization of platelets to the lung microvasculature after endotoxin is species-dependent and perhaps dose-dependent (Meyrick and Brigham 1983). While endotoxin does induce complement-dependent platelet aggregation, this process is also species-dependent as to whether the alternative or classical pathway is involved, and which portion of the endotoxin molecule is most important (Abdelnoor et al 1980; Harris et al 1982; Morrison and Oades 1979; Semeraro et al 1979).

Even in species in which platelet deposition occurs in the lungs after endotoxemia, it occurs in other organs as well, and there is no convincing evidence that it occurs in the lungs in preference to the other organs. There are, however, several mechanisms which might promote lung platelet deposition. The slowing of pulmonary blood flow noted by Bredenburg et al (1980) would be expected to promote platelet deposition as well as to enhance PMN localization by methods described by Martin et al (1982). Similarly, local lung synthesis of thromboxane A_2 would stimulate platelet localization. However, Thorne et al (1981) did not observe any decrement in lung localization of labeled platelets in dogs after pretreatment with acetylsalicylic acid. PMN-mediated injury to pulmonary endothelium would also be expected to enhance platelet deposition. Both endothelial detachment, as described by Harlan et al (1981), and the retraction of endothelial cells from one another, described by Shasby et al (1983), would expose subendothelial matrix and enhance platelet deposition. Intravenous aggregation of platelets with endotoxin would promote platelet emboli which would then localize to the lung as the first microvascular bed. All but the last of these mechanisms require some effect of endotoxemia on the lung microvasculature before lung platelet localization.

In summary, early lung localization of platelets after endotoxemia is species-dependent as are the direct and complement-mediated effects of endotoxin on platelets. Whether platelets localize to the lung in preference to other organs after endotoxin is not certain, and what mechanism might be responsible for this is not known either.

6. MECHANISMS OF MICROVASCULAR INJURY PRODUCED BY POLYMORPHONUCLEAR LEUKOCYTES

Several potential mechanisms have been identified for PMN to produce the microvascular consequences of endotoxin in the lung. Both oxygen radical and proteolytic enzyme release from PMN have been suggested as important factors for the increase in lung microvascular permeability observed after endotoxemia. Lucht and coworkers (1983) demonstrated that pretreatment of sheep with acetylcysteine blunted the increase in lymph protein clearance expected after endotoxin infusion. Since acetylcysteine is believed to work through enhancing the activity of the glutathione cycle, an important cellular defense against oxidant injury, these results are consistent with an oxidant mechanism of microvascular injury. More evidence for PMN oxidant-mediated processes comes from the report of Yamada et al (1981), which showed enhanced ^{51}Cr release from cultured endothelial cells exposed to PMN and endotoxin, and inhibition of this effect by the oxygen radical scavengers superoxide dismutase and catalase. The reports of Shasby and coworkers (1982, 1983) on PMA also support a role for PMN oxygen radicals in PMN-mediated increases of microvascular permeability. Both in isolated perfused rabbit lungs and with monolayers of endothelial cells cultured on micropore filters, normal PMN and PMA increased endothelial permeability, while PMA and chronic granulomatous disease (CGD) PMN did not. Since the only recognized defect in CGD-PMN is the inability to make oxygen radicals, these studies strongly supported a role of PMN oxygen radicals in causing the increased permeability.

Oxygen radicals also seem to be important in other models of acute increases in lung microvascular permeability. Till and coworkers (1982) reduced the increase in albumin in rat lungs after intravascular activation of complement with superoxide dismutase and catalase. This study supported previous studies from the same laboratory which indicated that oxygen metabolites have a role in altering lung microvascular permeability (Johnson and Ward 1981) in the setting of acute inflammation. However, all the studies are based on complement-dependent acute inflammation. Flick et al (1981b) also suggested a role for PMN oxygen radicals in the increase of microvascular permeability after microembolization. Direct evidence for participation of intravascularly generated oxygen radicals in this process comes from Tate et al (1982), who showed that when radicals were chemically generated within the vasculature, there was an increase in alveolar capillary permeability and in pulmonary vascular resistance.

Release of PMN proteolytic enzymes has also been suggested as a mechanism for microvascular injury produced by PMN after endotoxemia. Many of the early histologic studies previously cited described release of

PMN lysosomes in the lung microvasculature after endotoxemia. Demling et al (1981b) noted an increase in sheep lung lymph β-glucuronidase content after endotoxin infusion. Harlan et al (1982) described detachment of endothelial cells by complement- and fMLP-stimulated PMN. The detachment could be inhibited by antiproteases, and the authors demonstrated proteolysis of fibronectin. While they attributed the detachment primarily to proteolytic processes, the antiproteases inhibit the release of oxidants by PMN as well. The potential for proteases to increase the permeability of the microvasculature of the lung was also demonstrated by Spragg and coworkers (1982), who observed lung edema after infusing neutrophil elastase into isolated perfused lungs. However, Johnson and Ward (1981) were not able to blunt a complement-initiated PMN-dependent acute injury to rat lungs with antiproteases. Since Harlan was able to block the protease-mediated detachment of endothelial cells with serum, it is possible that endogenous antiproteases limit the effect of the proteolytic enzymes. Since antiproteases are inactivated by oxidants, these two mechanisms do have the potential to amplify each other. Indeed, Cochrane et al (1983) demonstrated oxidant inactivation of intrabronchial and/or intra-alveolar antiproteases but not intravascular antiproteases in patients with acute respiratory insufficiency.

Other potential mechanisms exist to explain increased permeability of lung microvasculature after endotoxin, especially since PMN depletion does not completely prevent the increase in permeability. Other PMN-dependent possibilities include PMN release of leukotrienes and oxidant stimulation of arachidonate metabolism (Ogletree and Brigham 1982). Busch et al (1974) suggested that fibrin fragments might increase microvascular permeability directly and that PMN proteases could produce some of these fragments from deposited fibrin.

In addition to the changes in microvascular permeability, some of the changes in sheep lung mechanics after endotoxemia are also blunted by granulocyte depletion (Hinson et al 1983). Since the changes in lung mechanics are totally abolished with cyclooxygenase inhibitors, PMN cyclooxygenase products may be involved. Alternatively, PMN may initiate cyclooxygenase activity in other cells.

In summary, the release of oxidants and proteolytic enzymes by PMN is a likely mechanism for the PMN-mediated increases in microvascular permeability following endotoxemia. Inhibition of the oxidant effects of PMN has successfully blunted injury of lung microvasculature initiated by endotoxin and several other stimuli, but similar inhibition by antiproteases has not been demonstrated. PMN might also contribute to changes in microvascular permeability via the release of metabolites of arachidonate and through the generation of proteolytic fragments of fibrin. Changes in lung mechanics after endotoxin seem to be dependent on products of cyclooxygenase, and while PMN contribute to these changes, there is no evidence yet for direct participation by PMN cyclooxygenase.

7.   MECHANISMS OF PLATELET-MEDIATED EFFECTS

As noted in Section 3, the contribution of platelets to the pulmonary consequences of endotoxemia remains less clear than the contribution of PMN. Platelets are certainly a rich source for thromboxane, and thromboxane levels are increased after endotoxemia and may contribute to the effects of endotoxin on vasoconstriction and lung mechanics. The platelet-dependent slowing of the pulmonary circulation in dogs injected with endotoxin observed by Bredenberg et al (1980) could be due, at least in part, to thromboxane release, although the authors offered no data to support this hypothesis. Heffner et al (1983) addressed this question directly by both inhibiting thromboxane synthesis and by adding a thromboxane receptor antagonist; they demonstrated that platelet thromboxane release was responsible for the pulmonary pressor response of isolated rabbit lungs exposed to AGECP (platelet activating factor). While both of these studies support a role for thromboxane released from platelets in the early pulmonary pressor response following endotoxemia, McDonald and coworkers did not find that platelet depletion altered either the pressor response or thromboxane release in sheep (McDonald et al 1983) given zymosan-activated plasma. They concluded that PMN caused endothelium to release thromboxane, although the exact mechanism for this was not stipulated. Similarly, Snapper and coworkers (1983) found no effect of platelet depletion on the responses of hemodynamics and airways of sheep given endotoxin.

Other studies in which platelet aggregation was induced with collagen (Bo and Hognestad 1972), protamine (Radergan and McAslan 1972) or ADP (Minnear et al 1982) also support a role for platelet cyclooxygenase products, in that these seem to be able to alter lung vasopressor responses and lung mechanics. All of these platelet stimuli caused decreased lung compliance, but only the protamine and collagen caused pulmonary vasoconstriction. Since the collagen was given to cats, protamine to dogs, and ADP to sheep, this may represent interspecies differences in the response. Nevertheless, since the early response of vasoreaction and lung mechanics to endotoxin seems to be mediated by cyclooxygenase-dependent pathways, these studies support the potential role for platelets as contributors to these responses.

Serotonin is another vasoactive and inflammatory molecule released from platelets that could contribute to the responses of hemodynamics and airways to endotoxin. However, Murphy and coworkers (1981) found no evidence for significant release of platelet serotonin in dogs given intravenous endotoxin nor could they demonstrate changes in lung mechanics after directly injecting serotonin in a dose greater than that released from the platelets. While Murphy's report did not support an important role for massive platelet serotonin release in causing the changes in lung mechanics after endotoxin, Boogaerts and coworkers (1982) identified platelet serotonin as an agent that increased PMN adherence to endothelium and potentiated PMN endothelial injury, which suggests that platelet serotonin may be important in endotoxemia.

Because data on whether or not platelets contribute to the pulmonary consequences of endotoxemia are conflicting, it is difficult to be more than speculative about mechanisms. However, both cyclooxygenase products of platelets and platelet serotonin could play important roles in the lung's response to endotoxemia.

8.   FUTURE DIRECTIONS

While an important role for PMN-dependent processes in mediating the pulmonary consequences of endotoxemia is established, exactly what those processes are is much less certain. An important role for oxygen radicals released by the PMN is suggested by several studies with other models but is not yet certain for endotoxin. Whether PMN proteases are important in the response is also not yet certain. While PMN depletion reduces the quantity of cyclooxygenase metabolites and their effects, it is not clear if these active metabolites come from PMN or from other cells.

While the studies with PMN depletion demonstrate an important role for PMN in the lung's response to endotoxin, it also seems that there are some PMN-independent effects of endotoxin on the lung. It is necessary to be certain that these are PMN-independent since in the depletion studies PMN are never totally absent. It is possible that platelets could be part of the PMN-independent process although previous studies argue against this. However, almost all of them involve platelet depletion and the effects of the PMN that are still present may mask the effects of platelet depletion. It is also possible that endotoxin has direct effects independent of normal circulating cells.

The importance of PMN adherence to the endothelium also needs to be explored closely. Endotoxin causes more severe endothelial injury than activated complement and endotoxin causes a more prolonged PMN adherence to endothelium than does complement. With the recent observations on the importance of adherence to PMN activation (Dahinden et al 1983), this aspect of PMN function deserves more attention.

Finally, most of the animal models in which the effect of endotoxemia on the lung were studied show a readily reversible change in lung function. Certainly, the clinical condition of respiratory failure following gram-negative bacteremia is not so readily reversed. What are the factors responsible for this? Is it just a quantitative difference in the amount of injury or are other systems involved? There has been very little mechanistic research done in the clinical setting, and very few attempts at clinical correlation of animal experiments have been made.

REFERENCES

Abdelnoor AM, Kassem H, Bikhazi AB, Nowotny A (1980), Effect of gram-negative bacterial lipopolysaccharide-derived polysaccharides, glycolipids, and lipopolysaccharides on rabbit and human platelets in vitro. *Immunobiology 157*, 145-153.

Ambrus CM, Ambrus JL, Johnson GC, Packman EW, Chernick WS, Back N, Harrison JWE (1954). Role of the lungs in regulation of the white blood cell level. *Am. J. Physiol. 178*, 33-44.

Bachofen M, Weibel ER (1977) Alterations of the gas exchange apparatus in adult respiratory insufficiency associated with septicemia. *Am. Rev. Respir. Dis. 116*, 589-615.

Bessa SM, Dalmasso AP, Goodale RL Jr (1974) Studies on the mechanism of endotoxin-induced increase of alveolocapillary permeability. *Proc. Soc. Exp. Biol. Med. 147*, 701-705.

Bierman HR, Kelly KH, King FW, Petrakis NI. (1951) The pulmonary circulation as a source of leukocytes and platelets in man. *Science 114*, 276-277.

Binder AS, Kageler W, Perel A, Flick MR, Staub NC (1980) Effect of platelet depletion on lung vascular permeability after microemboli in sheep. *J. Appl. Physiol. 48*, 414-420.

Binder AS, Nakaharak K, Ohleuda K, Kageler W, Staub NC (1979) Effect of heparin of fibrinogen depletion on lung fluid balance in sheep after emboli. *J. Appl. Physiol. 47*, 213-219.

Bo G, Hognestad J (1972) Effects on the pulmonary circulation of suddenly induced intravascular aggregation of blood platelets. *Acta Physiol. Scand. 85*, 523-531.

Boogaerts MA, Yanada O, Jacob HS, Moldow CF (1982) Enhancement of granulocyte-endothelial cell adherence and granulocyte-induced cytotoxicity by platelet release products. *Proc. Natl Acad. Sci. 79*, 7019-7023.

Braude AI, Siemienski J, Williams D, Sanford JP (1953) Respiratory distress and cyanosis is a frequent manifestation of clinical bacteremic shock. *Univ. Mich. Med. Bull. 19*, 23-28.

Bredenberg CE, Taylor GA, Webb WR (1980) The effect of thrombocytopenia on the pulmonary and systemic hemodynamics of canine endotoxin shock. *Surgery 87*, 59-68.

Brigham KL, Woolverton WC, Blake LH, Staub NC (1974) Increased sheep lung vascular permeability caused by pseudomonas bacteremia. *J. Clin. Invest. 54*,

792-804.

Brigham KL, Bowers RE, Haynes J (1979) Increased sheep lung vascular permeability caused by *E. coli* endotoxin in unanesthetized sheep. *Circ. Res. 45*, 292-297.

Busch C, Lindquist O, Saldeen T (1974) Respiratory insufficiency in the dog induced by pulmonary microembolism and inhibition of fibrinolysis. *Acta Chir. Scand. 140*, 255-266.

Coalson JJ, Hinshaw LB, Guenter CA (1970) The pulmonary ultrastructure in septic shock. *Exp. Mol. Pathol. 12*, 84-103.

Cochrane CG, Spragg R, Revak SD (1983) Pathogenesis of the adult respiratory distress syndrome. Evidence of oxidant activity in bronchoalveolar lavage fluid. *J. Clin. Invest. 71*, 754-761.

Colucci M, Balconi G, Lorenzet R, Pietra A, Locati D, Donati MB, Sermeraro N (1983) Cultured human endothelial cells generate tissue factor in response to endotoxin. *J. Clin. Invest. 71*, 1893-1896.

Craddock PR, Fehr J, Dalmasso A, Brigham K, Jacob HS (1977) Hemodialysis leukopenia: pulmonary vascular leukostasis resulting from complement activation by dialyzer cellophane membrane. *J. Clin. Invest. 59*, 879-888.

Dahinden C, Galanos C, Fehr J (1983) Granulocyte activation by endotoxin *J. Immunol. 130*, 857-868.

Demling RH, Smith M, Gunther R, Flynn JT, Gee MH (1981a) Pulmonary injury and prostaglandin production during endotoxemia in conscious sheep. *Am. J. Physiol. 240*, H348-H353.

Demling RH, Proctor R, Grossman J, Dey N, Starling J (1981b) Lung injury and lung lysosomal enzyme release during endotoxemia. *J. Surg. Res. 30*, 135-141.

Esbenshade AM, Newman JH, Lams PM, Jolles H, Brigham KL (1982) Respiratory failure after endotoxin infusion in sheep: lung mechanics and lung fluid balance. *J. Appl. Physiol. 53*, 967-976.

Flick MR, Perel A, Staub NC (1981a) Leukocytes are required for increased lung microvascular permeability after microembolization in sheep. *Circ. Res. 48*, 344-351.

Flick MR, Perel A, Staub NC (1979) Increased lung vascular permeability after microemboli in unanesthetized sheep requires circulating leukocytes. *Physiologist 22*, 39-48.

Flick MR, Hoeffel J, Staub NC (1981b) Superoxide dismutase prevents increased lung vascular permeability after air emboli in unanesthetized sheep. *Fed. Proc. 40*, 405.

Fowler AA, Hamman RF, Good JT, Benson KN, Baird M, Eberle DJ, Petty TL, Hyers TM (1983) Adult respiratory distress syndrome: risk with common predispositions. *Ann. Intern. Med. 98*, 593-597.

Freudenberg MA, Galanos C (1978) Interaction of lipopolysaccharides and lipid A with complement in rats and its relation to toxicity. *Infect. Immun. 19*, 875-882.

Garner R, Chater BV, Brown DL (1974) The role of complement in endotoxin shock and disseminated intravascular coagulation: experimental observations in the dog. *Br. J. Haematol. 28*, 393-401.

Gaynor E (1973) The role of leukocytes in endotoxin-induced vascular injury. *Blood 41*, 797-807.

Goodman ML, Way BA, Irwin JW (1979) The inflammatory response to endotoxin. *J. Pathol. 128*, 7-14.

Harlan JM, Killen PD, Harker LA, Striker GE, Wright DG (1981) Neutrophil-mediated endothelial injury in vitro. Mechanisms of cell detachment. *J. Clin. Invest. 68*, 1394-1403.

Harris RH, Schmeling JW, Fletcher RJ, Romwell PW (1982) Endotoxin interaction with canine platelets fails to stimulate thromboxane production. *Proc. Soc. Exp. Biol. Med. 169*, 397-400.

Heffner JE, Shoemaker SA, Canham EM, Patel M, McMurtry IF, Morris HG, Repine JE (1983) Acetyl glyceryl ether phosphorylcholine-stimulated human platelets cause pulmonary hypertension and edema in isolated rabbit lungs. Role of thromboxane A_2. *J. Clin. Invest. 71*, 351-357.

Heflin AC, Brigham KL (1981) Prevention by granulocyte depletion of increased vascular permeability of sheep lung following endotoxemia. *J. Clin. Invest. 68*, 1253-1260.

Hinshaw LB, Kuida H, Gilbert RP, Yisscher MB (1957) Influence of perfusate characteristics on pulmonary vascular response to endotoxin. *Am. J. Physiol. 191*, 292-301.

Hinson JM Jr, Hutchison AA, Ogletree ML, Brigham KL, Snapper JR (1983) Effect of granulocyte depletion on altered lung mechanics after endotoxemia in sheep. *J. Appl. Physiol. 55*, 92-99.

Hoover RL, Folger R, Haering WA, Ware BR, Karnovsky MJ (1980) Adhesion of leukocytes to endothelium: roles of divalent cations, surface change, chemotactic agents and substrate. *J. Cell Sci. 45*, 73-86.

Horn REG, Collins RD (1968) Studies on the pathogenesis of the generalized Shwartzman reaction. The role of granulocytes. *Lab. Invest. 18*, 101-107.

Hosea S, Brown E, Hammer C, Frank M. (1980) Role of complement activation in a model of adult respiratory distress syndrome. *J. Clin. Invest. 66*, 375-382.

Hüttemeier PC, Watkins WD, Peterson MB, Zapol WM (1982) Acute pulmonary hypertension and lung thromboxane release after endotoxin infusion in normal and leukopenic sheep. *Circ. Res. 50*, 688-694.

Johnson A, Malik A (1980a) Effect of granulocytopenia on extravascular lung water content after microembolization. *Am. Rev. Respir. Dis. 122*, 561-566.

Johnson A, Malik AB, (1980b) Effect of defibrinogenation on extravascular fluid accumulation after pulmonary microembolization in dogs. *J. Appl. Physiol. 49*, 841-845.

Johnson KJ, Ward PA (1981) Role of oxygen metabolites in immune complex injury of lung. *J. Immunol. 126*, 2365-2369.

Kux M, Coalson JJ, Maisson WH, Guenter CA (1972) Pulmonary effects of E. coli endotoxin: role of leukocytes and platelets. *Ann. Surg. 175*, 26-34.

Lentnek AL, Schreiber AD, MacGregor RR (1976) The induction of augmented granulocyte adherence by inflammation. *J. Clin. Invest. 57*, 1098-1103.

Loyd JE, Newman JH, English D, Ogletree ML, Meyrick B, Brigham KL (1983) Lung vascular effects of phorbol myristate acetate in awake sheep. *J. Appl. Physiol. 54*, 267-276.

Lucht WD, Bernard GR, Niecdermeyer MC, Snapper JR, Brigham KL (1983) Effect of n-acetyl cysteine on the pulmonary response to endotoxin in awake sheep (Abstract). *Fed. Proc. 42*, 4776.

Malik AB, Johnson A, Tahamont MV (1982) Mechanisms of lung injury after intravascular coagulation. *Ann. N.Y. Acad. Sci. 384*, 213-234.

Malik AB, Vander Zee H (1978) Mechanisms of pulmonary edema induced by microembolization in dogs. *Circ. Res. 42*, 72-79.

Martin BA, Wright JL, Thommasen H, Hogg JC (1982) Effect of pulmonary blood flow on the exchange between the circulating and marginating pool of polymorphonuclear leukocytes in dog lungs. *J. Clin. Invest. 69*, 1277-1285.

McDonald JWD, Ali M, Morgan E, Townsend ER, Cooper JD (1983) Thromboxane synthesis by sources other than platelets in association with complement-induced pulmonary leukostasis and pulmonary hypertension in sheep. *Circ. Res. 52*, 1-6.

McKay DG, Margarethen W, Cassovy I (1966) An electron microscopic study of effects of bacterial endotoxin on the blood-vascular system. *Lab. Invest. 15*, 1815-1829.

McKay DG, Margarethen W, Cassovy I (1967) An electron microscope study of endotoxin shock in rhesus monkeys. *Surg. Gynecol. Obstet. 125*, 825-832.

Meyrick B, Brigham KL (1983) Acute effects of *Escherichia coli* endotoxin on the pulmonary microcirculation of anesthetized sheep. Structure:function relationship. *Lab. Invest. 48*, 458-470.

Minnear FL, Moon DG, Kaplan JE, Malik AB (1982) Effect of ADP-induced platelet aggregation on lung fluid balance in sheep. *Am. J. Physiol. 242*, H645-H651.

Morrison DC, Oades ZG (1979) Mechanisms of lipopolysaccharide-initiated rabbit platelet responses. II. Evidence that lipid A is responsible for binding of lipopolysaccharide to the platelet. *J. Immunol. 122*, 753-758.

Murphy TL, Allison RC, Weisman IM, McCaffree DR, Gray BA (1981) Role of platelet serotonin in the canine pulmonary response to endotoxin. *J. Appl. Physiol. 50*, 178-184.

Northrup WF, Murphy EW (1976) Pulmonary and systemic capillary permeability to protein following endotoxemia. *Surg. Forum. 27*, 65-67.

O'Flaherty JT, Kreutzer DL, Ward PA (1978) The influence of chemotactic factors on neutrophil adhesiveness. *Inflammation 3*, 37-48.

Ogletree MC, Brigham KL (1982) Effects of cyclooxygenase inhibitors on pulmonary vascular responses to endotoxin in unanesthetized sheep. *Prostaglandins Leukotrienes Med. 8*, 489-502.

Oseas R, Yang HH, Bachner RL, Boxer LA (1981) Lactoferrin: a promotor of polymorphonuclear leukocyte adhesiveness. *Blood 57*, 939-945.

Pietra GG, Szidon JP, Carpenter HA, Fishman AP (1974) Bronchial venular leakage during endotoxin shock. *Am. J. Pathol. 77*, 387-406.

Radegran K, McAslan C (1972) Circulatory and ventilatory effects of induced platelet aggregation and their inhibition by acetylsalicylic acid. *Acta Anaesthesiol. Scand. 16*, 76-84.

Sacks T, Moldow CF, Craddock PR, Bowers TK, Jacobs HS (1978) Oxygen radicals mediate endothelial cell damage by complement stimulated granulocytes. An in vitro model of immune vascular damage. *J. Clin. Invest. 61*, 1161-1167.

Schneider RC, Zapol WM, Carvalho AC (1980) Platelet consumption and sequestration in severe acute respiratory failure. *Am. Rev. Respir. Dis. 122*, 445-451.

Semeraro N, Colucci M, Vermylen J (1979) Complement-dependent and complement-independent interactions of bacterial lipopolysaccharides and micropeptides with rabbit and human platelets. *Thrombus Haemostas 41*, 392-406.

Shasby DM, VanBenthuysen KM, Tate RM, Shasby SS, McMurtry IF, Repine JE (1982) Granulocytes mediate acute edematous lung injury in rabbits and in isolated rabbit lungs perfused with PMA: role of O_2 radicals. *Am. Rev. Respir. Dis. 125*, 443-447.

Shasby DM, Shasby SS, Peach MJ (1983) Granulocytes and phorbol myristate acetate increase permeability to albumin of cultured endothelial monolayers and isolated perfused lung. *Am. Rev. Respir. Dis. 127*, 72-78.

Smith ME, Gunther R, Gee M, Flynn J, Demling RH (1981) Leukocytes, platelets, and thromboxane A_2 in endotoxin-induced lung injury. *Surgery 90*, 102-107.

Snapper JR, Hinson JM Jr, Lefferts P, Ogletree ML, Brigham KL (1983) Effect of anti-platelet antibody (APA) on the awake sheep and platelet depletion on the sheep's response to endotoxemia (Abstract). *Fed. Proc. 42*, 4777.

Snyderman R, Shin HS, Phillips JK, Gewurz H, Morganhagen SE (1969) A neutrophil chemotactic factor derived from C's upon interaction of guinea pig serum with endotoxin. *J. Immunol. 103*, 413-422.

Spagnuolo PJ, Ellner JJ (1980) Comparative stimulation of granulocyte adherence and chemotaxis by bacterial products. *Infect. Immun. 27*, 519-524.

Spagnuolo PJ, Ellner JJ, Hassid A, Dunn MJ (1981) Thromboxane A_2 mediates augmented polymorphonuclear leukocyte adhesiveness. *J. Clin. Invest. 66*, 406-414.

Spragg RG, Lonky SA, Loomis WH, Marsh J, Abraham JL (1982) Human neutrophil elastase causes acute high permeability edema in isolated perfused rabbit lung (Abstract). *Am. Rev. Resp. Dis. 126*, 276.

Staub NC, Schultz EL, Albertine KH (1982) Leukocytes and pulmonary microvascular injury. *Ann. N.Y. Acad. Sci. 384*, 332-343.

Stetson CA Jr (1951) Studies on the mechanism of the generalized Shwartzman phenomenon. Certain factors involved in the production of hemorrhagic necrosis. *J. Exp. Med. 93*, 489-505.

Stetson CA Jr, Good RA (1951) Studies of the mechanism of the Shwartzman phenomenon. Evidence for the participation of polymorphonuclear leukocytes in the phenomenon. *J. Exp. Med. 93*, 49-65.

Tahamont M, Malik A (1983) Granulocytes mediate the increase in pulmonary vascular permeability after thrombin embolism. *J. Appl. Physiol. 54*, 1489-1495.

Tate RM, VanBenthuysen KM, Shasby DM, McMurtry IF, Repine JE (1982) Oxygen-radical-mediated permeability edema and vasoconstriction in isolated perfused rabbit lungs. *Am. Rev. Resp. Dis. 126*, 802-806.

Thomas J, Land T, Good RA (1952) Studies on the generalized Shwartzman reaction. I. General observations concerning the phenomenon. *J. Exp. Med. 96*, 605-614.

Thorne LJ, Keuenzig M, Schwartz SI (1981) Effects of acetylsalicylic acid and prostaglandin on endotoxin induced pulmonary platelet sequestration. *Acta Chir. Scand. 147*, 161-162.

Till GO, Johnson KJ, Kunkel R, Ward PA (1982) Intravascular activation of complement and acute lung injury. Dependency on neutrophils and toxic oxygen metabolites. *J. Clin. Invest. 69*, 1126-1135.

Tomashefski JF, Davies P, Boggie C, Greene R, Zapol WM, Reid LM (1983) The pulmonary vascular lesions of the adult respiratory distress syndrome. *Am. J. Pathol. 112*, 112-126.

Ulevitch RJ, Cochran CG, Henson PM, Morrison DC, Doe WF (1975) Mediation systems in bacterial lipopolysaccharide-induced hypotension and disseminated intravascular coagulation. I. Role of complement *J. Exp. Med. 142*, 1570-1590.

Ulevitch RJ, Cochrane CG, Bangs K, Herman CM, Fletcher JR, Rice CL (1978) The effect of complement depletion on bacterial lipopolysaccharide (LPS)-induced hemodynamic and hematologic changes in the rhesus monkey. *Am. J. Pathol. 92*, 227-240.

Vaage J, Nicolaysen G, Wooler BA (1976) Aggregation of blood platelets and increased hydraulic conductivity of pulmonary exchange vessels. *Acta Physiol. Scand. 98*, 175-184.

Walker RI, Parker GA (1980) Clearance of inflammatory cells from the microcirculation of endotoxin resistant C3H/HeJ mice. *Can. J. Microbiol. 26*, 725-727.

Walker RI, Casey LC, Ramwell PW, Fletcher JR (1982) Association of prostacyclin production with resistance of C3H/HeJ mice to endotoxin shock. *Adv. Shock Res.* 7, 125-132.

Wedmore CV, Williams TJ (1981) Control of vascular permeability by polymorphonuclear leukocytes in inflammation. *Nature (London) 289*, 646-650.

Weiss SJ, Young J, LoBuglio AF, Slivka A, Nimeh NF (1981) Role of hydrogen peroxide in neutrophil-mediated destruction of cultured endothelial cells. *J. Clin. Invest. 68*, 714-721.

Yamada O, Moldow CF, Sacks T, Craddock PR, Boogaerts MA, Jacob HS (1981). Deleterious effects of endotoxin on cultured endothelial cells. An in vitro model of vascular injury. *Inflammation 5*, 115-125.

Handbook of Endotoxin, Vol. 2: Pathophysiology of Endotoxin
L.B. Hinshaw, editor
© Elsevier Science Publishers B.V., 1985
ISBN 0 444 90385 2
$0.85 per article per page (transactional system)
$0.20 per article per page (licensing system)

CHAPTER 6

Role of complement in endotoxin shock

DOUGLAS P. FINE

1. INTRODUCTION

Since 1914, it has been suggested that complement activation is an integral part of most if not all acute infections (Gunn 1914). On the basis of experimental studies of endotoxemia in dogs, Spink and colleagues more recently postulated that many of the clinical manifestations of gram-negative sepsis and endotoxemia might be mediated by complement activation (Spink and Vick 1961; Spink et al 1964). It will be the purpose of this chapter to review the data generated since original studies in order to assess how well this hypothesis has fared and to suggest paths for future investigation.

2. REVIEW OF COMPLEMENT SYSTEM

Complement is the generic name for a group of plasma proteins which circulate generally in inactive precursor form and which can, upon interaction with appropriate stimuli, be converted into potent biological mediators. This system was first identified at the end of the 19th century as a heat-labile activity of nonimmune serum which mediated or 'complemented' the bactericidal effect of antibodies (Rapp and Borsos 1970). The multiplicity of components was only gradually recognized. Not until the 1960s were all nine components of the classical pathway identified, and the alternative pathway only in the late 1960s, although Pillemer and associates had clearly recognized an alternative pathway in 1954 (Pillemer et al 1954).

2.1. The classical pathway

The bactericidal and hemolytic system first called 'complement' is now more properly referred to as the classical pathway. Some authorities would now limit that term to the first three components of this system (C1, C4, and C2), segregating them from C3 and the terminal components or membrane attack mechanism, which are shared with the alternative pathway.

The classical pathway is activated by immune complexes. Following specific interaction with antigen, the immunoglobulin Fc fragment is so altered as to expose sites which can bind the first component of complement, C1. Two contiguous Fc fragments are required for this step; thus, two IgG molecules but only one IgM molecule are necessary for C1 activation. C1 is a trimolecular complex at physiological calcium concentrations; in the absence of calcium the complex dissociates into the subunits C1q, C1r, and C1s. C1q is the portion of the complex that binds to Fc fragments, following which internal rearrangement results in exposure of the active esterase site on the C1s subunit. The active C1s, also referred to as C1 esterase, can then cleave C4 into its two subunits, C4a and C4b. C4a dissociates as a fluid-phase phlogistic particle; the larger C4b can attach to the antigen surface and participate with C1s in the similar cleavage of C2 into the surface-active C2a and the fluid-phase C2b. These interactions can be represented by the equations

$$(1) \qquad \mathrm{Ag + Ab \rightarrow AgAb}$$

$$(2) \qquad \mathrm{AgAb + C1 \rightarrow AgAb\overline{C1}}$$

$$(3) \qquad \mathrm{AgAb\overline{C1} + C4 \rightarrow AgAb\overline{C14b} + C4a}$$

$$(4) \qquad \mathrm{AgAb\overline{C14b} + C2 \rightarrow AgAb\overline{C14b2a} + C2b.}$$

The $\mathrm{C\overline{4b2a}}$ complex expresses the enzymatic activity for the cleavage of C3 into the active subunits C3a, a fluid-phase phlogistic fragment, and the membrane-active C3b. This reaction may be expressed by the equation

$$(5) \qquad \mathrm{C\overline{4b2a} + C3 \rightarrow C\overline{4b2a3b} + C3a.}$$

The classical pathway C3 convertase ($\mathrm{C\overline{4b2a}}$) does not require immunoglobulin or C1 for the conversion of C3; indeed, multiple convertases can be generated on the same antigen by a single antigen-antibody-C1 complex. Likewise, C3b need not bind to the $\mathrm{C\overline{4b2a}}$ complex, and multiple C3b molecules can be generated by a single convertase.

The $\mathrm{C\overline{4b2a3b}}$ complex is the C5 convertase, cleaving that molecule into the powerful mediator of inflammation, C5a, and the membrane-active C5b. The latter molecule, in complex with C6 and C7, binds at still another site on the antigen surface to initiate the membrane attack mechanism, as shown in equations (6) and (7):

$$(6) \qquad \mathrm{C\overline{4b2a3b} + C5 \rightarrow C\overline{4b2a3b} + C\overline{5b} + C5a}$$

$$(7) \qquad \mathrm{C\overline{5b} + C6 + C7 \rightarrow C\overline{5b67} + C8 + 6C9 \rightarrow C\overline{5b678(6C9)}.}$$

As the last equation implies, as many as six C9 molecules may be incorporated into this complex. With the formation of the complete complex, a transmembrane channel can form in the lipid bilayer (if this reaction occurs on such a membrane surface, e.g., an erythrocyte or a gram-negative bacterium), the result of which would be free diffusion of small molecules and osmotic damage.

As will be discussed, some substances are capable of activating this pathway by direct interaction with C1q. The alternative pathway was recognized as a system by which C3 could be activated directly, bypassing C1, C4, and C2.

2.2. The alternative pathway

If the classical pathway can be said to be specifically activated, by virtue of the specificity inherent in antibody recognition, a hallmark of the alternative pathway is its nonspecificity. The proteins of this system include C3, factor B (B), factor D (D), and properdin (P). They interact in an autocatalytic fashion, after which, in the presence of an appropriate surface, the interaction accelerates. The mechanisms and functions of this pathway have been the subject of an excellent review (Fearon and Austen 1980).

These proteins in fluid phase undergo slow spontaneous interaction, whereby the enzymatically active D cleaves small amounts of C3:

$$(8) \qquad C3 + B + \overline{D} + P \rightarrow C\overline{3b} + C3a + B + \overline{D} + P.$$

Several fates may befall the nascent C3b, including binding to beta-1-H globulin and subsequent inactivation by the C3b inactivator; however, if the molecule interacts with an appropriate activator surface (characteristically poor in sialic acid), it is protected from its natural inactivators and preferentially binds factor B, in a magnesium-dependent reaction. The B can then be cleaved by D, with release of a small soluble fragment, Ba:

$$(9) \qquad C3bB + \overline{D} \rightarrow C\overline{3bBb} + Ba + \overline{D}.$$

Properdin can bind to this complex and further stabilize and protect the enzyme. The C3bBbP then is a powerful C3 and, with accumulation of additional C3b molecules, C5 convertase:

$$(10) \qquad \overline{C3bBbP} + C3 \rightarrow \overline{C3bBbP3b} + C3a$$

$$(11) \qquad \overline{C3bBbP3b} + C5 \rightarrow \overline{C3bBbP3b} + C5b + C5a.$$

Immunoglobulin is clearly not required for this process, thus the only specificity is the presence of an appropriate surface. Many microbes and other cells possess such surface characteristics as to make them potent activators of the alternative pathway. It should be noted, however, that immunoglobulins may participate in the activation of the alternative pathway to enhance it and may thereby introduce immunological specificity. The mechanism of this antibody activity is not yet clear.

2.3. Biological effects of complement activation

Potential biological consequences of complement activation include lytic damage to membranes, mediation of attachment of particles to C3b receptors on phagocytic (opsonization) and other cells, virus neutralization, and elaboration of inflammatory molecules which interact with a variety of mammalian tissues and cells.

With activation of the membrane attack mechanism through either the classical or alternative pathway, a transmembrane channel is produced which permits osmotic lysis of cells. C3b, deposited on a particle (e.g., a bacterium) through either pathway, can function as a ligand to bind the particle to polymorphonuclear leukocytes, macrophages, primate erythrocytes, B lymphocytes, and renal glomerular epithelial cells. This binding can, especially in concert with immunoglobulin, stimulate phagocytosis. By these two mechanisms and also by direct surface accretion, the complement proteins may mediate virus neutralization.

The soluble fragments elaborated upon cleavage of complement proteins (C4a, C2b, Bb, C3a, and C5a) have the potential of playing important roles in the inflammatory response. For C2b and Bb, this role is largely speculative. For C5a and C3a, important roles are better established. C5a, and to a lesser extent perhaps C3a, is an attractive stimulus to leukocytes, a chemotactic factor. Moreover, C5a appears to be able to increase leukocyte adhesiveness and aggregation. These activities, which may be very important in the in vivo effects of endotoxin, are discussed in Chapter 5.

Both C5a and C3a are also anaphylatoxins, that is, they induce the classical manifestations of anaphylaxis – vasodilatation, increased vascular permeability, and smooth muscle contraction (Hugli and Müller-Eberhard 1978). These effects are largely due to stimulation of histamine release by basophils and mast cells, though there may be some direct smooth muscle contraction induced by C5a and C3a. Because anaphylatoxic activity is rapidly destroyed, these molecules are probably active only at short distances. Nevertheless, it is this activity of complement which has suggested that the system may directly mediate some of the clinical manifestations of endotoxin shock.

3. COMPLEMENT ACTIVATION BY ENDOTOXIN IN VITRO

Using endotoxin prepared from *Salmonella typhi*, Muschel and colleagues demonstrated that serum complement could be consumed (Muschel et al

1964). They suggested that antibody was not required for this reaction since absorption of serum with *S. typhi* did not alter the complement consumption. Furthermore, the endotoxin could activate complement in the immunoglobulin-deficient serum of newborn piglets. The authors further demonstrated that human serum was more reactive with endotoxin than was guinea pig serum, which was in turn more active than rabbit serum. Endotoxins prepared by the Boivin method were more anticomplementary than those prepared by the Westphal method. Several points about this investigation should be emphasized. First, it emphasized the differences between species which were to render extrapolation of data from one species to another quite dangerous. Secondly, these investigators recognized that different preparations of endotoxins could give different results. Finally, they presaged an important interest in the role of antibody in complement activation by endotoxin, an interest which was crucial to the clear delineation of the alternative pathway (see Section 3.1.).

The transmembrane channels of the membrane attack mechanism can be visualized by electron microscopy and appear as doughnut-shaped 'holes.' Bladen and colleagues demonstrated identical lesions on *Escherichia coli* membranes after incubation with antibody and complement (Bladen et al 1966). Likewise, *Veillonella alcalescens* organisms or lipopolysaccharide derived from those organisms displayed similar lesions when reacted with guinea pig serum (Bladen et al 1967). In the latter study, they further demonstrated that complement depletion abrogated the ability of serum to induce these lesions and also the bactericidal capacity of the serum. Finally, they confirmed that endotoxin consumed complement in vitro. Gewurz and colleagues confirmed the surface lesions and the complement consumption produced by endotoxin (Gewurz et al 1968a). The complement components lost were primarily C3,5–9, and the earlier components of the classical pathway were relatively spared. Endotoxin was a more powerful activator of complement than zymosan, sheep erythrocytes, or heat-aggregated gamma globulins. The authors were able to demonstrate, however, that in immune serum, all components, including C1, C4, and C2, were consumed. The investigators attempted to evaluate the relation of complement activation to toxicity of the lipopolysaccharide in vitro. Endotoxin chemically altered to the endotoxoid retained antigenicity but lost both toxicity and complement-fixing activity. This concomitant loss of both properties suggested a close relationship.

3.1. Pathways of complement activation by endotoxin

In the context of the understanding of the complement system in the mid-1960s, the observations that antibody might not be required for complement activation and that only the terminal components might be consumed were surprising and controversial. Pillemer and colleagues had suggested ten years earlier that gram-negative bacterial cell walls and endotoxins could react with properdin and consume C3 (Pillemer et al 1955), but this notion had lain dormant for some time.

Gewurz et al further evaluated the role of antibody (Gewurz et al 1968b). Endotoxin consumed complement, generated chemotactic and anaphylatoxic activities, and developed typical complement 'holes' in precolostral piglet serum and immunoglobulin-deficient sera from pigs, cows, and man. When serum from individuals with Swiss-type lymphopenic agammaglobulinemia was used, however, none of these effects could be demonstrated. These last results could perhaps be explained by the fact that such serum is also deficient in C1q and properdin. Nevertheless, the study demonstrated that the interactions of complement and endotoxin could take place at low antibody concentrations.

With the discovery of a strain of guinea pigs deficient in C4 and the identification of the alternative pathway, it became possible to study the interactions of complement and endotoxin more definitively. Lipopolysaccharide prepared from *V. alcalescens* by phenol-water extraction consumed complement poorly in C4-deficient serum, results which indicated a requirement for an intact classical pathway (Snyderman et al 1971). A dose of 10 µg endotoxin consumed 32% of C3,5–9 activity in normal guinea pig serum, none in C4-deficient serum; 100 µg consumed 83% and 34% respectively. The investigators concluded that C4 was required, certainly for optimal complement activation. Using endotoxin-coated sheep erythrocytes, Phillips et al demonstrated a requirement for immunoglobulin and the early components C1, C4, and C2 (Phillips et al 1972). Fine confirmed those observations using a different endotoxin preparation (*Salmonella enteritidis* instead of *S. typhi* and *V. alcalescens*) but also found evidence that soluble lipopolysaccharide, as distinct from cell-bound lipopolysaccharide, could activate the alternative pathway as well (Fine 1974). It thus seemed likely that endotoxins could activate both complement pathways, that antibody could modulate this activation, and that different preparations or states of endotoxins would activate complement in different ways.

3.2. Components of the endotoxin molecule and their interactions with complement

Galanos et al (1971) evaluated complement activation by a variety of endotoxins from *Salmonellae* and *Escherichia coli*. Most importantly, they used lipopolysaccharide preparations with varying amounts of polysaccharide, that is, rough and smooth endotoxins. All these preparations were toxic but only some were anticomplementary; in contrast, lipid A preparations from all were both toxic and anticomplementary. In subsequent studies with lipopolysaccharide preparations from *Salmonella minnesota*, it was demonstrated that lipid A preparations were more efficient at activating C3 than were rough lipopolysaccharide preparations which were in turn more active than smooth preparations (Dierich et al 1973). Next purified C1 was found to react directly with lipopolysaccharides in a serum-free system (Loos et al 1974). Furthermore, these investigators demonstrated that purified, isolated C1q could bind to lipid A in the absence of immunoglobulin.

These interactions were further explored by Morrison and Kline (1977) in studies with a variety of lipid A and lipopolysaccharide preparations. They confirmed that efficiency of complement activation in whole serum (i.e., the amount of lipopolysaccharide required to consume a given amount of complement) was inversely related to the amount of polysaccharide contained on the molecule. Thus, polysaccharide appeared to modulate complement activation by lipid A. However, the picture was actually more complex. Using *E. coli* lipopolysaccharides varying only in the amounts of polysaccharide, the authors could not relate complement activation with the amount by weight of lipid A. Lipid A was then shown to consume components of the classical pathway (C1, C4, and C2) as well as C3; antibody appeared not to be required. In contrast, complete lipopolysaccharides consumed primarily C3, with very little consumption of the early components. Furthermore, in C2-deficient serum, lipopolysaccharide but not lipid A could activate C3. Thus, Morrison and Kline concluded that the polysaccharide moiety modulated the direct classical pathway activation by lipid A but that the polysaccharide could also activate the alternative pathway.

Cooper and Morrison (1978) confirmed the observations of Loos et al (1974) that lipid A could bind isolated C1q in an immunoglobulin-independent fashion. This binding could, when the C1q was part of the intact C1 macromolecule, lead to direct activation of the C1 esterase, C1s.

Thus, complement can be activated in at least three ways by endotoxin: an antibody-independent activation of C1 by the lipid A portion, an antibody-dependent activation of the classical pathway, and activation of the alternative pathway by the polysaccharide portion (Morrison and Ulevitch 1978). Which mechanisms are active in whole serum or in vivo will be affected by the source and nature of the endotoxin and by the immunological status of the host.

3.3. In vitro effects of complement activation on other biological systems

Endotoxin activation of complement has been demonstrated to generate activities which might be expected to have important biological effects.

Chemotactic activities, probably due to generation of C5a, have been demonstrated (Gewurz et al 1968b; Snyderman et al 1968). Important to considerations of any role in endotoxin shock have been demonstrations of the elaboration of anaphylatoxins (Gewurz et al 1968b; Lichtenstein et al 1969).

In addition, endotoxins have been shown to interact in a complex fashion with a variety of host tissues and cells, including platelets, neutrophils, and endothelial cells. In many of these situations, investigators have suggested roles for complement. These interactions are reviewed in Chapters 5 and 8 and also in Volume 3 of this series, and are beyond the scope of this chapter. The reader is also referred to two excellent reviews (Mergenhagen et al 1969; Morrison and Ulevitch 1978).

4. STUDIES IN EXPERIMENTAL ANIMALS

There is an extensive body of information relating to the effects of endotoxin in experimental animals. In this section, having accepted that endotoxin can activate and interact with complement in complex ways, we will focus on evidence that complement is or is not involved in producing the syndrome of endotoxin shock. Because of potentially important interspecies differences, the discussion will be divided on that basis.

4.1. Canine model

Spink and colleagues measured complement levels sequentially following endotoxin injections (Spink et al 1964). Complement levels fell abruptly and were minimal by 10 minutes; plasma histamine rose even more abruptly and was maximal by 30–60 seconds; hypotension accompanied these changes. Lethality could be related to the degree of hypocomplementemia. These authors suggested that the interaction of endotoxin, antibody, and complement resulted in histamine release and shock. Other investigators have subsequently confirmed that endotoxin injection in dogs will produce both hypotension and hypocomplementemia which are temporally related (Aasen et al 1978; Füst and Keresztes 1969).

Füst et al (1977) attempted to define a causal relationship between complement activation on the one hand and hypotension and lethality on the other by manipulating the endotoxin molecule. Irradiation of the endotoxin molecule with ^{60}Co altered its complement activating properties: the irradiated molecule was deficient in classical pathway activation though it retained ability to activate the alternative pathway and to bind C1. Injection of this preparation produced less hypotension and hypocomplementemia than did the untreated endotoxin. The irradiated endotoxin was also less lethal to mice.

An alternative approach is to manipulate the complement system. From and colleagues depleted dogs of complement by the injection of cobra venom factor (a potent activator of the alternative pathway, which depletes C3 and the later components) and evaluated the subsequent hypotensive effect of endotoxin injection. Complement depletion abrogated the initial (2–5 min) hypotension but had no effect on the second, more gradual and more pronounced fall in mean arterial blood pressure. Neither could the authors demonstrate any amelioration of mortality rates by cobra factor treatment (From et al 1970). In contrast, Garner et al (1974) demonstrated a profoundly protective effect of complement depletion. In their model, endotoxin injection was followed immediately by a biphasic hypotension, the latter phase being gradual, progressive, and fatal. Cobra factor-treated animals had no hypotension at all and they did not die. No explanation for the discrepancy between these two studies can be given, except for the fact that endotoxin preparations and animal breeds were not the same.

4.2. Rabbit model

Endotoxin injection produces hypocomplementemia in rabbits as well (Gilbert and Braude 1962; Pearlman et al 1963). Furthermore, there is some correlation with lethality in that nonlethal doses produce less complement depletion than do lethal doses (Gilbert and Braude 1962), although, of course, these two could be independent dose effects.

Ulevitch and colleagues studied a variety of endotoxins in sublethal doses (Ulevitch et al 1975). Cobra venom treatment or congenital C6-deficiency had no effect on hypotension or other manifestations of endotoxemia (disseminated intravascular coagulation, leukopenia, or the major phase of thrombocytopenia). With lethal doses, Ulevitch and Cochrane (1978) were able to demonstrate that complement depletion (cobra venom factor) abrogated the initial hypocomplementemia seen in the first few minutes after endotoxin injection. However, the later hypotension (maximal at 2 hrs), acidosis, and mortality appeared to have no relation to complement.

A metabolite of prostacyclin, 6-oxo-prostaglandin-F1-alpha, has been shown to be increased during the hypotensive phase of endotoxemia. The rise in this metabolite and the hypotension were inhibited by complement depletion (Rampart et al 1981). These investigators postulated that complement activation was necessary for the hypotension, which was related to prostacyclin.

Others have suggested that complement, or some of the complement components, might actually play a protective role during endotoxemia. In one study (Johnson and Ward 1971), an ordinarily sublethal dose of *E. coli* endotoxin, 100 µg, produced lethal disease in 11 of 12 congenitally C6-deficient rabbits. There was some suggestion that the C6-deficient serum was less effective at detoxifying endotoxin in vitro (mouse lethality assay, chick embryo lethality) (Johnson and Ward 1972).

4.3. Guinea pig model

May and colleagues used the C4-deficient strain of guinea pigs to analyze the protective role of complement (May et al 1972), as first suggested by Johnson and Ward (1971). C4-deficient animals or normal guinea pigs depleted of C3 and later components by cobra venom factor had accelerated mortality following endotoxin injection. Cobra factor-treated C4-deficient guinea pigs had markedly accelerated mortality; they were all dead within 4 hours, whereas normal animals did not begin dying until 8 hours. May et al (1972) then examined the effect of variously treated sera on endotoxin subsequently injected into cobra factor-treated, C4-deficient pigs. Endotoxin previously incubated with normal guinea pig serum was more slowly lethal than endotoxin incubated with saline, cobra factor-treated serum, or C4-deficient serum. These data, therefore, also strongly suggest that in guinea pigs, as in rabbits (Johnson and Ward 1971, 1972), serum is protective against lethal effects of endotoxin. Drake et al (1976) have confirmed these observations. They depleted guinea pigs of C3 and terminal components with cobra venom factor and antiserum to C3. The average time to death was shortened in C3-deficient animals, the degree of shortening being related to the residual C3 concentration.

4.4. Other animals

In a cat model of endotoxin shock, Kitzmiller et al (1972) demonstrated that endotoxin produced a moderate hypocomplementemia and concomitant hypotension. The initial phases of hypotension, but not the later phases or mortality, were inhibited by prior decomplementation. Their results were similar to those obtained in the canine model by most investigators.

Other investigators have confirmed that complement activation can be documented to follow endotoxin injection in rats (Füst and Fóris 1970) and newborn pigs (Miler et al 1966). In the latter study, results suggested that the hypocomplementemia required antibody to endotoxin.

5. STUDIES IN MAN

McCabe (1973) measured C3 levels by radial immunodiffusion in blood obtained at the time of onset of fever from patients later proven to have gram-negative bacteremia. Bacteremic patients had essentially the same mean value as did a control group of noninfected patients, but the range of values was considerably greater. Values in some bacteremic patients were more than 2 standard deviations below the control mean. When the bacteremic group was subdivided into those without complications, those manifesting shock, and those dying, the latter two groups had markedly decreased mean C3 levels and the range of values was skewed downward, whereas the uncomplicated bacteremic patients had a normal mean and a widened but not skewed range. Thus, hypocomplementemia correlated not only with bacteremia but with complications and worsened prognosis. These observations do not imply any causal relationship, although that would be a possible explanation. A major determinant of complications and prognosis is a dose of organisms or endotoxin; likewise, the degree of complement activation can also be related to dose effects (Snyderman et al 1971). Thus, hypocomplementemia likely reflects greater bacteremia and endotoxemia.

These observations were extended to include measurements of additional complement proteins (Fearon et al 1975). Patients with gram-negative bacteremia had levels of C1q and C1s below normal and normal or supranormal levels of most other components. However, patients in whom shock developed had low levels of C3 and all patients with gram-negative bacteremia had low levels of properdin. Of 19 patients with shock, 11 had properdin levels more than 2 standard deviations below normal. These results suggested preferential activation of the alternative pathway. Other investigators have also confirmed the highly variable changes in serum complement levels during gram-negative bacteremia and the tendency of patients with shock to have the lowest levels (Füst et al 1976).

In another study, patients scheduled to undergo genitourinary manipulation were evaluated for changes in peripheral vascular resistance, complement, and prekallikrein. Eight patients had no decline in resistance; of these, none had bacteremia or endotoxemia (as measured by the Limulus lysate test), one had decrease in prekallikrein, and one had a decline of more than 10% in factor B levels. Six of 10 patients with decreased vascular resistance had bacteremia (one patient, *Staphylococcus aureus*) or endotoxemia (positive Limulus lysate). Seven of these 10 had decreased prekallikrein postoperatively and 4 had decreased C3 or factor B levels (Robinson et al 1975). Leaving aside the question of the validity of the Limulus test (see chapter by Jorgensen in Volume 4), this study suggests a correlation among the findings of complement activation, kallikrein activation, and endotoxin-induced hypotension. As with other studies in man, however, no causal relationship can be inferred and the results are intriguing only.

6. CONCLUSIONS

It has been amply demonstrated in vitro and in vivo in a number of species that endotoxins of various gram-negative bacilli can activate the complement system by the classical and the alternative pathways. Antibody-dependent and -independent mechanisms have been demonstrated. The activation of complement by endotoxins is generally dose-dependent. Potentially phlogistic complement molecules are released during activation. These molecules include C5a and C3a, both of which can cause histamine release by basophils, which in turn can lead to vasodilatation. These molecules thus have the theoretical potential, if activated in sufficient quantity, to produce hypotension. Thus, it is possible that complement activation by endotoxin is important in the syndrome of endotoxin shock.

Ulevitch and Cochrane (1978) have suggested that there are multiple species of endotoxin released into the circulation during infection with gram-negative bacteria. Some of these forms of endotoxin are rapidly cleared by the reticuloendothelial system or inactivated by various biological protection systems (perhaps including complement). Some endotoxin species may bind to host cells to produce more prolonged and probably ultimately more serious effects, and some species interact immediately with proteolytic enzyme systems to generate short-lived mediators of inflammation. This interpretation is most consistent with the data as reviewed here.

There does appear to be a very early, transient, relatively benign hypotensive response to endotoxin injection which can be related to complement activation; the kinin system may or may not be involved in this reaction. The later, prolonged and potentially fatal hypotension appears not to be related to complement activation. Having questioned any major mediating role for complement in endotoxin shock, one must point out that contradictory data have not been explained away. In particular, one must point to the studies of Garner et al (1974) in the canine model.

It also seems clear that, in some animal models, primarily rabbits and guinea pigs, the complement system has a major protective effect against otherwise lethal doses of endotoxins. Whether these observations are species-specific has not been determined. It should be mentioned that so far nothing has suggested that people with congenital deficiencies of C3 or C5 or C6 are unduly susceptible to the hypotensive or lethal effects of endotoxin.

The role of complement in mediating or, alternatively, protecting against endotoxin shock remains unclear. The bulk of evidence does not favor a major role. Other aspects of endotoxemia and gram-negative sepsis, particularly pulmonary manifestations, may of course be more clearly related.

7. PROPOSED FUTURE WORK

There does not appear to be any need for additional studies demonstrating that endotoxin can activate complement or that complement activation can be observed during endotoxemia and endotoxin shock. These points have been well made. The essential questions to be answered are: (a) Does complement activation play a causal role in the syndrome of endotoxin shock? (b) Does the complement system protect against some of the lethal effects of endotoxin? (c) Does the final manifestation of endotoxemia represent the balance between both effects of complement? (d) If complement plays an important role in endotoxin shock, does it do so by interaction with other host inflammatory systems? (e) Can the complement system be manipulated in a clinically relevant way in order to improve the condition of the patient?

Some observations need to be further tested and discrepancies or contradictions need to be evaluated. The studies of Garner et al (1974) stand in contrast to the general experience that complement depletion affects endotoxin shock in relatively modest ways. These studies should be confirmed or explained. Further attention should be directed to possible protective roles of complement (Drake et al 1976; Johnson and Ward 1971, 1972). Are these observations peculiar to the rabbit and guinea pig models?

In general, more careful analysis is needed of the variations which can be attributed to the source and preparation of the lipopolysaccharide, the dose administered, and the animal species used. Much of the discrepancy in the literature can probably be explained by such analysis.

Multidisciplinary approaches are clearly needed. It is highly unlikely that any effects of endotoxin are directly attributable to the complement acting independently of other host systems, such as the kallikrein system, coagulation and fibrinolytic systems, lymphocytes, or polymorphonuclear leukocytes and macrophages. Major advances in our understanding of the syndrome of endotoxin shock are most likely to come from studies by investigative groups capable of evaluating the syndrome from multiple viewpoints simultaneously.

Finally, as pharmacological manipulation of complement and other systems becomes available, it may be possible to provide more definitive analyses of the various factors involved in endotoxin shock. Development of pharmacological agents capable of modulating the complement system is currently an area in need of greater attention.

REFERENCES

Aasen AO, Mellbye OJ, Ohlsson K (1978) Complement activation during subsequent stages of canine endotoxin shock. *Scand. J. Immunol. 8*, 509-513.
Bladen HA, Evans RT, Mergenhagen SE (1966) Lesions in *Escherichia coli* mem-

branes after actions of antibody and complement. *J. Bacteriol. 91*, 2377-2381.

Bladen HA, Gewurz H, Mergenhagen SE (1967) Interactions of the complement system with the surface and endotoxic lipopolysaccharides of *Veillonella alcalescens. J. Exp. Med. 125*, 767-786.

Cooper NR, Morrison DC (1978) Binding and activation of the first component of human complement by the lipid A region of lipopolysaccharides. *J. Immunol. 120*, 1862-1868.

Dierich MP, Bitter-Suermann D, König W, Hadding U, Galanos C, Rietschel ET (1973) Analysis of bypass activation of C3 by endotoxic LPS and loss of this potency. *Immunology 24*, 721-733.

Drake WP, Pokorney DR, Kopyta LP, Mardiney MR Jr (1976) In vivo decomplementation of guinea pigs with cobra venom factor and anti-C3 serum: analysis of the requirement of C3 and C5 for the mediation of endotoxin-induced death. *Biomed. Express (Paris) 25*, 91-94.

Fearon DT, Austen KF (1980) The alternative pathway of complement – a system for host resistance to microbial infection. *N. Engl. J. Med. 303*, 259-263.

Fearon DT, Ruddy S, Schur PH, McCabe WR (1975) Activation of the properdin pathway of complement in patients with Gram-negative bacteremia. *N. Engl. J. Med. 292*, 937-940.

Fine DP (1974) Activation of the classic and alternate complement pathways by endotoxin. *J. Immunol. 112*, 763-769.

From AHL, Gewurz H, Gruninger RP, Pickering RJ, Spink WW (1970) Complement in endotoxin shock: effect of complement depletion on the early hypotensive phase. *Infect. Immun. 2*, 38-41.

Füst G, Fóris G (1970) Newer data on the role of complement system in toxicity of endotoxin. *Experientia 26*, 1362-1363.

Füst G, Keresztes M (1969) Effect of endotoxin on the serum level of complement components. II. Effect of endotoxin on dog serum complement levels in vivo and in vitro. *Acta Microbiol. Acad. Sci. Hung. 16*, 135-147.

Füst G, Petrás G, Ujhelyi E (1976) Activation of the complement system during infections due to Gram-negative bacteria. *Clin. Immunol. Immunopathol. 5*, 293-302.

Füst G, Bertók L, Juhász-Nagy S (1977) Interactions of radio-detoxified *Escherichia coli* endotoxin preparations with the complement system. *Infect. Immun. 16*, 26-31.

Galanos C, Rietschel ET, Lüderitz O, Westphal O (1971) Interaction of lipopolysaccharides and lipid A with complement. *Eur. J. Biochem. 19*, 143-152.

Garner R, Chater BV, Brown DL (1974) The role of complement in endotoxin shock and disseminated intravascular coagulation: experimental observations in the dog. *Br. J. Haematol. 28*, 393-401.

Gewurz H, Mergenhagen SE, Nowotny A, Phillips JK (1968a) Interactions of the complement system with native and chemically modified endotoxins, *J. Bacteriol. 95*, 397-405.

Gewurz H, Snyderman R, Shin HS, Lichtenstein L, Mergenhagen SE (1968b) Complement (C′) consumption by endotoxic lipopolysaccharide (LPS) in immunoglobulin deficient sera. *J. Clin. Invest. 47*, 39a.

Gilbert VE, Braude AI (1962) Reduction of serum complement in rabbits after injection of endotoxin. *J. Exp. Med. 116*, 477-490.

Gunn WC (1914) The variation in the amount of complement in the blood in some acute infectious diseases and its relation to the clinical features. *J. Pathol. Bacteriol. 19*, 155-181.

Hugli TE, Müller-Eberhard HJ (1978) Anaphylatoxins: C3a and C5a. *Adv. Immunol. 26*, 1-53.

Johnson KJ, Ward PA (1971) Protective function of C6 in rabbits treated with bacterial endotoxin. *J. Immunol. 106*, 1125-1127.

Johnson KJ, Ward PA (1972) The requirement for serum complement in the detoxification of bacterial endotoxin. *J. Immunol. 108*, 611-616.

Kitzmiller JL, Lucas WE, Yelenosky PF (1972) The role of complement in feline endotoxin shock. *Am. J. Obstet. Gynecol. 112*, 414-421.

Lichtenstein LM, Gewurz H, Adkinson NF Jr, Shin HS, Mergenhagen SE (1969) Interactions of the complement system with endotoxic lipopolysaccharide: the generation of an anaphylatoxin. *Immunology 16*, 327-336.

Loos M, Bitter-Suermann D, Dierich M (1974) Interaction of C1, C2, and C4 with different preparations of bacterial lipopolysaccharides and with lipid A. *J. Immunol. 112*, 935-940.

McCabe WR (1973) Serum complement levels in bacteremia due to Gram-negative organisms. *N. Engl. J. Med. 288*, 21-23.

May JE, Kane MA, Frank MM (1972) Host defense against bacterial endotoxemia. Contribution of the early and late components of complement to detoxification. *J. Immunol. 109*, 893-895.

Mergenhagen SE, Snyderman R, Gewurz H, Shin HS (1969) Significance of complement to the mechanism of action of endotoxin. *Curr. Topics Microbiol. Immunol. 50*, 37-77.

Miler I, Tlaskalova H, Kostka J, Jilek M (1966) Effect of endotoxin on the complement level in sera of precolostral newborn pigs. *Folia Microbiol. 11*, 475-478.

Morrison DC, Kline LF (1977) Activation of the classical and properdin pathways of complement by bacterial lipopolysaccharides (LPS). *J. Immunol. 118*, 362-368.

Morrison DC, Ulevitch RJ (1978) The effects of bacterial endotoxins on host mediation systems. *Am. J. Pathol. 93*, 525-618.

Muschel IH, Schmoker K, Webb PM (1964) Anticomplementary action of endotoxin. *Proc. Soc. Exp. Biol. Med. 117*, 639-643.

Pearlman DS, Sauers JB, Talmage DW (1963) The effect of adjuvant amounts of endotoxins on the serum hemolytic complement activity in rabbits. *J. Immunol. 91*, 748-756.

Phillips JK, Snyderman R, Mergenhagen SE (1972) Activation of complement by endotoxin: a role for gamma-2 globulin, C1, C4 and C2 in the consumption of terminal complement components by endotoxin-coated erythrocytes. *J. Immunol. 109*, 334-341.

Pillemer L, Blum L, Lepow IH, Ross OA, Todd EW, Wardlaw AC (1954) The properdin system and immunity. I. Demonstration and isolation of a new serum protein, properdin, and its role in immune phenomena. *Science 120*, 279-285.

Pillemer L, Schoenberg MD, Blum L, Wurz L (1955) The properdin system and immunity. II. Interaction of the properdin system with polysaccharides. *Science 122*, 545-549.

Rampart M, Bult H, Herman AG (1981) Complement activation and blood levels of 6-oxo-prostaglandin F1-alpha during endotoxin-induced hypotension in rabbits. *Arch. Int. Pharmacodyn. 249*, 328-329.

Rapp HJ, Borsos T (1970) *Molecular Basis of Complement Action*. Appleton-Century-Crofts, New York.

Robinson JA, Klodnycky ML, Loeb HS, Racic MR, Gunnar RM (1975) Endotoxin, prekallikrein, complement and systemic vascular resistance. Sequential measurements in man. *Am. J. Med. 59*, 61-67.

Snyderman R, Gewurz H, Mergenhagen SE (1968) Interaction of the complement system with endotoxic lipopolysaccharide. Generation of a factor chemotactic for polymorphonuclear leukocytes. *J. Exp. Med. 128*, 259-275.

Snyderman R, Gewurz H, Mergenhagen SE, Jensen J (1971) Effect of C4 depletion on the utilization of the terminal components of guinea-pig complement by endotoxin. *Nature (London) New Biol. 231*, 152-154.

Spink WW, Vick J (1961) A labile serum factor in experimental endotoxin shock: cross-transfusion studies in dogs. *J. Exp. Med. 114*, 501-508.

Spink WW, Davis RB, Potter R, Chartrand S (1964) The initial stage of canine endotoxin shock as an expression of anaphylactic shock: studies on complement titers and plasma histamine concentrations. *J. Clin. Invest. 43*, 696-704.

Ulevitch RJ, Cochrane CG (1978) Role of complement in lethal bacterial lipopolysaccharide-induced hypotensive and coagulative changes. *Infect. Immun. 19*, 204-211.

Ulevitch RJ, Cochrane CG, Henson PM, Morrison DC, Doe WF (1975) Mediation systems in bacterial lipopolysaccharide-induced hypotension and disseminated intravascular coagulation. I. The role of complement. *J. Exp. Med. 142*, 1570-1590.

Handbook of Endotoxin, Vol. 2: Pathophysiology of Endotoxin
L.B. Hinshaw, editor
© Elsevier Science Publishers B.V., 1985
ISBN 0 444 90385 2
$0.85 per article per page (transactional system)
$0.20 per article per page (licensing system)

CHAPTER 7

Adrenergic aspects of endotoxin shock*

H. RICHARD ADAMS, SETH D. IZENBERG, AND CHARLES
R. BAXTER

1. INTRODUCTION

Despite over three decades of intensive clinical and experimental study, the precise involvement of sympathoadrenal mechanisms in the mammalian response to bacterial endotoxins has remained an unsettled issue. Conflicting conclusions derived by different investigative groups no doubt reflect the myriad of animal species, endotoxin sources, and experimental protocols used over the years. In addition, however, recent major discoveries in autonomic physiology-pharmacology have shown clearly that control mechanisms operative in adrenergic neurohumoral transmission are exceedingly more complex than originally surmised (Shepherd and Vanhoutte 1981). Furthermore, these emerging concepts have considerable application to the interpretation of data derived from experimental study of sympathoadrenal involvement in endotoxin shock and other forms of shock.

The α-class of adrenergic receptors, for example, actually comprises two distinct subtypes now classified as α_1 and α_2. The two types of α-adrenergic receptors can subserve opposing physiologic responses in the same organ, and yet both types are susceptible to pharmacologic blockade by those α-receptor antagonists used historically and also currently in shock research (Adams 1983a). Accordingly, if specific effector responses associated with the α_1- and α_2-receptor subtypes are not distinguished by the investigator, interpretation of data may be oversimplified at best or incorrect at worse.

This chapter addresses these types of problems through, first, a brief overview of general ideas about sympathoadrenal involvement in endotoxin shock, and secondly, a more detailed synthesis of current concepts about physiologic control of communication pathways between noradrenergic

* The work presented in this chapter was supported in part by National Institutes of Health grant number GM-21681, and by postdoctoral fellowship HL-06796 awarded to S. Izenberg.

neurons and effector cell elements. Lastly, these concepts are incorporated into a summary discussion of selected reports that typify experimental evidence imputing adrenergic mechanisms in the pathogenesis of endotoxin shock.

2. OVERVIEW OF SYMPATHOADRENAL INVOLVEMENT IN ENDOTOXIN SHOCK

The sympathetic division of the autonomic nervous system has long been implicated in both the physiologic-compensatory phase and the pathophysiologic-decompensatory phase of the hemodynamic response evoked by gram-negative septicemia and its experimental counterpart, endotoxin shock. This important concept evolved along several lines of evidence, including observations that either surgical or chemical sympathectomy or pharmacologic blockade of sympathetically-mediated effector cell responses altered the outcome of experimental endotoxicosis in different mammalian species (Bolton and Atuk 1978; Filkins 1979; Palmerio et al 1963; Pardini et al 1982). Importantly, since drugs that blunted α-adrenergic responses also prevented or delayed death in shock, it was concluded that α-receptor activation by endogenous adrenergic stimulants must be an underlying participant in progression of endotoxin shock to the irreversible stage (Lillehei 1979; Nickerson and Gourzis 1962).

Indeed, it is now generally accepted that circulatory-metabolic hypodynamics of shock activate autonomic reflex adjustments that entail an acceleration of efferent impulse traffic over sympathetic vasomotor neurons with a concomitant increase in catecholamine release from the adrenal glands. Resulting increments in the availability of norepinephrine and epinephrine lead to α-mediated vasoconstriction, thereby promoting an increase in peripheral vascular resistance. This response would tend to oppose the early hypotensive effects of endotoxin and no doubt represents an endogenous compensatory attempt to improve perfusion pressure and circulation to vital organs. If unabated, however, α-mediated constriction of nutrient blood vessels can lead to intensification of the stagnant hypoxia and attendant metabolic consequences already underway as part of the shock syndrome. This self-potentiating or decompensatory α-mediated sequence is thought to be manifested especially in viscerocutaneous tissues (Lillehei 1979), and recent evidence suggests its concurrent operation even in more vital regions such as the myocardium (Birinyi et al 1977).

Many details remain unresolved, however, and we believe that part of the problem can be traced to the complexity of cell-to-cell interchange between the sympathetic nervous system and its innervated organs. A clear under-

146

standing of physiologic regulation of information transmission across noradrenergic neuroeffector junctions is essential for the definition of pathophysiologic derangement of these processes during endotoxin shock.

3. PHYSIOLOGIC ASPECTS OF ADRENERGIC MECHANISMS

3.1. Noradrenergic neurohumoral transmission

Historically, the principal events responsible for information transmission across noradrenergic neuroeffector junctions were believed to comprise the biosynthesis and storage of norepinephrine in the neuron terminals, its exocytotic discharge from the neuron, and its activation of receptors on the innervated cells (Rand et al 1980). The neuron endings served further by active reuptake of norepinephrine, by which the availability of transmitters at the postjunctional receptors was decreased. Pharmacologically, adrenergic agonists activated effector cell receptors and associated cellular changes, while adrenergic antagonists blocked the receptors and thereby prevented or inhibited receptor-linked effector cell responses.

This model, however, could not satisfactorily account for the puzzling observation that adrenergic agonists and antagonists also exerted specific influences on norepinephrine release from the neuron (Brown and Gillespie 1958; Langer 1974; Rand et al 1980). These pharmacological findings were explained by the presence of prejunctional adrenergic receptors that subserved physiologic modulation of mechanisms for the release of transmitters (Langer 1974, 1977). Identification of adrenoceptor populations on neuron endings not only refined the basic model of noradrenergic transmission, but it also prompted a reevaluation of complete pharmacodynamic profiles of drugs that interact with adrenergic receptors.

3.2. Adrenergic receptors

The basic physiologic function of adrenergic receptors is to recognize and interact with norepinephrine and epinephrine, the endogenous adrenergic mediators. Interaction of agonists with either the α- or β-class of adrenoceptors activates a complex series of intracellular events. These perturbations culminate in characteristic changes in physiologic activity of the effector cell, depending upon the particular tissues and prevalence of α- or β-receptor populations. In vascular smooth muscle, for example, α-receptor activation by norepinephrine or epinephrine leads to active vasoconstriction.

Adrenergic blocking drugs also exert their pharmacologic effects through interaction with either the α- or β-class of receptors. In contrast to agonist activity, however, binding of antagonist to the receptor does not elicit receptor activation, but makes critical binding sites of the receptor inaccessible to other ligands. Adrenergic mediators are thereby prevented from affixing to the receptor, and their physiologic-pharmacologic effects are abolished or attenuated. In vascular smooth muscle contracted by epinephrine or norepinephrine, for example, α-antagonists evoke passive vasodilation because of withdrawal of α-vasoconstrictor tone.

However, a unimodal description of drug interaction with effector cell receptors is no longer sufficient to explain fully the pharmacodynamics of adrenergic drugs. Direct and indirect influences of these compounds on other adrenoceptor populations, both α and β, should also be considered.

3.2.1. *α- and β-adrenoceptors of noradrenergic nerve terminals*

Studies in the last decade have clearly established the existence of receptor-coupled mechanisms operating within autonomic nerve endings, but uncertainties remain about precise physiologic roles of all prejunctional receptor types purportedly discovered. Relative to peripheral noradrenergic axons, there is convincing evidence that norepinephrine release is susceptible to acceleration or deceleration by a wide range of locally produced and circulating neurotransmitters, hormones, and autacoids. Some of these factors have been reviewed (Langer 1977; Rand et al 1980; Shepherd and Vanhoutte 1981) and are listed in Table 1 along with associated receptors and effects on norepinephrine discharge from the neuron.

Although the significance was not fully appreciated at first, studies in 1958 by Brown and Gillespie showed that an α-adrenergic blocking drug (phenoxybenzamine) could enhance release of norepinephrine from sympathetic neurons within the spleen. Further investigation provided the related finding that an α-agonist could actually suppress release of norepinephrine from the neuron. Thus evolved the concept of prejunctional α-adrenoceptors that subserve an important autoinhibitory effect on norepinephrine release mechanisms (Langer 1974; Table 1). This response is mediated through the calcium-dependent link in the stimulus–secretion coupling mechanism of the axon terminal, and it results in decreased exocytosis of norepinephrine storage vesicles (Exton 1981; Graham 1981). The physiologic function of α-mediated prejunctional events is envisioned as a local servo-mechanism through which norepinephrine can govern its own release once a threshold intrajunctional concentration of transmitter has been exceeded.

Epinephrine can also activate prejunctional α-receptors with a potency approximately equal to that of norepinephrine (Rand et al 1980). Interestingly, however, low concentrations of epinephrine, equivalent perhaps to

148

TABLE 1   *Prejunctional receptors of noradrenergic neuroeffector junctions*

Receptor	Endogenous ligand	Influence on norepinephrine release
α_2-Adrenoceptor	Norepinephrine	Inhibitory
β_2-Adrenoceptor	Epinephrine	Facilitatory
Muscarinic cholinergic	Acetylcholine	Inhibitory
Nicotinic cholinergic	Acetylcholine	Facilitatory
Dopaminergic	Dopamine	Inhibitory
H_2-Histaminergic	Histamine	Inhibitory
Angiotensin	Angiotensin II	Facilitatory
Serotonergic	Serotonin	Inhibitory
Opioid	Enkephalins	Inhibitory
Prostaglandin (?)	Prostaglandin E_2	Inhibitory
Prostaglandin (?)	Prostaglandin $F_{2\alpha}$	Facilitatory

those associated with increased release from the adrenals, may actually enhance norepinephrine discharge. This facilitatory action is shared by the β-agonist isoprenaline and prevented by β-antagonists (Adler-Graschinsky and Langer 1975). Thus, noradrenergic nerve endings also seem to possess β-receptors which subserve a facilitatory effect on transmitter release, precisely the opposite response as that associated with prejunctional α-receptors (Adler-Graschinsky and Langer 1975; Rand et al 1979).

3.2.2. *'Cotransmitter' theory*

Norepinephrine itself seems to have only slight influence on prejunctional β-receptors, apparently because this receptor population is representative of β_2 rather than β_1 subtype (Stjärne and Brundin 1976). Thus, the α-controlled autoinhibitory cycle probably dominates during usual noradrenergic transmission. Under some conditions, however, epinephrine arising from the adrenals may be taken up by noradrenergic nerve endings and subsequently released as a cotransmitter along with resident norepinephrine (Rand et al 1979). Conversion of noradrenergic nerves to partly adrenergic nerves may set the stage for activation of the β_2 autofacilitatory loop owing to neuronal release of newly stored epinephrine. Figure 1 displays a model of noradrenergic neurotransmission incorporating prejunctional α- and β-receptor subtypes modulating mechanisms for norepinephrine release.

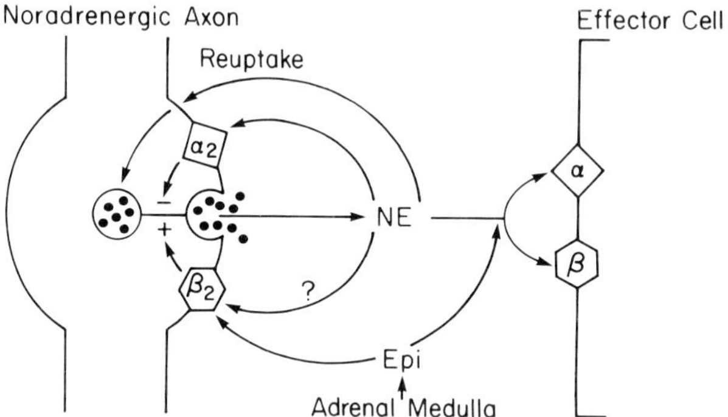

Fig. 1 *Schematic diagram of a peripheral noradrenergic neuroeffector junction with the axon terminal varicosity on the left and effector cell on the right. Norepinephrine (NE) released from the neuron interacts postjunctionally with α- or β₁-adrenoceptors of effector cells and interacts also with prejunctional α-adrenoceptors (α₂ subtype) to inhibit further release of NE. Prejunctional β-adrenoceptors (β₂ subtype) subserve a facilitatory effect on NE release but it is questionable (?) whether NE itself activates this facilitatory feedback loop. Epinephrine (Epi) from the adrenals can interact with prejunctional and postjunctional α- and β-adrenoceptors. NE is removed from the junctional cleft by diffusion, extraneuronal uptake, and active uptake (Reuptake) into the nerve terminal where it is metabolized or reincorporated into the transmitter storage vesicles. There is some evidence that Epi can be taken up by the noradrenergic neuron, stored, and subsequently released as a cotransmitter along with resident NE (not shown). Reprinted from Adams (1983a), with permission.*

3.2.3. α_1-α_2 *Adrenoceptor subtypes*

Prejunctional α-receptors respond differently to some agonists and antagonists when compared with the classical α-receptor of innervated cells. Differences in pharmacologic responsiveness led to the subdivision of α-adrenoceptors into α_1 and α_2 classes (Berthelson and Pettinger 1977; Langer 1974; U'Prichard and Snyder 1977), somewhat analogous to the original pharmacologic classification of α- and β-receptors by Ahlquist (1948) and the later demarcation of β_1 and β_2 subtypes by Lands et al (1967). Subsequent radioligand binding studies with membrane fractions indeed confirmed the existence of different adrenoceptor populations with characteristic orders of binding affinities for adrenergic compounds (Hoffman and Lefkowitz 1980; U'Prichard and Snyder 1979).

Classical α-receptors of effector cells are now placed in the α_1 subtype; they are blocked more potently by prazosin than yohimbine and generally show high responsiveness to the agonists phenylephrine and methoxamine, but less responsiveness to clonidine and corbadrine. Prejunctional α-receptors belong to the α_2 subtype (Fig. 1); they are blocked potently by yohimbine, and have high affinity for clonidine and corbadrine but low affinity for prazosin, phenylephrine, and methoxamine (Exton 1981; Hoffman and Lefkowitz 1980). It should be emphasized that α_2-receptors are not necessarily restricted to prejunctional elements; they are also present on some effector cell types (Exton 1981; U'Prichard and Snyder 1979).

Table 2 lists several tissues where α_1- and/or α_2-receptor subtypes have been identified, and Table 3 ranks the order of selectivity of several an-

TABLE 2 *Some tissue distributions of α-adrenoceptor subtypes*

Effector organ	Receptor subtype	Effector response
Noradrenergic nerve ending	α_2	Inhibition of norepinephrine release
Platelets	α_2	Aggregation
Kidney	α_2	Inhibition of renin release
Adipose tissue	α_2	Antilipolysis
Pancreatic islets	α_2	Inhibition of insulin secretion
Central nervous system	α_1 and α_2	Numerous
Vascular smooth muscle	α_1 and α_2	Vasoconstriction
Liver	α_1	Glycogenolysis
Uterus	α_1	Smooth muscle contraction
Vas deferens	α_1	Smooth muscle contraction

tagonists relative to their blocking activities at the α-receptor subtypes (Exton 1981; Graham 1981; Starke and Docherty 1980). These tabulations should be considered only as working models because ongoing research continually reveals important species and tissue exceptions to adrenoceptor distributions and associated effector mechanisms. Nevertheless, it now seems fairly clear that two distinct α-receptor subtypes are distributed in different tissues where they may subserve similar, dissimilar, or even opposing responses of effector organs.

3.3. Adrenoceptor-coupled effector mechanisms

3.3.1. β-Receptors

The biochemical pathways linking cell surface receptors with complex functional changes in intracellular organelles have been relatively well mapped for the β-adrenergic receptor–adenylate cyclase–cyclic AMP triad. Interac-

TABLE 3 *Relative order of selectivity of several α-antagonists for α_1- and α_2-adrenoceptor subtypes*

Prazosin	α_1
Corynanthine	
WB 4101	
Phenoxybenzamine	
Clozapine	
Phentolamine	
Piperoxan	
Tolazoline	
Yohimbine	
Rauwolscine	α_2

α_1-Blocking activity diminishes and α_2-blocking activity increases as the list is traversed from top to bottom. Source: Starke and Docherty (1980).

tion of agonist ligands with either the β_1- or β_2-adrenoceptor subtype elicits an increase in adenylate cyclase activity that, in turn, accelerates the synthesis of cyclic $3',5'$-adenosine monophosphate (cyclic AMP). With rare exceptions, cellular responses to β-receptor stimulation can be traced to this increase in cyclic AMP. However, although there is ample evidence for a cause–effect relationship between β-receptor stimulation and cyclic AMP synthesis, the biomolecular events responsible for the expression of this relationship are exceedingly complex.

The β-receptor-linked adenylate cyclase system actually comprises at least three principal and closely affiliated membrane proteins, namely, the β-adrenergic receptor itself, a guanine nucleotide-binding regulatory protein (referred to as G/F or N), and a catalytic protein (C) that is relatively inactive under basal conditions (Ross and Gilman 1980). The β-receptor does not interact directly with the catalytic unit; instead, the former communicates with the latter by using the guanine nucleotide-binding protein as an intermediary signal (Levitzki 1982; Northup et al 1982). The basic steps involved in this progression are outlined below and schematized in Figure 2.

The β-receptors (R_β) are localized in the plasma membrane with their ligand binding sites oriented externally; they serve by recognizing and binding the appropriate β-agonist or hormone (H), yielding a ligand receptor complex ($H \cdot R_\beta$). The latter combines with G/F_{GDP} to form an intermediate complex of $H \cdot R_\beta \cdot G/F_{GDP}$. However, interaction of G/F_{GDP} with the hormone-receptor complex promotes release of GDP, with its subsequent replacement by GTP (Fig. 2). The resulting ternary complex of $H \cdot R \cdot G/F_{GTP}$ dissociates to $H \cdot R_\beta$ and G/F_{GTP}, and the latter product now binds to the catalytic protein. This reaction yields the active form of adenylate cyclase

152

designated as $G/F_{GTP} \cdot C$, which catalyzes the conversion of adenosine triphosphate (ATP) to cyclic AMP. A GTPase activity resides with G/F, and activation of the catalytic protein is terminated by hydrolysis of G/F-bound GTP to yield G/F_{GDP} and inorganic phosphate (P_i). The C moiety is again relatively quiescent until reactivated by another cycle of $H \cdot R_\beta$ interaction (Fig. 2).

3.3.2. α-Receptors

Compared with the β-adrenoceptor-linked effector mechanism, much less is known about the intracellular messengers or signals generated through agonist binding with the α class of receptors. Interestingly, whereas β-receptor stimulation activates adenylate cyclase, there is increasing evidence that the α_2 subgroup of adrenergic receptors subserves an inhibitory influence on adenylate cyclase in some cell types. For example, the proaggregate effect of α-agonists in thrombocytes is believed to be mediated through an α_2-linked reduction in adenylate cyclase activity; this would impede platelet synthesis of cyclic AMP, a potent antiaggregate (Exton 1981). A guanine nucleotide-binding regulatory protein seems to be involved in α_2-inhibition of adenylate cyclase, just as a similar (or the same?) guanine nucleotide-binding protein

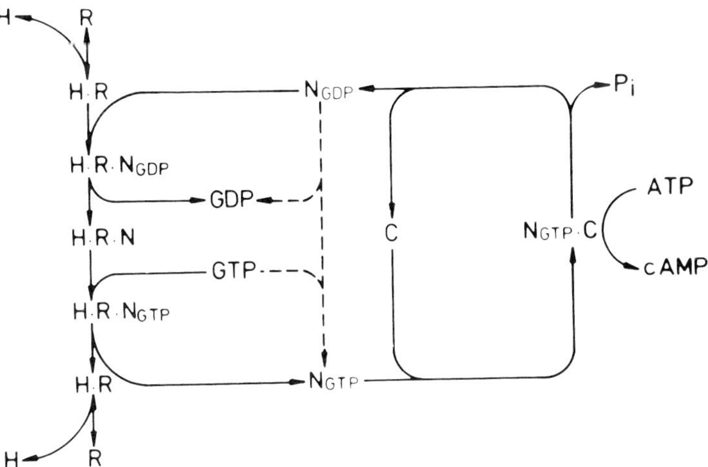

Fig. 2 *β-Adrenergic regulation of adenylate cyclase activity. H = hormone (β-adrenergic agonist); R = β-adrenoceptor; N = guanine nucleotide-binding regulatory protein (referred to as G/F in the text); C = catalytic subunit (adenylate cyclase). For details see text. Reprinted from Jakobs and Schultz (1982), with permission.*

is involved in β-receptor stimulation of the cyclase enzyme. The suggestion was made that α_2-agonists may activate a GTPase, thereby promoting breakdown of the active form of adenylate cyclase, that is, α_2-receptor activation leads to acceleration of the disassociation of $G/F_{GTP} \cdot C$ (active form) to G/F_{GDP} and C (inactive form) (Jakobs and Schultz 1982). Various data support this assumption, but isolation and reconstitution of the various components are needed before definitive conclusions can be reached about cause–effect relationships.

Relative to α_1-adrenoceptor subtype, stimulation of these receptors generally results in an increase in the ionized calcium (Ca^{2+}) concentration of the cytosol compartment. The mechanisms are still only partly understood, but there is evidence that α_1-receptor activation facilitates Ca^{2+} influx across the plasma membrane and also enhances release of this cation from intracellular storage sites. Increased free Ca^{2+} then serves as the primary intracellular signal linking α_1-receptor activation to those cellular changes subserved by this receptor type, for example stimulation of vascular smooth muscle contractile proteins.

The biochemical pathways from α_1-receptor stimulation to increased Ca^{2+} concentration are currently equivocal, but different investigators have engendered support for an association between α_1-receptors and increased turnover of membrane phosphatidylinositol. This putative cause–effect relationship exists in several cell types, including hepatocytes, adipocytes, pineal tissue, and vas deferens (Exton 1981; Jakobs and Schultz 1982; Rahwan 1983). Phosphatidylinositol (PI) is an acidic phospholipid component mainly of the inner leaflet of the plasma membrane bilayer. Metabolism of PI precedes the entry of Ca^{2+} into stimulated cells and may therefore serve as a universal biochemical event intrinsic to activation of Ca^{2+} passageways or 'Ca^{2+}-gates' through excitable membranes (Michell and Kirk 1981). The PI-dependent Ca^{2+}-gating mechanism is believed to involve the following sequence. A specific Ca^{2+}-dependent phospholipase C catalyzes the breakdown of PI into diacylglycerol and inositol phosphates, and resynthesis occurs through phosphorylation of diacylglycerol to phosphatidic acid which is then converted back to PI. This degradative-resynthesis cycle in some way promotes Ca^{2+} mobilization and is proposed to be dependent upon, or at least linked to, α-receptor stimulation. As pointed out by Jakobs and Schultz (1982), however, this theory has been challenged as it remains unclear how ligand binding to the receptor activates the phospholipase C or how the membrane PI-signal is transferred to internal Ca^{2+} stores.

Guanylate cyclase and cyclic guanosine 3′,5′-monophosphate (cyclic GMP) represent another enzyme-substrate pair purportedly associated with α_1-receptor mechanisms. However, whereas PI metabolism precedes increased Ca^{2+} concentrations, guanylate cyclase activation and increased

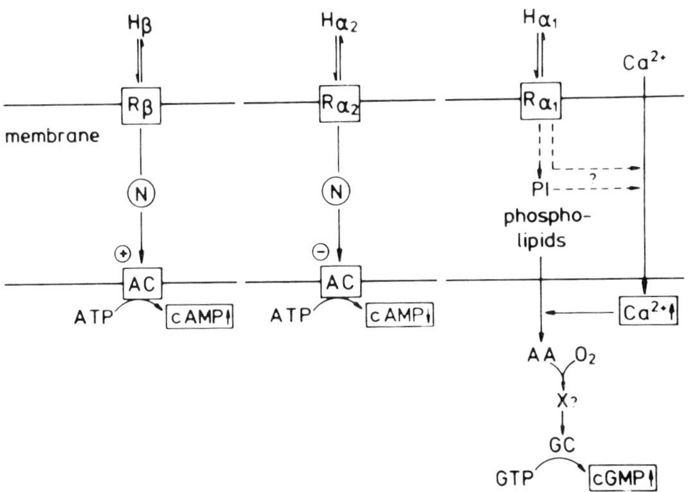

Fig. 3 *Intracellular signals and signal generating systems coupled to subtypes of adrenoceptors. H = hormone (adrenergic agonist); R = adrenoceptor; N = guanine nucleotide-binding regulatory protein; AC = adenylate cyclase; PI = changes in phosphatidylinositol metabolism (PI response); AA = arachidonic acid; GC = guanylate cyclase. Reprinted from Jakobs and Schultz (1982), with permission.*

synthesis of cyclic GMP seem to follow the increase in Ca^{2+}. Ca^{2+}-stimulated release of arachidonic acid or related compounds may precede activation of the guanylate cyclase (Spies et al 1980). The illustrations shown in Figure 3 summarize the salient features currently believed to typify receptor-linked mechanisms for the β-, α_1-, and α_2-adrenoceptor groups.

4. PATHOPHYSIOLOGIC ASPECTS OF ADRENERGIC EVENTS IN ENDOTOXIN SHOCK

With the preceding schema of physiologic control of sympathetic neuro-transmission and adrenoceptor-linked events in effector cells, it should be possible to construct a pathophysiologic sequence that would incorporate sympathoadrenal participation into the endotoxin shock syndrome. As will be developed in the following sections, however, the precise story of endotoxin-induced changes in tissue disposition and cellular effects of catecholamines remains questionable.

4.1. Plasma and tissue catecholamines

Measurement of catecholamine concentrations in plasma and other tissues represents one of the earliest approaches taken by investigators in the search for adrenergic involvement in the pathogenesis of sepsis and endotoxemia. Results from numerous studies have provided the general observations that plasma concentrations of norepinephrine and epinephrine are elevated during endotoxin shock, while peripheral organ stores of norepinephrine undergo partial depletion (Rosenberg et al 1959; Spink et al 1966; Zetterström et al 1964; see Rosenberg et al 1961 for early references, and Pardini et al 1982 for newer references). The importance of these changes to endotoxin pathogenesis has been debated, and part of the controversy revolves perhaps around the relative insensitivity of the older catecholamine assay techniques and the study of sedated or anesthetized animals in the experimental setup.

Rosenberg et al (1959, 1961) used an early fluorimetric assay technique for measuring catecholamines and they observed that intravenous injection of a lethal quantity of dried *E. coli* organisms resulted in pronounced increments in plasma epinephrine and norepinephrine in morphine-sedated dogs. Plasma epinephrine increased from a basal (preshock) value of about 1.5 ng/ml to almost 30 ng/ml within 5 minutes after injection of the endotoxin source, representing about a 20-fold increase; whereas plasma norepinephrine values increased during the same interval from a basal level of 2.2 ng/ml to 27 ng/ml, representing about a 12-fold increase. Spink et al (1966), using a trihydroxyindole assay method for catecholamines, found that basal (preshock) plasma concentrations of epinephrine and norepinephrine in pentobarbital-anesthetized dogs varied considerably from of 0.2 to 1.0 ng/ml and from 0.04 to 1.6 ng/ml, respectively. Despite such variability in their control values, Spink et al (1966) reported that endotoxin-induced hypotension caused plasma epinephrine to rapidly become several times higher. In contrast to the earlier findings by Rosenberg et al (1961), Spink et al (1966) concluded that the early plasma catecholamine response to endotoxin was selective for epinephrine because it occurred without significant changes in plasma norepinephrine concentration.

Vick (1965) used a bioassay system for assessing the influence of endotoxin shock on circulating catecholamines and other vasoactive agents. This system was based on contractile responses of an isolated saphenous vein preparation superfused with donor blood from pentobarbital-anesthetized, heparinized, endotoxin-shocked dogs. Vick (1965) observed a biphasic contractile response of the blood-perfused venous smooth muscle after administration of endotoxin to the donor dog. The early contractile response (1–10 min post endotoxin) was blocked by antihistamines, while only the delayed contractile response (20–120 min post endotoxin) was prevented by either adrenalectomy or by the α-adrenoceptor-blocking drug phentolamine. Unlike Spink et al (1966) and Rosenberg et al (1961), Vick (1965) therefore concluded that the early circulatory response to endotoxin was dominated by histamine-dependent events rather than by a change in plasma catecholamines.

With the advent of more sensitive and more reliable catecholamine assay techniques, a clearer picture seems to be emerging. Feuerstein et al (1981), for example, recently applied an extremely sensitive radioenzymatic and chromatographic assay system to the measurement of plasma catecholamines in pentobarbital-anesthetized cats injected with *E. coli* endotoxin. From a basal (preshock) value of 42 ± 9 pg/ml ($\overline{X}\pm SE$), plasma epinephrine rapidly increased to 295 ± 87 pg/ml within 20 minutes following endotoxin injection; high plasma epinephrine values (> 200 ng/ml) were sustained throughout the remainder of this 2-hour experiment. On the other hand, plasma norepinephrine concentrations slowly but progressively increased from a basal level of 339 ± 37 pg/ml to a maximum of 1988 ± 276 pg/ml after 2 hours. The epinephrine and norepinephrine concentrations measured by Feuerstein et al (1981) were much lower and displayed considerably less intragroup variability than values obtained from most earlier reports, and this no doubt reflects the superior sensitivity of the radioenzymatic assay. On the negative side, however, this study lacked an important control, that is, these investigators did not report values for plasma catecholamines in pentobarbital-anesthetized cats for a 2-hour sham-shock interval. Hence, it is difficult to discern from their data whether changes in plasma catecholamines were due strictly to endotoxin alone, to the combination of endotoxin and anesthesia (a plausible explanation), or even to anesthesia alone. This same criticism of a lack of appropriate anesthesia control applies to many studies of plasma catecholamines in endotoxin shock (e.g., Prager et al 1975; Spink et al 1966). General anesthesia affects central and peripheral autonomic control centers and can modify the sympathoadrenal response of animals to shock-inducing stimuli (Farnebo et al 1979; Parker and Adams 1978).

Another important variable pertains to other components of experimental protocols commonly used in studies of the interactions between endotoxin and catecholamines. Heparinization is an extremely common and often routine procedure in endotoxin experiments with intact animals, and yet Devereux et al (1977) recently reported that heparin pretreatment itself can suppress the magnitude of the endotoxin-induced increase in plasma norepinephrine concentrations observed in pentobarbital-anesthetized dogs. This effect seemed specific for norepinephrine because heparin pretreatment did not significantly alter the increase in plasma epinephrine concentrations evoked by endotoxin in the same dogs. Unfortunately, this was a brief report without any detail about the intralaboratory reliability of the catecholamine assay procedures used by this group. Moreover, sham-shock controls for the influence of anesthesia were not reported.

Jones and Romano (1983) applied the radioenzymatic catecholamine assay for measuring plasma epinephrine and norepinephrine in unanesthetized rats, thereby circumventing secondary effects of general anesthesia. These investigators found basal plasma epinephrine and norepinephrine concentrations of about 100 and 250 pg/ml, respectively. Within 30 minutes after the injection of endotoxin, plasma epinephrine concentration became almost 30 times higher and norepinephrine about 5 times. Plasma catecholamines change little in sham-shock unanesthetized rats in this laboratory, and heparin is used only to maintain patency of in-dwelling vascular catheters (S.B. Jones, personal communication, 1983).

Zetterström et al (1964) isolated sympathetically innervated tissues from morphine-sedated dogs and found that lethal doses of *E. coli* endotoxin resulted in partial depletion of norepinephrine stores in heart (51% decrease), spleen (78% decrease) and liver (56% decrease). Essentially similar results have been observed in endotoxicosis caused by *E. coli* (Pohorecky et al 1972) and *S. enteritidis* (Pardini et al 1982) in the rat, and there is some evidence that more severe or late stages of endotoxin shock may be associated with quantitatively greater decrements in tissue norepinephrine content (Pardini et al 1982).

Depletion of peripheral stores of norepinephrine is often assumed to reflect increased release of physiologically active neurotransmitter to effector cell receptors subsequent to accelerated sympathetic neuronal discharge (Pardini et al 1982). However, there are important alternatives to explain the depletion of norepinephrine tissue stores. For example, reserpine is a drug that causes profound depletion of catecholamines from sympathetic neurons throughout the body and yet by no means causes increased transfer of norepinephrine to extraneuronal receptor sites. On the contrary, reserpine inhibits the uptake and storage of catecholamines in the intraneuronal storage vesicles and thereby causes norepinephrine leakage from the vesicles into the neuronal cytoplasm. Cytoplasmic norepinephrine is subsequently deaminated by monoamine oxidase into inactive metabolites prior to exit of the latter from the neuron. Reserpine is therefore classified as a sympatholytic drug because of its catecholamine-depleting action. Norepinephrine depletion observed during endotoxicosis could therefore theoretically reflect a decreased availability of norepinephrine for neuronal transfer to receptors instead of increased transfer to receptors as has been suggested. This problem was indirectly studied by Pohorecky et al (1972) and Rao et al (1972).

Pohorecky et al (1972) radioisotopically labeled tissue norepinephrine stores in rats by injecting tritiated norepinephrine (^{3}H-norepinephrine), and they concluded that endotoxin increases norepinephrine turnover rates. Endotoxin shock accelerated the loss of ^{3}H-norepinephrine from brain and various sympathetically-innervated peripheral tissues. This accelerated loss of parent compound was accompanied by an increase in both the deaminated (3,4-dihydroxymandelic acid) and o-methylated (normetanephrine) metabolites of ^{3}H-norepinephrine in the brain. Rao et al (1972) studied the distribution and fate of ^{3}H-norepinephrine in selected peripheral tissues of endotoxin-shocked dogs, and they too observed endotoxin-induced increases in ^{3}H-normetanephrine concentrations. Catechol-o-methyltransferase is principally an extraneuronal enzyme, and hence, an increase in ^{3}H-normetanephrine concentration suggests that at least part of the norepinephrine liberated during endotoxin shock is indeed released from neurons in a physiologically active form. On the other hand, monoamine oxidase enzyme is present within noradrenergic neurons. Therefore, the increased concentration of deaminated metabolites of ^{3}H-norepinephrine observed by Pohorecky et al (1972) suggests that another portion of the norepinephrine liberated during endotoxemia is degraded to inactive metabolite prior to its exit from the neuron.

The mechanism for the complex effects of endotoxin on norepinephrine turnover remains unknown, but it may involve decreased reuptake of norepinephrine by the neurons (Pardini et al 1982), reserpine-like effects on catecholamine storage mechanisms, modification of catecholamine synthesis, and/or a response to increased sympathetic nerve traffic (Pohorecky et al 1972). Detailed temporal studies of the complete catecholamine-metabolite profile in plasma, peripheral organs, and also urine are required before a composite picture can be drawn for the disposition and fate of catecholamines during endotoxin shock.

Epinephrine is not normally found in appreciable quantities within peripheral noradrenergic neurons, and this probably accounts for the general lack of interest shown by investigators in assessing the potential effects of endotoxin on peripheral tissue stores of epinephrine. As pointed out earlier in this chapter, however, there is some interesting evidence that epinephrine arising from the adrenal glands may be taken up by noradrenergic neurons and incorporated into the intraneuronal storage vesicles occupied by norepinephrine (Section 3.2.2.). The newly arrived epinephrine may then serve as a 'cotransmitter' and be released along with resident norepinephrine when the nerve is stimulated (Rand et al 1979). Epinephrine is more potent than norepinephrine at both α- and β-adrenoceptors, and partial conversion of noradrenergic neurons to adrenergic neurons could therefore result in an increase in effector cell response to nerve stimulation even if replacement of norepinephrine by epinephrine does not follow a 1:1 stoichiometric relationship. Furthermore, epinephrine released from the neuron may feedback and activate the β_2-autofacilitatory adrenoceptors on the neuron terminals and through this mechanism promote nerve-stimulated transmitter output (see

Figure 1; Section 3.2.1.). In this context, although different investigative groups have detected partial depletion of norepinephrine stores during endotoxin shock, the functional relevance of such partial depletion to the actual efficiency of sympathetic neurotransmission in these tissues was not determined. In view of the increase in circulating epinephrine occurring during endotoxin shock, we suggest endotoxicosis as a prime candidate for testing whether or not the epinephrine 'cotransmitter' mode can become a functional alternative of noradrenergic neurons.

There is inferential support for the cotransmitter theory in hemorrhagic shock, in that several groups have detected increased epinephrine content of peripheral tissues after bleeding of experimental animals (Glaviano and Coleman 1961; Gómez-Poviña and Canépa 1965). Farnebo et al (1979) reported that the cardiac epinephrine concentration (0.21 ± 0.012 nmol/kg) comprised only 3% of the total catecholamine content of control rat hearts, but that after 4 hours of hemorrhagic hypotension cardiac epinephrine concentration (2.96 ± 0.11 nmol/kg) became more than 10 times higher and now comprised 34% of the total catecholamine content in this organ. Furthermore, this epinephrine increment seemed to be localized exclusively to noradrenergic neurons because an increase in cardiac epinephrine did not occur when rats were subjected to chemical sympathectomy with 6-hydroxydopamine prior to bleeding. Again, however, the function relevance of this increase in epinephrine:norepinephrine ratio was not determined.

4.2. Adrenergic receptors and effector cell responses

Reports described in the preceding section have provided the general knowledge that experimental endotoxin shock results in relatively pronounced elevations of plasma catecholamine concentrations in concert with accelerated turnover and partial depletion of tissue norepinephrine reserves. Inherent to this line of observations is the trailing assumption that increased availability of adrenergic mediators results in enhanced stimulation of adrenoceptors, with the latter leading in turn to acceleration of receptor-coupled effector cell responses as a contribution to endotoxin pathogenesis.

Indeed, direct or indirect evidence has been garnered for endotoxin-induced dysfunctions of a large number of adrenoceptor-dependent events in various cell types, including the following examples of alterations of: sympathetic reflexes (Halinen 1976); cardiac inotropic responsiveness to β-adrenoceptor agonists and antagonists (Goodyer 1967; Parratt 1973); vascular contractile sensitivity to catecholamines (Gourzis et al 1961; Zweifach et al 1956); α-adrenoceptor-dependent contractile response of bladder (Nergårdh et al 1977) and bronchial smooth muscle (Simonsson et al 1973); α-adrenoceptor involvement in platelet aggregation during the Shwartzman reaction (Collins et al 1972; Thomas 1956); β-adrenoceptor-linked adenylate cyclase and/or cyclic AMP disposition in the heart (Romano and Jones 1983); α-adrenoceptor-dependent metabolic events in adipocytes (Spitzer 1974; Hikawyj-Yevich and Spitzer 1977); epinephrine-responsive adenylate cyclase in hepatocytes (Bitensky et al 1971), perhaps involving the guanine nucleotide-binding regulatory subunit of this enzyme complex (Donlon and Walker 1976); peripheral β-adrenoceptors involved in fever responses (Székely 1979); cyclic 3′,5′-guanosine monophosphate metabolism in hepatocytes (Graber et al 1979), and α- and β-adrenoceptor regulation of insulin and glucose metabolism (Agarwal and Lazar 1977; Cryer et al 1972; Filkins 1982; Hinshaw 1976; Yelich and Filkins 1982). Despite this plethora of experimental observations, it seems that a unified concept about the relevance of such changes to the pathogenesis of endotoxin shock has not received consensus support. Cause–effect relationships between the presence of an excessive amount of catecholamines and endotoxin lethality have been especially debated.

Some investigators have suggested that high plasma concentrations of catecholamines may foretell lethal actions of endotoxin (Jones and Romano 1983), while others have implied that lethal actions of endotoxin are actually accelerated in the absence of circulating catecholamines (Spink et al 1966). Hinshaw et al (1964) reported that they found no evidence to assign detrimental roles to the sympathoadrenal system in endotoxin shock, whereas Bolton and Atuk (1978) said they obtained direct evidence that an intact reactive sympathetic nervous system is essential for development of lethal endotoxicosis. McLean and Berry (1960) found that an α-adrenergic blocking drug (phenoxybenzamine) was ineffective against *E. coli* endotoxin lethality in mice, while a drug with β-receptor blocking properties (dichloroisoprenaline) was protective. Filkins (1979) later concluded just the opposite after he observed than an α-blocking agent (phentolamine) protected against *S. enteritidis* endotoxin in rats, whereas a β-blocker (propranolol) actually increased endotoxin lethality. A large number of studies have been designed in attempts to reconcile such divergent conclusions, and many of these studies have focused on drugs that either simulate or block interaction of adrenergic mediators with the α- or β-class of receptors. Representative examples of such studies are presented in the remainder of this chapter with emphasis on cardiac β-adrenoceptors and vascular α-adrenoceptors.

4.2.1. Cardiac β-adrenoceptors

Experiments with different mammalian species have provided evidence that endotoxin shock influences β-receptor-dependent events in the heart in at least two general ways. First, increased sympathetic drive to the cardiac β-receptors subserving positive inotropic activity seems to serve as an important compensatory stimulus for supporting myocardial contractile force and cardiac output during shock. Secondly, and somewhat paradoxically, the capability of the heart to respond to β-receptor stimulants is diminished.

Increased sympathetic drive to the heart during endotoxemia was unmasked by Goodyer (1967), in his studie of the nonselective β_1-β_2 blocking agent propranolol in urethane-anesthetized dogs. Propranolol had minimal cardiodynamic effects during basal conditions in normal dogs when the heart was not under sympathetic dominance. In endotoxic dogs, however, propranolol evoked substantial reductions in cardiac output. Goodyer (1967) not only uncovered dependency of the endotoxic heart upon increased sympathetic drive, but in doing so, he also revealed myocardial depression as an underlying component of endotoxin shock. Parratt (1973) reached similar conclusions after testing hemodynamic responses to another nonselective β_1-β_2 blocking agent, alprenolol, in his endotoxin shock model in pentobarbital-anesthetized cats. In fact, Parratt (1973) indicated that β-adrenoceptor blockade was potentially lethal under these circumstances because alprenolol rapidly precipitated irreversible right ventricular failure in 3 cats. Parratt (1973) also reported that the cardiodynamic effects of norepinephrine and epinephrine were markedly reduced 2–3 hours after endotoxin administration. Hinshaw and coworkers (Archer et al 1975) documented myocardial dysfunction in pentobarbital-anesthetized dogs within 4–6 hours following a LD_{70} dose of endotoxin, and they too observed decreased cardiodynamic effects of epinephrine.

Decreased cardiac responsiveness to catecholamines was also observed by Bhagat et al (1970) and Cavanagh et al (1970) in heart muscle preparations isolated from endotoxic guinea pigs and baboons, respectively. The β-selective agonist dobutamine, a synthetic catecholamine analog, was likewise less effective in eliciting cardiac stimulation when administered after endotoxin (McCaig and Parratt 1980). This series of related observations could be explained by a desensitization of cardiac β-receptors or a 'down-regulation' of their number, secondary to persistent receptor activation in response to excessive sympathetic neuronal stimulation and/or excessive release of adrenal catecholamines (Baumann et al 1981). However, other cardiac stimulants, whose mechanism of inotropic action does not depend upon β-adrenoceptors, for instance glucagon (an adenylate cyclase activator) and quazodine (a phosphodiesterase inhibitor), have also been shown to be less effective during endotoxin shock (Bower et al 1970; Parratt and Winslow 1974).

Since all of the preceding agents act partly or totally by increasing intracellular concentrations of cyclic AMP, an abnormality in generation of this second messenger might account for depressed responsiveness to these ino-

162

tropic stimulants during endotoxic shock (McCaig and Parratt 1980). Indeed, there is emerging evidence that endotoxin does effect cyclic AMP disposition in some cell types (see Bitensky et al 1971; Donlon and Walker 1976; Wolff and Cook 1975). Recently, Romano and Jones (1983) reported that cardiac tissue content of cyclic AMP was significantly less in endotoxic rats than in control rats within 1 hour after endotoxin injection, and remained lower in the shocked rats until death. Moreover, 45 seconds of exposure to isoprenaline increased cyclic AMP content in control heart by 22% but failed to increase cyclic AMP in tissue from shocked rats.

Studies of experimental cardiogenic shock by Baumann et al (1981) revealed a 74% loss and a 10 times lower binding affinity (K_D) of β-receptors in noninfarcted myocardium 3 days following partial coronary ligation in guinea pigs. Receptor changes were accompanied by a diminished inotropic responsiveness to isoprenaline in perfused hearts and a 70% reduction in isoprenaline stimulation of adenylate cyclase in a cardiac membrane preparation. All of these alterations were prevented by reserpine or by the β_1-selective blocking agent metoprolol. Baumann et al (1981) thus concluded that the abnormalities they detected in noninfarcted myocardium were the result of specific reversible damage of sarcolemmal β-receptors due to excessive levels of circulating catecholamines in the cardiogenic shock model. The situation seems to be more complex in endotoxic shock.

If decreased myocardial responsiveness to catecholamines and other activators of cyclic AMP during endotoxicosis is explained solely by a defect in cyclic AMP-dependent reactions, then agents which stimulate the heart through other pathways should not be affected by endotoxin shock. The positive inotropic response to calcium does not depend upon β-receptor-linked events, but is explained instead by a direct increase in transsarcolemmal Ca^{2+} influx and thus enhanced delivery and availability of this cation at the contractile proteins. McCaig and Parratt (1980) observed impaired inotropic responses of the heart to intravenous infusions of calcium in endotoxic cats, and Parker and Adams (1981) observed several calcium-dependent contractile irregularities in atrial myocardium from endotoxin-shocked guinea pigs. Hess and coworkers conducted a comprehensive series of studies and obtained clear biochemical evidence of altered myocardial calcium metabolism in dogs subjected to endotoxin shock (Hess 1979; Soulsby et al 1978).

Thus, although depressed myocardial responsiveness to β-adrenoceptor agonists is a characteristic of experimental endotoxin shock in several mammalian species, it remains to be established whether this deficiency is explained by: (a) decreased sensitivity of β-receptors to agonist ligands, (b) down-regulation in the number of β-receptors, (c) impaired linkage between receptor and adenylate cyclase, (d) decreased synthesis or enhanced degradation of cyclic AMP, (e) changes in basic calcium-dependent contractile processes distal to receptor-linked mechanisms, or most likely, (f) a combination of the above factors.

4.2.2. *Vascular and other α-adrenoceptors*

As discussed earlier in this chapter, excessive activation of α-adrenergic receptors subserving vascular smooth muscle contraction has been proposed as a component of endotoxin shock pathogenesis. Much of this concept was based on beneficial pharmacologic effects of α-blocking drugs in endotoxin-shocked animals.

4.2.2.1. *α-Adrenergic blocking drugs*

The use of drugs as experimental probes is based on the premise that specificity of drug action provides the means to chemically modulate only selected physiologic-pathologic events, which in turn leads to the delineation of particular functions/dysfunctions as important components of shock. A potential danger of this pharmacologic approach, however, is the problem – which is often overlooked – that many drugs assumed to be mechanistically specific actually exert multiple biologic influences (Adams 1983b). This criticism is valid irrespective of the routine classification of drug groups according to a single pharmacologic action, and it applies for those α-blocking drugs used historically and also currently in shock research (Adams 1983a).

Phenoxybenzamine and other haloalkylamines have antihistaminic, anticholinergic and antiserotonin activities in addition to their better recognized α-blocking actions. The H_1-blocking action of phenoxybenzamine is as potent in this regard as many of the clinically employed H_1-antihistamines. In fact, Iampietro et al (1963) concluded from their studies with dogs that beneficial hepatosplanchnic effects of phenoxybenzamine in endotoxin shock were due not to α-blockade but to the antihistaminic properties of the drug. In addition to subsidiary pharmacologic actions, it is also becoming increasingly evident that the net pharmacologic response to α-adrenoceptor blockade itself is more complex than originally assumed.

4.2.2.2. *Nonselective $α_1$-$α_2$ antagonists*

Clinical use of α-blocking agents has been limited by their paradoxical sympathomimetic activities. Treatment with phentolamine or phenoxybenzamine, for instance, can result in tachycardia and also lead to increased plasma catecholamines (Graham 1981; Saeed et al 1982). These responses are believed to be instigated along the following progression: binding of antagonist to vascular α-receptors, block of α-vasoconstrictor tone, vascular dilation, hypotension, activation of baroreceptor reflex mechanisms, and increased impulse traffic over sympathetic nerves and increased release of adrenal catecholamines.

As pointed out in Table 3, however, we now know that phenoxybenzamine and especially phentolamine are actually nonselective α_1-α_2 antagonists (Starke and Docherty 1980). That is, they also block α_2-receptors of noradrenergic nerve endings in addition to α-receptors of effector cells. Since prejunctional α_2-receptors subserve the previously discussed feedback inhibition of transmitter release, their blockade by phentolamine or phenoxybenzamine would free or disinhibit the noradrenergic neuron from a resident suppressor system.

An important effect of α_2-blockade would therefore be an augmentation of the net quantity of catecholamines mobilized and released by reflex autonomic adjustments. This activity would not be fully expressed at α-receptors because of the original α-blocking action of the drugs. In the heart, however, the increased availability of catecholamines can explain the β-receptor (β_1-subtype) stimulation and attendant tachycardia associated with the use of nonselective α-blockers (Saeed et al 1982).

This complex sequence prompts the following question: would increased availability of catecholamines in response to prejunctional α_2-blockade be sufficient to activate other β-receptor populations as well? If so, then the net pharmacologic response to nonselective α-blocking agents must entail an active, albeit indirect, β-selective sympathomimetic component in tissues other than just the heart. This hypothesis was tested by Saeed et al (1982) during a comparative examination of the hemodynamic effects of phentolamine and prazosin.

4.2.2.3. *Phentolamine and prazosin*

Since prazosin is a selective α_1-antagonist (Starke and Docherty 1980; Table 3), it leaves prejunctional α_2-receptors virtually unoccupied. Thus distinct hemodynamic differences should exist between the effects of prazosin and phentolamine if prejunctional α_2-blocking action and indirect β-stimulatory effects of the latter are important.

When the two drugs were tested in conscious dogs, Saeed et al (1982) observed equivalent decrements in peripheral vascular resistance and blood pressure. Phentolamine also evoked significant increases in heart rate, cardiac output, oxygen consumption, and plasma concentrations of norepinephrine and epinephrine. Prazosin did not produce these sympathomimetic responses despite equivalent reductions in blood pressure.

Not surprisingly, perhaps, β-adrenergic blockade with nadolol almost completely abolished phentolamine-induced vasodilation without affecting prazosin-induced vasodilation (Saeed et al 1982). β-Blockade also abolished the cardiac and metabolic (increased oxygen consumption) effects of phentolamine but did not blunt the capability of phentolamine to increase plasma catecholamines.

It seems clear, therefore, that nonselective α_1-α_2 antagonists do not simply prevent α-controlled events in effector organs. Because of their prejunctional α_2-blocking action, they also free mechanisms for catecholamine release from a resident inhibitory control. This action can elicit a substantial increase in circulating catecholamines (Saeed et al 1982) and perhaps even a relatively greater increment in norepinephrine at effector cell receptors adjacent to or near noradrenergic nerve endings (Silverberg et al 1978). As a result, β-receptors and associated cellular changes are activated in the heart and vasculature, and perhaps in muscle, fat, liver, and other tissues as well.

Indeed, the vascular response to nonselective α-antagonists may paradoxically depend as much or more on β_2-mediated vasodilation and metabolic vasodilation than on direct withdrawal of α-mediated constriction of blood vessels. Even metabolic responses to nonselective α-blockers seem secondary, at least in part, to β-receptor-coupled events instigated by increased catecholamines (Saeed et al 1982). These pharmacodynamic complexities prompted Saeed et al (1982) to issue a crucial warning: 'The combination of pre- and postsynaptic effects invalidates the use of phentolamine in the assessment of α-adrenoceptor-mediated vasoconstrictor tone.'

4.2.2.4. *α-Antagonists and shock*

Numerous investigators have assessed that phentolamine and phenoxybenzamine improve hemodynamics and prolong survival in experimental shock via block of α-vasoconstrictor tone. The obvious question that now arises, however, is: do nonselective α-blockers prove helpful in shock because of α-vasoconstrictor block (which Saeed et al (1982) imply is an invalid or incomplete interpretation) or do their indirect β-stimulating properties play an important or even dominant role?

Some insight into this issue may be inferred from data reported by Filkins (1979). In this study, phentolamine exerted a marked protective effect against *S. enteritidis* endotoxin shock in rats. Phentolamine also blunted the development of endotoxic hypoglycemia and protected against endotoxic depression of hepatic gluconeogenesis. In contrast, β-adrenergic blockade with propranolol sensitized rats to the lethality of shock, and accentuated both the hypoglycemic response and the depressed hepatic gluconeogenesis in shock. Importantly, concurrent treatment with propranolol and phentolamine resulted in increased mortality in shock, similar to the sensitization phenomenon seen with propranolol alone (Filkins 1979). Thus, β-blockade-induced sensitization to endotoxin lethality was dominant to α-blockade protection. Alternatively, however, it may be that β-blockade was not simply dominant, but that it actually prevented the protective effects of phentolamine because the latter involved indirect activation of β-receptors.

As a related example, Nagasawa et al (1975) studied cardiogenic shock in dogs and found that phentolamine increased cardiac index, coronary blood flow, and mean arterial blood pressure while lowering systemic vascular resistance and coronary arterial resistance. Nagasawa et al (1975) believed that phentolamine improved cardiac performance by lowering impedance to left ventricular ejection via α-vasoconstrictor blockade. Alternatively, however, the pharmacologic responses to phentolamine in these dogs are not incompatible with those resulting from β-receptor stimulation in the heart and vasculature. Few attempts have been made in the past to ascertain whether or not beneficial effects of α-blockers in shock depended in part upon β-receptor-coupled events.

5. SUMMARY

Despite hundreds of research reports dealing with adrenergic aspects of endotoxin shock, only general conclusions about this problem can be advanced at this time. Most studies show that endotoxemia results in relatively pronounced increments in circulating catecholamines with moderate decrements in norepinephrine neuronal stores in the heart and other sympathetically innervated tissues (Section 4.1). Sympathoadrenal activation is not specific to endotoxicosis, but reflects an endogenous compensatory effort common to different forms of hypodynamic circulatory states. Increased sympathetic drive to the heart represents an important reserve mechanism for maintaining cardiac pump performance in the early and perhaps intermediate shock periods (Section 4.2.1). On the other hand, however, prolonged delivery of catecholamines to the myocardium does not prevent development of spontaneous heart failure, but may lead instead to reduced sensitivity of the cardiac β_1-adrenoceptors (Baumann et al 1981) and possibly to impairment of contractile-dependent mechanisms downstream from the receptor–adenylate cyclase–cyclic AMP triad (Section 4.2.1.). Reflex activation of vascular α-adrenoceptors produces vasoconstriction and an increase in peripheral vascular resistance, again reflecting a common survival mechanism for improving perfusion pressure and circulation to vital organs. Intense prolonged vasoconstriction no doubt is detrimental to membrane and cell viability in the hypoperfused regions. Also, indirect or preliminary evidence has been presented for adrenoceptor-dependent changes in a myriad of other cells, tissues, and organs as putative characteristic effects of endotoxin (Section 4.2.).

The preceding scenario seems to be generally supported by most investigative groups in this field; it should, however, be emphasized that a consensus has not been reached about the actual relevance of sympathoadrenal activation to endotoxin pathogenesis. In essence, and on the basis of the data available, investigators have been unable to agree on whether activation of adrenoceptor-dependent mechanisms is simply a peripheral accompaniment or a true obligatory participant in the endotoxin shock syndrome. Opposing opinions about this problem are replete throughout the literature, and we believe that this controversy reflects the basic complexity of endotoxicosis as well as the often overlooked multifaceted nature of sympathetic neuroeffector communication. Indeed, recent discoveries in adrenergic physiology-pharmacology have prompted new questions about standard interpretations of old data, and future study with α_1-, α_2-, β_1-, and β_2-selective adrenergic drugs should help clarify some of the issues addressed in this overview of adrenergic aspects of endotoxic shock.

REFERENCES

Adams HR (1983a) Pharmacologic problems in circulation research: alpha adrenergic blocking agents. *Circ. Shock 10*, 215-223.

Adams HR (1983b) Pharmacologic problems in shock research. In: Reichard SM, Reynolds DG, Adams HR (Eds), *Advances in Shock Research, Vol 10*, pp 3-8. Allen R. Liss, New York.

Adler-Graschinsky E, Langer SZ (1975) Possible role of a β-adrenoceptor in the regulation of noradrenaline release by nerve stimulation through a positive feedback mechanism. *Br. J. Pharmacol. 53*, 43-50.

Agarwal MK, Lazar G (1977) Metabolic basis of endotoxicosis. *Microbios 20*, 183-214.

Ahlquist RP (1948) Study of adrenotropic receptors. *Am. J. Physiol. 153*, 586-600.

Archer LT, Black MR, Hinshaw LB (1975) Myocardial failure with altered response to adrenaline in endotoxin shock. *Br. J. Pharmacol. 54*, 145-155.

Baumann G, Reiss G, Erhardt WD, Felix SB, Ludwig L, Blutel G, Blomer H (1981) Impaired beta-adrenergic stimulation in the uninvolved ventricle post-acute myocardial infarction: reversible defect due to excessive circulating catecholamine-induced decline in number and affinity of beta receptors. *Am. Heart J. 101*, 569-581.

Berthelson S, Pettinger WA (1977) A functional basis for classification of alpha-adrenergic receptors. *Life Sci. 21*, 595-606.

Bhagat B, Cavanagh D, Merrild BN, Rana MW, Rao PS (1970) Noradrenaline and tyramine action on isolated atrial muscle of endotoxin-treated guinea pigs. *Br. J. Pharmacol. 29*, 688-695.

Birinyi F, Hacket DB, Mikat E (1977) Effects of α-adrenergic blockade on coronary blood flow of dogs in hemorrhagic shock. *Circ. Shock 4*, 297-303.

Bitensky MW, Gorman RE, Thomas L (1971) Selective stimulation of epinephrine-response adenyl cyclase in mice by endotoxin. *Proc. Soc. Exp. Biol. Med. 138*,

773-775.

Bolton WK, Atuk NO (1978) Study of chemical sympathectomy in endotoxin-induced lethality and fibrin deposition. *Kidney Int. 13*, 262-270.

Bower MG, Okude S, Jolley WB, Smith LL (1970) Hemodynamic effects of glucagon following hemorrhagic and endotoxic shock in the dog. *Arch. Surg. (Chicago) 101*, 411-419.

Brown GL, Gillespie JS (1958) The output of sympathetic transmitter from the spleen of the cat. *J. Physiol. (London) 138*, 81-102.

Cavanagh D, Rao PS, Sutton DMC, Bhagat DB, Bachmann F (1970) Pathophysiology of endotoxin shock in the primate. *Am. J. Obstet. Gynecol. 108*, 705-711.

Collins AD, Henson EC, Izard SR, Brunson JG (1972) Norepinephrine, endotoxin shock, and the generalized Shwartzman reaction. *Arch. Pathol. Lab. Med. 93*, 82-88.

Cryer PE, Coran AG, Sode J, Herman CM, Horwitz DL (1972) Lethal Escherichia coli septicemia in the baboon: alpha adrenergic inhibition of insulin secretion and its relationship to the duration of survival. *J. Lab. Clin. Med. 79*, 622-638.

Devereux DF, Michas CA, Rice S (1977) Heparin pretreatment suppresses norepinephrine concentrations in dogs in endotoxic shock. *Clin. Chem. 23*, 1346-1347.

Donlon MA, Walker RI (1976) Adenylate cyclase activity of mouse liver membranes after incubation with endotoxin and epinephrine. *Specialia 32*, 179-181.

Exton JH (1981) Molecular mechanisms involved in α-adrenergic responses. *Mol. Cell. Endocrinol. 23*, 233-264.

Farnebo LO, Hallman H, Hamberger B, Jonsson G (1979) Catecholamines and hemorrhagic shock in awake and anesthetized rats. *Circ. Shock 6*, 109-118.

Feuerstein G, Dimicco JA, Ramu A, Kopin IJ (1981) Effect of indomethacin on the blood pressure and plasma catecholamine responses to acute endotoxaemia. *J. Pharm. Pharmacol. 33*, 576-579.

Filkins JP (1979) Adrenergic blockade and glucoregulation in endotoxin shock. *Circ. Shock 6*, 99-107.

Filkins JP (1982) Effect of phentolamine on insulin levels and insulin responsiveness in endotoxicosis. *Adv. Shock Res. 7*, 185-190.

Glaviano VV, Coleman B (1961) Tissue levels of norepinephrine and epinephrine in hemorrhagic shock. *Proc. Soc. Exp. Biol. Med. 107*, 761-764.

Gómez-Poviña OA, Canépa JF (1965) Catecholamine content of the ventricular myocardium in dogs following hemorrhagic hypotension. *J. Surg. Res. 5*, 341-345.

Goodyer AVN (1967) Left ventricular function and tissue hypoxia in irreversible hemorrhagic and endotoxin shock. *Am.J. Physiol. 212*, 444-450.

Gourzis JT, Hollenberg MW, Nickerson M (1961) Involvement of adrenergic factors in the effects of bacterial endotoxin. *J. Exp. Med. 144*, 593-604.

Graber SE, Bomboy JD Jr, Salmon WD Jr, Krantz SB (1979) Evidence that endotoxin is the cyclic 3':5'-GMP-producing factor in erythropoietin preparations. *J. Lab. Clin. Med. 93*, 25-31.

Graham RM (1981) The physiology and pharmacology of alpha and beta blockade. *Cardiovas. Med. Spec. Suppl.* 7-21.

Halinen MO (1976) Initial effects of endotoxin on cardiovascular reflex functions and circulation in dogs. *Acta Physiol. Scand. Suppl.* 439.

Hess ML (1979) Subcellular function in the acutely failing myocardium. *Circ. Shock* 6, 119-136.

Hikawyj-Yevich I, Spitzer JA (1977) The role of adrenergic receptors and Ca^{2+} in the action of endotoxin on human fat cells. *J. Surg. Res. 23*, 233-238.

Hinshaw LB (1976) The role of glucose in endotoxin shock. *Circ. Shock 3*, 1-10.

Hinshaw LB, Brake CM, Emerson TE Jr, Jordan MM, Masucci FD (1964) Participation of sympathoadrenal system in endotoxin shock. *Am. J. Physiol. 207*, 925-930.

Hoffman BB, Lefkowitz RJ (1980) Radioligand binding studies of adrenergic receptors: new insights into molecular and physiological regulation. *Ann. Rev. Pharmacol. Toxicol. 20*, 571-608.

Iampietro P, Hinshaw LB, Brake CM (1963) Effect of an adrenergic blocking agent on vascular alterations associated with endotoxin shock. *Am. J. Physiol. 204*, 611-614.

Jakobs HH, Schultz G (1982) Signal transformation involving α-adrenoceptors. *J. Cardiovasc. Pharmacol. 4*, S63-S67.

Jones SB, Romano FD (1983) Plasma catecholamines in rats subjected to surgery and endotoxicosis (Abstract). *Fed. Proc. 43*, 980.

Lands AM, Arnold A, McAuliff JP, Luduena FP, Brown TC (1967) Differentiation of receptor systems activated by sympathomimetic amines. *Nature (London) 214*, 597-598.

Langer SZ (1974) Presynaptic regulation of catecholamine release. *Biochem. Pharmacol. 23*, 1793-1800.

Langer SZ (1977) Presynaptic receptors and their role in the regulation of transmitter release. *Br. J. Pharmacol. 60*, 481-497.

Levitski A (1982) Activation and inhibition of adenylatecyclase by hormones: mechanistic aspect. *Trends Pharmacol. Sci. May*, 203-208.

Lillehei RC (1979) History of vasodilation in treating shock and low flow states. In: Lefer AM, Saka TM, Mela LM (Eds), *Advances in Shock Research, Vol 1*, pp 1-17. Allen R. Liss, New York.

Michell RH, Kirk CJ (1981) Why is phosphatidylinositol degraded in response to stimulation of certain receptors? *Trends Pharmacol. Sci. 2*, 86-89.

McCaig DJ, Parratt JR (1980) Reduced myocardial response to calcium during endotoxin shock in the cat. *Circ. Shock 7*, 23-30.

McLean RA, Berry LJ (1960) Adrenergic inhibition and lethal effect of bacterial endotoxin in mice. *Proc. Soc. Exp. Biol. Med. 105*, 91-95.

Nagasawa K, Vyden JK, Forrester JS, Groseth-Dittrich MF, Corday E, Swan JHC (1975) Effect of phentolamine on cardiac performance and energetics in acute myocardial infarction. *Circ. Shock 2*, 5-11.

Nergårdh A, Boréus LO, Holme T (1977) The inhibitory effect of coli-endotoxin on alpha-adrenergic receptor functions in the lower urinary tract. *Scand. J. Urol. Nephrol. 11*, 219-224.

Nickerson M, Gourzis JT (1962) Blockade of sympathetic vasoconstriction in the treatment of shock. *J. Trauma 2*, 399-411.

Northup JK, Smigel MD, Gilman AG (1982) The guanine nucleotide activating site of the regulatory component of adenylate cyclase. *J. Biol. Chem. 257*, 11416-

11423.

Palmerio C, Zetterström B, Shammash J, Euchbaum E, Frank E, Fine J (1963) Denervation of the abdominal viscera for the treatment of traumatic shock. *N. Engl. J. Med. 269*, 709-716.

Pardini GJ, Jones SB, Filkins JP (1982) Contribution of depressed reuptake to the depletion of norepinephrine from rat heart and spleen during endotoxin shock. *Circ. Shock 9*, 129-143.

Parker JL, Adams HR (1978) The influence of chemical restraining agents on cardiovascular function: a review. *Lab. Anim. Sci. 28*, 575-583.

Parker JL, Adams HR (1981) Contractile dysfunction of atrial myocardium from endotoxin-shocked guinea pigs. *Am. J. Physiol. 240*, H954-H962.

Parratt JR (1973) Myocardial and circulatory effects of E. coli endotoxin; modification of responses to catecholamines. *Br. J. Pharmacol. 47*, 12-25.

Parratt JR, Winslow E (1974) The haemodynamic effects of quazodine, a cardiac stimulant, in experimental E. coli endotoxin shock in the cat. *Clin. Exp. Pharm. Physiol. 1*, 31.

Pohorecky L, Wurtman RJ, Taam D, Fine J (1972) Effects of endotoxin on monoamine metabolism in the rat. *Proc. Soc. Exp. Biol. Med. 140*, 739-746.

Prager RL, Dunn EL, Seaton JF (1975) Increased adrenal secretion of norepinephrine and epinephrine after endotoxin and its reversal with corticosteroids. *J. Surg. Res. 18*, 371-375.

Rahwan RG (1983) Mechanisms of action of membrane calcium channel blockers and intracellular calcium antagonists. *Med. Res. Rev. 3*, 21-42.

Rand MJ, Majewski H, McCullock MW, Story DF (1979) An adrenalin-mediated positive feedback look in sympathetic transmission and its possible role in hypertension. In: *Proceedings of the Symposium on Presynaptic Receptors*. IUPHAR Satellite Meetings, Paris. Pergamon Press, Oxford.

Rand JJ, Majewski H, Medgett IC, McCulloch MW, Story DF (1980) Prejunctional receptors modulating autonomic neuroeffector transmission. *Circ. Res. 46*, 170-176.

Rao PS, Bhagat BD, Cavanagh D (1972) Effect of endotoxin on hemodynamics and norepinephrine metabolism in the dog. *Proc. Soc. Exp. Biol. Med. 141*, 412-418.

Romano FD, Jones SB (1983) Desensitization of the endotoxic rat myocardium to adrenergic stimulation (Abstract). *Fed. Proc. 43*, 980.

Rosenberg JC, Lillehei RC, Moran WH, Zimmerman B (1959) Effect of endotoxin on plasma catecholamines and serum serotonin. *Proc. Soc. Exp. Biol. Med. 102*, 335-337.

Rosenberg JC, Lillehei RC, Longerbeam J, Zimmerman B (1961) Studies on hermorrhage and endotoxin shock in relation to vasomotor changes in endogenous circulating epinephrine, norepinephrine and serotonin. *Ann. Surg. 154*, 611-628.

Ross EM, Gilman AG (1980) Biochemical properties of hormone-sensitive adenylate cyclase. *Ann. Rev. Biochem. 49*, 533-564.

Saeed M, Sommer O, Holtz J, Bassenge E (1982) α-Adrenoceptor blockade by phentolamine causes β-adrenergic vasodilation by increased catecholamine release due to presynaptic α-block. *J. Cardiovas. Pharmacol. 4*, 44-52.

Shepherd JT, Vanhoutte PM (1981) Local modulation of adrenergic neurotransmission. *Circulation 64*, 655-666.

Silverberg AB, Shah SD, Haymond MW, Cryer PE (1978) Norepinephrine: hormone and neurotransmitter in man. *Am. J. Physiol. 243*, E252-E256.

Simonsson BG, Svedvyr N, Skoogh BE, Andersson R, Bergh NP (1973) *In vivo* and *in vitro* studies on alpha receptors in human airways; potentiation with bacterial endotoxins. *Chest 63, Suppl*, 3S-4S.

Soulsby ME, Bruni FD, Looney TJ, Hess ML (1978) Influence of endotoxin on myocardial calcium transport and the effect of augmented venous return. *Circ. Shock 5*, 23-34.

Spies C, Schultz KD, Schultz G (1980) Inhibitory effects of mepacrine and eicosatetraynoic acid on cyclic GMP elevations caused by calcium and hormonal factors in rat ductus deferens. *Naunyn Schmiedebergs Arch. Pharmakol. 311*, 71-77.

Spink WW, Reddin J, Zak SJ, Peterson M, Starzecki B, Seljeskog E (1966) Correlation of plasma catecholamine levels with hemodynamic changes in canine endotoxin shock. *J. Clin. Invest. 45*, 78-85.

Spitzer JA, Kovach AGB, Rosell S, Sandor P, Spitzer JJ, Storch R (1974) Influence of endotoxin on adipose tissue metabolism. *Adv. Exp. Med. Biol. 33*, 337-344.

Starke K, Docherty JR (1980) Recent developments in α-adrenoceptor research. *J. Cardiovas. Pharmacol. 2, Suppl 3*, S269-S286.

Stjärne L, Brundin J (1976) β_2-adrenoceptors facilitating noradrenaline secretion from human vasoconstrictor nerves. *Acta Physiol. Scand. 97*, 888-933.

Székely M (1979) Central and peripheral adrenergic mechanisms in endotoxin fever of newborn guinea pigs. *Acta Physiol. Acad. Sci. Hung. 54*, 257-263.

Thomas L (1956) The role of epinephrine in the reactions produced by the endotoxins of gram-negative bacteria. *J. Exp. Med. 104*, 865-880.

U'Prichard DC, Snyder SH (1977) [³H] epinephrine and [³H] norepinephrine binding to alpha noradrenergic receptors. *Life Sci. 20*, 527-533.

U'Prichard DC, Snyder SH (1979) Distinct alpha-noradrenergic receptors differentiated by binding and physiological relationships. *Life Sci. 24*, 79-88.

Vick JA (1965) Bioassay of the prominent humoral agents involved in endotoxin shock. *Am. J. Physiol. 209*, 75-78.

Wolff J, Cook GH (1975) Endotoxic lipopolysaccharides stimulate steroidogenesis and adenylate cyclase in adrenal tumor cells. *Biochim. Biophys. Acta 413*, 291-297.

Yelich MR, Filkins JP (1982) Insulin hypersecretion and potentiation of endotoxin shock in the rat. *Circ. Shock 9*, 589-603.

Zetterström BEM, Palmerio C, Fine J (1964) Changes in tissue content in catecholamines in traumatic shock. *Acta Chem. Scand. 128*, 13-19.

Zweifach B, Nagler AL, Thomas L (1956) The role of epinephrine in the reactions produced by the endotoxins of gram-negative bacteria. *J. Exp. Med. 104*, 881-896.

Handbook of Endotoxin, Vol. 2: Pathophysiology of Endotoxin
L.B. Hinshaw, editor
© Elsevier Science Publishers B.V., 1985
ISBN 0 444 90385 2
$0.85 per article per page (transactional system)
$0.20 per article per page (licensing system)

CHAPTER 8

Release and vascular effects of histamine, serotonin, angiotensin II and renin following endotoxin

THOMAS E. EMERSON Jr

1. INTRODUCTION

I have been assigned the interesting and welcome challenge of reviewing the roles of histamine, 5-hydroxytryptamine (5-HT, serotonin) and angiotensin II in the pathogenesis of endotoxin shock. I will first cover the cardiovascular actions of these naturally occurring vasoactive agents under normal conditions, which leads naturally to a better understanding of their action in shock. Because of space limitations, I have targeted the lung and heart for special consideration. These critical organs have in recent years been shown to be seriously affected during the early phase of circulatory shock as well as during chronic sepsis associated with high cardiac output. Also, it is well known that acutely induced septic and endotoxin shock are characterized by a decrease in cardiac output secondary to a decrease in venous return, but the role of cardiac depression in the pathogenesis of shock is not known (Goldfarb 1982).

I have also placed major emphasis upon the role of histamine in the pathogenesis of endotoxin shock, largely because it was one of the first vasoactive hormones to be considered as the primary 'shock toxin', and although its role in this capacity continues to be investigated, an unassailable conclusion one way or the other cannot presently be made.

2. HISTAMINE

2.1. General

The pharmacodynamics of histamine have been studied extensively since its synthesis in 1907 by Windaus and Vogt (1907), and its numerous biological activities have been the subject of many comprehensive reviews (Coode 1952; Kahlson and Rosengren 1968; Lindell and Westling 1966; Parott et al 1964; Porter and Mitchell 1972; Reite 1972; Rocha e Silva 1966); the review by Reite (1972) is particularly exhaustive and covers a wide range of biological activities of histamine in a large number of vertebrate and invertebrate species.

2.1.1. Storage

Histamine is stored in tissue mast cells and circulating basophils in plasma (Ackerman 1963; Riley and West 1953, 1966). In addition to stored histamine, the concept of 'induced histamine' was introduced by Schayer (1962, 1963). While stored (bound or performed) histamine is generally a rapid 'one-shot' release event, Schayer (1960a, 1960b) proposes that induced histamine may be continually involved in regulation of the microvasculature under normal and abnormal conditions, including endotoxin shock. While this hypothesis has received little support, it should not be ruled out (Altura and Zweifach 1967; Kahlson and Rosengren 1968). Additional studies to test this interesting, unique hypothesis are clearly needed.

2.1.2. Release

Histamine is readily released by a number of agents (Bloom and Haegermark 1965; Fawcett 1954; Mota 1966; Paton 1957; Riley 1959), including 5-hydroxytryptamine (Feldberg and Smith 1953) and purines (Moulton et al 1957; Spector and Willoughby 1957). Of particular relevance to the present topic, endotoxin from gram-negative microorganisms causes a large, explosive histamine release (Hinshaw et al 1960, 1961a, 1961b, 1962a, 1962b; Reddin et al 1966; Vick 1964; Vick et al 1971) as early as 30–60 seconds after intravenous injection in adult dogs (Reddin et al 1966; Vick et al 1971). A high plasma histamine concentration is maintained for at least 10 minutes (Vick et al 1971), but at 60 and 180 minutes of shock it does not differ from preshock control values (Reddin et al 1966). It is also of interest that plasma volume expanders such as dextran are capable of releasing histamine (Halpern 1956; Silk 1966; Walton et al 1959) and that intravenous (Halpern 1956) or intraperitoneal (Silk 1966) administration of dextran results in profound shock in some rodent species.

174

2.2. Cardiovascular effects

2.2.1. General

Effects of histamine on various cardiovascular functions were described in the early 1900s (Barger and Dale 1910; Burn and Dale 1926; Dale and Laidlaw 1910, 1911, 1918-1919; Dale and Richards 1918).

Effects of histamine on cardiovascular dynamics are profound. For example, a 5-minute intravenous infusion at 50 mg/min causes a dramatic decrease in venous return and hence cardiac output, total peripheral resistance, and arterial blood pressure in the open-chest anesthetized dog (Emerson 1968). An equivalent infusion in the eviscerated dog, however, causes only a small decrease in venous return and cardiac output but still causes a striking decrease in total peripheral resistance and arterial blood pressure (Emerson 1968). The pooling of large amounts of blood in the hepatosplanchnic viscera is dependent on localized venous constriction, particularly in the liver (Andrews et al 1955; Hinshaw et al 1960, 1966). Hepatic 'sphincter-type' veins in the dog are extremely sensitive to injected (Brake et al 1964) or endogenously released (Hinshaw et al 1962a) histamine. That evisceration does not completely prevent the fall in venous return during steady-state histamine infusion probably reflects intravascular pooling of blood and loss of plasma water. These perturbations are secondary to venous constriction and hence increased venous transmural and capillary hydrostatic pressures, respectively (Haddy 1960a). The decrease in total peripheral resistance during histamine infusion should increase venous return and cardiac output (Guyton and Sagawa 1961). However, factors which oppose an increase, namely pronounced constriction of outflow veins, predominate, and venous return decreases. It is perhaps instructive to compare the action of other potent vasodilator agents on venous return and cardiac output. The action of histamine on venous return is in striking contrast to that of bradykinin (Emerson 1967) and acetylcholine (Emerson 1968).

While histamine, bradykinin, and acetylcholine each dilate precapillary resistance vessels and lower total peripheral resistance and arterial blood pressure (Haddy and Scott 1966), only bradykinin and acetylcholine increase venous return. The differential effects on veins (Haddy 1960a; Haddy and Scott 1966), particularly hepatic sphincter-type veins (Hinshaw et al 1966), appear to be responsible for this difference.

A recent study by Olson (Olson 1982; Olson et al 1983) demonstrates the effect of long-term infusion of histamine on the cardiopulmonary system in the dog. Intravenous infusion of histamine base (10 mg/kg/min) for 90 minutes caused a decrease in cardiac output (due to a decrease in stroke volume which was proportionately greater than the increase in heart rate) and a fall in arterial blood pressure (due to both a decrease in cardiac output and total peripheral resistance). Postcapillary pulmonary vascular resistance increased markedly and remained elevated throughout the infusion period, while pulmonary artery wedge pressure and mean pulmonary artery pressure markedly decreased. Total pulmonary blood volume decreased strikingly, but pulmonary extravascular fluid volume remained unaltered during the 90-minute infusion period.

2.3. Effects on organs

2.3.1. General

Intravenous administration of histamine decreases vascular resistance, particularly precapillary resistance, and increases blood flow in most peripheral organs including the heart (Giles et al 1977), kidney (Altura and Zweifach 1967; Hinshaw et al 1962a), forelimb (Grega et al 1972; Haddy et al 1972; Marciniak et al 1978), equine digit (Robinson et al 1975, 1976), brain (Edvinsson and Owman 1975; Tindall and Greenfield 1973), stomach (Jacobson 1963), intestine (Jacobson 1968; Texter 1963), and liver, except that of the dog (Andrews et al 1955; Lautt 1977). On the other hand, intravenous infusion of histamine directly and/or indirectly causes active vasoconstriction of postcapillary resistance vessels in most organs (Grega et al 1980; Haddy et al 1972; Hinshaw et al 1962; Marciniak et al 1977; Staszewska-Barczak and Vane 1965). The venoconstrictor effect of histamine is particularly prominent in the lungs of most species (Brody and Stemmler 1968; Glazier and Murray 1971; Gilbert et al 1958; Olson 1982; Olson et al 1983), in the dog liver (Andrews et al 1955; Hinshaw et al 1960), and in the intestine (Jacobson 1968; Texter 1963).

Histamine is a notorious edemagenic agent in most vascular beds studied and causes this untoward effect through every conceivable mechanism: it increases capillary hydrostatic pressure via simultaneous dilation of precapillary resistance vessels and constriction of postcapillary resistance vessels, increases microvascular surface area, and increases microvascular permeability to protein (Haddy et al 1972, 1976; Kozlowski et al 1981; Olson et al 1983).

While the increases in capillary microvascular surface area and hydrostatic pressure are important contributing factors, it now seems clear that the majority of peripheral edema produced by histamine results from an increase in microvascular permeability to protein (Korthuis et al 1982; Kozlowski et al 1981).

Two organs which have been increasingly implicated in the pathology of irreversible shock are the lung and heart. Some effects of histamine on these organs are reviewed below.

2.3.2. Lung

For a more comprehensive review of this subject, the reader is referred to the excellent article by Olson (1982). Results from this recent and extensive study demonstrate that intravenous infusion of histamine (10 mg/kg/min) for 90 minutes, with or without α-(phentolamine) or β-(propranolol)receptor blockade, does not alter lung extravascular thermal volume or the wet-to-dry lung weight ratio. These and other data lead to the conclusion that histamine has little if any effect on lung microvascular permeability to protein or extravascular water.

Conclusions resulting from the Olson studies (1982, 1983) are well supported by earlier work. Addition of histamine to lung perfusate resulted in no increase in dog alveolar capillary permeability to radioactive albumin (Goetzman and Visscher 1969). No increase in lung weight was observed during pulmonary arterial infusion of histamine in isolated fetal, neonatal, or adult dog lungs (Grega et al 1971). Numerous other studies also demonstrated that intravenous histamine infusion exerts little if any effect on pulmonary microvascular permeability to protein or lung water (Drake and Gabel 1980; Gabbiani et al 1972; Pietra et al 1971, 1979).

On the other hand, Brigham and Owen (1975) demonstrated a 2–6-fold increase in pulmonary lymph flow with no change in the lymph/plasma protein ratio during histamine infusion in sheep, which led them to conclude that histamine increased pulmonary vascular permeability. Others have also reported increases in pulmonary microvascular permeability and lymph protein clearance in sheep during either bronchial or pulmonary blood infusion of histamine (Nakahara et al 1979). However, Drake et al (1981) recently presented data suggesting that caudal mediastinal lymph contains appreciable amounts of nonpulmonary lymph, which cast doubts on conclusions drawn from sheep studies. However, assuming the sheep pulmonary data are valid, the effects of histamine on pulmonary microvascular permeability and lung water in this species are not striking.

The contrast between the data obtained from studies with sheep and those from studies with other species may be related to the greater sensitivity of sheep lungs to histamine and sympathetic nerve activity. This phenomenon is apparently related to a more intense sympathetic innervation of sheep pulmonary vessels (Hebb 1969) and greater concentration of the epinephrine precursor dopamine in tissue mast cells (Falck et al 1964).

2.3.3. Heart

Histamine is readily taken up by myocardial tissue (Cross 1973) and is released by various stimuli (Capurro and Levi 1975; Levi 1972). It has a potent dilator effect on coronary arteries which is mediated via stimulation of H_1 and H_2 receptors (Giles et al 1977). However, in spite of this potent vasodilatory effect, histamine does not appear to be involved in the reactive hyperemia response in coronary blood vessels (Giles et al 1977). Histamine causes a positive inotropic effect in both isolated and intact hearts in a variety of animal species (Giles et al 1977; McNeill and Muschek 1972; McNeill and Verma 1974; Pöch and Kukovetz 1967; Trendelenburg 1960). In this regard, it is of interest that the inotropic effect of histamine apparently acts through the same mechanism proposed for the catecholamines, viz, by activation of phosphorylase and increased production of cyclic adenosine 3′,5′-monophosphate (McNeill and Verma 1974).

2.4. Endotoxin shock and lung function: effects of histamine

The lung is adversely affected during endotoxin shock (Ayres 1977; Kerstein et al 1982; Webb 1977), exhibiting an early increase in vascular resistance and microvascular permeability to protein (Brigham et al 1979; Demling and Flynn 1982; Hirose et al 1978). The Adult Respiratory Distress Syndrome is often associated with circulatory shock of various etiologies including sepsis, hemorrhage, burn, and trauma (Ayres 1977; Webb 1977). However, this horrendous and usually irreversible condition may be a direct result of treatment regimens, especially those involving overvigorous transfusion with crystalloid solutions (Carlson 1977).

Since histamine causes mild to moderate (Brigham and Owen 1975) or no pulmonary edema (Olson 1982; Olson et al 1983) in normal lungs, it appears unlikely that this vasoactive amine is primarily involved in pulmonary dysfunction of endotoxin shock or the Adult Respiratory Distress Syndrome. More likely candidates for this distinction might be pulmonary prostaglandins (especially thromboxane A_2) whose syntheses are increased during endotoxin shock (Hirose et al 1978; Massion and Kux 1973; Parratt and Sturgess 1977; Reeves et al 1972). However, involvement of these agents during endotoxin shock has been investigated primarily to explain the early pulmonary hypertension characteristic of this condition; relatively few data are available concerning effects of prostaglandins on pulmonary microvascular permeability to protein (Demling and Flynn 1982).

2.5. Endotoxin shock and heart function: effects of histamine

In a recent review, Goldfarb (1982) emphasizes that studies are equally divided between confirming or rejecting the hypothesis that cardiac contractility is depressed during circulatory shock of various etiologies.

Archer et al (1975, 1978, 1982) documented myocardial failure in dogs during *E. coli* endotoxin or bacterial shock. Decreased cardiac function has also been reported in dogs during *Pseudomonas aeruginosa* bacteremia (Postel and Schloerb 1972) and in man during sepsis and/or septic shock (Cann et al 1972; Cerra et al 1978; Siegel et al 1967; Weisul et al 1975; Woo et al 1979).

In the Archer et al (1982) study cited above, myocardial performance was evaluated by pressure–function curve tests in hearts from small adult 'donor' dogs whose hearts were transferred 2 hours after initiation of shock to an extracorporeal circuit of a 'support' dog. Three to five hours after *E. coli* infusion, marked myocardial depression occurred in 75% of the hearts tested; increased left ventricular end diastolic pressures and depressed peak positive and negative dP/dt occurred at every aortic pressure tested. Also, myocardial efficiency and power were depressed, oxygen uptake was increased and coronary blood flow was unchanged.

Mechanisms of cardiac dysfunction and factors which contribute to this disorder are not well understood and could be related to either hypoperfusion problems (Hinshaw et al 1973, 1974a, 1974b), circulating mediators (e.g., myocardial depressant factor) (Lefer 1970), a combination of both, or possibly other factors. However, coronary hyperperfusion for 4 hours (cardiac afterload = 50 mmHg) in isolated working hearts resulted in changes of myocardial substrate utilization different from changes which occur during experimental hemorrhagic or endotoxin shock (Spitzer et al 1975). The endotoxin molecule does not appear to have a direct cardiotoxic effect (Parker and Adams 1979).

We do have some hard facts with respect to pathophysiological occurrences within the myocardium. Hearts exhibiting dysfunction during endotoxin shock have accumulation of fluid between myofibrillar elements (Coalson et al 1972; Mela et al 1974) and dogs with *Pseudomonas aeruginosa* bacteremia and cardiac depression exhibit severe interstitial myocardial edema (Postel and Schloerb 1972) which may account for some cardiac depression (Cohn et al 1972; McLaurin et al 1973). Coalson et al (1972) reported both edema and rupture of myocardial mitochondria during endotoxin shock in dogs.

However, not all news is bad! Archer et al (1982) noted that oxygen uptake and carbon dioxide production in hearts depressed by live *E. coli* were elevated, indicating a productive rather than wasteful use of oxygen (Mela et al 1971). Mela et al (1974) also demonstrated normal functioning of mitochondria in dog hearts during endotoxemia.

Recent studies by Parker and Adams (1985) demonstrate very early (within 2 hrs) and direct myocardial depression in guinea pigs administered LD_{25} doses of *E. coli* endotoxin. The cardiac preparations used in these studies were contracting independently of depressive or supportive factors present in the intact animal.

Recent studies by Hess and colleagues (Hess et al 1977, 1980; Krause and Hess 1979; Soulsby et al 1978), and others (Estes et al 1979) suggest a causal relationship between histamine release and myocardial sarcoplasmic reticulum dysfunction during endotoxin shock in the dog. The most recent study by Hess et al (1980) showed an uncoupling of calcium transport from ATPase activity. In contrast, Soulsby et al (1978) and Estes et al (1979) reported a parallel depression of ATPase activity and calcium transport. An improved isolation procedure and elimination of 'nonspecific' ATPase activity in the more recent study may explain this difference (Hess et al 1980).

The Hess 'Cardiac Histamine Hypothesis' is based primarily on the observations that bolus histamine infusion causes a small but significant depression of both calcium transport and ATPase activity, with no uncoupling, and that pretreatment of animals with the histamine blocker diphenhydramine prevents some of the changes in myocardial sarcoplasmic reticular function, particularly in the subendocardial area during endotoxin shock (Hess et al 1980). However, it is difficult to reconcile this hypothesis with the fact that histamine causes coronary artery dilation and has a positive inotropic effect in normal animals (Giles et al 1977; McNeill and Muschek 1972; McNeill and Verma 1974; Pöch and Kukovetz 1967; Trendelenburg 1960). If a vasoactive hormone causes myocardial depression during shock, a more likely candidate would appear to be vasopressin. Plasma vasopressin concentrations are markedly elevated during endotoxin or live *E. coli* bacterial shock in the baboon and dog (Wilson et al 1981), and infusion of vasopressin into the left atrium causes depressed myocardial function in nonshocked isolated dog hearts (Wilson et al 1980).

2.6. Hypothesis 1: histamine is the primary 'shock toxin'

Histamine is released in an explosive manner, similar to an anaphylactic reaction, following intravenous administration of endotoxin from gram-negative microorganisms (Hinshaw et al 1960, 1961a, 1961b, 1962b; Reddin et al 1966; Vick 1964; Vick et al 1971). Plasma histamine concentrations reach high levels within 30–60 seconds of shock induction and are sustained for at least 10 minutes (Reddin et al 1966; Vick et al 1971).

However, histamine levels then return to normal, a fact which impugns a primary role for histamine in the pathogenesis of endotoxin shock. Hinshaw et al (1962b) investigated this problem in part, and their data suggest that histamine may participate in the progression of shock by triggering the sustained release of epinephrine-like agents (Haddy 1960a), which in turn superimpose their effects on those of histamine (Hinshaw et al 1962b).

The studies of Schayer (1960a, 1960b) suggest an additional mechanism for a continual action of histamine as shock progresses. Although endotoxin molecules are rapidly cleared from the plasma by the reticuloendothelial system (Buchanan and Filkins 1976; Zlydaszyk and Moon 1976), the activity of histidine decarboxylase was found to increase progressively and markedly for a period of 6 hours following an intravenous injection of *E. coli* endotoxin in the mouse (Schayer 1960b). Specific tissues in which this activity increased included skin, lung, liver, spleen, and carcus. Since histamine production parallels histidine decarboxylase activity, production of 'induced' histamine presumably increases as shock progresses and peaks when the animal becomes moribund (Schayer 1960a).

Significant increases in histidine decarboxylase activity in mouse skin have been shown following intraperitoneal injection of histamine, serotonin, histamine plus epinephrine, serotonin plus epinephrine or epinephrine alone (Schayer 1960a). Epinephrine plasma concentration is elevated throughout shock. The rapid release of histamine evidently acts indirectly by stimulating epinephrine release which in turn stimulates the activity of histidine decarboxylase (Schayer 1960b).

Plasma histamine concentration is not increased after 60 minutes of endotoxin shock (Reddin et al 1966). This may be explained on the basis that induced histamine is both produced and acts at the microvascular level. Actually, sites of 'induced' histamine production are not clearly defined, although their close proximity to microcirculatory vessels has been proposed (Schayer 1960a, 1960b, 1962, 1963). Also, a role of 'induced' histamine in either maintenance of normal hemostasis or in the pathogenesis of endotoxin shock has not been widely accepted but appears to be an area deserving additional investigation.

An interesting study by Jordan et al (1965) compares cardiovascular and tissue pathological effects of intravenously administered histamine and endotoxin in the dog. Three groups were studied: (a) a priming histamine dose of 1 mg was injected and an intravenous histamine infusion was then adjusted to maintain systemic arterial blood pressure at 50 mmHg for 240 minutes; (b) 20 mg of histamine was injected intravenously at time 0 and 2 hours later; (c) *E. coli* endotoxin was injected intravenously at 1–5 mg/kg.

In group *a*, continuous infusion of histamine resulted in maintained alterations of all cardiovascular parameters, which were similar to those produced by endotoxin shock (except for maintenance of the increased portal vein pressure). Autopsy and histological examination after 240 minutes of histamine infusion showed gastric mucosal erosions, loss of surface epithelium, lesions in the intestinal mucosa (especially with the villi tips), edema of the gallbladder, passive congestion of the liver with perivascular edema and hemorrhage, and mild to moderate congestion of the kidney and spleen.

Lesions in group *c* (endotoxin shock) were similar to those described above for the histamine infusion group.

In the group of dogs receiving histamine as an intravenous bolus injection (group *b*), cardiovascular alterations were transient and gastric mucosal lesions were not observed, although the other lesions were. The lesions were more variable and tended to be less severe.

The authors interpret their data as support for the histamine shock toxin hypothesis, suggesting that both an initial (bound) and continuing histamine release (possibly 'induced' histamine?) occurs during endotoxin shock.

A more recent study by Sleeman et al (1971) attacked this problem by comparing the effects of *E. coli* endotoxin on serum enzyme activity to the effects produced by a single, large intravenous injection of histamine. Initial cardiovascular and physiological abnormalities produced by endotoxin and histamine were similar. However, these variables returned to control within 3 hours in the histamine injection group but not in the endotoxin group. Significantly, the initial similarities could not be equated to tissue damage or cellular permeability changes. The measured variables in the histamine group were not significantly different from the saline control group.

On the other hand, induced rather than bound histamine could be responsible for the prolonged abnormalities of the cardiovascular system and the hepatosplanchnic pathology. On the negative side of this argument is the report by Schayer (1960a) that the mechanism for increasing the activity of histidine decarboxylase and hence synthesis of induced histamine may not exist in the intestine, brain, heart, and kidney; no increase in histidine decarboxylase activity was measured in these tissues during endotoxin shock in mice (Schayer 1960a).

While the present reviewer would tend to assign a positive, favorable influence to this lack of activity, it has been argued that the resulting inadequate amounts of histamine fail to balance the damaging vasoconstrictor effects of increased plasma levels of epinephrine (Schayer 1960a).

2.7. Hypothesis 2: histamine is not the primary 'shock toxin'

There is an impressive body of data showing that released (bound) histamine is not necessary for either the long-term cardiovascular/metabolic perturbations or lethality of endotoxin shock. To wit, prevention or amelioration of the initial effects of histamine on the cardiovascular system (Brake et al 1964; Hinshaw et al 1962a) does not increase survival (Hinshaw et al 1965).

While the sudden and severe hypotension associated with bolus endotoxin injection does not occur following slow infusion of endotoxin (Emerson and Gill 1967) or injection of live *E. coli* bacteria (Emerson and Kelly 1967), there is no difference in the ultimate outcome, viz., cardiovascular/metabolic alterations and mortality are similar in all three groups (Emerson and Gill 1967; Emerson and Kelly 1967). Looking at this same problem from a different angle, Jacobson et al (1964) found no increase in plasma histamine levels in adult dogs made hypotensive by sublethal doses of endotoxin. It may also be instructive to compare the response of adults to those of neonate and infant animals.

Reddin et al (1966) reported that while endotoxin injections were lethal in both puppies and adult dogs, the puppies exhibited no sudden decrease in arterial blood pressure. Plasma histamine concentration did not increase in the puppies during an observation period of 1–180 minutes post endotoxin, whereas it increased strikingly in the adults (0–0.24 µg/ml by 1 min and 0.27 µg/ml by 5 min). The same difference in arterial blood pressure responses between puppies and adults was also reported earlier by Hinshaw et al (1962a).

Other studies which do not support the histamine 'shock toxin' hypothesis are those which suggest a protective role for histamine during endotoxin shock in mice and adult dogs (Fox and Lasker 1962; Markley et al 1971).

It might also be appropriate to mention here that the much touted role of histamine in promoting loss of vascular fluid during shock does not hold up well under close examination. As discussed in Section 2.3.2, a recent study in unstressed dogs clearly demonstrates that long-term intravenous infusion of histamine does not cause pulmonary edema (Olson 1982; Olson et al 1983). Also, long-term intravenous infusion of histamine at rates sufficient to markedly lower arterial blood pressure has little effect on capillary filtration in dog forelimb, skin, or skeletal muscle (Grega et al 1980; Marciniak et al 1977). Histamine does decrease the ratio between pre- and postcapillary resistance, which alone would increase capillary hydrostatic pressure and fluid filtration. However, this decrease in the ratio between pre- and postcapillary resistance is counteracted by a proportionate or proportionately greater decrease in arterial and venous blood pressures, two other major determinants of capillary blood volume and hence hydrostatic pressure.

Although the histamine 'shock toxin' hypothesis is not supported by these data, it should not be completely ruled out because of the relatively uninvestigated role of induced histamine (Schayer 1960a, 1960b) in the pathogenesis of endotoxin shock (Schayer 1962, 1963).

3. 5-HYDROXYTRYPTAMINE (5-HT; SEROTONIN)

3.1. General

A serum vasoconstrictor agent, which later turned out to be serotonin, was known to exist since 1868 (cited by Heinzelman and Weisblat 1951) but not identified as 5-hydroxytryptamine until the late 1940s (Rapport et al 1948;

Rapport 1949). Serotonin was first synthesized in 1951 (Hamlin and Fischer; Heinzelman and Weisblat), and since this time research concerning its various biological activities has been extensive. Included in these investigations are its possible involvement in pulmonary hypertension associated with certain disorders, its ability to promote platelet and white blood cell aggregation, and its role as a neurotransmitter.

3.1.1. Storage

Serotonin is synthesized in the central nervous system (Stacey 1959) and intestine (Humphrey and Toh 1954) from tryptophan (Warner 1967) and is distributed throughout the body, primarily by platelets (Stacey 1959; Whelan 1959). Serotonin is taken up actively by platelets against a several hundred-fold concentration gradient and its uptake is energy- and temperature-dependent (Burningham et al 1966; Weissbach and Redfield 1960). While over half of the plasma serotonin is in platelets (Stacey 1959; Whelan 1959), it is also localized in mast cells (Stacey 1959) and some exists as 'free' serotonin in plasma (Robertson and Andrews 1961).

In addition to plasma serotonin, large amounts of this amine are located throughout the brain and central nervous system (Stacey 1959) and the gastrointestinal tract (Reid and Rand 1951; Rosenberg et al 1961). Serotonin has also been found in the uterus and placenta (Klinge et al 1964; Reid and Rand 1951), spleen (Sivo Sankar et al 1961), pancreas (Tobe et al 1966), kidney (Whelan 1959), adrenal medulla (Stacey 1959) and pineal gland (Giarman and Freedman 1960).

3.1.2. Release

Serotonin is released from platelets by a variety of factors (Reid 1952; Westerholm 1966), including endotoxin from gram-negative bacteria (Kobold et al 1964).

3.2. Cardiovascular effects

The effects of serotonin on the circulatory system are many and varied, and the specific responses to this amine depend on the species of animal and doses given; also, response variations have been noted within the same species under similar conditions (Kabins et al 1959; Page and McCubbin 1956). A comparison of the systemic blood pressure response to serotonin of the rabbit with that of the dog shows how striking species variations can be. Arterial blood pressure changes in the dog are inconsistent and unimpressive during intravenous infusion of serotonin (Daugherty et al 1968), but in the rabbit, serotonin causes severe and prolonged hypotension, similar in fact to that caused by bradykinin (Cuevas and Fine 1973). In the dog, serotonin cannot accurately be described as a pressor or depressor agent since it may increase, decrease, or have little effect on systemic arterial blood pressure (Haddy 1960b). Alterations in systemic arterial blood pressure are unimpressive, particularly compared to chose caused by histamine, catecholamines, or angiotensin II. For example, large amounts infused intravenously have little effect on systemic arterial blood pressure (Daugherty et al 1968), venous return, or cardiac output (Emerson 1968).

3.3. Effects on organs

3.3.1. General

Peripheral vascular effects of serotonin are also unique and interesting. As with systemic arterial blood pressure, serotonin may increase, decrease, or have little effect on total vascular resistance in the in vivo perfused dog forelimb (Daugherty et al 1968) or gracilis muscle (Emerson et al 1973). Directional vascular resistance changes during intra-arterial serotonin administration in skeletal muscle (Emerson et al 1973) or other organs (Haddy 1960b) can be predicted and depend upon the existing level of vascular resistance; to wit, if the initial resting resistance is low, high, or in between, serotonin causes constriction, dilation, or no effect, respectively. This unusual phenomenon has been explained on the basis of differential effects of serotonin on small and large arteries (Daugherty et al 1968; Haddy and Scott 1966). On the other hand, serotonin always causes constriction in forelimb skin (Daugherty et al 1968; Emerson et al 1973) and always causes an increase in tension in isolated arterial vessel strips from both skin and skeletal muscle (Iturri and Emerson 1971). Serotonin apparently has little effect on microvascular permeability to protein or transvascular fluid movement (Daugherty et al 1968).

3.3.2. Lung

Serotonin administration produces a marked pressor response in the pulmonary circulatory system (MacCanon and Horvath 1954) with a rapid increase in pulmonary artery pressure (MacCanon and Horvath 1954; Maxwell et al 1959; Reid and Rand 1951) which is caused, at least in part, by pulmonary arteriolar and venous constriction (Kabins et al 1959; Noble and Nanson 1959). This increase in pulmonary vascular resistance during serotonin administration is transient in nature (Haddy and Scott 1966; Schneider et al 1967) and is most likely due to both active vasoconstriction of pulmonary vessels (Hinshaw et al 1957) and passive factors, particularly aggregation of platelets and leukocytes (Hissen and Swank 1965; Swank et al 1963; Swank and Edwards 1968).

3.3.3. Heart

Serotonin has a direct inotropic effect on the intact dog heart and isolated cat papillary muscle (Buccino et al 1967) and increases right ventricular work (Maxwell et al 1959). It also causes tachycardia in the dog (Maxwell et al 1959; Noble and Nanson 1959) and reportedly increases cardiac output in the rat (Takacs and Vajda 1963).

3.4. Activity during endotoxin shock

This subject was reviewed by the present author a decade ago (Emerson 1972). Since then the status of serotonin in the shock scenario has not changed significantly. It is clear that serotonin is released during the early phase following endotoxin administration. The in vivo and in vitro work of Kobold et al (1964) demonstrated an early release of serotonin from platelets in whole blood and from splanchnic visceral organs following endotoxin administration. Davis et al (1960, 1961) noted that plasma serotonin levels rise suddenly following intravenous injections of endotoxin in dogs, peak at about 30 seconds, and return to near control within 5 minutes. Rosenberg et al (1959, 1961) and Jacobson et al (1964) have shown that serum and plasma serotonin concentrations decrease markedly from 5 minutes post endotoxin injection.

Considering its transient increase in plasma following endotoxin shock induction and its relatively minor effects on cardiovascular function, the role of this plasma amine in the pathogenesis of endotoxin shock appears to be minimal in most species. In further support of this view are studies by Cuevas and Fine (1973) which suggest that serotonin per se does not produce lasting tissue injury, except in rabbits; and by Markley et al (1971), which proposes a protective role for serotonin against burn, tourniquet, or endotoxin shock in mice. It may also be relevant that Sibbald et al (1980) found no significant difference in platelet numbers or serum serotonin levels across the pulmonary vascular bed in septic patients with the Adult Respiratory Distress Syndrome, which led them to conclude that platelet sequestration and/or serotonin were probably not major factors in this condition.

Finally, serotonin may be involved in the transient pulmonary hypertension characteristic of the early phase of endotoxin shock, due to an active (Hinshaw et al 1957) and/or passive increase in vascular resistance as a result of platelet-leukocyte aggregation (Swank et al 1963, 1964; Swank and Edwards 1968); these transient perturbations could conceivably play a role in lung injury sometimes associated with endotoxin shock (Demling and Flynn 1982). This possible role of serotonin in the pathogenesis of endotoxin shock has not been thoroughly explored.

4. ANGIOTENSIN II AND RENIN

4.1. Formation

Plasma angiotensin II results from a reaction cascade, beginning with the release of renin from the kidneys which cleaves plasma angiotensinogen to form the relatively inactive decapeptide angiotensin I (Skeggs et al 1954). Angiotensin I is subsequently converted to the vasoactive polypeptide angiotensin II by a converting enzyme in the lung.

It is significant that circulatory shock and alterations in autonomic nerve activity are potent stimuli for renin release and the ensuing increase in plasma angiotensin formation (Brown et al 1966; Jakschik et al 1974; Morton et al 1977; Thames et al 1977).

4.2. Cardiovascular effects

4.2.1. General

A pressor substance associated with the kidney was recognized as early as 1898 (Tigerstedt and Bergman), but it was not until 1940 that Page and Helmer in the USA and Braun-Menendez et al in Venezuela independently identified the active pressor substance. This was named angiotensin by the former investigators and hypertensin by the latter. However, in 1958 Braun-Menendez and Page jointly proposed the 'bastard' word angiotensin, a name which holds to date.

4.2.2. Specific

Intravenous infusion (5 min) of angiotensin II in open-chest dogs causes a steady-state decrease in venous return and cardiac output, a proportionately greater increase in total peripheral resistance, and hence a large increase in mean systemic arterial blood pressure (Emerson 1966). In these animals, angiotensin II does not alter pulmonary artery or left atrial pressures; pulmonary vascular resistance tends to increase on the average, but this increase is not significant (Emerson 1966). Similar effects of angiotensin on dog arterial blood pressure, cardiac output, pulmonary artery pressure, and left atrial pressure have also been reported by McCaa et al (1967) following intravenous bolus injection. The chronotropic effect attributed to angiotensin is apparently not direct but it is mediated via the release of catecholamines from myocardial adrenergic nerve endings (Krasney et al 1967).

Angiotensin II is an extremely potent vasoconstrictor in peripheral organs (Daugherty et al 1968; Emerson et al 1965), with its site of action being predominantly the precapillary resistance vessels. Angiotensin II has relatively minor effects on postcapillary resistance vessels (Emerson et al 1965). Hence, this polypeptide decreases microvascular hydrostatic pressure and would not be expected to increase microvascular filtration during circulatory shock.

An interesting property of angiotensin II is the rapid development of tachyphylaxis during intravenous infusion or repeated bolus injections (Emerson et al 1965; Khairallah et al 1966), which is perhaps fortunate considering the severe ischemia of most peripheral organs which would occur during long-term exposure to high plasma levels of angiotensin II. This phenomenon was first shown in 1898 by Tigerstedt and Bergman as a decreased pressor response to repeated injections of crude renal extracts. This self-inhibition has subsequently been demonstrated in many organs and species, including perfused isolated hindlimbs and kidneys of rats, isolated rabbit heart, superior cervical ganglion of cats, guinea pig innervated vas deferens, innervated spleen of cats, perfused iris-ciliary body and enucleated whole eyes, adrenal medulla, isolated dog coronary arteries and isolated perfused human umbilical arteries (see Khairallah et al 1966 for specific citations). Work by Page's group (Khairallah et al 1966) demonstrated tachyphylaxis to angiotensin and some analogs on spirally cut arterial strips from cat, dog, sheep, and rat, and on venous blood vessel strips from rabbit and cat; however, arterial strips from rabbit and guinea pig did not develop tachyphylaxis.

4.2.3. *Activity during endotoxin shock*

Plasma concentrations of renin and angiotensin II increase early and remain elevated throughout the course of endotoxin shock (Hall and Hodge 1971; Isakson et al 1977; White et al 1967). While angiotensin is a potent vasoconstrictor, tachyphylaxis is apparently protective against long-term ischemia; indeed, the renal vascular and systemic pressor responses to infused angiotensin II wane early during endotoxin shock and are essentially lost within 30–60 minutes (Emerson et al 1964). Renin and angiotensin II plasma concentrations are also increased during hemorrhagic shock (Yamashita et al 1977).

Finally, even though angiotensin II decreases venous return, cardiac output, and blood flow through peripheral organs, there is little convincing evidence that it plays a major role in the pathogenesis of endotoxin shock. To the contrary, angiotensin II has been proposed by several investigators to be an effective adjunctive therapeutic agent during circulatory shock of various etiologies (Beaty et al 1982; Del Greco and Johnson 1961; Derrick et al 1962; Nassif et al 1963). However, its use for treatment of shock has been questioned by others, particularly in preference to norepinephrine or metaraminol (Udhoji and Weil 1964).

5. SUMMARY

The release and involvement of histamine, serotonin and renin-angiotensin II during endotoxin shock have been discussed, and the role of histamine and the heart-lung systems in the pathogenesis of endotoxin shock has been emphasized.

With specific reference to the vasoactive hormones covered in this review, involvement of 'induced' histamine in the pathogenesis of endotoxin shock has not been adequately investigated. This vasoactive agent has a particularly detrimental effect on microvascular fluid fluxes, increasing vascular fluid filtration via increasing microvascular permeability, hydrostatic pressure and surface area. Its continual production at discrete microvascular sites should not be ignored.

Serotonin does not appear to play a major role in the pathogenesis of endotoxin shock, although it could be responsible for some of the early pulmonary problems characteristic of this condition (e.g., platelet and leukocyte aggregation).

Angiotensin II is a potential candidate for detrimental action during endotoxin shock, with particular reference to its potent constriction of precapillary resistance vessels. However, after reviewing the literature relative to the activity of angiotensin during endotoxin shock, especially the rapid development of tachyphylaxis, its involvement in the pathogenesis of endotoxin shock seems unlikely.

6. GLOBAL EVALUATION OF ENDOTOXIN SHOCK AND AREAS FOR FUTURE WORK

An effective therapeutic regimen for endotoxin shock has yet to be developed, largely because we do not understand the basic underlying mechanisms of irreversibility. It seems likely that an effective pre- and/or very early post-treatment therapeutic regimen against endotoxin shock will be developed in the foreseeable future, but it is equally likely that an effective mid to late post-treatment regimen will not. Once the pathological alterations in systems such as the cardiovascular, hormonal, metabolic, immunological, etc., are firmly established and cell damage occurs, it is simply not possible to support life long enough for the body to repair itself. Also, certain cells such as those of the central nervous system cannot be rejuvinated or replaced. In fact, once this stage is reached the wisdom of maintaining life by heroic efforts and sophisticated life support systems must be questioned; at this point, recovery or death seems to be more in the 'hands of God' than 'Man.'

During the past decade, emphasis has progressively shifted from solving the elusive problem of treating endotoxin shock towards preventing its occurrence. Clinically, low cardiac output septic (endotoxin) shock is often preceded by a period of lingering high cardiac output sepsis and the attendant perturbations involving all major physiological systems.

While the cardiovascular perturbations during chronic gram-negative sepsis and classical septic (endotoxin) shock are exactly opposite, many of the metabolic ones are similar. With this brief background, perhaps the most important unanswered question is: why do patients pass from high cardiac output sepsis to low cardiac output septic (endotoxin) shock, that is, what is the mechanism? Is this transition the result of peripheral factors which decrease cardiac output via a decrease in venous return, or is it the result of cardiac depression secondary to a plasma mediator, a loss of myocardial contractile units (myocardial protein 'wasting'?) in the face of increased cardiac work, or a combination of these factors? Other important unanswered questions relating to chronic sepsis involve: (a) the inability of apparently normal cells to adequately utilize plasma substrates (e.g., free fatty acids, amino acids) and the increased formation and storage of triglycerides; (b) disseminated intravascular coagulation and pseudohemophilia; (c) inadequate immunological function, including severely decreased plasma levels of fibronectin; and (d) insulin resistance (glucose intolerance).

In addition to the constant danger of passing from high cardiac output sepsis to low cardiac output septic shock, there is the unpredictable but sudden development of the Adult Respiratory Distress Syndrome.

I believe the final resolution of this problem will involve a multifactorial approach. Dr. Charles Brake, a deceased former colleague, Dr. Lerner Hinshaw, my former mentor, and I spent many hours during the early 1960s discussing and trying to predict the ultimate therapeutic regimen. Dr. Brake steadfastly favored a 'cocktail' containing several different products, each targeted against a specific abnormality. Indeed, my belief in this concept has grown steadily; correcting one abnormality or another alone cannot be nearly as efficacious as correcting several major ones simultaneously. For example, a super 'shock cocktail' containing antibiotics, immune globulins and fibronectin to eliminate the infection and shore up the immunological system, antithrombin to ameliorate the clotting abnormalities, a proteinase inhibitor to fight the development of the Adult Respiratory Distress Syndrome and an effective substrate for metabolism (e.g., intravenous lipid solution) might effectively promote recovery and prevent the transition from high cardiac output sepsis to low cardiac output septic shock.

I am optimistic that an effective therapeutic regimen for high cardiac output sepsis will be developed in the not too distant future, but do not hold the same optimism for the middle- to late-phase low cardiac output endotoxin shock. In terms of recovery from this condition, I believe a quote favored by the late Dr. Richard Lillehei that 'the spirit is willing but the flesh is weak' is still valid. Furthermore, based upon state-of-the-art treatments available now or on the horizon for classical septic (endotoxin) shock, the best advice for the physician shopping for a good treatment is perhaps 'caveat emptor'.

This should not necessarily be taken as bad news or as a negative outlook, because development of an efficacious regimen for high cardiac output sepsis will in most cases make the problem of endotoxin shock a moot point.

ACKNOWLEDGEMENTS

My sincere appreciation goes to Ms. Julie Tast for the many hours spent on this manuscript, including word processing, proofing, and editing. In short, many thanks for a superior effort beyond the call of duty. Mr. John Lurton and Mr. Michael Everett, Associate Research Physiologists, also deserve credit for assisting in editing this manuscript.

My gratitude also goes to Mrs. Wen Ng, Director of Library Services at The Cutter Group of Miles Laboratories, Inc., and her dedicated staff, Mrs. Shelly Monson and Mrs. Beatrice Yuan, for their truly excellent job of verifying the 200-plus references in this review.

REFERENCES

Ackerman GA (1963) Cytochemical properties of the blood basophilic granulocyte. *Ann. N.Y. Acad. Sci. 103*, 376-393.
Altura BM, Zweifach BW (1967) Endogenous histamine formation and vascular reactivity. *Am. J. Physiol. 212*, 559-564.
Andrews WHH, Heckler R, Maegraith BG, Ritchie HD (1955) The action of adrenaline, L-noradrenaline, acetylcholine and other substances on the blood vessels of the perfused canine liver. *J. Physiol. (London) 128*, 413-434.
Archer LT, Black MR, Hinshaw LB (1975) Myocardial failure with altered response

to adrenaline in endotoxin shock. *Br. J. Pharmacol. 54*, 145-155.

Archer LT, Beller BK, Drake JK, Whitsett TL, Hinshaw LB (1978) Reversal of myocardial dysfunction in endotoxin shock with insulin. *Can. J. Physiol. Pharmacol. 56*, 132-138.

Archer LT, Benjamin BA, Beller-Todd BK, Brackett DJ, Wilson MF, Hinshaw LB (1982) Does LD_{100} E. coli cause myocardial failure? *Circ. Shock 9*, 7-16.

Ayres SM, (1977) The shock lung. In: Thompson WL (Moderator), *The Organ in Shock*, pp. 24-31. Scope Publication, Kalamazoo.

Barger G, Dale HH (1910) Die physiologische Wirkung einer Secalebase und deren Identifizierung als Imidazolylathylamin. *Zentralbl. Physiol 24*, 885-889.

Beaty O III, Sloop CH, Schmid HE Jr (1982) Action of angiotensin II on renal blood flow and function during hemorrhagic shock. *Circ. Shock 9*, 117-128.

Bloom GD, Haegermark O (1965) A study of morphological changes and histamine release increase by compound 48/80 in rat peritoneal mast cells. *Exp. Cell. Res. 40*, 637-654.

Brake CM, Emerson TE Jr, Wittmers LE, Hinshaw LB (1964) Alteration of vascular responses to endotoxin by adrenergic blockade. *Am. J. Physiol. 207*, 149-151.

Braun-Menendez E, Fasciolo JC, Leloir LF, Munoz JM (1940) The substance causing renal hypertension. *J. Physiol. (London) 98*, 283-298.

Braun-Menendez E, Page IH (1958) Suggested revision of nomenclature–angiotensin. *Science 127*, 242.

Brigham KL, Bowers RE, Haynes J (1979) Increased sheep lung vascular permeability caused by *Escherichia coli* endotoxin. *Circ. Res. 45*, 292-297.

Brigham KL, Owen PJ (1975) Increased sheep lung vascular permeability caused by histamine. *Circ. Res. 37*, 647-657.

Brody JS, Stemmler EJ (1968) Differential reactivity in the pulmonary circulation. *J. Clin. Invest. 47*, 800-808.

Brown JJ, Davies DL, Lever AF, Robertson JIS, Verniory A (1966) The effect of acute hemorrhage in the dog and man on plasma renin concentration. *J. Physiol. (London) 182*, 649-663.

Buccino RA, Covell JW, Sonnenblick EH, Braunwald E (1967) Effects of serotonin in the contractile state of the myocardium. *Am. J. Physiol. 213*, 483-486.

Buchanan BJ, Filkins JP (1976) Hypoglycemic depression of RES function. *Am. J. Physiol. 231*, 265-269.

Burn JH, Dale HH (1926) The vasodilator action of histamine and its physiological significance. *J. Physiol. (London) 61*, 185-214.

Burningham RA, Arimura GK, Yunis AA (1966) Effect of monase and related compounds on uptake of 5-hydroxytryptamine by platelets. *Proc. Soc. Exp. Biol. Med. 122*, 711-714.

Cann M, Stevenson T, Fiallos E, Thal AP (1972) Depressed cardiac performance in sepsis. *Surg. Gynecol. Obstet. 134*, 759-763.

Capurro N, Levi R (1975) The heart as a target organ in systemic allergic reactions; comparison of cardiac anaphylaxis in vivo and in vitro. *Circ. Res. 36*, 520-528.

Carlson RW (1977) Oncotic and hydrostatic forces in perfusion failure and volume loading. In: Thompson WL (Moderator), *The Organ in Shock*, pp 50-56. Scope Publication, Kalamazoo.

Cerra FB, Hassett J, Siegel JH (1978) Vasodilator therapy in clinical sepsis with low output syndrome. *J. Surg. Res. 25*, 180-183.

Coalson JJ, Woodruff HK, Greenfield LJ, Guenter CA, Hinshaw LB (1972) Effects of digoxin on myocardial ultrastructure in endotoxin shock. *Surg. Gynecol. Obstet. 135*, 908-912.

Cohn PF, Liedtke AJ, Serur J, Sonnenblick EH, Urschel CW (1972) Maximal rate of pressure fall (peak negative dp/dt) during ventricular relaxation. *Cardiovasc. Res. 6*, 263-267.

Coode CF (1952) Histamine in blood. *Physiol. Rev. 32*, 47-65.

Cross SAM (1973) Distribution of histamine, burimamide and metiamide and their interactions as shown by autoradiography. In: Wood CJ, Simkins MA (Eds), *International Symposium on Histamine H2-Receptor Antagonists*, pp 73-83. Deltakas, London.

Cuevas P, Fine J (1973) Production of fatal endotoxin shock by vasoactive substances. *Gastroenterology 64*, 285-291.

Dale HH, Laidlaw PP (1910) The physiological action of β-iminazolylethylamine. *J. Physiol. (London) 41*, 318-344.

Dale HH, Laidlaw PP (1911) Further observations on the action of β-iminazolylethylamine. *J. Physiol. (London) 43*, 182-195.

Dale HH, Laidlaw PP (1918-1919) Histamine shock. *J. Physiol. (London) 52*, 355-390.

Dale HH, Richards AN (1918) The vasodilator action of histamine and of some other substances. *J. Physiol. (London) 52*, 110-165.

Daugherty RM Jr, Scott JB, Emerson TE Jr, Haddy FJ (1968) Comparison of IV and IA infusion of vasoactive agents on dog forelimb blood flow. *Am. J. Physiol. 214*, 611-619.

Davis RB, Meeker WR, McQuarrie DC (1960) Immediate effects of intravenous endotoxin on serotonin concentrations and blood platelets. *Circ. Res. 8*, 234-239.

Davis RB, Meeker WR, Bailey WL (1961) Serotonin release by bacterial endotoxin. *Proc. Soc. Exp. Biol. Med. 108*, 774-776.

Del Greco F, Johnson DC (1961) Clinical experience with angiotensin II in the treatment of shock. *J. Am. Med. Assoc. 178*, 994-999.

Demling RH, Flynn JT (1982) Humoral factors and lung injury during shock, trauma and sepsis. In: Cowley RA, Trump BF (Eds), *Pathophysiology of Shock, Anoxia, and Ischemia*, pp 395-407. Williams & Wilkins, Baltimore-London.

Derrick JR, Anderson JR, Roland BJ (1962) Adjunctive use of a biologic pressor agent, angiotensin, in management of shock. *Circulation 25*, 263-267.

Drake RE, Gabel JC (1980) Effect of histamine and alloxan on canine pulmonary vascular permeability. *Am. J. Physiol. 239*, H96-H100.

Drake R, Adair T, Traber D, Gabel J (1981) Contamination of caudal mediastinal node efferent lymph in sheep. *Am. J. Physiol. 241*, H354-H357.

Edvinsson L, Owman C (1975) A pharmacologic comparison of histamine receptors in isolated extracranial and intracranial arteries in vitro. *Neurology 25*, 271-276.

Emerson TE Jr (1966) Effect of angiotensin, norepinephrine, epinephrine, and vasopressin on venous return. *Am. J. Physiol. 210*, 933-942.

Emerson TE Jr (1967) Changes of venous return and other hemodynamic parameters during bradykinin infusion. *Am. J. Physiol. 212*, 1455-1460.

Emerson TE Jr (1968) Effects of acetylcholine, histamine, and serotonin infusion on venous return in dogs. *Am. J. Physiol. 215*, 41-48.

Emerson TE Jr (1972) Participation of endogenous vasoactive agents in the pathogenesis of endotoxin shock. *Adv. Exp. Med. Biol. 23*, 25-45.

Emerson TE Jr, Hinshaw LB, Brake CM (1965) Vascular effects of angiotensin and norepinephrine in the dog, cat and monkey. *Am. J. Physiol. 208*, 260-264.

Emerson TE Jr, Gill CC (1967) Effects of slow intravenous endotoxin infusion on hemodynamics and survival in dogs. *J. Appl. Physiol 22*, 874-877.

Emerson TE Jr, Kelly FC (1967) Effect of living E. coli cells on hemodynamics and mortality in the dog. *J. Appl. Physiol. 23*, 609-612.

Emerson TE Jr, Brake CM, Hinshaw LB (1964) Renal vascular response to angiotensin and other vasoactive agents in endotoxin shock. *Am. J. Physiol. 207*, 1260-1264.

Emerson TE Jr, Meier PD, Daugherty RM Jr (1973) Dependency of skeletal muscle vascular response to serotonin upon the level of vascular resistance. *Proc. Soc. Exp. Biol. Med. 142*, 1185-1188.

Estes JE, Farley PE, Goldfarb RD (1979) Shock serum reduces the ability of cardiac sarcoplasmic reticulum to accumulate calcium. *Circ. Shock 6*, 187.

Falck B, Nystedt T, Rosengren E, Stenflo J (1964) Dopamine and mast cells in ruminants. *Acta Pharmacol. 21*, 51-58.

Fawcett DW (1954) Cytological and pharmacological observations on the release of histamine by mast cells. *J. Exp. Med. 100*, 217-224.

Feldberg W, Smith AN (1953) Release of histamine by tryptamine and 5-hydroxytryptamine. *Br. J. Pharmacol. 8*, 406-411.

Fox CL Jr, Lasker SE (1962) Protection by histamine and metabolites in anaphylaxis, scalds and endotoxin shock. *Am. J. Physiol. 202*, 111-113.

Gabbiani G, Badonnel MC, Gervasoni C, Portmann B, Majno G (1972) Carbon deposition in bronchial and pulmonary vessels in response to vasoactive compounds. *Proc. Soc. Exp. Biol. Med. 140*, 958-962.

Giarman NJ, Freedman DX (1960) Serotonin content of the pineal glands of man and monkey. *Nature (London), 186*, 480-481.

Gilbert RP, Hinshaw LB, Kuida H, Visscher MB (1958) Effects of histamine, 5-hydroxytryptamine and epinephrine on pulmonary hemodynamics with particular reference to arterial and venous segment resistances. *Am. J. Physiol. 194*, 165-170.

Giles RW, Heise G, Wilcken DEL (1977) Histamine receptors in the coronary circulation of the dog. *Circ. Res. 40*, 541-546.

Glazier JB, Murray JF (1971) Sites of pulmonary vasomotor reactivity in the dog during alveolar hypoxia and serotonin and histamine infusion. *J. Clin. Invest. 50*, 2550-2558.

Goetzman BW, Visscher MB (1969) The effects of alloxan and histamine on the permeability of the pulmonary alveolar-capillary barrier to albumin. *J. Physiol. (London) 204*, 51-61.

Goldfarb RD (1982) Cardiac mechanical performance in circulatory shock: a critical review of methods and results. *Circ. Shock 9*, 633-653.

Grega GJ, Daugherty RM Jr, Scott JB, Radawski DP, Haddy FJ (1971) Effect of pressure, flow and vasoactive agents on vascular resistance and capillary filtration in canine fetal, newborn, and adult lung. *Microvasc. Res. 3*, 297-307.

Grega GJ, Kline RL, Dobbins DE, Haddy FJ (1972) Mechanisms of edema formation by histamine administered locally into canine forelimbs. *Am. J. Physiol. 223*, 1165-1171.

Grega GJ, Marciniak DL, Jandhyala BS, Raymond RM (1980) Effects of intravenously infused histamine on canine forelimb transvascular protein efflux following adrenergic receptor blockade. *Circ. Res. 47*, 584-591.

Guyton AC, Sagawa K (1961) Compensations of cardiac output and other circulatory functions in areflex dogs with large A-V fistulas. *Am. J. Physiol. 200*, 1157-1163.

Haddy FJ (1960a) Effect of histamine on small and large vessel pressures in the dog foreleg. *Am. J. Physiol. 198*, 161-168.

Haddy FJ (1960b) Serotonin and the vascular system. *Angiology 11*, 21-24.

Haddy FJ, Scott JB (1966) Cardiovascular pharmacology. *Ann. Rev. Pharmacol. 6*, 49-76.

Haddy FJ, Scott JB, Grega GJ (1972) Effects of histamine on lymph protein concentration and flow in the dog forelimb. *Am. J. Physiol. 223*, 1172-1177.

Haddy FJ, Scott JB, Grega GJ (1976) Peripheral circulation: fluid transfer across the microvascular membrane. *Int. Rev. Physiol. 9*, 63-109.

Hall RC, Hodge RL (1971) Vasoactive hormones in endotoxin shock: a comparative study in cats and dogs. *J. Physiol. (London) 213*, 69-84.

Halpern BN (1956) Histamine release by long-chain molecules. In: Wolstenholme GEW, O'Connor CM (Eds), *Ciba Foundation Symposium on Histamine*. Little, Brown and Co., Boston.

Hamlin KE, Fischer FE (1951) The synthesis of 5-hydroxytryptamine. *J. Am. Chem. Soc. 73*, 5007-5008.

Hebb C (1969) Motor innervation of the pulmonary vessels of mammals. In: Fishman AP, Hecht HH (Eds), *The Pulmonary Circulation and Interstitial Space*, pp 195-222. University of Chicago Press, Chicago.

Heinzelman RV, Weisblat DI (1951) The synthesis of the blood serum vasoconstrictor principle serotonin creatinine sulfate. *J. Am. Chem Soc. 73*, 5514-5515.

Hess ML, Soulsby ME, Davis JA, Briggs FN (1977) The influence of venous return on cardiac mechanical and sarcoplasmic reticulum function during endotoxemia. *Circ. Shock 4*, 143-152.

Hess ML, Krause SM, Kornwatana P (1980) Myocardial failure and excitation–contraction uncoupling in canine endotoxin shock: role of histamine and the sarcoplasmic reticulum. *Circ. Shock 7*, 277-287.

Hinshaw LB, Kuida H, Gilbert RP, Visscher MB (1957) Influence of perfusate characteristics on pulmonary vascular response to endotoxin. *Am. J. Physiol. 191*, 293-295.

Hinshaw LB, Vick JA, Carlson CH, Fan YL (1960) Role of histamine in endotoxin shock. *Proc. Soc. Exp. Biol. Med. 104*, 379-381.

Hinshaw LB, Jordan MM, Vick JA (1961a) Histamine release and endotoxin shock in the primate. *J. Clin. Invest. 40*, 1631-1637.

Hinshaw LB, Jordan MM, Vick JA (1961b) Mechanism of histamine release in endotoxin shock. *Am. J. Physiol. 200*, 987-989.

Hinshaw LB, Emerson TE, Iampietro PF, Brake CM (1962a) A comparative study of the hemodynamic actions of histamine and endotoxin. *Am. J. Physiol. 203*, 600-606.

Hinshaw LB, Vick JA, Jordan MM, Wittmers LE (1962b) Vascular changes associated with the development of irreversible endotoxin shock. *Am. J. Physiol. 202*, 103-110.

Hinshaw LB, Brake CM, Emerson TE Jr (1965) Biochemical and pathological alterations in endotoxin shock. In: Mills LC, Moyer JH (Eds), *Shock and Hypotension: Pathogenesis and Treatment,* pp 431-441. Grune & Stratton, New York.

Hinshaw LB, Reins DA, Hill RJ (1966) Response of isolated liver to endotoxin. *Can. J. Physiol. Pharmacol. 44*, 529-541.

Hinshaw LB, Archer LT, Black MR, Greenfield LJ, Guenter CA (1973) Prevention and reversal of myocardial failure in endotoxin shock. *Surg. Gynecol. Obstet. 136*, 1-11.

Hinshaw LB, Archer LT, Black MR, Elkins RC, Brown PP, Greenfield LJ (1974a) Myocardial function in shock. *Am. J. Physiol. 226*, 357-366.

Hinshaw LB, Archer LT, Spitzer JJ, Black MR, Peyton MD, Greenfield LJ (1974b) Effects of coronary hypotension and endotoxin on myocardial performance. *Am. J. Physiol. 227*, 1051-1057.

Hirose T, Ikeda T, Aoki E, Hara N (1978) The protective effect of PGI2 on increased lung vascular permeability caused by endotoxin in dogs. *Nippon Kyobu Shikkan Gakkai Zasshi 16*, 410-417.

Hissen W, Swank RL (1965) Screen filtration pressure and pulmonary hypertension. *Am. J. Physiol. 209*, 715-722.

Humphrey JH, Toh CC (1954) Absorption of serotonin (5-hydroxytryptamine) and histamine by dog platelets. *J. Physiol. (London), 124*, 300-304.

Isakson PC, Shofer F, McKnight RC, Feldhaus RA, Raz A, Needleman P (1977) Prostaglandins and the renin-angiotensin system in canine endotoxemia. *J. Pharmacol. Exp. Ther. 200*, 614-622.

Iturri S, Emerson TE Jr (1971) Responses of arterial vessel strips from skin and muscle to serotonin. *Proc. Soc. Exp. Biol. Med. 137*, 416-420.

Jacobson ED (1963) Effects of histamine, acetylcholine and norepinephrine on gastric vascular resistance. *Am. J. Physiol. 204*, 1013-1017.

Jacobson ED (1968) The gastrointestinal circulation. *Annu. Rev. Physiol. 30*, 133-146.

Jacobson ED, Mehlman B, Kalas JP (1964) Vasoactive mediators as the 'trigger mechanism' of endotoxin shock. *J. Clin. Invest. 43*, 1000-1013.

Jakschik BA, Marshall GR, Kourik JL, Needleman P (1974) Profile of circulating vasoactive substances in hemorrhagic shock, and their pharmacological manipulation. *J. Clin. Invest. 54*, 842-852.

Jordan MM, Holmes DD, Hinshaw LB (1965) Pathophysiological comparisons of histamine and endotoxin shock. *J. Trauma 5*, 726-736.

Kabins SA, Molina C, Katz LN (1959) Pulmonary vascular effects of serotonin (5-hydroxytryptamine) in dogs: its role in causing pulmonary edema. *Am. J. Physiol. 197*, 955-958.

Kahlson G, Rosengren E (1968) New approaches to the physiology of histamine.

Physiol. Rev. 48, 155-196.

Kerstein MD, Kohler J, Gould S, Moseley P (1982) Pulmonary extraction of biogenic amines during septic shock. *Am. Surg. 48*, 552-554.

Khairallah PA, Page IH, Bumpus FM, Turker RK (1966) Angiotensin tachyphylaxis and its reversal. *Circ. Res. 19*, 247-254.

Klinge E, Penttila O, Tissari A (1964) The content of 5-hydroxytryptamine in human uterine and placental tissue, and their 5-hydroxytryptophan decarboxylase and monoamine oxidase activities in normal and toxemic pregnancies. *Acta Obstet. Gynecol. Scand. 43*, 107-114.

Kobold EE, Lovell R, Katz W, Thal AP (1964) Chemical mediators released by endotoxin. *Surg. Gynecol. Obstet. 118*, 807-813,

Korthuis RJ, Wang CY, Scott JB (1982) Transient effects of histamine on microvascular fluid movement. *Microvasc. Res. 23*, 316-328.

Kozlowski T, Raymond RM, Korthuis RJ, Wang CY, Grega EJ, Robinson WE, Scott JB (1981) Microvascular protein efflux: interaction of histamine and H_1 receptors. *Proc. Soc. Exp. Biol. Med. 166*, 263-270.

Krasney JA, Thompson JL, Lowe RF (1967) Cardiac effects of angiotensin injections into perfused right coronary artery. *Am. J. Physiol. 213*, 134-138.

Krause S, Hess ML (1979) Diphenhydramine protection of the failing myocardium during gram-negative endotoxemia. *Circ. Shock 6*, 75-87.

Lautt WW (1977) Hepatic vasculature: a conceptual review. *Gastroenterology 73*, 1163-1169.

Lefer AM (1970) Role of a myocardial depressant factor in the pathogenesis of circulatory collapse. *Fed. Proc. 29*, 1836-1847.

Levi R (1972) Effects of exogenous and immunologically released histamine on the isolated heart: a quantitative comparison. *J. Pharmacol. Exp. Ther. 182*, 227-238.

Lindell SE, Westling H (1966) In: Eicher O, Farah A (Eds), *Handbook of Experimental Pharmacology, Vol 18(1)*, pp 734-788. Springer, Berlin.

MacCanon DM, Horvath SM (1954) Some effects of serotonin in pentobarbital anesthetized dogs. *Am. J. Physiol. 179*, 131-134.

Marciniak DL, Dobbins DE, Maciejko JJ, Scott JB, Haddy FJ, Grega GJ (1977) Effects of systemically infused histamine on transvascular fluid and protein transfer. *Am. J. Physiol. 223*, H148-H153.

Marciniak DL, Dobbins DE, Maciejko JJ, Scott JB, Haddy FJ, Grega GJ (1978) Antagonism of histamine edema formation by catecholamines. *Am. J. Physiol. 234*, H180-H185.

Markley K, Smallman E, Thornton SW (1971) Protection against burn, tourniquet and endotoxin shock by histamine, 5-hydroxytryptamine and 5-hydroxytryptamine derivatives. *Br. J. Pharmacol. 42*, 13-24.

Massion WH, Kux M (1973) Protective effects of various inhibitor agents against pulmonary damage following shock. In: Haberland GL, Elberfeld W, Lewis DH (Eds), *New Aspects of Trasylol Therapy, The Shock Lung*, pp 191-198. F.K. Schattauer Verlag, Stuttgart.

Maxwell GM, Castillo CA, Clifford JE, Crumpton CW, Rowe GG (1959) Effect of serotonin (5-hydroxtryptamine) on the systemic and coronary vascular bed of the dog. *Am. J. Physiol. 197*, 736-738.

McCaa RE, Richardson TQ, Langford HG, Douglas BH (1967) Circulatory changes following angiotensin adminstration. *Am. J. Physiol. 212*, 565-568.

McLaurin LP, Rolett EL, Grossman W (1973) Impaired left ventricular relaxation during pacing-induced ischemia. *Am. J. Cardiol 32*, 751-757.

McNeill JH, Muschek LD (1972) Histamine effects on cardiac contractility, phosphorylase and adenyl cyclase. *J. Mol. Cell. Cardiol 4*, 611-624.

McNeill JH, Verma SC (1974) Blockade by burimamide of the effects of histamine and histamine analogs on cardiac contractility, phosphorylase activation and cyclic adenosine monophosphate. *J. Pharmacol. Exp. Ther. 188*, 180-188.

Mela L, Bacalzo LV, Miller LD (1971) Defective oxidative metabolism of rat liver mitochondria in hemorrhagic and endotoxin shock. *Am. J. Physiol. 220*, 571-577.

Mela L, Hinshaw LB, Coalson JJ (1974) Correlation of cardiac performance, ultrastructural morphology and mitochondrial function in endotoxemia in the dog. *Circ. Shock 1*, 265-275.

Morton JJ, Semple PF, Ledingham IM, Stuart B, Tehrani MA, Garcia AR, McGarritz G (1977) Effect of angiotensin-converting enzyme inhibitor (SQ 20881) on the plasma concentration of angiotensin I, angiotensin II, and arginine vasopressin in the dog in hemorrhagic shock. *Circ. Res. 41*, 301-308.

Mota I (1966) Release of histamine from mast cells. In: Eichler O, Farah A (Eds), *Handbook of Experimental Pharmacology, Vol 18(1)*, pp 569-636. Springer, Berlin.

Moulton R, Spector WG, Willoughby DA (1957) Histamine release and pain production by xanthosine and related compounds. *Br. J. Pharmacol. 12*, 365-370.

Nakahara K, Ohkuda K, Staub NC (1979) Effect of infusing histamine into pulmonary or bronchial artery on sheep pulmonary fluid balance. *Am. Rev. Respir. Dis. 120*, 875-882.

Nassif AC, Nolan MD, Corcoran AC (1963) Angiotensin II in treatment of hypotensive states. *J. Am. Med. Assoc. 183*, 751-754.

Noble JG, Nanson EM (1959) An investigation of the effects of 5-hydroxytryptamine on the cardiovascular system of the dog. *Ann. Surg. 150*, 846-853.

Olson NC (1982) I. *Effects of Brain or Carotid Body Hypoxia and/or Hypercapnia on Pulmonary Hemodynamics: II. Effects of Histamine on Lung Water and Hemodynamic Before and After Adrenergic Blockade.* Ph.D. Dissertation, Department of Physiology, Michigan State University, East Lansing, MI.

Olson NC, Robinson NE, Scott JB (1983) Effects of histamine on lung water and hemodynamics before and after b-blockade. *J. Appl. Physiol. 54*, 967-971.

Page IH, Helmer OM (1940) A crystalline pressor substance (angiotensin) resulting from the reaction between renin and renin-activator. *J. Exp. Med. 71*, 29-42.

Page IH, McCubbin JW (1956) Arterial pressure response to infused serotonin in normotensive dogs, cats, and hypertensive dogs and man. *Am. J. Physiol. 184*, 265-270.

Parker JL, Adams HR (1979) Myocardial effects of endotoxin shock: characterization of an isolated heart muscle model. *Adv. Shock Res. 2*, 163-173.

Parker JL, Adams HR (1985) Development of myocardial dysfunction in endotoxin shock. *Am. J. Physiol.*, in press.

Parott JL, Laborde-Burtin C, Saindelle A (1964) The allergic terrain. *Ann. Allergy*

22, 511-610.

Parratt JR, Sturgess RM (1977) The possible roles of histamine, 5-hydroxytryptamine and prostaglandin F_2alpha as mediators of the acute pulmonary effects of endotoxin. *Br. J. Pharmacol. 60*, 209-219.

Paton WD (1957) Histamine release by compounds of simple chemical structure. *Pharmacol. Rev. 9*, 269-328.

Pietra GG, Szidon JP, Leventhal MM, Fishman AP (1971) Histamine and interstitial pulmonary edema in the dog. *Circ. Res. 29*, 323-337.

Pietra GG, Magno M, Johns L (1979) Morphological and physiological study of the effect of histamine on the isolated perfused rabbit lung. *Lymphology 12*, 165-176.

Pöch G, Kukovetz WR (1967) Drug-induced release and pharmacodynamic effects of histamine in the guinea pig heart. *J. Pharmacol. Exp. Ther. 156*, 522-527.

Porter JF, Mitchell RG (1972) Distribution of histamine in human blood. *Physiol. Rev. 52*, 361-381.

Postel J, Schloerb PR (1972) Cardiac depression in bacteremia. *Ann. Surg. 186*, 74-82.

Rapport MM (1949) Serum vasoconstrictor (serotonin). The presence of creatinine in the complex. A proposed structure of the vasoconstrictor principle. *J. Biol. Chem. 180*, 961-969.

Rapport MM, Page IH, Green AA (1948) Crystalline serotonin. *Science 108*, 329-330.

Reddin JL, Starzecki B, Spink WW (1966) Comparative hemodynamic and humoral responses of puppies and adult dogs to endotoxin. *Am. J. Physiol. 210*, 540-544.

Reeves JT, Daoud FS, Estridge M (1972) Pulmonary hypertension caused by minute amounts of endotoxin in calves. *J. Appl. Physiol. 33*, 739-743.

Reid G (1952) Circulatory effects of 5-hydroxytryptamine. *J. Physiol (London) 118*, 435-453.

Reid G, Rand M (1951) Physiologic actions of partially purified serum vasoconstrictor serotonin. *Aust. J. Exp. Biol. Med. Sci. 29*, 401-415.

Reite OB (1972) Comparative physiology of histamine. *Physiol. Rev. 52*, 778-819.

Riley JF (1959) *The Mast Cells*. Livingstone, Edinburg.

Riley JF, West GB (1953) The presence of histamine in tissue mast cells. *J. Physiol. (London) 120*, 528-537.

Riley JF, West GB (1966) The occurrence of histamine in mast cells. In: Eichler O, Farah A (Eds), *Handbook of Experimental Pharmacology, Vol. 18(1)*, pp 116-135. Springer, Berlin,

Robertson JI, Andrews TM (1961) Free serotonin in human plasma. Quantitative and qualitative estimation. *Lancet 1*, 578-580.

Robinson NE, Jones GA, Scott JB, Dabney JM (1975, 1976) Effects of histamine and acetylcholine on equine digital lymph flow and composition. *Proc. Soc. Exp. Biol. Med. 149*, 805-807.

Rocha e Silva M (1966) Action of histamine upon the circulatory apparatus. In: Eichler O, Farah A (Eds), *Handbook of Experimental Pharmacology, Vol 18(1)*, pp. 238-293. Springer, Berlin.

Rosenberg JC, Lillehei RC, Moran WH, Zimmerman B (1959) Effect of endotoxin on plasma catecholamines and serum serotonin. *Proc. Soc. Exp. Biol. Med. 102*, 335-337.

Rosenberg JC, Lillehei RC, Longerbeam J, Zimmerman B (1961) Studies on hemorrhagic and endotoxin shock in relation to vasomotor changes and endogenous circulating epinephrine, norepinephrine and serotonin. *Ann. Surg. 154*, 611-628.

Schayer RW (1960a) Relationship of induced histidine decarboxylase activity and histamine synthesis to shock from stress and from endotoxin. *Am. J. Physiol. 198*, 1187-1192.

Schayer RW (1960b) Relationship of stress-induced histidine decarboxylase to circulatory homeostasis and shock. *Science 131*, 226-227.

Schayer RW (1962) Evidence that induced histamine is an intrinsic regulator of the microcirculatory system. *Am. J. Physiol. 202*, 66-72.

Schayer RW (1963) Induced synthesis of histamine, microcirculatory regulation and the mechanism of action of the adrenal glucocorticoid hormones. *Prog. Allergy 7*, 187-211.

Schneider CL, Meyers NL, Chaplick MJ, Engstrom RM (1967) Transient pulmonary arteriolar constriction, then increased perfusion, following serotonin, and following fibrination. *Bibl. Anat. 9*, 65-70.

Sibbald W, Peters S, Lindsay RM (1980) Serotonin and pulmonary hypertension in human septic ARDS. *Crit. Care Med. 8*, 490-494.

Siegel JH, Greenspan M, Del Guercio LRM (1967) Abnormal vascular tone, defective oxygen transport, and myocardial failure in human septic shock. *Ann. Surg. 165*, 504-517.

Silk MR (1966) The effect of dextran and hydroxyethyl starch on renal hemodynamics. *J. Trauma 6*, 717-723.

Sivo Sankar DV, Sankar DB, Phipps E, Gold E (1961) Effect of administration of LSD on serotonin levels in the body. *Nature (London) 191*, 499-500.

Skeggs LT, Marsh WH, Kahn JR, Shumway NP (1954) The existence of two forms of hypertension. *J. Exp. Med. 99*, 275-282.

Sleeman HK, Lamborn PB, Diggs JW, Emery CE (1971) Effects of endotoxin and histamine on serum enzyme activity. *Proc. Soc. Exp. Biol. Med. 138*, 536-541.

Soulsby ME, Bruni FD, Looney TJ, Hess ML (1978) Influence of endotoxin on myocardial calcium transport and the effect of augmented venous return. *Circ. Shock 5*, 23-34.

Spector WG, Willoughby DA (1957) Capillary-permeability factors, nucleosides, and histamine release. *J. Pathol. 73*, 133-139.

Spitzer JJ, Bechtel A, Archer LT, Black MR, Greenfield LJ, Hinshaw LB (1975) Effects of coronary hypotension on myocardial substrate utilization. *Am. J. Physiol. 228*, 365-368.

Stacey RS (1959) 5-Hydroxytryptamine and other pharmacologically active substances in the central nervous system. *Acta Physiol. Pharmacol. Neerl. 8*, 222-239.

Staszewska-Barczak J, Vane JR (1965) The release of catecholamines from the adrenal medulla by histamine. *Br. J. Pharmacol. 25*, 728-742.

Swank RL, Edwards MJ (1968) Microvascular occlusion by platelet emboli after transfusion and shock. *Microvasc. Res. 1*, 15-22.

Swank RL, Fellman JH, Hissen WW (1963) Aggregation of blood cells by 5-hydroxytryptamine (serotonin). *Circ. Res. 13*, 392-400.

Swank RL, Hissen W, Fellman J (1964) 5-Hydroxytryptamine (serotonin) in acute hypotensive shock. *Am. J. Physiol. 207*, 215-222.

Takacs L, Vajda V (1963) Effect of serotonin on cardiac output and organ blood flow of rats. *Am. J. Physiol. 204*, 301-303.

Texter EC Jr (1963) Small intestinal blood flow. *Am. J. Dig. Dis. 8*, 587-613.

Thames MD, Jarecki M, Donald DE (1977) Neural control of renin secretion in anesthetized dogs. *Circ. Res. 42*, 237-245.

Tigerstedt R, Bergman PG (1898) Niere und Kreislauf. *Scand. Arch. Physiol. 8*, 223-271.

Tindall GT, Greenfield JC Jr (1973) The effects of intra-arterial histamine on blood flow in the internal and external carotid artery of man. *Stroke 4*, 46-49.

Tobe T, Fujiwara M, Tanaka C (1966) Distribution of serotonin (5-hydroxytryptamine) in the human gastrointestinal tract. *Am. J. Gastroenterol. 46*, 34-37.

Trendelenburg U (1960) The action of histamine and 5-hydroxytryptamine on isolated mammalian atria. *J. Pharmacol. Exp. Ther. 130*, 450-460.

Udhoji VN, Weil MH (1964) Circulatory effects of angiotensin, levarterenol and metaraminol in the treatment of shock. *New Engl. J. Med. 270*, 501-505.

Vick JA (1964) Trigger mechanism of endotoxin shock. *Am. J. Physiol. 206*, 944-946.

Vick JA, Mehlman B, Heiffer MH (1971) Early histamine release and death due to endotoxin. *Proc. Soc. Exp. Biol. Med. 137*, 902-906.

Walton RP, Richardson JA, Thompson WL (1959) Hypotension and histamine release following intravenous injection of plasma substitutes. *J. Pharmacol. Exp. Ther. 127*, 39-45.

Warner RR (1967) Current studies and implications of serotonin in clinical medicine. *Adv. Intern. Med. 13*, 241-282.

Webb WR (1977) Lung perfusion and oxygen uptake. In: Thompson WL (Moderator), *The Organ in Shock*, pp 16-23. Scope Publication, Kalamazoo.

Weissbach H, Redfield BG (1960) Factors affecting the uptake of 5-hydroxytryptamine by human platelets in an inorganic medium. *J. Biol. Chem. 235*, 3287-3291.

Weisul JP, O'Donnell TF Jr, Stone MA, Clowes GHA Jr (1975) Myocardial performance in clinical septic shock: effects of isoproterenol and glucose potassium insulin. *J. Surg. Res. 18*, 357-363.

Westerholm B (1966) Dextran-induced release of 5-hydroxytryptamine from rabbit platelets. *Acta Physiol. Scand. 67*, 236-242.

Whelan RF (1959) Cardiovascular effects of serotonin in man. *Am. Heart J. 58*, 933-934.

White FN, Gold EM, Vaughn DL (1967) Renin-aldosterone system in endotoxin shock in the dog. *Am. J. Physiol. 212*, 1195-1198.

Wilson MF, Brackett DL, Archer LT, Hinshaw LB (1980) Mechanisms of impaired cardiac function by vasopressin. *Ann. Surg. 191*, 494-500.

Wilson MF Brackett DJ, Tompkins P, Benjamin B, Archer LT, Hinshaw LB (1981) Elevated plasma vasopressin concentrations during endotoxin and E. coli shock. *Adv. Shock Res. 6*, 15-26.

Windaus A, Vogt W (1907) Synthese des Imidazolylathylamins. *Chem. Ber. 40*, 3691-3695.

Woo P, Carpenter MA, Trunkey D (1979) Ionized calcium: the effect of septic shock in the human. *J. Surg. Res. 26*, 605-610.

Yamashita M, Oyama T, Kudo T (1977) Effect of the inhibitor of angiotensin I converting enzyme on endocrine function and renal perfusion in hemorrhagic shock. *Can. Anaesth. Soc. J. 24*, 695-701.

Zlydaszyk JC, Moon RJ (1976) Fate of ^{51}Cr-labeled lipopolysaccharide in tissue culture cells and livers of normal mice. *Infect. Immun. 14*, 100-105.

Handbook of Endotoxin, Vol. 2: Pathophysiology of Endotoxin
L.B. Hinshaw, editor
© Elsevier Science Publishers B.V., 1985
ISBN 0 444 90385 2
$0.85 per article per page (transactional system)
$0.20 per article per page (licensing system)

CHAPTER 9

The role of arachidonic acid metabolites in endotoxin shock I: lipoxygenase products

JAMES R. PARRATT

1. INTRODUCTION

There is little direct evidence to date that lipoxygenase products are involved in the pathophysiological consequences of endotoxin administration. This, of course, is not to imply that they are not involved; indeed, some of the indirect evidence suggests an active and important accessory role for these products in at least some forms of endotoxin-induced cellular injury. The reasons for the lack of conclusive, or even convincing, evidence at this time include the relatively recent chemical characterization of these products, the problems involved in their quantitative estimation in biological fluids, the lack of specific inhibitors of the several enzymes involved, or of suitable antagonists at the receptor level, and the complex interrelationships between these substances and those derived from the cyclooxygenase pathway of arachidonic acid metabolism. This review has two aims. First, to anticipate further work relating to these substances in shock by summarizing the pharmacological effects of the administration of these products with special reference to respiratory and cardiovascular functions. Second, to evaluate the current evidence, such as it is, for the role of lipoxygenase products in shock induced by endotoxin.

2. THE LIPOXYGENASE PATHWAY OF ARACHIDONIC ACID METABOLISM

Arachidonic acid is a 20-carbon polyunsaturated fatty acid constituent of all membrane phospholipids (Lands and Samuelsson 1968) which is obtained directly from the diet, or from the essential dietary fatty acid, linoleic acid, by desaturation. Chemically it is an eicosatetraenoic acid; the 'eicosa' indicates a 20-carbon fatty acid, 'tetraenoic' indicates it has four double bonds. There are two major enzyme systems that act on arachidonic acid to give rise to highly potent, pharmacologically active substances which almost certainly play major roles in the pathophysiology of endotoxin shock; these enzyme systems are cyclooxygenase (which catalyzes oxygenation and cyclization reactions to give rise to the prostaglandins, prostacyclin and thromboxanes discussed in Chapter 10) and the more recently described series of lipoxygenase enzymes, which give rise to noncyclic products, such as the leukotrienes. These form the subject of this chapter.

The fact that arachidonate metabolizing enzymes are present in nearly all mammalian cells suggests that the rate-limiting step in the production of

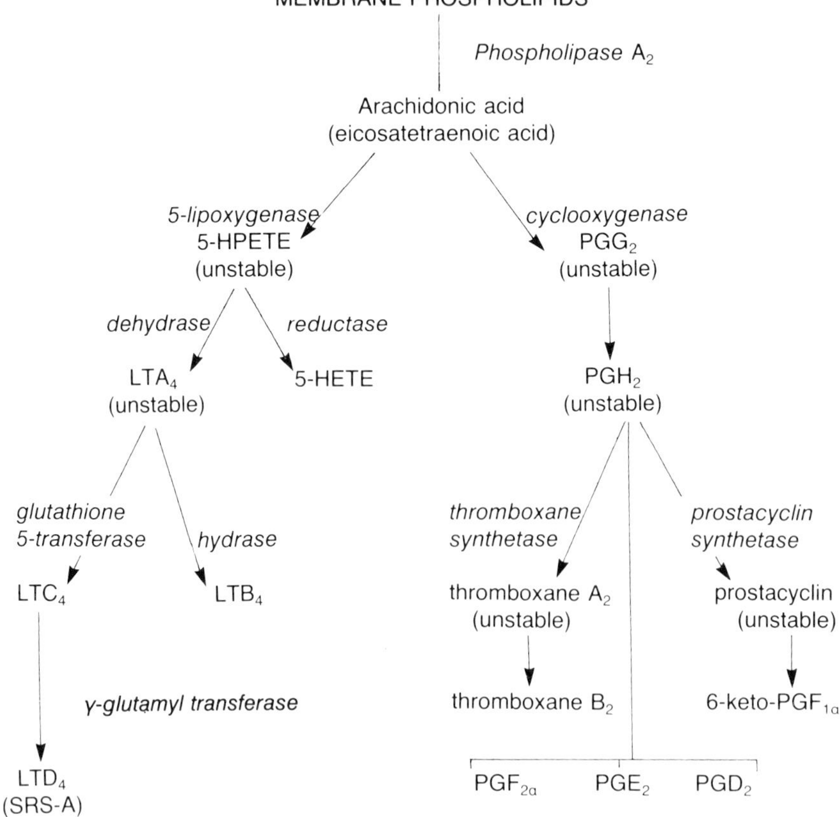

Fig. 1. *A simplified, schematic account of the metabolism of arachidonic acid to cyclooxygenase products (prostaglandins, thromboxane A$_2$, prostacyclin) and 5-lipoxygenase products (leukotrienes A$_4$, B$_4$, C$_4$, D$_4$).*

arachidonic acid metabolites is the availability of free arachidonic acid. The essential step is the activation of phospholipase A_2 (Flower and Blackwell 1976). This is achieved by even minor disturbances of the cell membrane (disorientation and translocation of membrane phospholipids), by more severe cellular trauma, by inflammatory mediators such as plasma kinins and by interaction of IgE with specific allergens. Phospholipase A_2 is inhibited by steroidal antiinflammatory drugs such as dexamethasone, probably because these steroids induce a release of the 'intermediary inhibitor' macrocortin (Flower and Blackwell 1979).

The main aspects of the arachidonic acid lipoxygenase and cyclooxygenase cascades are illustrated schematically in Figure 1. Unlike the ubiquitous cyclooxygenase enzyme, there are at least two different biologically important lipoxygenase enzymes, a 5-lipoxygenase (found, for example in leukocytes and in the lungs) and a 12-lipoxygenase (found, for example, in platelets). Peroxidation at the 5- or 12-position gives rise to 5-hydroperoxy-6,8,11,14-eicotetraenoic acid (5-HPETE) and to 12-hydroperoxy-5,8,10,14-eicotetraenoic (12-HPETE) respectively. These are further metabolized to hydroxyeicosatetraenoic acids (HETEs) and, in leukocytes, 5-lipoxygenase gives rise to a family of leukotrienes (Fig. 2).

2.1. Leukotrienes

Leukotrienes were named because of their proposed leukocyte origin (leuko) and because three of their four double bonds are present in a conjugated triene, that is, these three double bonds occur on alternate carbon atoms (e.g. 7, 9 and 11 in LTD_4). The subscript '4' refers to the total number of double bonds in the molecule (e.g. LTB_4, LTC_4 etc.). The unstable leukotriene A is a 5,6-epoxide of arachidonic acid which can be hydrolyzed to the 5,6-dihydroxy acid or converted by a hydrolase enzyme (for example in granulocytes) to LTB_4 (5,12-DHETE) which has several naturally occurring isomers. An alternative pathway for LTA_4 (for example in the lung) gives rise to leukotriene C_4 by the addition of glutathione to the epoxide by glutathione-5-transferase. The removal of glutamic acid from LTC_4 by glutamyl transpeptidase gives rise to a cysteinyl-glycinyl derivative (leukotriene D_4) which can be further metabolized to the cysteine derivative, leukotriene E_4.

It is likely that mast cells (including those in human lung and derived from bone marrow), basophils, neutrophils and macrophages are the major sources of leukotrienes. In some inflammatory reactions a specialized type of macrophage produces especially large amounts of leukotrienes. The localization of cells with the ability to synthesize leukotrienes has recently been investigated using immunocytochemical fluorescence techniques (Murphy et al 1983).

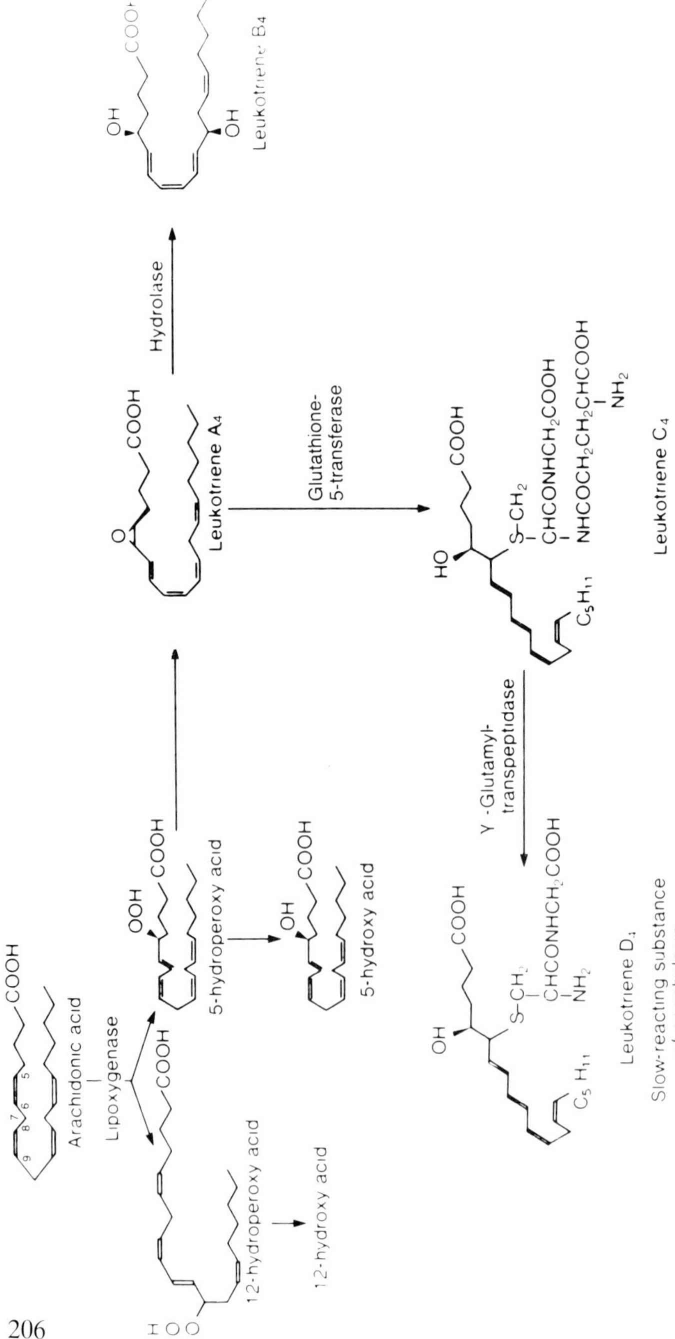

Fig. 2. *Formation of 12-HETE, 5-HETE and the leukotrienes (reproduced from Zeitlin 1981, with permission).*

Most of the work on the activation and chemical characterization of leukotrienes involved murine mastocytoma cells, which were stimulated with the calcium ionophore A 23187 after pretreatment with cysteine. 5-Lipoxygenase requires calcium for its activation (probably at the first step of the conversion of arachidonic acid to 5-hydroperoxyarachidonate) and, like LTA-synthetase (dehydrase) and LTB-hydrolase, is a soluble enzyme. No significant conversion of LTA_4 to LTC_4 occurs in plasma membrane preparations whereas the enzymes that convert LTC_4 to LTD_4 are localized in these membranes. This suggests that glutamic acid is removed from LTC_4 as it passes through the membrane to the outside of the cell and that the functional significance of LTC_4 may be intracellular and that of LTD_4 extracellular (Jakschik and Kuo 1983).

2.2. Platelet lipoxygenase products

Platelets contain a separate (12-)lipoxygenase enzyme system which synthesizes 12-HETE rather than 5-HETE from arachidonic acid. Formation of this monohydroxy acid is preceded by abstraction of hydrogen and peroxidation at the 12-position to give rise to the unstable hydroperoxy intermediate (12-HPETE); in platelets 12-HPETE peroxidase generates the stable hydroxy acid (12-HETE). Much less is known about the biological activity of such 12-lipoxygenase platelet products although there is some evidence that 12-HPETE inhibits both platelet aggregation, induced by an analog of PGH_2, and platelet thromboxane synthetase enzyme, and that it activates C-5 lipoxygenase (suggesting a possibly important interaction between 12- and 5-lipoxygenases). 12-HETE has been shown to possess chemotactic activity (Turner et al 1975).

Considerable advances have recently been made in the measurement of leukotriene levels in tissues and biological fluids. This is of vital importance to the elucidation of their role in, for example, the pathophysiology of shock induced by endotoxin because the classic methods of estimating leukotriene (SRS-A) activity involved bioassay on the guinea pig ileum, a method which lacks specificity since many different lipoxygenase (and cyclooxygenase) products have similar activity on this preparation. There are two available techniques for the measurement of leukotriene levels:

a. *Reverse-phase, high pressure liquid chromatography (RP-HPLC).* This technique can be used to analyze leukotrienes, but whilst it has proved useful in structural determination of lipoxygenase products, it is laborious and lacks the sensitivity required to measure the submicrogram quantities normally found in biological systems.

b. *Leukotriene-specific radioimmunoassays.* Defining a role for leukotrienes in shock states requires assay techniques that are specific, sensitive and simple, and these conditions can be fulfilled by the use of radioim-munoassay techniques. The double-antibody method, for example, has the advantage of good separation of labeled and unlabeled antigens (low nonspecific binding), the ability to analyze active substances in small (1 ml) amounts of blood, rapidity and reproducibility. Levine et al (1981) raised antibodies in rabbits by immunizing with 11-trans LTD_4 coupled with bovine albumin; these antibodies reacted in radioimmunoassays with LTC_4, LTD_4 and LTE_4 but not with other leukotrienes or with related compounds. Young and colleagues (Hayes et al 1983; Young et al 1982) raised antibodies in rabbits by conjugating the leukotriene with keyhole limpet hemocyanin (KLH) via a bifunctional coupling agent (6-N-maleimidohexanoic acid chloride). This method is described in detail by Hayes et al (1983).

2.3. Lipoxygenase inhibitors

A great many different drugs inhibit lipoxygenase. These include a number of nonsteroidal antiinflammatory drugs (NSAIDs, especially naproxen, BW755C, indometacin, isoxicam) which are discussed below, the antifilarial drug diethylcarbamazine (Hagmann and Keppler 1982; Piper and Temple 1981) and various analogs of LTA_4 (e.g. 5,6-methano-LTA; Letts and Piper 1983b) and of prostacyclin (e.g. 6,9-deepoxy-6,9-(phenylimino)-$\Delta^{6,8}$-prosta-glandin I, U-60,257 and its methylester U-56,467; Bach et al 1983; Hedqvist and Dahlen 1983a; Johnson et al 1983). U-60,257 is probably the most selec-tive lipoxygenase inhibitor yet described. For example, using ionophore-chal-lenged rat peritoneal mononuclear cells the ID_{50} dose was 4.57 µM, and at this concentration there was no alteration in thromboxane formation (used as an index of cyclooxygenase activity). Like FPL-55712 (Casey et al 1983), the Upjohn prostacyclin analog also competitively and selectively inhibits the effects of LTD on the guinea pig ileum (Bach et al 1983). The exact site of inhibition of U-60,257 on the lipoxygenase pathway has yet to be eluci-dated but an effect on the glutathione 5-transferase reaction (the last step in the formation of LTC/D; Fig. 1) has been recently suggested (Bach et al 1983). It would be of considerable interest to use this compound as a tool to examine the possible role of leukotrienes in shock induced by endotoxin; it certainly effectively inhibits the pulmonary changes (increased airways resis-tance, reduced lung compliance) induced in sensitized monkeys by *Ascaris* antigen (Johnson et al 1983).

The drug most used as a tool to examine the effects of lipoxygenase inhib-ition is the Wellcome compound BW755C (3-amino-1-(m-(trifluoro-methyl)phenyl)-2-pyrazoline) which is equiactive in inhibiting cyc-looxygenase and lipoxygenase (Higgs et al 1980). The evidence for this is that BW755C inhibits similarly carrageenin-induced prostaglandin synthesis and leukocyte infiltration in rabbits; the former is mediated by cyclooxygenase products, the latter probably mainly by lipoxygenase products (Higgs et al 1982). Similarly BW755C reverses indometacin-induced increases in leukocyte accumulation following carrageenin administration in rats (Higgs et al 1980); the reason being that indometacin, by selectively inhibiting cyclooxygenase, 'turns on' the lipoxygenase pathway, shunting arachidonic acid metabolism towards leukotrienes. Antiinflammatory corticosteroids, such as dexamethasone, by interfering with phospholipase activity, would prevent arachidonic acid release from phospholipids and would, indirectly, inhibit the generation of both cyclooxygenase and lipoxygenase products. Both prostaglandin production (and hence their biological effects, such as pain and vasodilatation) and leukotriene production (and hence leukocyte migration and accumulation) would be equally inhibited by this procedure.

At the time this chapter was completed (September 1983) no work had been published on the possible effects of BW755C on shock induced by endotoxin. Studies in the author's own laboratory, however, suggest that this approach (inhibition of *both* cyclooxygenase and lipoxygenase pathways) is much more effective at increasing survival than is inhibition of cyclooxygenase (or selectively, thromboxane synthetase) alone. These results are summarized below. One important study (Mullane and Moncada 1982) has demonstrated that BW755C is effective in limiting the development of myocardial ischemic damage following acute coronary artery occlusion (followed by reperfusion) in anesthetized dogs. In contrast, indometacin (which in the doses used inhibited only cyclooxygenase) did not limit myocardial ischemic damage. These results again suggest that in damage resulting from tissue hypoperfusion (whether induced in a major artery by occlusion or in smaller vessels by platelet aggregation and endothelial swelling) it may be important to inhibit both pathways.

2.4. Cyclooxygenase inhibition and the lipoxygenase pathway

Much confusion has arisen, and is likely to continue to arise, because of the possibility that classical inhibitors of cyclooxygenase, such as the nonsteroidal antiinflammatory drugs, might also inhibit the lipoxygenase pathway. This evidence can be summarized as follows:

2.4.1. *High-dose indometacin inhibits lipoxygenase*

Inhibition of lipoxygenase by high-dose indometacin has been shown in rabbit and rat leukocytes incubated with carbon-labeled arachidonic acid; the metabolites (hydroxy-hepta-decatrienoic acid, HHT, as an index of cyclooxygenase activity), and 5-hydroxy-eicosatetraenoic acid (5-HETE, as an index of lipoxygenase activity) were separated by thin layer chromatography and HPLC and quantified by liquid scintillation (Randall et al 1980). In rabbit leukocytes there was an 80% inhibition of HHT production at 1 µM indometacin, at which concentration 5-HETE production was actually increased by 20% (presumably because of the greater availability of arachidonic acid to the uninhibited lipoxygenase pathway). At 100 µM indometacin there was a 50% inhibition of 5-HETE production, with complete inhibition of HHT by concentrations between 1 and 10 µM. Similar results resulted from studies using rat leukocytes; at a concentration of 100 µM there was a 90% inhibition of HHT (and 11- and 15-HETE production) and a 20% inhibition of 12-HETE production. In contrast, BW755C influenced both pathways similarly (40% inhibition of both HHT and 12-HETE at 10 µM). The 'selectivity' also depends upon the biological system; for example in horse platelets (Higgs and Flower 1981) both BW755C and indometacin are equiactive at inhibiting cyclooxygenase (70% inhibition at 10 µg/ml) whereas at the same concentrations BW755C completely inhibits lipoxygenase and indometacin hardly inhibits this enzyme at all (< 10%).

Using a model in which leukocyte migration and prostaglandin synthesis were measured in rabbit skin, Higgs et al (1982) showed that a dose of 50 mg/kg BW755C inhibited equally prostaglandin synthesis and leukocyte migration; a dose of 1 mg/kg indometacin inhibited prostaglandin synthesis to the same extent (50%) and actually *increased* migration.

2.4.2.   Inhibition by NSAIDs of leukocyte accumulation in inflammatory exudates

Many other nonsteroidal antiinflammatory drugs probably inhibit both pathways. This statement is based, not on measurements of lipoxygenase products but on studies involving inhibition by these drugs of leukocyte accumulation in inflammatory exudates induced, for example, by carrageenin in rats (Higgs et al 1980; Higgs and Flower 1981). More recently Sircar et al (1983) have compared the ability of a variety of NSAIDs to inhibit soybean lipoxygenase (E.C. 1.13.11.12) using linoleic acid as substrate. In the Higgs and Flower studies, low doses of indometacin, acetylsalicylic acid, sodium salicylate, flurbiprofen and phenylbutazone significantly *increased* leukocyte migration (by 20–70% of control values). This enhancement could be explained by substrate diversion toward the production of chemotactic lipoxygenase products. At higher doses (Table 1) all of these drugs inhibited leukocyte migration, presumably by lipoxygenase inhibition. Naproxen, benoxaprofen, acetylsalicylic acid and ibuprofen showed the least selectivity (ratio of less than 10); indometacin and ketoprofen the most (i.e. markedly

TABLE 1   *Ratio between inhibition of prostaglandin (PG) synthesis and of leukocyte migration (LM) by various nonsteroidal antiinflammatory drugs following carrageenin-induced inflammation in rats*

Drug	Inhibition of leukocyte migration	Inhibition of PG synthesis	Ratio inhibition LM/PG synthesis
BW755C	8.0	21.0	0.4
Naproxen	5.1	1.1	4.6
Benoxaprofen	7.8	1.2	6.5
Ibuprofen	45.0	5.6	8.0
*Acetylsalicylic acid	130.0	15.0	8.7
*Flurbiprofen	0.47	0.032	14.7
Diclofenac	3.0	0.2	15.0
*Phenylbutazone	52.0	2.8	18.6
*Indometacin	5.7	0.15	38.0
*Ketoprofen	2.5	0.017	147.1

*Indicates evidence of potentiation of leukocyte migration at lower doses.
Drugs were administered at ED_{50} doses (mg/kg). Inhibition of PG synthesis is indicative of cyclooxygenase inhibition, while inhibition of LM is probably an index of lipoxygenase inhibition. Data taken from Higgs et al (1980) and Higgs and Flower (1981).

preferential inhibition of cyclooxygenase). These in vivo results correlate reasonably well with the in vitro studies involving inhibition of soybean lipoxygenase (Sircar et al 1983). In these, naproxen, BW755C, and indometacin were the most active inhibitors (compare Table 1), meclofenamic acid, phenylbutazone and benoxaprofen had intermediate potency, and ibuprofen was relatively weak.

There is one practical application of the above results in the context of endotoxin shock. This concerns the debate regarding the effect of NSAIDs on survival following endotoxin administration. There is no doubt that these drugs reduce, for example, the early pulmonary changes (pulmonary hypertension, increase in airways resistance, reduced lung compliance) that result from endotoxin administration (e.g. Parratt and Sturgess 1974, 1976; Parratt et al 1982; Parratt 1983a), mainly because these drugs prevent the release of thromboxane A_2 and $PGF_{2\alpha}$ (Ball et al 1983; Coker et al 1983; Parratt et al 1982). There is, however, considerable controversy regarding the value of these drugs (or selective inhibitors of thromboxane synthesis) at increasing overall survival. One reason could be that, in the doses used, NSAIDs inhibit not only cyclooxygenase but have varied effects on lipoxygenase (lower doses diverting arachidonic acid down this pathway; higher doses of some of these drugs inhibiting the pathway). If leukotriene generation is an important determinant of survival, then survival might be increased with doses of NSAIDs that inhibit lipoxygenase as well as cyclooxygenase but decreased by lower doses. The results with BW755C in a conscious rat model (Furman et al, unpublished results, see below) would support the concept that inhibition of both pathways is important.

2.5. Nafazatrom

One other approach which might be valuable in examining the possible role of leukotrienes in endotoxin shock is the use of nafazatrom (3-methyl-1-[2-(2-naphthyloxy)-ethyl]-2 pyrazolin-5-one). This, like BW755C, has been found to reduce the extent of myocardial ischemic damage in rats and rabbits subjected to acute coronary artery ligation (Fiedler 1983a, 1983b, 1984) and to reduce the severity and incidence of reperfusion arrhythmias in greyhound dogs subjected to coronary artery occlusion and subsequent release (Coker and Parratt 1984). There are a number of possible explanations for these protective effects which include potentiation of ischemia-induced prostacyclin release (Coker and Parratt 1984), probably mediated by inhibition of 15-hydroxyprostaglandin dehydrogenase (Wong et al 1982) and an anti-thrombotic action (Seuter et al 1979). Recently, however, nafazatrom has been shown to inhibit cellular lipoxygenase activity in leukocytes (Busse et al 1982; Mardin and Busse 1983). The precise sites of action have yet to be elucidated, but two possibilities are inhibition of 5-HPETE formation (through an 'antioxidase' effect of nafazatrom) and stimulation of the conversion of 15-HPETE to 15-HETE by acting as a reducing cofactor. The combination of the properties of lipoxygenase inhibition and prostacyclin potentiation offers interesting possibilities in shock states; such studies with endotoxin are in progress.

2.6. Leukotriene antagonists

Most studies involving antagonism of the biological effects of leukotrienes at the receptor level have used the Fisons compound FPL55712, which was first described as an SRS-A antagonist by Augstein et al in 1973. This drug also selectively inhibits the lipoxygenase pathway (Casey et al 1983). FPL55712 antagonizes, in a dose-related and selective manner, a number of the biological effects of purified leukotrienes including increased capillary permeability in guinea pig skin (LTB_4, LTC, LTD; Morley et al 1981), increased microvascular permeability in the guinea pig trachea (LTD_4; Woodward et al 1983) and the pulmonary responses (bronchoconstriction, cough) induced by aerosols of LTC in normal human volunteers (Holroyde et al 1981). In guinea pig isolated Langendorff perfused hearts, FPL55712 fully antagonized the coronary vasoconstrictor effects of LTC_4 but was much less effective against LTD_4-induced coronary vasoconstriction (Letts and Piper 1982a, 1982b). In contrast, the marked reduction in myocardial contractile force that resulted from the administration of both these leukotrienes in isolated guinea pig and rat hearts was markedly inhibited by FPL55712 (Letts and Piper 1982a, 1982b; Letts et al 1983).

A similar antagonism by FPL55712 of the marked coronary vasoconstriction that results from the local (intracoronary) injection of LTD_4 has been described in both anesthetized sheep (Michelassi et al 1982) and pigs (Boyd et al 1983). In the latter study the coronary vasoconstrictor effects of both $PGF_{2\alpha}$ and vasopressin were also reduced by FPL55712. This emphasizes that care has to be taken in interpreting results from studies in which this antagonist is used. It is clear that the drug does interfere with leukotriene-mediated vasoconstrictor (and other) responses but that very large molar ratios ($FPL55712:LTD_4 = 300:1$) are required to achieve substantial blockade. At these concentrations, specificity is sometimes lost (Feniuk et al 1983). Further, it is also clear from the Michelassi study (Michelassi et al 1982) that the coronary vasoconstrictor response to LTD_4 is easier to antagonize with FPL55712 than is the reduction in systolic shortening. This, and other evidence, suggests that a single receptor site is unlikely. Indeed, it has been suggested (Fleisch et al 1982; Krell et al 1983) that in smooth muscle there are three distinct receptors for LTD, two present in trachea and smooth muscle, and one in the ileum. At present, evidence arising from the inhibition of a response by FPL55712 suggests, but does not necessarily prove, that that particular response is mediated by the release of leukotrienes. Such evidence certainly needs to be substantiated by other approaches such as the specific inhibition of synthesis and/or the estimation of leukotriene levels. We badly need selective and potent leukotriene antagonists.

2.7. Mediation of some of the biological effects of leukotrienes by activation of the cyclooxygenase pathway

There is considerable evidence that some of the effects of leukotrienes may be indirectly mediated through the release of cyclooxygenase products. This applies especially to effects on the airways. Thus, in vitro, the contractile effects of LTC_4 and LTD_4 on strips of guinea pig lung parenchyma are, in some hands (Piper et al 1981), but not others (Austen et al 1983) prevented by indometacin. Austen's group found that although the effects of LTB_4 in this preparation are prevented by indometacin (and thus involve cyclooxygenase products, principally TXA_2), those of LTC_4, LTD_4 and LTE_4 are not (except in high concentrations). One reason for this discrepancy is suggested by experiments using this same preparation by Hedqvist and Dahlen (1983b). They showed that the response to LTC_4 was not influenced by indometacin unless the preparation was superfused at a high flow and LTC_4 was added as a bolus. In this case indometacin reduced the response by 50%, but by the use of this technique the sensitivity of the preparation to LTC_4 was decreased by 100-fold. The thromboxane A_2 antagonist EP 045 also reduces leukotriene mediated contractions in this preparation (Barnett and Piper 1983).

There is direct evidence for the release of cyclooxygenase products by leukotrienes. Thromboxane A_2 is released by LTC_4 and LTD_4 from guinea pig isolated perfused lungs (Coleman et al 1983a, 1983b; Piper et al 1981), from chopped lungs (Coleman et al 1983a), from trachea (Cuthbert and Gardiner 1983) and from guinea pig lung parenchyma (Zijlstra et al 1983). This release is not secondary to smooth muscle contraction (Coleman et al 1983a).

In vivo indometacin and large doses of acetylsalicylic acid have been reported to inhibit LTC_4- and/or LTD_4-induced bronchospasm in guinea pigs (Piper et al 1981; Schiantarelli et al 1981; Vargaftig et al 1981) although this has been recently disputed.

All of the above studies were carried out in guinea pigs, and there is some recent evidence that there may be marked species differences in what mediates the pulmonary effects of leukotrienes. For example, LTD_4-induced contraction of isolated human bronchial tissue is not prevented by indometacin (Jones et al 1982), LTD_4-induced contractions of cynomolgus and squirrel monkey lung parenchyma are not antagonized by meclofenamic acid and LTD_4 does not generate TXB_2 from monkey lung parenchyma or trachea (Weichman et al 1983). The route of administration may also be important; even in guinea pigs the aerosol administration of LTD_4 (in contrast to intravenous administration) does not elevate plasma prostaglandin (6-keto-$PGF_{1\alpha}$) or thromboxane levels, and meclofenamic acid does not antagonize LTD_4-induced decreases in lung compliance or increases in airways resistance (Weichman et al 1983).

Ogletree et al (1982a) have reported that LTD_4 injected intravenously in adult sheep with chronic lung lymph fistulas increased pulmonary arterial pressure and released TXA_2 into lung lymph, an effect which was not seen in sheep pretreated with sodium meclofenamate. These results also indicate that some of the pulmonary effects of leukotrienes may be mediated by the local release of TXA_2. The mechanism of this release and the source of the TXA_2 are unclear.

The situation with regard to other body systems is not very much clearer. In guinea pig and rat isolated perfused hearts neither indometacin, nor the selective thromboxane inhibitor dazoxiben, modify LTD_4-induced decreases

214

in coronary flow (Letts and Piper 1982a, 1982b; Letts et al 1983) although the responses to LTC_4 are inhibited by indometacin (Letts et al 1983). In the same preparation both LTC_4 and LTD_4 released 6-keto-$PGF_{1\alpha}$ (the main metabolite of prostacyclin) but not thromboxane into the venous effluent (Terashita et al 1981). In vivo neither ibuprofen (in sheep) nor indometacin (in pigs) modify LTD_4-induced coronary vasoconstriction (Boyd et al 1983; Michelassi et al 1982), and under these conditions LTD_4 does not release TXB_2 or 6-keto-$PGF_{1\alpha}$ into coronary sinus blood (Michelassi et al 1982).

It is difficult to come to any firm general conclusions regarding the release of cyclooxygenase products by leukotrienes. The main evidence for such a release comes from one species (guinea pigs) and one organ (lungs). The possibility should, however, be borne in mind that under other conditions (endotoxin shock?), actions attributed to leukotrienes might, in part, be mediated by prostaglandins and thromboxane.

3. THE BIOLOGICAL EFFECTS OF LIPOXYGENASE PRODUCTS WHICH ARE RELATED TO ACTIONS OF ENDOTOXIN

The purpose of this part of the review is to summarize those biological actions of lipoxygenase products, mainly leukotrienes, which are similar to those that result from endotoxin administration. This does not of course imply that these particular effects of endotoxin are mediated, even partly, through the endogenous release of lipoxygenase products; it simply summarizes what actions in physiological systems these products might be expected to have if they were released during endotoxemia.

3.1. Lipoxygenase products and pulmonary function

There is considerable experimental evidence, in a number of animal species and man, that leukotrienes have marked effects on pulmonary function which are in many ways similar to those of endotoxin itself. These pulmonary actions of leukotrienes have been reviewed (Hedqvist and Dahlen 1983a). LTC_4, LTD_4 and LTE_4 are all potent bronchoconstrictors both in vivo and in vitro; this bronchoconstriction is slow in onset and is of a longer duration than that produced by histamine, which is much less potent (by at least three orders of magnitude) in this respect than, for example, LTC_4. In vivo the effect on pulmonary compliance is generally more pronounced than effects on airways resistance (e.g. Drazen et al 1980; Hedqvist et al 1980; Smedegård et al 1982) and in this respect differs from the response of airways to histamine. For example, in anesthetized monkeys (Smedegård et al 1982) nebulized LTC_4 induced a long-lasting increase in transpulmonary pressure and in pulmonary artery pressure. Hypoxemia was pronounced and long lasting despite attempts to combat atelectasis. Under these conditions pulmonary (airways) resistance was very slightly affected. In the same study histamine, also given by aerosol, resulted in marked but reversible increases in pulmonary arterial pressure, and the main airways effects were on resistance rather than compliance.

In human volunteers inhalation of LTC_4 (Holroyde et al 1981; Weiss et al 1982) and LTD_4 (Griffin et al 1983) induces bronchoconstriction (assessed from measurements of maximum expiratory flow) and again, this is much more pronounced and prolonged than that resulting from histamine inhalation. As in experimental animal studies, LTC_4 and LTD_4 also have a much slower onset of action than does histamine. The effects of leukotrienes and HETEs resemble in some respects the pathophysiological features of asthma (bronchoconstriction, mucosal edema, increased mucus production and impaired clearance), and SRS-A-like activity has been identified in the sputum and blood of asthmatics (Morris et al 1983). Nevertheless the role of lipoxygenase products either in asthma or as mediators of the airways obstruction which is a common feature of chronic bronchitis and cystic fibrosis is unproven (O'Driscoll and Kay 1982). There is as yet no evidence (Griffin et al 1983) that asthmatic subjects are hyperresponsive to leukotrienes; this is again in contrast to histamine.

Leukotrienes also induce contraction of various isolated smooth muscle preparations derived from bronchi and lung parenchyma (Dahlen et al 1980, 1983; Drazen et al 1980, 1983; Piper et al 1981). Again, these contractions are characterized by a slow onset (an effect is often not immediately detectable and a peak response is not reached until several minutes have passed), by resistance to washout (persistence) and by the development of tachyphylaxis (Dahlen et al 1983). LTC_4, LTD_4 and LTE_4 are equipotent, at least on the guinea pig lung strip, and are some 5000 times more potent, on a molar basis, than histamine. The fact that lung parenchymal strips are much more sensitive to leukotrienes than tracheal smooth muscle suggests a predominant effect on the smooth muscle cells in terminal airways.

Although the pulmonary effects of leukotrienes are now well characterized, there is still doubt as to the extent of the indirect contribution of other smooth muscle-contracting substances released by lipoxygenase products, a possibility that is suggested by the slow onset of action. Contractions evoked by leukotrienes in the guinea pig lung strip and trachea, and in human bronchial smooth muscle are unaffected by drugs that block muscarinic, histamine and 5-hydroxytryptamine receptors (Dahlen et al 1980), but there is good evidence in some experimental situations for a major contribution by products derived from arachidonic acid by cyclooxygenase, principally thromboxane A_2. For example, contractions induced by LTC_4 and

216

LTD$_4$ in guinea pig lung parenchyma are reduced (or prevented) not only by FPL55712 but also by indometacin (Piper et al 1981). This effect of indometacin might, however, depend on the experimental conditions; Hedqvist and Dahlen (1983b) have shown that whilst indometacin does inhibit the increase in tension induced in lung parenchyma by LTC$_4$ in perfusion studies, it has no effect when the LTC$_4$ contraction is already well developed. 'This makes it essential to define the experimental conditions, since the interpretation of observed pulmonary reactions to leukotrienes will depend on whether direct or indirect mechanisms are prevailing' (Hedqvist and Dahlen 1983a).

In vivo the situation is just as confused. Indometacin greatly attenuates the bronchoconstrictor effects of intravenously administered LTC$_4$ but not that of LTC$_4$ given by aerosol, when the response is often enhanced (Hedqvist and Dahlen 1983a, 1983b). On the other hand, large doses of acetylsalicylic acid (20 mg/kg) suppress the bronchoconstrictor effects of both the intravenous and aerosol administration of LTC$_4$ and LTD$_4$ in guinea pigs (Vargaftig et al 1981), and indometacin inhibits the effects of intra-arterial injections of LTD$_4$ and LTC$_4$ in the same species (Piper et al 1981). Sodium meclofenamate also abolishes the acute increase in pulmonary arterial pressure and the reduced lung compliance that result from the intravenous administration of LTD$_4$ in unanesthetized sheep (Ogletree et al 1982a). This evidence suggesting a mediatory role, in the pulmonary effects of leukotrienes, of released cyclooxygenase products, is supported by studies in which the release of thromboxane A$_2$ has been demonstrated both in vitro (Piper et al 1981 in perfused guinea pig lungs) and in vivo (Ogletree et al 1982b in unanesthetized sheep) following leukotriene administration.

These findings suggest that leukotrienes are potent bronchoconstrictors in their own right, perhaps acting on specific receptors, but also that secondarily released cyclooxygenase products, especially TXA$_2$, play a major role. This is especially well illustrated by the Vanderbilt University study in anesthetized sheep referred to above. In this preparation, LTD$_4$ injected intravenously increased pulmonary artery pressure and lung lymph flow and decreased lung compliance. Only the increases in pulmonary artery pressure and the reduced lung compliance were prevented by a dose of sodium meclofenamate that also prevented leukotriene-mediated TXA$_2$ release. The increase in lung lymph flow and also the delayed increases in left atrial pressure and in aortic pressure were unaffected by pretreatment with sodium meclofenamate. The conclusion reached was that 'some, but not all (of the cardiac, lung vascular and airways) LTD$_4$ responses may be mediated by the local release of TXA$_2$' (Ogletree et al 1982a). The similarities will be noted, in this same model, between the modification, by inhibitors of cyclooxygenase, of the effects of leukotriene D$_4$ and those of endotoxin itself.

The possibility that part of the bronchoconstrictor effects of leukotrienes might be mediated by effects on the efferent or afferent components of the irritant receptor pathway has been recently investigated in anesthetized guinea pigs in which airways resistance and dynamic compliance were measured (Advenier et al 1983). The effects of LTD_4, administered intravenously, on airways resistance were increased by the β-adrenoceptor blocking drug propranolol and by hexamethonium (suggesting a catecholamine involvement), and effects on both airways resistance (increased by LTD_4) and lung compliance (reduced by LTD_4) were significantly attenuated by atropine. This suggests that LTD_4 induces a cholinergic reflex, possibly by stimulation of irritant receptors in the bronchi or within the pulmonary circulation. A recent study of LTD_4 in dogs with hyperactive airways (Hirshman et al 1983) comes to the same conclusion; atropine completely prevented the increase in airways resistance and the reduced lung compliance induced by LTD_4 (5 μg/ml by aerosol). There is other evidence in man that leukotrienes stimulate irritant receptors and the possibility has also been raised that a reflex component is also involved in the bronchoconstriction produced by LTC_4 and LTD_4 in man (Holroyde et al 1981).

3.2. Lipoxygenase products and the inflammatory response

The main features of the inflammatory response are vasodilatation, an increase in vascular permeability, and cellular, particularly leukocyte, infiltration; similar responses occur in responses to the administration of endotoxin.

The most pronounced cutaneous effects of LTC_4 and LTD_4 are a reduction in blood flow and the exudation of plasma, especially in the presence of PGE_1 and PGE_2 (Piper et al 1981). For example, a combination of LTD_4 and PGE_1 resulted in a 30-fold increase in the amount of plasma exuded in guinea pig skin as compared to PGE_1 alone; LTC_4, LTD_4 and LTE_4 are equiactive in promoting plasma exudation and are 1000 times more active than histamine. This effect of cysteinyl-containing leukotrienes appears to be a direct action on the endothelial lining and does not require the presence of polymorphonuclear cells or the release of prostaglandins, although, in this respect, synergism between leukotrienes and prostaglandins is of considerable importance (Williams et al 1983).

Leukotriene B_4 also elicits plasma exudation but is much less potent than LTC_4 and unlike the other leukotrienes, this occurs only after some delay. This effect may be secondary to the pronounced cytotactic activity of LTB_4 and the rapid involvement of polymorphonuclear leukocytes (PMN) at the inflammatory site. There appear to be two possible explanations for the delayed increase in vascular permeability induced by LTB_4 (Smith 1981, 1982). It might result from the release of enzymes from PMN, which could degrade substances present at the junction between endothelial cells, or it might result from leukocytes which lodge at endothelial cell junctions and which then swell and force open the junctions mechanically.

This action on leukocyte behavior and function is the most pronounced biological effect of LTB_4 (Smith 1981, 1982). It causes leukocytes, but not platelets or lymphocytes, to adhere to each other (aggregation) and indeed this is used as a sensitive bioassay for LTB_4. This leukotriene also stimulates the random migration of leukocytes (chemokinesis) and the directed migration of leukocytes (leukocyte chemotaxis); in this respect LTB_4 is considerably more potent than either 3-HETE, other mono-HETEs or the natural dihydroxy-HETE (DHETE) isomers (Higgs et al 1981). These effects of LTB_4 can also be demonstrated in vivo. Thus intraperitoneal or intradermal injections of LTB_4, or its introduction into the aqueous humor, induces leukocyte migration and adherence of white cells to the walls of small blood vessels (particularly venules) with subsequent neutrophil penetration through the wall and migration into the extravascular space (Bray et al 1981; Smith 1982). The result is a marked decrease in the number of granulocytes in the postcapillary venules and a transient but profound neutropenia. It is this marked cytotactic activity and the possible subsequent enzyme release, that accounts for the effect of LTB_4 on vascular permeability; depletion of circulating PMN (for example, with nitrogen mustard) prevents this effect.

The effects of leukotrienes outlined above have been mainly described in preparations such as the hamster cheek pouch, mesentery and in skin. Similar leukotriene-induced inflammatory responses have also recently been described in the lung (Woodward et al 1983) and this, and the fact that a profound neutropenia is also found in Adult Respiratory Distress Syndrome ('shock lung') raises the possibility that leukotrienes might be mediators of pulmonary damage in this situation, especially as leukotrienes can be generated from mast cells present in lung (MacGlashan et al 1982). Woodward et al (1983) administered LTC_4, LTD_4 and LTE_4 locally and measured the vascular permeability in the trachea of anesthetized guinea pigs. All three leukotrienes increased microvascular permeability (as assessed by measurements of extravascular content of labelled serum albumin), an effect significantly reduced by FPL55712 but not by mepyramine or by indometacin. The leukotrienes were 100–1000 times more potent at increasing tracheal extravascular albumin content than was histamine.

3.3. Cardiovascular effects of lipoxygenase products

Very little detailed information is at present available with regard to the effects of leukotrienes on the heart and circulation. In conscious rats and guinea pigs (with indwelling, externalized intra-arterial catheters) LTD_4, administered either intra-arterially (rats) or intravenously (guinea pigs) in doses ranging from 0.2 to 20 µg/kg, caused short-lasting, dose-dependent increases in systemic arterial pressure (Lux et al 1983a, 1983b; Zukowska-Grojec et al 1982, 1983a, 1983b). This was more marked in spontaneously hypertensive rats (SHR) than in normotensive Wistar-Kyoto (WKY) strain rats and was accompanied by tachycardia and by increases in plasma levels of noradrenaline, adrenaline and dopamine (Zukowska-Grojec et al 1982, 1983a, 1983b). In SHR, but not WKY rats, and in guinea pigs, LTD_4 administration resulted in a prolonged systemic hypotension and a decrease in heart rate following the transient period of hypertension and tachycardia.

A study by Feniuk et al (1983) describes the effects of LTC_4, LTD_4 and LTE_4 in vagotomized cats anesthetized with chloralose. All three leukotrienes, when administered intravenously, caused dose-related increases in diastolic blood pressure (of about 50 mmHg with 5 µg/kg of LTC_4 and LTD_4 and with 10 µg/kg of LTE_4). Heart rate was usually decreased but changes in aortic flow were variable. The hypertensive response was unaffected by phentolamine, mecamylamine, saralasin or mepyramine and only slightly reduced (25%) by FPL55712. Whether this inability to reduce the pressor response of leukotrienes with this compound in vivo is due to a difference in receptor type or to the rapid clearance of the antagonist (Mead et al 1981) is not clear.

The initial systemic hypertension induced by LTD_4 probably results from a combination of a direct peripheral vasoconstrictor effect (Pfeffer et al 1983) and a centrally-mediated increase in sympathoadrenalmedullary activity. This is consistent with (a) the increased levels of plasma catecholamines observed in conscious rats and (b) the abolition, in some experiments, by anesthesia of the hypertensive effect of LTD_4 (Drazen et al 1980). The acute hypertensive effect is unaffected by pretreatment with indometacin.

The delayed systemic hypotension in the conscious guinea pig is unaffected by methylatropine, inhibition of cyclooxygenase, naloxone or by antioxidants (Lux et al 1983b); it is, however, prevented (and reversed) by the intracerebroventricular administration of thyrotropin releasing hormone (TRH). TRH has been found to be beneficial in various low perfusion states (including shock induced by spinal cord injury) and was originally presumed to act by physiological antagonism of endogenous opioids at cardioregulatory centres in the anterior hypothalamus, the nucleus tractus solitarius or the nucleus ambiguus. The observation (Lux et al 1983a) that TRH, but not naloxone, prevents systemic hypotension induced by LTD suggests a potential non-endorphin link between the cardiovascular effects of this leukotriene and the centrally active peptide TRH. On the basis of these studies Lux et al suggest that such a link might have implications for the development of new therapeutic approaches to shock.

220

It is possible that there is a cardiac component in the LTD_4-induced delayed systemic hypotension described above. LTD_4 administration in conscious SHR rats often resulted in myocardial ischemia, in bradyarrhythmias and in conduction defects; about 25% of the animals exhibited particularly severe ischemic changes and A-V block and died before the end of the experiment (Zukowska-Grojec et al 1982, 1983b). Since the electrocardiographic changes often occurred before the onset of systemic hypotension, this suggests direct effects of LTD on the coronary circulation and/or conduction. This is supported by the following:

a. LTD_4, LTE_4, and especially LTC_4, cause dose-dependent reductions in coronary flow and in myocardial contractility in Langendorff preparations of guinea pig and rat isolated hearts (Kennedy et al 1983; Letts and Piper 1982; Letts et al 1983; Terashita et al 1981).

b.   Although LTC_4 and LTD_4 release 6-keto-$PGF_{1\alpha}$ (the major metabolite of prostacyclin) into the coronary perfusate (Terashita et al 1981), there was no evidence of thromboxane release (Letts et al 1983; Terashita et al 1981) as one might perhaps expect in the absence of platelets. The effects of these leukotrienes on the coronary circulation were unaffected by prior treatment with dazoxiben, a specific inhibitor of thromboxane synthetase, but were prevented by the leukotriene antagonist FPL55712 (Letts et al 1983). Since, however, this drug in similar concentrations also antagonizes the coronary vasoconstrictor effects of the thromboxane-mimetic U 44619 (Kennedy et al 1983), it may be of limited value in elucidating the role of endogenous leukotrienes in the heart.

c.   In vivo the intracoronary administration of LTD_4 causes profound dose-related coronary constriction in anesthetized sheep (Michelassi et al 1982), and both LTC_4 and LTD_4 produce coronary vasoconstriction in anesthetized dogs (Letts et al 1983; Woodman and Dusting 1983) and pigs (Boyd et al 1983; Letts et al 1983). In sheep, coronary vasoconstriction is accompanied by impaired regional ventricular wall motion and by T-wave inversion characteristic of acute myocardial ischemia. Leukotrienes are clearly very active coronary vasoconstrictors in all these species, with detectable effects in concentrations as low as 2×10^{-11} molar of LTD_4 (Michelassi et al 1982).

d. LTD_4 does not cause the myocardial release of either TXB_2 or 6-keto-$PGF_{1\alpha}$ (the main metabolites of TXA_2 and prostacyclin respectively) as evaluated from coronary sinus sampling and radioimmunoassay. The coronary effects of LTD_4 and LTC_4 are unaffected by the prior administration of inhibitors of cyclooxygenase (ibuprofen, Michelassi et al 1983; indometacin, Boyd et al 1983; Letts et al 1983; Woodman and Dusting 1983) but are prevented by the prior intracoronary administration of FPL55712 (Boyd et al 1983). These findings suggest that leukotriene-induced vasoconstriction in vivo is independent of the production of any cyclooxygenase metabolites of arachidonic acid (e.g. thromboxane A_2) and is due to interaction with leukotriene receptors in the coronary circulation. There is some evidence in vitro that some of the effect of leukotrienes on the heart could involve cyclooxygenase products (Letts and Piper 1982, 1983a; Terashita et al 1981).

The possibility that leukotrienes have direct depressant effects on the myocardium has been examined in isolated cardiac muscle preparations. Neither LTC_4 nor LTD_4 influence spontaneous rate or contractile force of guinea pig isolated paired or right atrial preparations (Letts and Piper 1982; Terashita et al 1981), or modify contractile force of electrically-driven right ventricular muscle strips from the same species (Letts and Piper 1982). On the other hand, leukotrienes (LTC_4, LTD_4, LTE_4) do depress contractility in electrically-paced left atrial and right ventricular papillary muscles of guinea pigs and in specimens of human right atrial appendage obtained following surgery (Burke et al 1982).

Much less is known about the effects of the leukotrienes on other vascular beds. It does seem, however, that the predominant effect is vasoconstriction. Besides the pronounced effects on the coronary circulation outlined above, LTC_4 and LTD_4 induce vasoconstriction in isolated perfused rat kidneys (Rosenthal and Pace-Asciak 1983) in the skin (see Section 3.2) and in the pulmonary circulation (Ogletree et al 1982a; Smedegård et al 1982). The effects of leukotrienes on some regions of extralobar pulmonary arteries are enhanced by histamine (Hand et al 1983).

4. EVIDENCE FOR THE PARTICIPATION OF LIPOXYGENASE PRODUCTS IN SHOCK INDUCED BY ENDOTOXIN

There are two main studies to be considered:

4.1. Evidence for a role of leukotrienes in the lethal effects of endotoxin in mice sensitized by D-galactosamine

The amino-sugar D-galactosamine reduces, by several orders of magnitude, the lethal dose of endotoxin in mice (Galanos et al 1979), an effect reversed by uridine. The explanation for this 'sensitization' is that the amino sugar acts as a specific hepatotoxin by inducing a rapid depletion of certain uracil nucleotides (Keppler et al 1974), with subsequent necrosis of liver parenchymal cells (Keppler et al 1968). Mice were given an intravenous injection of purified lipopolysaccharide (LPS), derived from *S. abortus equi*, together with D-galactosamine; this resulted in all the mice dying by 72 hours. Neither lipopolysaccharide nor D-galactosamine given separately resulted in any deaths. The effects of the repeated administration of either an inhibitor of lipoxygenase (diethylcarbamazine; Matthews and Murphy 1982; Orange et al 1968; Piper and Temple 1981), a 'selective' leukotriene antagonist (FPL55712; Augstein et al 1973) or a specific inhibitor of calmodulin (calmidazolium; Van Belle 1981) were examined against lethality resulting from the LPS–D-galactosamine combination (Hagmann and Keppler 1982; Hagmann et al 1983). All three procedures reduced lethality from 100% to 0% (Table 2) and this suggested to the authors 'a crucial role for leukotrienes in endotoxic actions'.

The mechanism proposed for this leukotriene involvement in shock induced by endotoxin is outlined in Figure 3. The LPS–D-galactosamine combination increases calcium uptake by the liver (Fritz and Keppler 1982), as does LPS alone, although over a longer time period. The accumulation of hepatic calcium results in a loss of liver glycogen since an elevated cytoplasmic Ca^{2+} leads to a stimulation of glycogen phosphylase and to an inactivation of glycogen synthetase by a kinase reaction controlled by calmodulin. Glycogen loss is inhibited by phenathiazines such as trifluoperazine and promethazine, which can act as calmodulin antagonists, and by the more specific calmidazolium. Leukotriene B_4 is a potent endogenous calcium ionophore (Serhan et al 1982) and further, the lipoxygenase pathway is stimulated by calcium and by specific calcium ionophores such as A 23187. If leukotriene B_4 was produced under these conditions, it could enhance its own production by a positive feed-back mechanism. The increased sensitivity of mice given D-galactosamine to endotoxin is explained by interference with hepatic leukotriene elimination.

Clearly, the working hypothesis formulated above needs more rigorous testing. It would be important to determine, for example, whether the effects

TABLE 2 *Lethal effects of purified lipopolysaccharide (LPS) (S. abortus equi) and D-galactosamine, and modification by drugs*

Treatment*	Lethality at 72 hrs (5%)
LPS alone	0
D-galactosamine	0
Lipopolysaccharide + D-galactosamine	100
LPS + D-galactosamine + FPL 55712	0
LPS + D-galactosamine + diethylcarbamazine	0
LPS + D-galactosamine + calmidazolium	0
LPS + D-galactosamine + promethazine/trifluoperazine	0

*See original reference for details of doses. Administered by intravenous injection in mice.

Adapted from Hagmann and Keppler (1982) and Fritz and Keppler (1982).

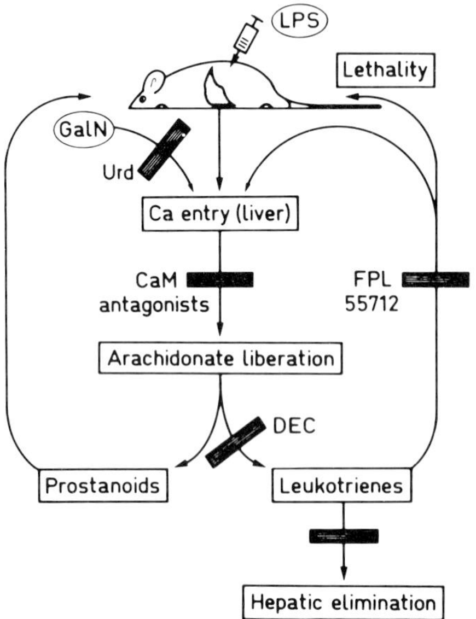

Fig. 3.   *Possible mechanisms, suggested from the work of Professor Keppler's group, for the role of leukotrienes in endotoxin-induced lethality. CaM = calmodulin; GalN = D-galactosamine; DEC = diethylcarbamazine; Urd = uridine; LPS = lipopolysaccharide.*

of larger doses of endotoxin (without D-galactosamine sensitization) are prevented by inhibitors of leukotriene synthesis and by other leukotriene antagonists. Unfortunately neither diethylcarbamazine nor FPL55712 are really specific enough for such studies. Diethylcarbamazine, which is a widely used antifilarial drug with particularly marked effects against *Wuchereria bancrofti*, *Loa Loa* and *Onchocerca volvulus*, has pharmacological effects other than that of lipoxygenase inhibition. For example, it stimulates autonomic ganglia (Abaitey and Parratt 1976, 1977), inhibits the immunological release of histamine (Orange et al 1971) and antagonizes the effects on smooth muscle of a variety of spasmogens (acetylcholine, histamine, potassium, prostaglandin $F_{2\alpha}$; Abaitey and Parratt 1976, 1977), some of which are also released by endotoxin. It is probably also dangerous to assume that the Fison compound FPL55712 is acting solely through leukotriene antagonism. For example, in isolated perfused hearts the coronary vasoconstriction and reduced myocardial contractility induced by vasopressin and by the thromboxane A_2 mimetic U-44619 (Coleman et al 1981) and by leukotriene C_4 are inhibited to a similar degree. FPL55712 therefore appears to be of limited value as a tool for elucidating the role of endogenous leukotrienes (Kennedy et al 1983).

Although the experiments of Keppler's group can be explained by leukotriene mediation, it is too early to accept the hypothesis as proven. It would be important, for example, to determine whether LPS lethality can be modified by procedures which modify the cyclooxygenase pathway, by specific inhibitors (and antagonists) of thromboxane A_2, and by combined cyclooxygenase and lipoxygenase inhibition (e.g. with BW755C). In other words, how specific is the protection afforded by procedures which apparently modify leukotriene synthesis and actions?*

4.2. Evidence from studies involving release of lipoxygenase products by endotoxin and from lipoxygenase administration

This evidence, which is again inconclusive, comes from studies in unanesthetized sheep. Sheep respond to minute amounts of endotoxin (e.g. 0.4 µg/kg) with a pronounced acute pulmonary hypertension followed by a prolonged period of increased pulmonary vascular permeability, with a high flow of protein-rich lung lymph; this can result ultimately in pulmonary edema and in respiratory death (Brigham et al 1979; Esbenshade et al 1982). There is no evidence that lipoxygenase products are involved in the acute pulmonary hypertensive response (discussed below), but it is possible that they are involved in the later increase in permeability. The evidence for this possibility will be summarized below.

4.2.1. Release of 5-hydroxyeicosatetraenoic acid (5-HETE) from lungs by endotoxin

5-HETE concentrations were measured (by a sensitive, stable isotope dilution technique using a combination of gas chromatography and mass spectrometry) in the lung lymph of unanesthetized sheep administered *E. coli* endotoxin (Ogletree et al 1982b). The resting (pre-endotoxin) lymph 5-HETE concentration was 1.7 ± 0.3 ng/ml. This was not altered (concentration 2.2 ± 0.5 ng/ml) in the early stage of endotoxemia (30–60 minutes after the start of a 15-minute endotoxin infusion), during which time pulmonary artery pressure was greatly elevated (from 16 ± 1 to 56 ± 4 cmH$_2$O) and lung lymph

---

* Keppler's group have recently shown (Hagmann et al 1984) that endotoxin from *Salmonella minnesota* greatly increased the biliary excretion of LTC$_4$-like material (measured by radioimmunoassay). Further, there was no protection in their mouse model with either indometacin or ibuprofen.

flow markedly increased (from 9.2 ± 1.3 to 33.8 ± 3.0 ml/hr). However, during the 2–2.5 hours following the administration of endotoxin the lymph 5-HETE concentration was significantly elevated (to 5.5 ± 2.0 ng/ml) at a time when pulmonary artery pressure was declining (to 25 ± 3 cmH$_2$O) and when lymph flow remained elevated (30.8 ± 1.4 ml/hr). Within 4 hours of endotoxin administration, lymph 5-HETE levels had returned to pre-endotoxin levels. This paper is the first to demonstrate that lipoxygenase products derived from arachidonic acid are released by endotoxin. The time course of release is somewhat different to that of cyclooxygenase products in that it reaches a peak following the initial period of severe pulmonary hypertension but before the steady-state period with high flow of protein-rich lung lymph. As is discussed below, the pulmonary release of cyclooxygenase products is explosive and is correlated with the rapid development of pulmonary hypertension.

The possible relevance of this 5-HETE release is that it might also reflect increased leukotriene production which, in turn, might contribute to the increased pulmonary vascular permeability. The effects of leukotrienes on the pulmonary circulation, on airways resistance and lung compliance have been discussed above.

There are several possible sources of the 5-HETE released by endotoxin. These include lymphocytes, macrophages and PMN (Borgeat et al 1976). There is a good deal of evidence that PMN play an important role in mediating pulmonary vascular injury. For example, depletion of PMN inhibits the increase in pulmonary vascular permeability induced by endotoxin (Heflin and Brigham 1981). It is thus possible that endotoxin simultaneously enhances endogenous arachidonate availability and stimulates lipoxygenase, probably through the complement system.

4.2.2. *Similarity between the pulmonary effects of endotoxin and those of lipoxygenase administration*

Smith et al (1981a, 1981b) compared the pulmonary vascular responses of unanesthetized sheep to a 5-hour infusion of soybean lipoxygenase with those of endotoxin. Lipoxygenase administration markedly increased lymph flow and pulmonary artery pressure and slightly decreased the lymph/plasma protein ratio. The effects on permeability were similar, although somewhat slower in onset, to those induced by *Escherichia coli* endotoxin, and there were similar effects on the lymph concentrations of β-glucuronidase (increased) on white blood cells (decreased) on the hematocrit and on body temperature (both increased). In this study therefore there were clearly similarities, especially between endotoxin- and lipoxygenase-induced 'permeability'. The authors are admirably reluctant to press these similarities too far. Thus 'there is a possibility that the soybean protein initiated a hypersensitivity response which could have evoked systems other than lipoxygenase activity. Due to these uncertainties this paper should be viewed as a preliminary work designed to arouse curiosity' (Flynn, personal communication, 1983).

4.2.3. Effects of inhibitors of cyclooxygenase on the pulmonary responses to endotoxin

There is good evidence in several species that cyclooxygenase inhibitors markedly modify the pulmonary responses to endotoxin (Parratt and Sturgess 1974, 1975; Parratt et al 1982). This in itself might indicate that lipoxygenase products are much less important than cyclooxygenase products in mediating endotoxin-induced pulmonary damage. In the unanesthetized sheep model, however, endotoxin causes a reproducible two-phase pulmonary vascular response characterized by an acute severe pulmonary hypertension (Phase 1) accompanied by arterial hypoxemia, and a somewhat delayed increase in pulmonary vascular permeability with a high flow of protein rich lymph (Phase 2; Brigham et al 1979). Ogletree and Brigham (1982) have recently shown that only the Phase 1 (pulmonary hypertension) response is prevented by inhibitors of cyclooxygenase such as indometacin and meclofenamate. The delayed increases in lung vascular permeability and in lymph protein clearance were exaggerated by cyclooxygenase inhibition. This could be explained on the basis of diversion of arachidonic acid metabolism through the lipoxygenase pathway.

In the cat, and probably also in the sheep models of endotoxin shock, the evidence is almost overwhelming that the initial pulmonary effects of endotoxin are mediated through products of the cyclooxygenase pathway of arachidonic acid metabolism. This evidence has been recently reviewed (Parratt et al 1982; Parratt 1983a). It is briefly that:

a.   The acute pulmonary effects of endotoxin (increase in pulmonary artery pressure and in airways resistance; reduced lung compliance) are prevented by a variety of cyclooxygenase inhibitors (indometacin, flurbiprofen, meclofenamate).

b.   These effects are also completely abolished by prior treatment with the prostaglandin receptor antagonist polyphloretin.

c.   The early pulmonary release of prostaglandin (PG)$F_{2\alpha}$ and TXA_2 can be demonstrated, and this correlates with the degree of pulmonary hypertension.

d.   The specific inhibition of TXA_2 synthesis by dazoxiben prevents endotoxin-induced pulmonary hypertension but not the increase in intratracheal pressure.

Taken together these results suggest an important mediatory role for cyclooxygenase products, especially $PGF_{2\alpha}$, PGD_2 and TXA_2. Most of the above evidence is against a contribution by products of the lipoxygenase pathway. This does not of course mean that activation of the lipoxygenase pathway might not play an important role in the secondary (delayed) effects of endotoxin on lung vascular permeability.

4.3. Recent evidence for a possible involvement of leukotrienes in endotoxin shock

Clearly this is an inopportune time to review this subject and the likelihood is that, in such a rapidly developing area, the possible role (if any) of leukotrienes in low perfusion states will be greatly clarified by the time this chapter appears in print. There are certainly as yet unpublished data available. One recent study (Furman, McKechnie and Parratt) demonstrates that combined cyclooxygenase and lipoxygenase inhibition with BW755C somewhat increases survival in conscious rats infused with *E. coli* endotoxin. The model is similar to that described by Adeleye et al (1981, 1982). Rats with indwelling arterial and venous catheters were given endotoxin by continuous infusion in a dose of 2.4 mg/kg/hr for 4 hours (10 mg/kg total dose). The metabolic effects include an initial hyperglycemia followed by a marked, sustained hypoglycemia (without hyperinsulinemia) and a severe lactic acidemia. The mortality is 75% assessed at 24 or 48 hours. This mortality is not influenced by specific inhibitors of thromboxane synthetase (e.g. dazoxiben; Furman et al 1984) but is reduced by methylprednisolone (McKechnie et al, unpublished observations). Survival in this model is somewhat increased by BW755C (50 mg/kg orally in a single dose given before the start of the endotoxin infusion) and the metabolic effects of endotoxin (plasma glucose, lactic acid) were not observed in rats so treated.

At the present time it is not possible to come to any definite conclusion regarding the role of lipoxygenase products of arachidonic acid in shock induced by endotoxin. The present author's view is that they might well prove to be important especially in the delayed effects (increase in pulmonary vascular permeability; ultimate multi-organ failure) and that either phospholipase inhibition or interference with both cyclooxygenase and lipoxygenase pathways of arachidonic acid metabolism offer promise in the treatment of endotoxin-induced shock. Whether this results in new ways of treating the patient with septic shock with high circulating levels of endotoxin is for future determination. It should, however, be borne in mind that arachidonic acid derivatives are only some of many potentially dangerous endogenous mediators released in shock (Parratt 1983b) and that ultimately

prophylaxis of patients at greatest risk (and hence prevention of the endotoxin-induced neurohumoral cascade) offers the best chance of success.

ACKNOWLEDGEMENTS

I am grateful to a number of people who helped in providing information for the
writing of this chapter. I especially thank Professor Dietrich Keppler (Freiburg) for
up-to-date information and for providing the figure of his mouse endotoxin studies,
to Dr. Rod Flower for much help over the question of lipoxygenase inhibition by
NSAIDs and to Professor M.J.H. Smith and Drs. Ogletree and Flynn for much
helpful correspondence. My own colleague Dr. Jack Zeitlin helped to set this subject
against arachidonic acid metabolism in general.

REFERENCES

Abaitey AK, Parratt JR (1976) Cardiovascular effects of diethylcarbamazine citrate.
    Br. J. Pharmacol. 56, 219-228.
Abaitey AK, Parratt JR (1977) The effects of diethylcarbamazine citrate on smooth
    muscle. *J. Pharm. Pharmacol. 29*, 428-432.
Adeleye GA, Al-Jibouri L, Furman BL, Parratt JR (1981) Endotoxin-induced
    metabolic changes in the conscious, unrestrained rat: hypoglycaemia and elevated
    blood lactate concentrations without hyperinsulinaemia. *Circ. Shock. 8*, 543-550.
Adeleye GA, Furman BL, Parratt JR (1982) Possible involvement of opiate receptors
    in the response of conscious rats to *Escherichia coli* endotoxin; protective effect
    of naloxone. *J. Physiol. 329*, 31P.
Advenier C, Cerrina J, Duroux P, Floch A, Pradel, J, Rennier A (1983) Sodium
    cromoglycate, verapamil and nicardipine antagonism to leukotriene D_4 bron-
    choconstriction. *Br. J. Pharmacol. 78*, 301-306.
Augstein J, Farmer JB, Lee TB, Sheard P, Tattersall ML (1973) Selective inhibitor
    of slow reacting substance of anaphylaxis. *Nature (London) New Biol. 245*, 215-
    217.
Austen KF, Corey EJ, Drazen JM, Leitch AG (1983) The effect of indomethacin on
    the contractile response of the guinea-pig lung parenchymal strip to leukotrienes
    B_4, C_4, D_4 and E_4. *Br. J. Pharmacol. 80*, 47-53.
Bach MK, Brashler JR, Fitzpatrick FA, Griffin RL., Iden SS, Johnson HG, McNee
    ML, McGuire JC, Smith HW, Smith RJ, Sun FF, Wasserman MA (1983) *In vivo*
    and *in vitro* actions of a new selective inhibitor of leukotriene C and D synthesis.
    In: Samuelsson B, Paoletti R, Ramwell P (Eds), *Advances in Prostaglandin,
    Thromboxane and Leukotriene Research, Vol 11*, pp 39-44. Raven Press, New
    York.
Ball HA, Parratt JR, Zeitlin IJ (1983) Effect of dazoxiben, a specific thromboxane
    synthetase inhibitor, on the pulmonary response to *E. coli* endotoxin in anaes-
    thetized cats. *Br. J. Clin. Pharmacol. 15, Suppl. 1*, 127-135S.

Barnett K, Piper P (1983) The effect of EP 045, a thromboxane A_2 antagonist, on leukotriene-induced contractions of parenchymal strips and human bronchus. *Br. J. Pharmacol. 79*, 215P.

Borgeat P, Hamberg M, Samuelsson B (1976) Transformation of arachidonic acid and homo-γ-linoleic acid by rabbit polymorphonuclear leukocytes. *J. Biol. Chem. 251*, 7816-7820.

Boyd L, Ezra D, Feuerstein G, Goldstein RE (1983) Effects of FPL-55712 or indomethacin on leukotriene-induced coronary constriction in the intact pig heart. *Eur. J. Pharmacol. 89*, 307-311.

Bray MA, Ford-Hutchinson AW, Smith MJH (1981) Leukotriene B_4: an inflammatory mediator *in vivo. Prostaglandins 22*, 213-222.

Brigham KL, Bowers RE, Haynes J (1979) Increased sheep lung vascular permeability caused by *E. coli* endotoxin. *Circ. Res. 45*, 292-297.

Burke JA, Levi R, Guo ZG, Corey EJ (1982) Leukotrienes Csub 4, Dsub 4 and Esub 4: effects on human and guinea-pig cardiac preparations in vitro. *J. Pharmacol. Exp. Ther. 221*, 235-241.

Busse WD, Mardin M, Gruetzmann R, Dunn JR, Theodoreau M, Sloane BF, Honn KV (1982) Nafazatrom (BAY g 6575) an inhibitor of cellular lipoxygenase activity. *Fed. Proc. 41*, 1717.

Casey FB, Appleby BJ, Buck DC (1983) Selective inhibition of the lipoxygenase metabolic pathway of arachidonic acid by the SRS-A antagonist, FPL 55712. *Prostaglandins 25*, 1-11.

Coker SJ, Parratt JR (1984) The effects of nafazatrom on arrhythmias and prostanoid release during coronary artery occlusion and reperfusion in anaesthetized greyhounds. *J. Mol. Cell. Cardiol. 16*, 43-52.

Coker SJ, Hughes B, Parratt JR, Rodger IW, Zeitlin IJ (1983) The release of prostanoids during the acute pulmonary response to *Escherichia coli* endotoxin in anaesthetized cats. *Br. J. Pharmacol. 78*, 561-570.

Coleman RA, Humphrey P, Kennedy I, Levy GP, Lumley GP (1981) Comparison of the actions of U-46619, a prostaglandin H_2-analogue, with those of prostaglandin H_2 and thromboxane A_2 on some isolated smooth muscle preparations. *Br. J. Pharmacol. 73*, 773-778.

Coleman RA, Kennedy I, Sheldrick RLG (1983a) Comparison of the release of thromboxane A_2 (TxA_2) from guinea-pig isolated perfused whole and superfused chopped lungs. *Br. J. Pharmacol. 79*, 357P.

Coleman RA, Kennedy I, Sheldrick RLG (1983b) Release of thromboxane A_2 from guinea-pig isolated perfused whole lung by leukotriene D_4; some unexplained observations. *Br. J. Pharmacol. 79*, 358P.

Cuthbert NJ, Gardiner PJ (1983) Leukotriene induced release of cyclooxygenase products from the guinea-pig trachea. In: Piper PJ (Ed), *Leukotrienes and Other Lipoxygenase Products*, pp 290-294. John Wiley & Sons, Chichester.

Dahlen S-E, Hedqvist P, Hammarström S, Samuelsson B (1980) Leukotrienes are potent constrictors of human bronchi. *Nature (London) 288*, 484-486.

Dahlen S-E, Hedqvist P, Hammarström S (1983) Contractile activities of several cysteine-containing leukotrienes in the guinea pig lung strip. *Eur. J. Pharmacol. 86*, 207-215.

Drazen JM, Austen KF, Lewis RA, Clark DA, Goto G, Marfat A, Cirey EJ (1980) Comparative airway and vascular activities in leukotrienes C-1 and D *in vivo* and *in vitro*. *Proc. Natl. Acad. Sci. 77*, 4354-4358.

Esbenshade AM, Newman JH, Lams PM, Jolles H, Brigham KL (1982) Respiratory failure after endotoxin infusion in sheep: lung mechanics and lung fluid balance. *J. Appl. Physiol. 53*, 967-976.

Feniuk K, Kennedy I, Whelan CJ (1983) Cardiovascular actions of leukotrienes. In: Piper PJ (Ed), *Leukotrienes and Other Lipoxygenase Products*, pp 108-112. John Wiley & Sons, Chichester.

Fiedler VB (1983a) Reduction of myocardial infarction and dysrhythmic activity by nafazatrom in the conscious rat. *Eur. J. Pharmacol. 88*, 263-267.

Fiedler VB (1983b) The effects of oral nafazatrom (BAY g 6575) on canine coronary artery thrombosis and myocardial ischemia. *Basic Res. Cardiol. 78*, 266-280.

Fiedler VD (1984) Reduction of acute myocardial ischemia in rabbit hearts by nafazatrom. *J. Cardiovasc. Pharmacol. 6*, 318-324.

Fleisch JH, Rinkema LE, Baker SR (1982) Evidence for multiple leukotriene D_4 receptors in smooth muscle. *Life Sci. 31*, 577-581.

Flower RJ, Blackwell GJ (1976) The importance of phospholipase A_2 in prostaglandin biosynthesis. *Biochem. Pharmacol. 25*, 285-291.

Flower RJ, Blackwell GJ (1979) Anti-inflammatory steroids induce biosynthesis of a phospholipase A_2 inhibitor which prevents prostaglandin generation. *Nature (London) 278*, 456-459.

Fritz MM, Keppler DOR (1982) Liver calcium and endotoxin action. *Naturwissenschaften 69*, 147.

Furman BL, McKechnie K, Parratt JR (1984) Failure of drugs which selectively inhibit thromboxane synthesis to modify endotoxin shock in conscious rats. *Br. J. Pharmacol. 82*, 289-294.

Galanos C, Freudenberg MA, Reutter W (1979) Galactosamine-induced sensitization to the lethal effects of endotoxin. *Proc. Natl Acad. Sci. 76*, 5339-5343.

Griffin M, Weiss JW, Keitch AG, McFadden ER, Corey EJ, Austen KF, Drazen JM (1983) Effects of leukotriene D on the airways in asthma. *N. Engl. J. Med. 308*, 436-439.

Hagmann W, Keppler D (1982) Leukotriene antagonists prevent endotoxin lethality. *Naturwissenschaften 69*, 594-595.

Hagmann W, Fritz MM, Keppler D (1983) Leukotrienes and endotoxin lethality. In: Piper PJ (Ed), *Leukotrienes and Other Lipoxygenase Products*, pp 128-133. John Wiley & Sons, Chichester.

Hagmann W, Denzlinger C, Keppler D (1984) Role of peptide leukotrienes and their hepatobiliary elimination in endotoxin action. *Circ. Shock 14*, 223-235.

Hand JM, Will JA, Buckner CK (1983) Enhancement by histamine of the contractile effects of leukotriene D_4 on extralobar pulmonary arteries isolated from the guinea-pig. *Eur. J. Pharmacol. 87*, 341-343.

Hayes EC, Lombardo DL, Girard Y, Maycock AL, Rokach J, Rosenthal AS, Young RN, Zweerink HJ (1983) Leukotriene specific radioimmunoassays. In: Piper PJ (Ed), *Leukotrienes and Other Lipoxygenase Products*, pp 26-43. John Wiley & Sons, Chichester.

Hedqvist P, Dahlen S-E (1983a) Leukotrienes as mediators of allergic and inflammatory reactions. In: Piper PJ (Ed), *Leukotrienes and Other Lipoxygenase Products*, pp 134-151. John Wiley & Sons, Chichester.

Hedqvist P, Dahlen S-E (1983b) Pulmonary and vascular effects of leukotrienes imply involvement in asthma and inflammation. In: Samuelsson B, Paoletti R, Ramwell P (Eds), *Advances in Prostaglandin, Thromboxane and Leukotriene Research, Vol 11*, pp 27-32. Raven Press, New York.

Hedqvist P, Dahlen S-E, Gustafson L, Hammarström S, Samuelsson B (1980) Biological profile of leukotrienes C_4 and D_4. *Acta Physiol. Scand. 110*, 331-333.

Heflin AC, Brigham KL (1981) Prevention by granulocyte depletion of increased lung vascular permeability of sheep lung following endotoxemia. *J. Clin. Invest. 68*, 1253-1260.

Higgs GA, Flower RJ (1981) Anti-inflammatory drugs and the inhibition of arachidonate lipoxygenase. In: Piper PJ (Ed), *SRS-A and Leukotrienes*, pp 197-207. John Wiley & Sons, Chichester.

Higgs GA, Eakins KE, Mugridge KG, Moncada S, Vane JR (1980) The effects of non-steroid anti-inflammatory drugs on leukocyte migration in carrageenin-induced inflammation. *Eur. J. Pharmacol. 66*, 81-86.

Higgs GA, Salmon JA, Spayne JA (1981) The inflammatory effects of hydroperoxy and hydroxy acid products of arachidonate lipoxygenase in rabbit skin. *Br. J. Pharmacol. 74*, 429-433.

Higgs GA, Bax CMR, Moncada S (1982) Inflammatory properties of lipoxygenase products and the effects of indomethacin and BW755C on prostaglandin production, leukocyte migration, and plasma exudation in rabbit skin. In: Samuelsson B, Paoletti R (Eds), *Leukotrienes and Other Lipoxygenase Products*, pp 331-339. Raven Press, New York.

Hirshman CA, Darnell M, Brugman T, Peters J (1983) Airways constrictor effects of leukotriene D_4 in dogs with hyperactive airways. *Prostaglandins 25*, 481-490.

Holroyde MC, Altounyan REC, Cole M, Dixon M, Elliott EV (1981) *Pharmacology of Inflammation and Allergy*, pp 147-152. Les colloques de l'INSERM No 100.

Jakschik BA, Kuo CG (1983) Subcellular localization of leukotriene-forming enzymes. In: Samuelsson B, Paoletti R, Ramwell P (Eds), *Advances in Prostaglandin, Thromboxane and Leukotriene Research, Vol 11*, pp 141-145. Raven Press, New York.

Johnson HG, McNee ML, Bach MK, Smith HW (1983) The activity of a new, novel inhibitor of leukotriene synthesis in rhesus monkey *ascaris* reactors. *Int. Arch. Allergy Appl. Immunol. 70*, 169-173.

Jones TR, Davis C, Daniel EE (1982) Pharmacological study of the contractile activity of leukotrienes C_4 and D_4 on isolated human airway smooth muscle. *Can. J. Physiol. Pharmacol. 60*, 638-643.

Kennedy I, Whelan CJ, Wright G (1983) The actions of leukotrienes and FPL55712 on guinea-pig isolated heart; lack of specificity of FPL55712. *Br. J. Pharmacol. 79*, 218P.

Keppler D, Lesch R, Reutter W, Decker K (1968) Experimental hepatitis induced by D-galactosamine. *Exp. Mol. Pathol. 9*, 279-290.

Keppler D, Pausch J, Decker K (1974) Selective uridine phosphate deficiency induced by D-galactosamine in liver and reversed by pyrimidine nucleotide precursors. *J. Biol. Chem. 249*, 211-216.

Krell RD, Isai BS, Giles RE (1983) Pharmacologic aspects of leukotriene receptors. In: Piper PJ (Ed), *Leukotrienes and Other Lipoxygenase Products*, pp 222-240. John Wiley & Sons, Chichester.

Lands WEM, Samuelsson B (1968) Phospholipid precursors of prostaglandins. *Biochim. Biophys. Acta 164*, 426-429.

Letts LG, Piper PJ (1982) The actions of leukotrienes C_4 and D_4 on guinea-pig isolated hearts. *Br. J. Pharmacol. 76*, 169-176.

Letts LG, Piper PJ (1983a) Cardiac actions of leukotrienes B_4, C_4, D_4 and E_4 in guinea pig and rat *in vitro*. In: Samuelsson B, Paoletti R, Ramwell P (Eds), *Advances in Prostaglandin, Thromboxane and Leukotriene Research, Vol 11*, pp 391-395. Raven Press, New York.

Letts LG, Piper PJ (1983b) 5,6-Methano leukotriene A_4: a potent specific inhibitor for 5-lipoxygenase. In: Samuelsson B, Paoletti R, Ramwell P (Eds), *Advances in Prostaglandin, Thromboxane and Leukotriene Research, Vol 11*, pp 163-167. Raven Press, New York.

Letts LG, Piper PJ, Newman DA (1983) Leukotrienes and their action in the coronary circulation. In: Piper PJ (Ed), *Leukotrienes and Other Lipoxygenase Products*, pp 94-107. John Wiley & Sons, Chichester.

Levine L, Morgan RA, Lewis RA, Austin KF, Clark DA, Marfat A, Corey EJ (1981) Radioimmunoassay of the leukotrienes of slow reacting substance of anaphylaxis. *Proc. Natl Acad. Sci. 78*, 7692-7696.

Lux WE, Feuerstein G, Faden AI (1983a) Alteration of leukotriene D_4 hypotension by thyrotropin releasing hormone. *Nature (London) 301*, 822-824.

Lux WE, Feuerstein GZ, Smith GP, Faden AI (1983b) Effects of pharmacologic interventions on leukotriene D_4 hypotension in the unanaesthetised guinea-pig. In: Piper PJ (Ed), *Leukotrienes and Other Lipoxygenase Products*, pp 338-342. John Wiley & Sons, Chichester.

MacGlashan DW, Schleimer RP, Peters SP, Schulman ES, Adams GK, Newball HH, Lichtenstein LM (1982) Generation of leukotrienes by purified human lung mast cells. *J. Clin. Invest. 70*, 747-751.

Mardin H, Busse W-D (1983) Effect of nafazatrom on the lipoxygenase pathways in PMN leukocytes and RBL-1 cells. In: Piper PJ (Ed), *Leukotrienes and Other Lipoxygenase Products*, pp 263-274. John Wiley & Sons, Chichester.

Matthews WR, Murphy RC (1982) Inhibition of leukotriene biosynthesis in mastocytoma cells by diethylcarbamazine. *Biochem. Pharmacol. 31*, 2129-2132.

Mead B, Patterson LH, Smith DA (1981) The disposition of FPL 55712 acid (7-[3-(4-acetyl-3-hydroxy-2-propyl-phenoxy)-2-hydroxypropoxy]-4oxo-8-propyl-4H-benzopyran-2-carboxylic acid) in rat and dog. *J. Pharm. Pharmacol. 33*, 682-684.

Michelassi F, Landa L, Hill RD, Lowenstein E, Watkins WD, Petkau AJ, Zapol WM (1982) Leukotriene D_4: a potent coronary artery vasoconstrictor associated with impaired ventricular contraction. *Science 217*, 841-843.

Michelassi F, Landa L, Hill RD, Huttemeier P, Lowenstein E, Zapol WM, Watkins WD (1983) Leukotriene D_4: a potent sheep coronary vasoconstrictor not blocked by ibuprofen. In: Samuelsson B, Paoletti R, Ramwell P (Eds), *Advances in Pros-*

taglandin, Thromboxane and Leukotriene Research, Vol. *11*, pp 397-399. Raven Press, New York.

Morley I, Page C, Paul W (1981) Leukotrienes, SRS-A and the vascular manifestations of PCA. *Agents Actions 11*, 585-587.

Morris HR, Taylor GW, Clinton PM, Piper PJ, Tippins JR, Barnett K, Costello J, Dunlop L, Henderson A, Heaton R (1983) Measurement of leukotrienes in asthmatics. In: Samuelsson B, Paoletti R, Ramwell P (Eds), *Advances in Prostaglandin, Thromboxane and Leukotriene Research, Vol 11*, pp 221-228. Raven Press, New York.

Mullane KM, Moncada S (1982) The salvage of ischaemic myocardium by BW755C in anaesthetised dogs. *Prostaglandins 24*, 255-266.

Murphy RC, Hoffer B, Palmer M, Olsen L, Hayes EC, Rosenthal A, Young RN, Rokach J (1983) Localization of cells with leukotriene-like immunoreactivity during anaphylaxis: preliminary observations. In: Piper PJ (Ed), *Leukotrienes and Other Lipoxygenase Products*, pp 70-77. John Wiley & Sons, Chichester.

O'Driscoll BRC, Kay AB (1982) Leukotrienes and lung disease. *Thorax 37*, 241-245.

Ogletree ML, Brigham KL (1982) Effects of cyclooxygenase inhibitors on pulmonary vascular responses to endotoxin in unanesthetized sheep. *Prostaglandins Leukotrienes Med. 8*, 489-502.

Ogletree ML, Snapper JR, Brigham KL (1982a) Immediate pulmonary vascular and airways responses after intravenous leukotriene (LT) D_4 injections in awake sheep. *Physiologist 25*, 275.

Ogletree ML, Oates JA, Brigham KL, Hubbard WC (1982b) Evidence for pulmonary release of 5-hydroxyeicosatetraenoic acid (5-HETE) during endotoxemia in unanesthetized sheep. *Prostaglandins 23*, 459-468.

Orange RP, Valentine MD, Austen KF (1968) Inhibition of the release of slow-reacting substance of anaphylaxis in the rat with diethylcarbamazine. *Proc. Soc. Exp. Biol. Med. 127*, 127-132.

Orange RP, Austen WG, Austen KF (1971) Immunological release of histamine and slow reacting substance of anaphylaxis from human lung. *J. Exp. Med, 134*, 136s-148s.

Parratt JR (1983a) Pulmonary function in experimental endotoxin shock – an examination of the role of chemical mediators. In: Lewis DH, Hagland U (Eds), *Shock Research*, pp 129-145. Elsevier Science Publishers, Amsterdam.

Parratt JR (1983b) Neurohumoral agents and their release in shock. In: Altura BM, Lefer AM, Schumer W (Eds), *Handbook of Shock and Trauma*, pp 311-336. Raven Press, New York.

Parratt JR, Sturgess RM (1974) The effect of indomethacin on the cardiovascular and metabolic responses to E. coli endotoxin in the cat. *Br. J. Pharmacol. 50*, 177-184.

Parratt JR, Sturgess RM (1975) *E. coli* endotoxin shock in the cat; treatment with indomethacin. *Br. J. Pharmacol. 53*, 485-488.

Parratt JR, Sturgess RM (1976) The effect of a new anti-inflammatory drug, flurbiprofen, on the respiratory, haemodynamic and metabolic responses to *E. coli* endotoxin shock in the cat. *Br. J. Pharmacol. 58*, 547-551.

Parratt JR, Coker SJ, Hughes B, MacDonald A, Ledingham IMcA, Rodger IW, Zeitlin IJ (1982) The possible role of prostaglandins and thromboxanes in the pulmonary consequences of experimental endotoxin shock and clinical sepsis. In: McConn R (Ed), *Role of Chemical Mediators in the Pathophysiology of Acute Illness and Injury*, pp 195-217. Raven Press, New York.

Pfeffer MA, Pfeffer JM, Lewis RA, Braunwald E, Corey EJ, Austen F (1983) Systemic hemodynamic effects of leukotrienes C_4 and D_4 in the rat. *Am. J. Physiol. 244*, H628-633.

Piper PJ, Temple DM (1981) The effect of lipoxygenase inhibitors and diethylcarbamazine on the immunological release of slow reacting substance of anaphylaxis (SRS-A) from guinea-pig chopped lung. *J. Pharm. Pharmacol. 33*, 384-386.

Piper PJ, Samhoun MN, Tippins JR, Williams TJ, Palmer MA, Peck MK (1981) Pharmacological studies on pure SRS-A, and synthetic leukotrienes C_4 and D_4. In: Piper PJ (Ed), *SRS-A and Leukotrienes*, pp 81-89. John Wiley & Sons, Chichester.

Randall RW, Eakins KE, Higgs GA, Salmon JA, Tateson JE (1980) Inhibition of arachidonic acid cyclooxygenase and lipoxygenase activities of leukocytes by indomethacin and compound BW755C. *Agents Actions 10*, 553-555.

Rosenthal A, Pace-Asciak CR (1983) Potent vasoconstriction of the isolated perfused rat kidney by leukotrienes C_4 and D_4. *Can. J. Physiol. Pharmacol. 61*, 325-328.

Schianterelli P, Congrani S, Folco G (1981) Bronchospasm and pressor effects induced in the guinea-pig by leukotriene C_4 are probably due to release of cyclooxygenase products. *Eur. J. Pharmacol. 73*, 363-366.

Serhan CN, Firdovich J, Goetzel EJ, Dunhan PB, Weissman G (1982) Leukotriene B_4 and phosphatidic acid are calcium ionophores. *J. Biol. Chem. 257*, 4746-4752.

Seuter F, Busse WD, Mank K, Hoffmeister F, Möller E, Horstmann H (1979) The antithrombotic activity of BAY g 6575. *Arzneim. Forsch. 29*, 54-59.

Sircar JC, Schwender CF, Johnson EA (1983) Soybean lipoxygenase inhibition by nonsteroidal antiinflammatory drugs. *Prostaglandins 25*, 393-396.

Smedegård G, Hedqvist P, Dahlen S-E, Revanas B, Hammarström S, Samuelsson B (1982) Leukotriene C_4 affects pulmonary and cardiovascular dynamics in monkey. *Nature (London) 295*, 327-329.

Smith MJ, (1981) Leukotriene B_4 (Minireview). *Gen. Pharmacol. 12*, 211-216.

Smith MJ (1982) Biological activities of leukotriene B_4 (Isomer III). In: Samuelsson B, Paoletti R, Ramwell P (Eds), *Advances in Prostaglandin, Thromboxane and Leukotriene Research*, Vol 9, pp 283-292. Raven Press, New York.

Smith M, Zaiss C, Gunther R, Demling R (1981a) Lipoxygenase pathway of arachidonic acid and acute lung injury. *Circ. Shock 8*, 198-199.

Smith M, Gunther R, Zaiss C, Flynn J, Demling R (1981b) Pulmonary microvascular injury from lipoxygenase infusion: comparison with endotoxemia. *Circ. Shock 8*, 647-656.

Terashita A-I, Fukui H, Hirata M, Terao S, Ohkawa S, Nishikawa K, Kikuchi S (1981) Coronary vasoconstriction and PGI_2 release by leukotrienes in isolated guinea pig hearts. *Eur. J. Pharmacol. 73*, 357-361.

Turner SR, Trainer SA, Lynn WS (1975) Biogenesis of chemotactic molecules by the arachidonate lipoxygenase systems of platelets. *Nature (London) 257*, 680-681.

Van Belle H (1981) A potent inhibitor of calmodulin-activated enzymes. *Cell Calcium 2*, 483-494.

Vargaftig BB, Lefort J, Murphy R (1981) Leukotrienes C and D induce aspirin-sensitive bronchoconstriction in the guinea-pig. *Pharmacology of Inflammation and Allergy*, pp 153-159. Les Colloques de l'INSERM No. 100.

Weichman BM, Muccitelli RM, Tucker SS, Osborn RR, Gleason JG, Wasserman MA (1983) Relationship between prostaglandin and thromboxane production and the bronchoconstrictive property of LTD_4. In: Piper PJ (Ed), *Leukotrienes and Other Lipoxygenase Products*, pp 198-204. John Wiley & Sons, Chichester.

Weiss JW, Drazen JM, Coles N (1982) Bronchoconstrictor effects of leukotriene C in humans. *Science 216*, 196-198.

Williams TJ, Jose PJ, Wedmore CV, Peck MJ, Forrest MJ (1983) Mechanisms underlying inflammatory edema: the importance of synergism between prostaglandins, leukotrienes, and complement-derived peptides. In: Samuelsson B, Paoletti R, Ramwell P (Eds), *Advances in Prostaglandin, Thromboxane and Leukotriene Research, Vol 11*, pp 31-37. Raven Press, New York.

Wong PYK, Chao PHW, McGiff JC (1982) Nafazatrom (BAY g-6575), an antithrombotic and antimestatic agent inhibits 15-hydroxy prostaglandin dehydrogenase. *J. Pharmacol. Exp. Ther. 223*, 757-760.

Woodman OL, Dusting GJ (1983) Coronary vasoconstriction induced by leukotrienes in the anaesthetised dog. *Eur. J. Pharmacol. 86*, 125-128.

Woodward DF, Weichman BM, Gill CA (1983) The effect of synthetic leukotrienes on tracheal microvascular permeability. *Prostaglandins 25*, 131-142.

Young RN, Kakushima M, Rokach J (1982) Studies on the preparation of conjugates of leukotriene C_4 with proteins for development of an immunoassay for SRS-A. *Prostaglandins 23*, 603.

Zeitlin IJ (1981) Arachidonate metabolites in inflammation. *Clin. Rheum. Dis. 7*, 781-797.

Zijlstra FJ, Bonta IL, Vincent JE (1983) Tachyphylaxis of leukotriene C_4-induced release of thromboxane A_2 from guinea-pig lung parenchyma and isoproterenol inhibition of this release. In: Samuelsson B, Paoletti R, Ramwell P (Eds), *Advances in Prostaglandin, Thromboxane and Leukotriene Research, Vol 12*, pp 167-171. Raven Press, New York.

Zukowska-Grojec Z, Bayorh MA, Kopin IJ, Feuerstein G (1982) Leukotriene D_4: cardiovascular and sympathetic effects in spontaneously hypertensive rats (SHR) and Wistar-Kyoto (WKY) rats. *J. Pharmacol. Exp. Ther. 223*, 183-189.

Zukowska-Grojec Z, Bayorh MA, Yaar I, Kopin IJ, Feuerstein G (1983a) Leukotriene D_4: divergent cardiovascular and sympathetic effects of spontaneously hypertensive and normotensive Wistar-Kyoto rats. In: Samuelsson B, Paoletti R, Ramwell P (Eds), *Advances in Prostaglandin, Thromboxane and Leukotriene Research, Vol 11*, pp 407-412. Raven Press, New York.

Zukowska-Grojec Z, Bayorh MA, Faden AI, Kopin IJ, Feuerstein G (1983b) Pharmacological interventions in the cardiovascular effects of leukotriene D_4 in pithed spontaneous hypertensive rats. In: Piper PJ (Ed), *Leukotrienes and Other Lipoxygenase Products*, pp 113-120. John Wiley & Sons, Chichester.

Handbook of Endotoxin, Vol. 2: Pathophysiology of Endotoxin
L.B. Hinshaw, editor
© Elsevier Science Publishers B.V., 1985
ISBN 0 444 90385 2
$0.85 per article per page (transactional system)
$0.20 per article per page (licensing system)

CHAPTER 10

The role of arachidonic acid metabolites in endotoxin shock II: involvement of prostanoids and thromboxanes*

JOHN T. FLYNN

1. INTRODUCTION

Arachidonic acid was first isolated from liver lipids and characterized by Hartley in 1909 as a polyunsaturated lipid with the chemical formula $C_{20}H_{32}O_2$. Brown (1928) further purified this fatty acid, and in 1942 Mowry et al fully elucidated the stereochemistry of the compound as a 20 carbon, polyunsaturated fatty acid (i.e., all-*cis*-5,8,11,14-eicosatetraenoic acid). The physiologic significance of this essential fatty acid was only vaguely understood until Van Dorp et al (1964) and Bergstrom et al (1964) described the biosynthesis of bisenoic prostanoids from arachidonic acid. Further work in the field characterized both the enzymatic and nonenzymatic synthesis of at least 6 stable, biologically active arachidonate metabolites. These were the classical bisenoic prostaglandins of the A, B, C, D, E and F series. The biological activity of these materials had been indirectly shown at least 30 years before by Kurzrok and Lieb (1930), Von Euler (1934) and Goldblatt (1935), who demonstrated that seminal fluid was able to contract uterine smooth muscle. Von Euler (1935) further purified the active substances and observed that the extracts were able to produce vasodilation in vivo.

By the late 1960s, several techniques for the synthetic production of various prostaglandins had been developed. These technical achievements led directly to the plethora of papers in the early 1970s which described the physiologic and pharmacologic actions of the classical prostaglandins. Prominent among these actions were the ability of E- and A-type prostaglandins to produce vasodilation in several species and the ability of F-type prostaglandins to contract capacitance vessels (see review by Malik and McGiff 1976). It had also been established that various prostaglandins could inhibit

---

* The preparation of this review was supported in part by GM 28023 from the National Institute of General Medical Sciences, National Institutes of Health, U.S.A.

platelet aggregation both in vitro and in vivo. In addition to these hemodynamic effects, Hedqvist in 1969 reported his work on the involvement of E-type prostaglandins in the possible modulation of sympathetic nervous system activity. It was primarily the above stated biological activities of the newly rediscovered prostaglandins and their possible emergence as a modulatory system that attracted the attention of several investigators working in the field of circulatory shock.

This interest in the possible involvement of prostaglandins in shock was strengthened by another set of observations. In 1971, Vane demonstrated that nonsteroidal antiinflammatory drugs such as acetylsalicylic acid were potent inhibitors of prostaglandin biosynthesis. This finding was of seminal importance to the area of prostaglandins in circulatory shock since Northover and Subramanian (1962) had earlier reported that large doses of acetylsalicylic acid could significantly improve the rate of survival of dogs receiving a lethal dose of endotoxin. Thus in the early 1970s, the assumption that the mechanism of action of acetylsalicylic acid in endotoxin shock was that of inhibiting de novo prostaglandin biosynthesis was an initiating factor in the first phase of studies which were directly designed to investigate the role of prostaglandins in the shock state.

The basic hypothesis that was being tested stated that endogenously synthesized prostaglandin-like materials were formed and released by various tissues of the body in response to a shock-inducing stimulus such as the bolus administration of endotoxin. Once released, the prostaglandins could then act in one of three ways. First, the newly formed prostanoids could act in a beneficial manner within a tissue and be viewed as a positive cellular homeostatic mechanism. An example of this would be prostaglandin E-induced vasodilation which would increase the rate of tissue perfusion. Secondly, arachidonic acid metabolism could result in the production of metabolites which could produce detrimental effects, for example, vasoconstriction in the lung or mesenteric bed when inappropriate. The third possibility was that endogenously synthesized prostaglandins played no role in the pathophysiology of the shock state but rather represented a temporally related artifact of the disease process.

Ten years of subsequent research in both the field of circulatory shock and prostaglandin physiology have demonstrated that the involvement of the arachidonic acid cascade in the series of events associated with cellular injury and shock is extremely complex. Very few generalizations hold. We are no longer dealing with prostaglandins per se. Rather, there are 'eicosanoids', including the classical prostaglandins, prostacyclin, thromboxanes, various hydroxy and hydroperoxy fatty acids and the leukotrienes. Most tissues have been shown to possess different profiles of the enzymatic machinery required to synthesize each of these materials. In addition, each substance possesses its own unique set of physiologic characteristics. Because of this complexity of the arachidonic acid cascade, one can no longer provide a simple answer on the role of eicosanoids in shock. One must now focus on the specific injury model to be studied, the tissue of interest, the particular eicosanoids that may be involved, and those factors within the animal which may modify or modulate the eicosanoid involvement. Within this review, I will be discussing only the primary arachidonic acid metabolites which are products of the cyclooxygenase reaction and the roles which they may be playing in endotoxin shock. These metabolites include the A-, E- and F-series prostaglandins, thromboxane A_2 and prostacyclin. For a summary of lipoxygenase involvement in endotoxic shock the reader is referred to Chapter 9.

2. OBJECTIVES OF THE REVIEW

This review will deal with several different aspects of arachidonic acid metabolism by animals in endotoxin shock. Initial data will be presented related to the ability of gram-negative bacterial endotoxin in vivo to stimulate endogenous arachidonic acid metabolism. These data will be compared and contrasted with the ability of endotoxin to produce similar results in several in vitro models. Data will also be presented on the rate of clearance and catabolism of various prostaglandin-like materials during the shock state. Two other major aspects of the presentation will be discussions of the physiologic effects of enhancing the arachidonic acid cascade during endotoxin shock or conversely, the effects of blockade of arachidonic acid metabolism by nonsteroidal antiinflammatory and other drugs. In reviewing these latter two aspects of the presentation, emphasis will be placed, wherever possible, on the rate of animal survival as the end-point in evaluating whether any manipulation of the arachidonic acid cascade was of physiologic importance. In addition to survival, other physiologic areas to be emphasized will be vascular hemodynamics, pulmonary fluid dynamics, and the involvement of individual cells such as platelets, leukocytes or tissue-based cells such as macrophages, in the shock state.

As a preface to the discussion of the involvement of cyclooxygenase products in endotoxin shock, one should be cognizant of several concepts or common assumptions which bear upon the interpretation of experimental data in this area. One of these factors is the concept of modulation of physiologic function. A large body of literature is developing which suggests that the arachidonic acid cascade represents a fundamental mechanism which normally modulates the activity of many biological systems, that is, the eicosanoid system is not, in most cases, an effector mechanism. Rather, within a normal range of physiologic function, arachidonate metabolism may act to change the sensitivity or responsiveness of other homeostatic mechanisms. The modulation of renin release by granular cells in the renal afferent arteriole by E- and F-type prostaglandins and the modulation of antidiuretic hormone action on the distal nephron by E-type prostaglandins are examples of this facet of eicosanoid activity. When functioning in their normal ranges, these modulatory actions may be quite appropriate, but the shock state is often characterized by excessive or inappropriate events. It therefore becomes quite important when reviewing the endotoxin literature to attempt to determine the appropriateness of the eicosanoid involvement. A specific prostanoid may exert a beneficial action early in the course of endotoxic shock when the homeostatic systems of an animal are responsive, whereas the presence of that same prostanoid during a decompensatory period may prove highly deleterious to the animal. An attempt will be made in this review to assess the degree of injury of the shock model in order to evaluate the nature of eicosanoid involvement in a qualitative sense.

Comparisons will also be made between different responses of various species, between the bolus injection models and the septic models of endotoxin shock and between models of differing lethality. It is hoped that these comparisons will develop perspective in the relationships between arachidonic acid metabolism and the physiologic and pathophysiologic events of endotoxin shock.

In the course of these discussions, three assumptions which are often made in the interpretation of data in the eicosanoid/shock field will not be accepted. The first of these is the assumption that the entire cyclooxygenase system responds to a stimulus as a unit, that is, that all pathways are simultaneously activated. To the contrary, the production of several materials such as prostacyclin and thromboxane A_2 can proceed independently of each other once the endoperoxide intermediates become available. The second invalid assumption is the preconceived notion that some prostanoids are salutary in all situations whereas others are deleterious. Data will be presented which demonstrate that a single prostanoid may play both roles during the time course of the shock state. The third invalid assumption is that an investigator can interpret the role of endogenous arachidonic acid products in a shock model without attempting to measure endogenous prostanoid biosynthesis in either the qualitative or quantitative sense. Several reports have attributed the effects of nonsteroidal antiinflammatory drugs in endotoxin shock solely to the inhibition of prostanoid production. This was done without characterizing (a) whether there was a prostanoid production, (b) which eicosanoids might have been involved, (c) the time course of the prostanoid response, or (d) whether or not the dosage of inhibitor used was adequate to ablate any possible arachidonic acid metabolite involvement. Thus, inhibitor studies which lack proper prostanoid quantitation will be viewed with some degree of skepticism.

3. ENDOTOXIN-MEDIATED PROSTANOID BIOSYNTHESIS IN VIVO

3.1. Bolus endotoxin models

Much of the initial work in the area of prostanoids in endotoxic shock deals with the measurement of plasma concentrations of the classical prostaglandins (PG) in the bolus endotoxin model in vivo (Table 1). Kessler et al (1973) administered a 3 mg/kg bolus dose of endotoxin to dogs and were able to demonstrate the presence of lipid depressor materials in the plasma of the shocked animals. These lipids cochromatographed with authentic PGE and PGA standards. In that same year, Collier et al (1973) demonstrated the release of E-type prostaglandins by the canine kidney in response to endotoxin-mediated hypotension. This prostaglandin activity diminished after the administration of indometacin, a cyclooxygenase inhibitor.

These initial studies, carried out before specific or sensitive radioimmunological techniques were available, gave the first indication that there was enhanced arachidonic acid metabolism during the course of endotoxic shock. Later, Herman and Vane (1976b) and Isakson et al (1977) also demonstrated rapid PGE and PGF production by the canine kidney following endotoxin administration. Korbut et al (1975), using bioassay techniques, observed the appearance of an E-type prostaglandin in mixed venous blood of a cat endotoxin shock model. This PGE synthesis could be affected by indometacin. With the advent of specific radioimmunoassays for the classical prostaglandins, more precise data as to the time course of arachidonic acid metabolism during endotoxin shock were gained. Anderson et al (1975a) showed a time-dependent production of E- and F-series prostaglandins in the

TABLE 1   *Early studies on the in vivo release of prostaglandins A, E and F during endotoxin shock*

Species	Dose of endotoxin	Prostanoid	Prostanoid concentration in plasma	Reference
Dog	3 mg/kg i.v. bolus	A, E	increased	Kessler et al (1973)
Dog	1–7 mg/kg i.v. bolus	E	increased	Collier et al (1973)
Cat	10–15 mg/kg i.v. bolus	E	increased	Korbut et al (1975)
Dog	1 mg/kg i.v. bolus	E, F	increased	Anderson et al (1975a)
Calf	1 mg/kg i.v. bolus	E, F	increased	Anderson et al (1975b)
Dog	5 mg/kg i.v. bolus	E, F	increased	Herman and Vane (1976b)
Dog	2 mg/kg i.v. bolus	E	increased	Isakson et al (1977)

dog model. Portal venous plasma PGE and PGF concentrations were significantly elevated within 15 minutes after the bolus endotoxin administration, whereas renal venous plasma concentrations of PGE and PGF became significantly elevated only after 60 to 90 minutes. No changes were observed in aortic plasma prostaglandin values. These same investigators examined pulmonary arterial and venous concentrations of PGE and PGF compounds in the calf, a species which responds to small doses of endotoxin with intense pulmonary vasoconstriction (1975b). In these animals, Anderson et al observed significant increases in pulmonary venous PGF concentrations at 15 and 45 minutes after endotoxin administration. The PGE concentration varied slightly. The pulmonary venous PGF values correlated with endotoxin-mediated increases in pulmonary arterial pressure. Pretreatment of the calves with indometacin inhibited both the PGF production and the pulmonary pressor response to endotoxin. These early studies strongly supported the hypothesis that the bolus intravenous administration of gram-negative endotoxin was a potent stimulus for E- and F-series prostaglandin production in vivo. Furthermore, these endogenously synthesized lipids seemed to be involved in the pathophysiology of the model.

The discovery of thromboxanes by Hamburg et al (1975) and prostacyclin by Moncada et al (1976) further stimulated interest in the question of prostanoid biosynthesis during the shock state. Table 2 lists several representative papers which describe enhanced production of these materials following the bolus administration of endotoxin to intact animals. Bult et al (1978) reported that concentrations of 6-keto-$PGF_{1\alpha}$, a stable metabolite of prostacyclin, began to increase in arterial blood within 5 minutes after a massive intravenous dose of endotoxin in the rabbit. This time course is somewhat faster than that previously seen for the increase in the production of E- and F-type prostaglandins. Within 3 hours after endotoxin, the 6-keto-$PGF_{1\alpha}$ concentration had declined. Harris et al (1980) and Fletcher et al (1981) demonstrated time course differences in prostacyclin and thromboxane (TX) biosynthesis in a baboon bolus endotoxin model. Femoral and pulmonary arterial 6-keto-$PGF_{1\alpha}$ concentrations began to increase progressively within 15 minutes after the bolus endotoxin administration, peaked at 2 hours post-endotoxin and declined thereafter. In contrast, plasma TXB_2 concentrations increased rapidly and maximally at 15 minutes post-endotoxin and then declined within the next 45 minutes. Thus in the baboon LD_{70} model, thromboxane production is rapid, approximately 5 times greater than the maximal prostacyclin response, and of relatively short duration. In contrast, the prostacyclin response is gradual and of a much longer duration.

This gradual increase in prostacyclin synthesis may not occur in all species. Webb et al (1981) were unable to demonstrate any increased prostacyclin concentration in a porcine model when endotoxin was constantly infused at

0.001 to 1 µg/kg/min for 90 minutes. The thromboxane response in the pig was similar to that described previously in the baboon model and was not dose-dependent. These data suggest that at least two different mechanisms or tissues are involved in producing thromboxane and prostacyclin-like materials in response to endotoxin. Other species respond in a similar qualitative manner but with slightly different time courses. In the horse, there is a slow progressive increase in prostacyclin production on a temporal basis similar to rabbit and baboon. However, the thromboxane response was just as rapid but prolonged when compared to other species, remaining elevated 4 hours after endotoxin (Bottoms et al 1982). The rat also demonstrates a prolonged thromboxane response to a lethal dose of endotoxin (Cook et al 1980).

The data described above were gathered in relatively lethal models (LD_{70} to LD_{100}) of endotoxin shock. It is interesting to examine the effect of low, nonlethal bolus doses of endotoxin in several species. Hales et al (1981) administered 15 to 150 µg/kg (i.v. bolus) of endotoxin to healthy dogs and observed a dose-related prostanoid response. At the 15 µg/kg dose, both 6-keto-$PGF_{1\alpha}$ and thromboxane-like materials were modestly and equally elevated in venous and arterial plasma. However, at the 150 µg/kg dose, 6-keto-$PGF_{1\alpha}$ increased modestly, but TXB_2 concentrations were markedly elevated to concentrations which had been observed by others in the lethal models of endotoxin shock. Other than a transient increase in pulmonary vascular resistance, the dogs appeared hemodynamically stable. This suggests that low doses of endotoxin can stimulate the arachidonic acid cascade maximally and produce prostanoids in the same concentration ranges as those observed in the lethal endotoxin models. In other words, high rates of prostanoid production do not necessarily produce endotoxin shock in a normal, healthy animal which can adequately compensate for the biologic actions of the prostanoids, but those same prostanoids could have a much different effect in an animal in which compensatory mechanisms are impaired. It is critical to interpret prostanoid involvement in relation to the physiologic status of the animal.

A great deal of attention has been paid to the pulmonary production of prostanoids during endotoxin shock as perhaps representing the target area for the initial endotoxin–arachidonic acid interaction. In species other than the dog, the lung appears to be initially targeted by endotoxin (Gilbert 1960). Cefalo et al (1980) had initially reported a net production of PGE_2 but not $PGF_{2\alpha}$ across the lung of sheep receiving a sublethal dose of endotoxin. Our group (Demling et al 1981a) has documented both venous plasma and pulmonary lymph concentrations of PGE_2, $PGF_{2\alpha}$, 6-keto-$PGF_{1\alpha}$ and TXB_2 in the unanesthetized sheep receiving 2 to 10 µg/kg, intravenous boluses of endotoxin. In this LD_{33} model, we observed modest increases in the plasma concentrations of $PGF_{2\alpha}$, TXB_2, and 6-keto-$PGF_{1\alpha}$ but not PGE_2, within the first

TABLE 2   *In vivo release of thromboxane, prostacyclin and E, F-like materials during endotoxin administration*

Species	Dose of endotoxin	Area measured	Eicosanoid	Change	Reference
Rabbit	5 mg/kg i.v. bolus	arterial blood	6-keto-PGF$_{1\alpha}$	↑	Bult et al (1978)
Baboon	LD$_{70}$ i.v. bolus	arterial blood	6-keto-PGF$_{1\alpha}$ TXB$_2$	↑ ↑	Harris et al (1980) Fletcher et al (1981)
Pig	1 μg/kg/min constant infusion for 90 min	venous blood	6-keto-PGF$_{1\alpha}$ TXB$_2$	←→ ↑	Webb et al (1981)
Horse	125 μg/kg i.v. bolus (LD$_{100}$)	venous blood	6-keto-PGF$_{1\alpha}$ TXB$_2$	↑ ↑	Bottoms et al (1982)
Rat	20 mg/kg i.v. bolus	cardiac puncture	TXB$_2$	↑	Cook et al (1980)
Dog	15–150 μg/kg i.v. bolus (LD$_0$)	venous/arterial plasma	6-keto-PGF$_{1\alpha}$ TXB$_2$	↑ ↑	Hales et al (1981)
Sheep	300 μg/kg i.v. bolus	transpulmonary	PGE$_2$ PGF$_{2\alpha}$	↑ ←→	Cefalo et al (1980)
Sheep	2–10 μg/kg i.v. bolus (LD$_{33}$)	pulmonary lymph and venous plasma	PGE$_2$ PGF$_{2\alpha}$ 6-keto-PGF$_{1\alpha}$ TXB$_2$	←→ ↑ ↑ ↑	Demling et al (1981a)
Sheep	2 μg/kg i.v. bolus (LD$_{10}$)	pulmonary lymph and venous plasma	TXB$_2$	↑	Smith et al (1981)

TABLE 2 *(continued)*

Species	Dose of endotoxin	Area measured	Eicosanoid	Change	Reference
Sheep	1 µg/kg i.v. bolus (LD$_0$)	transpulmonary	TXB$_2$ 6-keto-PGF$_{1\alpha}$ (control)	↑ ↑	Hüttemeier et al (1982)
			TXB$_2$ 6-keto-PGF$_{1\alpha}$ (leukopenic)	← → ↑	
Rat	sepsis (LD$_{100}$)	venous plasma	TXB$_2$ 6-keto-PGF$_{1\alpha}$	↑ ↑	Butler et al (1982)

30 minutes after endotoxin administration. Pulmonary lymph concentrations showed parallel changes except that thromboxane concentrations were 2 times and prostacyclin and $PGF_{2\alpha}$ concentrations were 3 times higher than those seen in plasma. Following the initial spike-like prostanoid production soon after endotoxin administration, most plasma and lymph prostanoid concentrations decreased toward baseline values. However, both plasma and pulmonary lymph 6-keto-$PGF_{1\alpha}$ concentrations remained elevated. Concentrations of both TXB_2 and 6-keto-$PGF_{1\alpha}$ measured in prefemoral lymph demonstrated values which were identical to their respective plasma values. These data suggest that endotoxin administration to conscious sheep provokes a rapid pulmonary but not systemic biosynthesis of $PGF_{2\alpha}$, prostacyclin and thromboxane which is sensitively reflected in lymph draining the pulmonary interstitium. The exact locus in the lung from which the prostanoids are generated is not fully known. We have correlated pulmonary thromboxane production in an LD_{10} sheep endotoxin model to the degree of leukopenia, but not thrombocytopenia, in the animals (Smith et al 1981). In this regard, Hüttemeier et al (1982) have observed increased transpulmonary synthesis of both TXB_2 and 6-keto-$PGF_{1\alpha}$ in conscious sheep with normal leukocyte levels. In leukopenic sheep, the same dose of endotoxin resulted in significantly less thromboxane production while the endotoxin-induced 6-keto-$PGF_{1\alpha}$ production was unaltered. These data suggest the possibility that endotoxin initiates the pulmonary or pulmonary vascular-associated production of prostacyclin while simultaneously stimulating leukocyte production of thromboxane-like materials. This possibility, while attractive, does not preclude specific prostanoid production by other tissues in response to endotoxin. However, the above data do support the concept that endotoxin-related arachidonic acid metabolism is not necessarily global in nature but may be a tissue-specific event with the lung and related formed elements being a focal point during endotoxemia in most species.

3.2. Septicemic models

While the bolus or short-term infusion models of endotoxic shock have been well characterized as to their arachidonic acid profiles, the septic model is less well understood. Butler et al (1982) have developed a rat fecal peritonitis model which is associated with an 8.6-hour mean survival time. Zero percent animal survival was achieved 13.5 hours after the fecal suspension was administered. This is, therefore, a severe septicemia model. In these rats, venous plasma thromboxane concentrations increased from less than 0.2 ng/ml of venous plasma to approximately 1 ng/ml plasma within 15 to 20 minutes following inoculation. The TXB_2 concentration remained constant until 2 hours after inoculation when it increased again to approximately 1.6 ng/ml plasma at 6 hours. Changes in venous plasma concentrations of 6-keto-PGF$_{1\alpha}$ followed a similar biphasic time course but occurred on a much greater scale. From nondetectable values during the control period, the plasma 6-keto-PGF$_{1\alpha}$ concentration rapidly increased to 4 ng/ml after inoculation and remained at the same level for 3 hours. Afterwards, the 6-keto-PGF$_{1\alpha}$ concentration progressively increased to 9 ng/ml at 6 hours post-inoculation. When compared to the bolus model, the rat septic model appears to be predominantly a prostacyclin event with plasma concentrations progressively increasing to extremely high values until the death of the animal. The initial, predominantly thromboxane spike of the bolus model is not observed in the septic rat model.

In a clinical study, Reines et al (1982) examined 12 patients with septicemia. The average venous plasma concentration of TXB$_2$ in 8 patients who eventually succumbed to the sepsis was 10 times higher than plasma TXB$_2$ concentrations in either surviving septic patients or in a control population. The time course of TXB$_2$ in these patients was not reported, and the elevated value represented the average plasma concentration of all samples drawn between 4 and 24 hours after admission. Since patients died up to 7 days after the elevated thromboxane concentrations were observed, it was not felt that the thromboxane production represented an agonal event. These data do suggest, however, that a long-lived thromboxane production may be characteristic of human sepsis. This type of long-term prostanoid profile over 24 to 48 hours is also seen in other forms of global injury. For example, Harms et al (1981) have reported that pulmonary lymph 6-keto-PGF$_{1\alpha}$ concentrations increased progressively over a 12- to 24-hour period following burn injury in a sheep model. This value was still significantly elevated 48 hours post injury.

It therefore appears that prostanoid production can be progressive and sustained over relatively long periods of time in certain forms of injury including sepsis. The differences between the bolus endotoxin administration and the septic models of endotoxin shock seem to involve the initial-phase response. The initial 'spike' of prostanoid production seen in several species following the bolus administration of endotoxin may in fact be unnatural. Although this initial response can be potent enough to kill the animal, it does not appear to reflect the involvement of prostanoids in the temporal course of events associated with clinical sepsis. The initial-phase prostanoid response to bolus endotoxin administration is more related to an anaphylactic reaction. Interpretation of the prostanoid/endotoxic shock literature is dependent upon which of the phases of shock, initial or secondary, the investigator is addressing. Data for 3- to 4-hour bolus model studies should not be used to interpret the involvement of prostaglandin-like materials in the overall course of endotoxic shock.

3.3. Summary

Both the bolus and septic models of endotoxic shock are associated with increased blood and lymph concentrations of several different groups of prostanoids. In a given model, a generalized activation of the entire cyclo-oxygenase system does not necessarily occur. Specific subpathways such as thromboxane synthetase or prostacyclin synthetase may be preferentially activated depending upon tissue type and presumably, cofactor availability. These various pathways may be activated on different time courses depending upon species and the initial mechanism for inducing the shock state. These qualitative and quantitative differences aside, there is no doubt that endogenous arachidonic acid metabolism is altered during endotoxic shock in vivo in all species studied including man.

4. PROSTANOID CATABOLISM AND CLEARANCE IN VIVO AND IN VITRO

4.1 Catabolism and clearance in vivo

The rate of clearance of endogenous prostanoids from the circulation and their subsequent catabolism in vivo during the shock state is not well documented. During the previous discussion of changes in circulating prostanoid concentrations during endotoxic shock, rapid changes were ascribed to de novo prostanoid biosynthesis, and the term 'production' was used to signify increased rates of arachidonic acid metabolism. There are, however, situations where elevated blood prostanoid concentrations could be due to decreased rates of clearance or catabolism of one or more substances in addition to increased rates of production. Studies designed to evaluate the clearance of prostanoids from the circulation are difficult to interpret. For example, Oettinger et al (1980) measured venous and arterial $PGF_{2\alpha}$ concentrations across the human lung during sepsis. In control patients, there was net extraction of $PGF_{2\alpha}$ across the lungs whereas in the septic patients there was a positive relationship, that is, the lungs were adding $PGF_{2\alpha}$ to the blood. The measurement of 13,14-dihydro-15-keto-$PGF_{2\alpha}$ during the shock phase demonstrated a finite rate of $PGF_{2\alpha}$ metabolism. Control patient 13,14-dihydro-15-keto-$PGF_{2\alpha}$ values were not presented for comparison. During the recovery phase, the $PGF_{2\alpha}$ metabolite concentration was high while the pulmonary A-V difference was zero. These data demonstrate the time lags and other factors which are involved in interpreting prostanoid clearance data in vivo. Oettinger et al suggested that the data resulted from a decreased rate of $PGF_{2\alpha}$ catabolism and clearance across the lungs during the shock state. However, one could just as easily ascribe the enhanced arterial $PGF_{2\alpha}$ concentration during shock to enhanced biosynthesis since the V-A difference was positive, and suggest that the enhanced circulating 13,14-dihydro-15-keto-$PGF_{2\alpha}$ metabolite was due to increased catabolism of $PGF_{2\alpha}$ early in the recovery stage with decreased removal of the metabolite from the circulation. We must therefore separate actual catabolism of prostanoids from the clearance rate of the materials from the circulation. We have made this observation in splanchnic artery occlusion shock (Flynn et al 1975). The rate and pattern of tritiated $PGF_{2\alpha}$ catabolism was not affected in vivo by the shock state, but the rate of clearance of the labeled metabolites from the circulation was significantly depressed.

Fletcher and Ramwell (1977a) studied the percentage of lung extraction of $PGF_{2\alpha}$ in the endotoxic baboon model. In contrast to hemorrhagic shock, endotoxin shock resulted in a 25% to 35% increase in the ability of the lungs to remove or metabolize this prostanoid. This enhanced ability to catabolize or clear $PGF_{2\alpha}$ was observed between 15 minutes and 4 hours after the bolus administration of an LD_{100} dose of endotoxin. It appears that prostanoid catabolism during endotoxic shock in vivo is less affected than is the rate of clearance of the metabolites from the circulation. Thus, pulmonary catabolism of the classical prostaglandins, which is normally quite efficient (Ferreira and Vane 1967) would be expected to proceed during endotoxemia since the lungs always receive 100% of the cardiac output. However, since TXA_2 and prostacyclin degrade both enzymatically and non-enzymatically, and are not cleared by the lung under normal conditions (Armstrong et al 1978), the decreased rate of removal of these metabolites may have physiological significance. One metabolite of prostacyclin, 6-keto-PGE_1, has been shown to be a potent platelet anti-aggregatory agent, a gastrointestinal smooth muscle contracting substance, and an agent which will constrict bovine coronary arteries (Quilley et al 1980). The possible role of 6-keto-PGE_1 in endotoxin shock has not been documented.

4.2. Catabolism and clearance in vitro

Because of the difficulty in interpreting the in vivo metabolism of various prostanoids, several investigators have used in vitro techniques involving radiochromatographic procedures to document prostanoid catabolism during endotoxic shock. Table 3 lists pertinent data. Nakano and Prancan (1973) reported that PGE_1 metabolism was depressed 12 hours after endotoxin administration. Both lung and kidney from endotoxic rats showed decreased activity when compared to the rates found in control tissues. Splawinski et al (1979) reported a slight reduction in the rate of PGE_2 catabolism in hypothalamic tissue taken from rats with endotoxin-induced fever. In the

TABLE 3   *In vitro catabolism of radiolabeled prostanoids in tissues removed from endotoxemic animals*

Type of prostanoid	Bolus i.v. dose of endotoxin	Species	Tissue studied	Time after endotoxin tissues sampled	Rate of catabolism	Reference
PGE_1	10 mg/kg	rat	lung	12 hrs	decreased	Nakano and Prancan (1973)
PGE_1	10 mg/kg	rat	kidney	12 hrs	decreased	
PGE_2	10 µg/rat	rat	hypothalamus	2 hrs	decreased	Splawinski et al (1979)
PGE_2	20 µg/kg	rabbit	uterus	2 hrs	decreased	Harper et al (1980)
$PGF_{2\alpha}$	20 µg/kg	rabbit	uterus	6 hrs	decreased	
PGE_2	20 µg/kg	rabbit	oviduct	2–24 hrs	decreased	
$PGF_{2\alpha}$	20 µg/kg	rabbit	oviduct	2–24 hrs	decreased	
PGE_2	20 µg/kg	rabbit	lung	2–24 hrs	unaffected	
$PGF_{2\alpha}$	20 µg/kg	rabbit	lung	2–24 hrs	decreased	
PGE_2	200 µg/kg	rabbit	jejunum	2 hrs	decreased	Blackwell et al (1976)
PGE_2	200 µg/kg	rabbit	lung	2 hrs	decreased	
PGE_2	500 µg/kg	rabbit	kidney	2 hrs	unaffected	Flynn and Lefer (1977)
$PGF_{2\alpha}$	500 µg/kg	rabbit	kidney	2 hrs	unaffected	
PGE_2	500 µg/kg	rabbit	liver	2 hrs	unaffected	
$PGF_{2\alpha}$	500 µg/kg	rabbit	liver	2 hrs	unaffected	
PGE_2	500 µg/kg	rabbit	lung	2 hrs	unaffected	
$PGF_{2\alpha}$	500 µg/kg	rabbit	lung	2 hrs	unaffected	

rabbit, Harper et al (1980) reported decreased rates of $PGF_{2\alpha}$ catabolism between 2 and 24 hours after administration of a low dose of endotoxin. These decreases occurred in the uterus, oviducts and lung. In contrast, endotoxin did not affect PGE_2 metabolism by lung tissue although PGE_2 metabolism was significantly depressed in the reproductive tissues studied. In a slightly more severe rabbit model, Blackwell et al (1976) reported that PGE_2 metabolism was decreased in both jejunum and lung tissues removed from rabbits 2 hours after endotoxin administration.

In contrast to these findings, our group (Flynn and Lefer 1977) could find no significant change in the rates of metabolism of either PGE_2 or $PGF_{2\alpha}$ in liver, kidney or lung slices removed from rabbits 2 hours after the administration of 500 µg/kg of endotoxin. We also determined the patterns of metabolite formation for the 15-keto, the 13,14-dihydro, and the 15-keto-13,14-dihydro metabolites of each compound, and also the patterns of uncharacterized metabolites of higher polarity. No significant changes in any metabolite pattern were found. The study by Blackwell et al (1976) and our study are very comparable. For example, Blackwell et al report a rate of pulmonary PGE_2 metabolism of 762 picomoles of PGE_2 metabolized per mg protein per minute in control tissue whereas the value which we reported was 881 picomoles/mg/min. Despite this good agreement in the rate of prostanoid metabolism, the results are different. Harper et al (1980) had also reported opposing data for PGE_2 and $PGF_{2\alpha}$ metabolism in the rabbit lung in vitro. No studies to date have suggested that prostanoid catabolism is increased during endotoxic shock.

4.3. Summary

The literature suggests that both the clearance rate and the actual rate of prostaglandin catabolism may be altered during endotoxin shock. Some tissues may be affected more so than others. There are not enough data available to determine whether the observed decreases in prostanoid catabolism in specific tissues are dependent on either time after endotoxin or the dose of endotoxin employed, that is, the severity of the model. Different findings in several studies suggest that there are additional factors, such as cofactor availability, etc., which may bear upon the ability of tissues to effectively catabolize endogenous prostanoids. The available data do suggest that enhanced circulating prostanoid concentrations as measured during endotoxin shock may be due to both enhanced biosynthesis and decreased metabolism. As mentioned earlier, a decreased rate of clearance of specific prostanoid metabolites may have significance during endotoxin shock.

5. ENDOTOXIN-MEDIATED PROSTANOID BIOSYNTHESIS IN VITRO

5.1. Effect of endotoxin on nonphagocytic cells

There is no doubt that the administration of a bolus dose of endotoxin in vivo or the induction of septicemia is associated with enhanced activity of the arachidonic acid cascade. Not all prostanoids are synthesized in all models, but the rate of arachidonate metabolism in vivo is always enhanced. The mechanism of this enhanced activity in vivo is undetermined, but is generally assumed to follow the calcium-mediated phospholipase A_2 pathway which leads to the release of free arachidonic acid (see Ramwell et al 1977 for review). Although other mechanisms may be envisioned, the available data suggest that endotoxin-mediated prostanoid biosynthesis is dependent upon substrate availability. The questions arise as to the mechanism of the endotoxin effect and which tissues may be significantly involved. A number of investigators have attempted to answer these questions by comparing the arachidonate-related endotoxin response observed in vitro to that observed in vivo. Table 4 presents representative data.

Herman and Vane (1976a) first assessed the in vitro production of PGE_2 mediated by endotoxin. One hundred µg/ml of endotoxin added in vitro did not alter rabbit jejunal prostanoid biosynthesis in an isolated organ bath system. In contrast, if the jejunal tissues were removed from rabbits which had received endotoxin in vivo 30 minutes before they were killed, an increased rate of PGE_2 biosynthesis was observed. The rate of net PGE_2 production appeared to be higher than that which could have been attributed to a possible decrease in PGE_2 catabolism in the tissue. A similar jejunal response was reported by Peskar et al (1981). Bult and Herman (1978) reported a different set of findings in aortic rings subjected to endotoxin treatment in vitro. Using platelet aggregation as a bioassay for prostacyclin production, these investigators reported an apparent decrease in prostacyclin-like anti-aggregatory activity in tissues treated with doses of endotoxin of between 32 and 1000 µg/ml. In a similar study using ^{14}C-labeled arachidonic acid in isolated rabbit peritoneum (Bult et al 1979), this group of investigators observed a decreased rate of formation of 6-keto-$PGF_{1\alpha}$ and PGE_2 in the presence of endotoxin whereas the rate of production of hydroxy fatty acids was unaltered. Since endotoxin had no significant effect on the percentage of ^{14}C-arachidonic acid utilized by the system, these data suggest that endotoxin in vitro may depress cyclooxygenase but not lipoxygenase activity in rabbit peritoneum.

Feuerstein and Ramwell (1981) examined the effects of adding endotoxin in vivo 45 minutes prior to removing rat lungs and perfusing the tissue in

TABLE 4 *Prostanoid biosynthesis in vitro: comparison between in vivo and in vitro endotoxin administration*

Species	Tissue	Endotoxin adminis-tered	Dose of endotoxin	Time after endotoxin	Type of prostanoid	Response	Reference
Rabbit	jejunum	in vitro	100 µg/ml	30 min	PGE_2	unaffected	Herman and Vane (1976a)
Rabbit	jejunum	in vivo	100 µg/kg i.v. bolus	30 min	PGE_2	increased	
Rat	jejunum	in vivo	25 µg/kg i.v. bolus	2 hrs	PGE_2, PGD_2 6-keto-$PGF_{1\alpha}$	increased	Peskar et al (1981)
Rabbit	aortic rings	in vitro	32–1000 µg/ml	5 min	PGI_2	decreased	Bult and Herman (1978)
Rabbit	peritoneum	in vitro	1 mg/ml	20 min	PGI_2, PGE_2	decreased	Bult et al (1979)
Rat	lung	in vitro	50–400 µg/ml	30-min perfusion	PGE_2, $PGF_{2\alpha}$, TXB_2, 6-keto-$PGF_{1\alpha}$	decreased	Feuerstein and Ramwell (1981)
Rat	lung	in vivo	7 mg/kg i.v. bolus	45 min; 30-min in vitro perfusion	PGE_2, $PGF_{2\alpha}$, TXB_2, 6-keto-$PGF_{1\alpha}$	increased	
Rabbit	liver	in vitro	10 µg/ml	120 min	PGD_2, PGE_2, $PGF_{2\alpha}$, TXB_2, 6-keto-$PGF_{1\alpha}$ HETE	unaffected	Flynn (1982)
Rat	vena cava and aorta	in vitro	100–500 µg/ml	5 min	PGI_2	unaffected	Villa et al (1981)
Rat	vena cava and aorta	in vivo	0.25–1 mg/kg i.v. bolus	1 hr	PGI_2	increased	
Rat	mesentery	in vitro	2.5–150 µg/ml	30 min	PGE_2	increased	Kahn and Brachet (1981)

vitro for an additional 30 minutes. This procedure resulted in significant increases in PGE_2, $PGF_{2\alpha}$, TXB_2 and 6-keto-$PGF_{1\alpha}$ production. Adding doses of endotoxin in vitro to normal, perfused lungs was seen to decrease the spontaneous release of all prostanoids measured. There was a dose–response inhibition ranging between 50 and 400 µg endotoxin/ml perfusate. This in vitro inhibition of arachidonate metabolism was also observed in Ca^{2+} ionophore stimulated preparations. These data confirm the findings of Bult et al (1979). Our laboratory has used the isolated, perfused rabbit liver preparation to assess the effect of endotoxin in vitro on arachidonic acid metabolism by an intact system (Flynn 1982). Livers which were perfused in vitro for 30 minutes with 10 µg/ml endotoxin in Krebs-Henseleit-bicarbonate buffer had slices removed 90 minutes later. These slices were incubated with ^{14}C-arachidonic acid, and the metabolic profile determined by radio-chromatographic techniques. Endotoxin in vitro had no significant effect on either the pattern or magnitude of arachidonic acid metabolism in the liver slices. In vascular tissue, Villa et al (1981) reported that endotoxin in vivo, but not in vitro, induced prostacyclin production in rat vena cava and aorta. In contrast, Kahn and Brachet (1981) reported that isolated rat mesentery preparations responded in a dose-dependent manner to endotoxin administered in vitro. Endotoxin resulted in a 3-fold increase in PGE_2 production which was prevented by indometacin.

To summarize the above data, it appears that the endotoxin molecule does not directly provoke enhanced arachidonic acid metabolism in vitro in most tissues. In fact, endotoxin in vitro may decrease cyclooxygenase activity by some unknown mechanism. If endotoxin is added in vivo and the tissue sampled between 30 minutes and 2 hours after the stimulus, there is an enhanced rate of eicosanoid production. It thus appears that an intermediate mechanism is required to mediate the endotoxin induced activation of the arachidonic acid cascade. The observed ability of endotoxin to decrease cyclooxygenase activity in vitro appears to be a direct action of the endotoxin molecule.

The actions of endotoxin on blood platelets are very interesting. Data on these actions, and on actions of various phagocytic cells are summarized in Table 5. Prancan et al (1981) administered endotoxin to rabbits and harvested platelets 120 minutes later. These platelets synthesized significantly more TXB_2 than did control platelets. This in vivo endotoxin effect is similar to that previously described. Stuart (1981) investigated the effect of adding endotoxin in vitro to human platelets on the ability of the platelets to form prostanoids in response to thrombin. TXB_2 production in response to equal doses of agonist was enhanced by endotoxin treatment. This ability of endotoxin to enhance the response of platelets to agonists appears to be indirect since Bult and Herman (1979) could not demonstrate a direct action of endotoxin in stimulating guinea pig platelet TXB_2 production in the absence of an agonist compound. Harris et al (1982) observed identical findings when using canine platelet-rich plasma. These data suggest that platelets respond to endotoxin administration in vitro in a manner similar to lung, liver, mesentery, etc. That is, there is no general stimulation of the arachidonic acid cascade by endotoxin unless the endotoxin was administered in vivo. However, endotoxin in vitro does appear to be able to sensitize platelets to their natural agonists.

5.2. Effect of endotoxin on phagocytic cells

The responses of the nonphagocytic cells discussed so far contrast with those of macrophages and other cells with phagocytic functions. Kurland and Bockman (1978) demonstrated the ability of human monocytes and mouse peritoneal macrophages to synthesize PGE in vitro. The PGE production by both types of cells was significantly stimulated by the addition of endotoxin in vitro. In contrast, lymphocytes, granulocytes or nonadherent peritoneal cells did not demonstrate any in vitro stimulation of arachidonic acid metabolism by endotoxin. Tanigawa and coworkers (1982) reported that phagocytic mouse macrophages released prostaglandins in response to endotoxin in vitro whereas nonphagocytic lines of the same cell line did not respond to endotoxin in vitro with enhanced arachidonic acid metabolism. These findings suggest that phagocytic cells may be of prime importance in initiating prostanoid production in response to endotoxin. As additional evidence in this regard, Cook et al (1981a) have demonstrated both TXB_2 and 6-keto-$PGF_{1\alpha}$ biosynthesis by rat peritoneal macrophages in response to the in vitro administration of *Salmonella* endotoxin. These investigators have recently summarized their findings on the modulation of macrophage arachidonic acid metabolism during endotoxin shock (Cook et al 1982).

Feuerstein et al (1981) also reported on the ability of endotoxin to directly stimulate PGE_2, $PGF_{2\alpha}$, TXB_2 and 6-keto-$PGF_{1\alpha}$ synthesis by rat peritoneal macrophages and further suggested that transmethylation reactions were important in the process. In hepatic Kupffer cells from the rat, Bhatnagar and co-workers (1982) demonstrated a dose-dependent production of PGE_2 in response to endotoxin. In addition, increases in Kupffer cell cyclic AMP appeared to be prostaglandin dependent. An interesting study was reported by Wahl et al (1979), who isolated macrophages from both endotoxin-susceptible C3H/HeN mice, and C3H/HeJ mice which are resistant to the effects of endotoxin in vivo. Endotoxin when added in vitro to C3H/HeN-derived macrophages stimulated PGE_2 biosynthesis. In contrast, macrophages isolated from tolerant C3H/HeJ mice did not synthesize PGE_2 in response to

TABLE 5 *Effects of endotoxin on arachidonic acid metabolism of platelets and phagocytic cells*

Species	Cell type	Endotoxin adminis-tered	Dose of endotoxin	Time after endotoxin	Type of prostanoid	Response	Reference
Rabbit	platelets	in vivo	1.1 µg/kg/min (120 min)	120 min	TXB_2	increased	Prancan et al (1981)
Man	platelets	in vitro	1 µg/ml	2 min	TXB_2	increased response to agonist	Stuart (1981)
Guinea pig	platelets	in vitro			TXB_2	unaffected	Bult and Herman (1979)
Dog	platelets	in vitro	0.2–20 µg/ml	15 min	TXB_2	unaffected	Harris et al (1982)
Mouse	peritoneal macrophage	in vitro	0.1–100 µg/culture	48 hrs	PGE	increased	Kurland and Bockman (1978)
Man	monocytes	in vitro	0.1–100 µg/culture	48 hrs	PGE	increased	
Mouse	28–12 (Ara) phagocytes	in vitro	1 µg/ml	24 hrs	PGE	increased	Tanigawa et al (1982)
Mouse	28–12 (Ara) non-phago-cytes	in vitro	1 µg/ml	24 hrs	PGE	unaffected	
Rat	peritoneal macrophage	in vitro	0.1–200 µg/ml	24 hrs	TXB_2, 6-keto-$PGF_{1\alpha}$	increased	Cook et al (1981a)

TABLE 5   *(continued)*

Species	Cell type	Endotoxin adminis-tered	Dose of endotoxin	Time after endotoxin	Type of prostanoid	Response	Reference
Rat	peritoneal macrophage	in vitro	0.1 µg/ml	3 hrs	PGE_2, $PGF_{2\alpha}$, TXB_2, 6-keto-$PGF_{1\alpha}$	increased	Feuerstein et al (1981)
Rat	Kupffer cells	in vitro	30 µg/ml	24 hrs	PGE_2	increased	Bhatnagar et al (1982)
Mouse	C3H/HeN macrophages	in vitro	0.1–30 µg/ml	48 hrs	PGE_2	increased	Wahl et al (1979)
Mouse	C3H/HeJ macrophages	in vitro	0.1–30 µg/ml	48 hrs	PGE_2	unaffected	

TABLE 6 *The administration of exogenous prostanoids and arachidonic acid to animals in endotoxin shock*

Species	Dose of endotoxin	Type of prostanoid (pre or post treatment)	Dose of prostanoid	Effect	Rate of survival	Reference
Dog	4 mg/kg i.v. bolus	PGE_1 (pre and post)	0.25 mg/kg bolus	attenuated pulmonary vascular resistance changes	not assessed	Sorrells et al (1972)
Dog	0.75 mg/kg i.v. bolus	PGE_1 (pre and post)	0.3–3.0 µg/kg bolus + 1-hr infusion at 0.3–3.0 µg/kg/min	↑ MABP, CO 'lysosomal stabilization'	increased from 4 to 35 hours	Raflo et al (1973)
Dog	0.75 mg/kg i.v. bolus	$PGF_{2\alpha}$ (pre and post)	1–10 µg/kg bolus + 1-hr infusion at 1–10 µg/kg/min	↑ MABP, CO	increased from 4 to 35 hours	
Dog	1.5 mg/kg i.v. bolus	PGA_1 PGE_1 } (15 min post)	50 µg/kg bolus	↑ MABP, CO ↓ pH	no effect at 24 hrs	Shatney and Lillehei (1976)
Rabbit	0.75 mg/kg i.v. bolus	PGA_2 PGE_1 $PGF_{2\alpha}$ } (15 min post)	1 µg/kg/min for 105 min	↓ LDH ←→ cathepsin D ←→ GPT	no effect at 48 hrs	Flynn (1978)
Dog	40 µg/kg/min i.v. for 30 min	PGI_2 (pre and post)	0.03 µg/kg/min for 5 hrs	↓ PAP ↓ vascular permeability	not assessed	Hirose et al (1978)
Cat	5 mg/kg i.v. bolus	PGI_2 (30 min post)	0.75 nM/kg/min for 4.5 hrs	←→ MABP ↑ SMAF ↓ lysosomal enzyme activity in plasma	not assessed	Lefer et al (1980)

TABLE 6 *(continued)*

Species	Dose of endotoxin	Type of prostanoid (pre or post treatment)	Dose of prostanoid	Effect	Rate of survival	Reference
Dog	1 mg/kg i.v. bolus	PGI_2 (pre and post)	0.02 µg/kg/min for 4 hrs	↓ MABP. HR ←→ PVR	40% increase at 72 hrs	Fletcher and Ramwell (1980)
Dog	1.75 mg/kg i.v. bolus	PGI_2 (30 min post)	100 µg/kg/min for 3 hrs	↑ MABP ↑ CO ↓ cathepsin D ↓ norepinephrine	90% increase at 24 hrs	Krausz et al (1981)
Sheep	2 µg/kg i.v. bolus	PGI_2 (post)	100–200 µg/kg/min for 5 hrs	↓ PAP change ↑ CO ↓ permeability of lung micro-vasculature	not applicable	Demling et al (1981b)
Rat	20 mg/kg i.v. bolus	essential fatty acid deficiency		↓ TXB_2	increased survival at 48 hrs	Cook et al (1981b)
Rat	20 mg/kg i.v. bolus	essential fatty acid deficiency + arachidonic acid (pre)	100 mg i.v. bolus	↑ TXB_2	decreased survival at 48 hrs	
Rabbit	0.75 mg/kg i.v. bolus	arachidonic acid (post)	15 µg/kg/min for 105 min	←→ MABP ↑ CVP	increased survival at 48 hrs	Flynn (1978)

CO = cardiac output; CVP = central venous pressure; GPT = glutamic pyruvic transaminase; HR = heart rate; LDH = lactic dehydrogenase; MABP = mean arterial blood pressure; PAP = pulmonary arterial pressure; PVR = pulmonary vascular resistance; SMAF = superior mesenteric artery blood flow.

either endotoxin or lipid A. These data raise some very interesting questions in regard to the role of prostanoid production by macrophages and the pathophysiologic events of endotoxic shock.

5.3. Summary

The administration of endotoxin in vitro appears to stimulate endogenous prostanoid biosynthesis primarily in phagocytic cells such as monocytes and macrophages. Endotoxin has little if any effect in vitro on nonphagocytic cell types. When endotoxin is given to intact animals, there appears to be a complex interaction which induces nonphagocytic cells to increase their rate of prostanoid production. This enhanced response remains intact when these nonphagocytic cells are subsequently studied in vitro. The mechanisms of this in vivo interaction remain to be elucidated. The possible interactions between endotoxin, phagocytic cells and the arachidonate scheme during endotoxic shock is an area of significant interest. (See Cook et al 1982, for additional information.)

6. ADMINISTRATION OF EXOGENOUS PROSTANOIDS TO INTACT ANIMALS DURING ENDOTOXIN SHOCK

Many investigations into the involvement of the arachidonic acid cascade in endotoxic shock have been based upon the classical techniques of stimulation and ablation. Stimulation studies involving the administration of either arachidonic acid as supplemental substrate or the administration of stabilized prostanoids to animals in shock began in the early 1970s when these compounds became available for experimental use (Table 6).

6.1. Classical prostanoids

Sorrells et al (1972) were the first to administer E-type prostanoids to dog lungs in situ prior to the administration of gram-negative endotoxin. It was found that PGE_1 could significantly attenuate the pulmonary pressor response to bolus endotoxin administration in this model. The effects of both PGE_1 and $PGF_{2\alpha}$ administration in the canine endotoxin shock model were investigated by Raflo et al (1973). These investigators reported that both prostaglandins tested prolonged the survival time of the animals, and that mean arterial blood pressure and cardiac output were better maintained in the treated groups. In addition, it was suggested that PGE_1 was able to reduce the accumulation of lysosomal enzymes in blood. It is not clear from this study whether the observed increase in survival time resulted in a greater number of permanent survivors or whether it reflected a prolongation of the pathophysiologic process. Shatney and Lillehei (1976) reported that post-treating endotoxic dogs with bolus injections of PGA_1 and PGE_1 had no significant effect on 24-hour survival rates of the animals. This study differed from that of Raflo et al (1973) in two ways. First, Raflo and coworkers pretreated their animals with prostanoids while Shatney and Lillehei post-treated after the initial hemodynamic changes had occurred. Secondly, Raflo et al maintained a constant 60-minute infusion of both E and F type prosta-glandins whereas a single bolus injection of PGA_1 was used by the Minnesota group. These data suggest the importance of dose, duration of administration and time of therapeutic intervention in the endotoxic model in regard to the use of exogenous prostaglandin-like materials.

Our group (Flynn 1978) has studied the effects of the posttreatment of endotoxic rabbits with constant infusions of PGA_2, PGE_1, and $PGF_{2\alpha}$. These infusions were begun 15 minutes after the bolus administration of an LD_{60} dose of endotoxin and were carried out for 105 minutes at rates of 1 µg/kg/min for each prostanoid. While each of these three prostaglandins significant-ly attenuated the accumulation of lactic dehydrogenase activity in the plasma of the shocked animals, none were able to protect against increases in plasma activities of glutamic pyruvic transaminase or cathepsin D during the period of shock. In addition, none were able to significantly enhance the survival rate of the treated rabbits when compared to vehicle-treated controls. From the data presented above, it appears that prostaglandins of the A, E and F series may be of some survival value during endotoxin shock when adminis-tered before the endotoxin bolus and when maintained with a constant infu-sion protocol. When administered as a post-endotoxin treatment, these pros-taglandins do not appear to possess survival value, even when constantly infused. It is not known whether the infusion of higher doses of prostaglan-dins to shocked animals results in enhanced survival.

6.2. Thromboxanes and prostacyclin

In comparison to the classical prostaglandins, prostacyclin (PGI_2) appears to have a more salutary profile of activity during endotoxin shock. In a dog model of endotoxin-mediated pulmonary injury, Hirose et al (1978) infused prostacyclin in both pre- and posttreatment protocols. They demonstrated that PGI_2 pretreatment can inhibit the changes in pulmonary vascular resis-tance and capillary permeability which occurs in this model following endo-toxin administration. In addition, posttreatment with PGI_2 was shown to reverse the endotoxin-mediated increase in vascular permeability. Long-term survival was not assessed in this study. Lefer et al (1980) also posttreated following endotoxin with a prostacyclin infusion in the cat and reported no differences in mean systemic arterial pressure between treated and untreated animals. However, superior mensenteric artery blood flow was better maintained in the PGI_2-treated cats and plasma lysosomal enzyme activity was not as elevated. Long-term survival was not assessed.

Permanent animal survival in conjunction with a prostacyclin infusion in endotoxemic dogs was first assessed by Fletcher and Ramwell (1980). Dogs were given an LD_{50} dose of endotoxin and observed for 72 hours. A 4-hour prostacyclin infusion decreased mean arterial blood pressure, did not affect endotoxin-mediated increases in pulmonary vascular resistance, but did improve long-term survival by approximately 40%. The prostacyclin infusion had little effect on the thrombocytopenia or the leukopenia which are associated with endotoxin administration in the dog model. Twenty-four hour survival was assessed in a dog model by Krausz et al (1981). These investigators reported a prostacyclin-mediated 90% increase in the number of animals surviving after receiving 1.75 mg/kg endotoxin. They also reported that an infusion of prostacyclin maintained mean arterial blood pressure and cardiac output, and attenuated endotoxin-mediated increases in the plasma concentrations of cathepsin D, norepinephrine, and fibrin degradation products. In a separate study using the same model, these investigators (Utsunomiya et al 1982) observed that cardiac mitochondria isolated from PGI_2-treated endotoxemic dogs were structurally and functionally normal when compared to mitochondria isolated from vehicle-treated controls. It is not known whether this enhanced mitochondrial viability is due to prostacyclin acting directly on the organelle, or due to improved myocardial hemodynamics or decreased cardiac load. These long-term survival studies by Krausz et al, and Fletcher and Ramwell demonstrate that exogenous prostacyclin may be of therapeutic use in the bolus model of endotoxin shock. Additional studies with prostacyclin administration during sepsis are required. In addition, possible side effects of prostacyclin treatment must be characterized before a beneficial action of prostacyclin in endotoxin shock can be ascribed.

There has been a great deal of interest in the relationships between endotoxin, prostanoids and the lung. As already cited, Hirose et al (1978) examined the protective effects of prostacyclin on pulmonary microvascular integrity in the dog. Demling et al (1981b) further examined this relationship in the chronic, unanesthetized sheep model. In this species, very small amounts of endotoxin produce the classic, biphasic hemodynamic response. There is an initial increase in pulmonary vascular resistance which results in a decrease in the cardiac index with a resultant fall in mean arterial pressure. There is a parallel leukopenia. The pulmonary lymph flow rate increases due to the pulmonary hypertension. This initial phase response resolves in 1–2 hours. The secondary phase response is characterized by sustained high rates of pulmonary lymph flow with the lymph being protein-rich. This non-hemodynamic mediated lymph flow is assumed to reflect pulmonary microvascular injury. The infusion of prostacyclin in these endotoxin-treated sheep was shown to significantly attenuate the initial endotoxin-mediated changes in pulmonary vascular pressure, cardiac output and lung lymph flow rate. Demling and coworkers also discussed a problem with using prostacyclin as a therapeutic drug. Twenty-five percent of the prostacyclin-treated sheep developed significant degrees of lung vascular injury when the prostacyclin infusion was discontinued after 5 hours. Thus, prostacyclin may have to be withdrawn slowly when used as a therapeutic agent.

Assuming that prostacyclin does have a net beneficial action in this model, the mechanism of action may relate to its ability to either directly or indirectly affect the rate of production of pulmonary pressor agents such as thromboxanes. We have shown (Flynn and Demling 1982) that the infusion of exogenous prostacyclin at a rate of 100–200 µg/kg/min to endotoxin-treated sheep can significantly attenuate the endotoxin-evoked increases in the pulmonary lymph concentrations of TXB_2. Thus, prostacyclin may possibly be acting by directly improving regional hemodynamics, affecting platelet function, etc., and in doing so, indirectly decrease cellular injury. This would decrease the stimulus for thromboxane synthesis which may be involved in the pulmonary hemodynamic changes following endotoxin. While this is one possibility, the precise mechanism by which prostacyclin may provide a salutary effect during endotoxin shock and pulmonary injury remains to be elucidated.

6.3. Arachidonic acid

Another means of investigating the role of prostanoids in endotoxin shock is to enhance the activity of the arachidonic acid cascade by providing exogenous substrate. This has been done in two ways. Cook et al (1981b) produced essential fatty acid-deficient rats by dietary manipulation and found that these rats were less sensitive to endotoxin-mediated injury when compared to rats on a normal diet. If arachidonic acid was given to the deficient animals, endotoxin-related mortality increased from 24% to 100%. These investigators measured plasma thromboxane concentration in both the deficient and supplemented rats and suggested that TXA_2 was the agent responsible for the increase in mortality. TXB_2 concentrations were very low in the deficient rats after endotoxin whereas the TXB_2 concentration was significantly elevated in deficient animals receiving arachidonic acid plus endotoxin. These data suggest that thromboxane-like materials play a detrimental role in the rat endotoxic shock model. Prostacyclin production was not assessed. We have administered arachidonate as a 105-minute infusion to rabbits which had received an LD_{50} dose of endotoxin (Flynn 1978). The arachidonate dose of 15 μg/kg/min had no significant effects on mean arterial blood pressure although it significantly elevated central venous pressure. The arachidonic acid administration did significantly improve the rate of survival at 48 hours post endotoxin. The beneficial effect of arachidonate treatment was abolished by pretreatment of the animals with sodium meclofenamate. Prostanoid concentrations were not measured.

These data demonstrate that numerous factors bear upon whether arachidonic acid may prove deleterious or beneficial in endotoxic shock. While thromboxane production appears to be of significance in the essential fatty acid-deficient rat as described by Cook et al (1979, 1981b), it does not appear to be the primary agent in the rabbit model in that an assumed beneficial prostanoid was active. A salutary effect of thromboxane A_2 in the rabbit model has not been described. In addition, the constant infusion of arachidonic acid may produce a different metabolic profile when compared to the endogenous pattern of arachidonic acid metabolism. Finally, different species may possess different degrees of endotoxin sensitivity in regard to target tissues. The rat requires 10–20 times greater amounts of endotoxin when compared to the rabbit to produce the same degree of lethality. Thus, the interpretation of the role of the arachidonic acid cascade in endotoxic shock as assessed by substrate supplementation must await further studies.

6.4. Summary

The effects of prostanoid administration to a number of species of animals in endotoxin shock appear to depend on a number of factors. Treatment with A-, E- and F-series prostaglandins has been observed to increase the rate of survival only when given as a pretreatment in a relative high dose. Posttreatment does not appear to be effective, even when a constant-infusion protocol is followed. In contrast, prostacyclin administration produces an increase in animal survival when given as an infusion either before or after endotoxin administration. This beneficial prostacyclin effect has been observed in the bolus endotoxin model where there is an initial spike of endogenous prostanoid production which subsides. It is not known whether prostacyclin will maintain this possible therapeutic activity in the sepsis model as described by Butler et al (1982) where high rates of endogenous prostacyclin production are seen in the decompensatory phase of the model. Thus, a general statement regarding a therapeutic role of exogenously administered prostanoids cannot be made. The net effect of the intervention depends upon species, severity of the model, whether the animal is in a compensatory or decompensatory phase, the time of treatment, the prostanoid involved, and the dose administered. In addition, the sepsis model may differ significantly from the bolus model of endotoxic shock. Additional characterization is required.

7. INHIBITION OF PROSTANOID BIOSYNTHESIS IN VIVO DURING ENDOTOXIN SHOCK

7.1. Chemical inhibitors

In contrast to the stimulation studies discussed previously, a large volume of literature exists on the ablation of the cyclooxygenase system during endotoxic shock, primarily by nonsteroidal antiinflammatory drugs (NSAID). No attempt will be made to review this literature in its entirety. Rather, representative studies will be presented in an attempt to summarize this area of research. Studies involving the use of NSAIDs are difficult to interpret in regard to arachidonic acid metabolism during endotoxin shock in that most of the drugs used possess a spectrum of biological actions. That is, they are less than specific. A NSAID which improves survival during endotoxic shock is not necessarily acting via inhibition of prostaglandin cyclooxygenase.

7.1.1. Acetylsalicylic acid

Among the first of the NSAIDs to be effective in endotoxic shock was acetylsalicylic acid. As mentioned in the introduction of this paper, Northover and Subramanian (1962) were the first to test acetylsalicylic acid in a dog model (Table 7). While they noted an increase in survival, they were unsure of the mechanism. Hinshaw et al (1967) both pretreated and posttreated with acetylsalicylic acid in an LD_{80} dog model and reported increased survival rates for pretreated animals only. The possible mechanisms of action of acetylsalicylic acid as cited by the Hinshaw group included its ability to antagonize the vascular actions of histamine, catecholamines, serotonin, bradykinin and SRS, and to prevent platelet adhesiveness, to alter bleeding time, and to stimulate the pituitary-adrenal system. In contrast to these studies, Greenway and Murthy (1971) used acetylsalicylic acid to study endotoxin-mediated mesenteric vasoconstriction in the cat. Doses of acetylsalicylic acid of 10-100 mg/kg were given before the bolus injection of endotoxin, and the authors observed an attenuation of the response in the initial phase. However, the cats still experienced the secondary-phase cardiovascular collapse and died; acetylsalicylic acid did not affect the survival rate.

In a comparative study, Hall et al (1972) examined the effect of acetylsalicylic acid on endotoxin shock in the cat, dog and sheep models. Pretreatment with acetylsalicylic acid attenuated the hemodynamic responses of the

TABLE 7   *The use of acetylsalicylic acid in endotoxin shock*

Species	Dose of endotoxin	Dose of acetylsalicylic acid	Effect	Rate of survival	Reference
Dog	0.15 mg/kg i.v. bolus	200–220 mg/kg (pre)	$\uparrow$ MABP	increased at 35 hrs	Northover and Subramanian (1962)
Dog	0.40 mg/kg i.v. bolus	100 mg/kg (pre) 100 mg/kg (post)	improved hemo-dynamics $\uparrow$ MABP	increased at 4 days no effect	Hinshaw et al (1967)
Cat	3 mg/kg i.v. bolus	10–100 mg/kg (pre)	attenuated initial hemodynamic changes	no effect	Greenway and Murthy (1971)
Dog Cat Sheep	2.0 mg/kg i.v. bolus 2.0 mg/kg i.v. bolus 0.5 mg/kg i.v. bolus	50 mg/kg i.v. (pre) 50 mg/kg i.v. (pre) 5 mg/kg i.v. (pre)	abolished initial hemodynamic changes	not assessed	Hall et al (1972)
Rat	20 mg/kg i.v. bolus	3.7–30 mg/kg 30 min pre 100 mg/kg 30 min pre 100 mg/kg 24 hrs pre	$\downarrow$ TXB$_2$ $\leftarrow\rightarrow$ 6-keto-PGF$_{1\alpha}$ $\downarrow$ TXB$_2$ $\downarrow$ 6-keto-PGF$_{1\alpha}$ $\downarrow$ TXB$_2$ $\leftarrow\quad\rightarrow$ 6-keto-PGF$_{1\alpha}$	increased no effect increased	Halushka et al (1981)

MABP = mean arterial blood pressure.

initial phase in all three species. The second-phase hemodynamic and metabolic effects of endotoxin were affected by acetylsalicylic acid only in the dog. Thus the dog appears to be different from other species in several regards. The target organ of endotoxin is different in the dog with the canine endotoxin model being characterized by intense hepatic pooling (Gilbert 1960) rather than pulmonary hypertension. In addition, Hall et al (1972) noted that acetylsalicylic acid inhibited the endotoxin-mediated release of serotonin in the dog, but not the cat or sheep. These species differences are notable.

The complexity of using NSAIDs such as acetylsalicylic acid in endotoxin shock is shown in a study by Halushka et al (1981). In a rat endotoxin model, pretreatment with acetylsalicylic acid at doses between 3.75 and 30 mg/kg 30 minutes before endotoxin administration significantly improved 24-hour survival and decreased endogenous TXB_2 production following an LD_{90} dose of endotoxin. However, a dose of 100 mg/kg of acetylsalicylic acid 30 minutes before endotoxin administration did not enhance survival, even though thromboxane production was inhibited. If the pretreatment with 100 mg/kg of acetylsalicylic acid was administered 24 hours before endotoxin, then survival was enhanced. These data are difficult to interpret. Halushka et al also measured 6-keto-$PGF_{1\alpha}$ concentrations in the experimental animals. The differential effects of acetylsalicylic acid in their study can be partly explained by the differences in its efficacy in inhibiting thromboxane synthetase and prostacyclin synthetase. Thus, lower doses of acetylsalicylic acid changed the TXB/PGI_2 ratio in favor of PGI_2, while high doses inhibited both enzymes equally. This suggests that thromboxane played a detrimental role in this model while prostacyclin played a beneficial role. The treatment of animals with high-dose acetylsalicylic acid 24 hours before endotoxin administration would alter platelet TXB production only and enhance survival by removing a deleterious factor. This study demonstrates the importance of quantifying endogenous prostanoid production during inhibitor studies.

7.1.2. Indometacin

Indometacin is another NSAID which has been extensively used during endotoxin shock (Table 8). This drug was first used by Erdös et al (1967) in a canine model. Although survival rates were not assessed, pretreatment with indometacin appeared to significantly improve hemodynamics for 2 hours following endotoxin. Parratt and Sturgess compared pretreatment (1974) versus posttreatment (1975a) of the cat with indometacin during the shock state. It was found that pretreatment with the drug abolished the pulmonary pressor response to the bolus injection of endotoxin and that the decrease in mean arterial blood pressure was attenuated. Survival at 6 hours post endoto-

TABLE 8   *Effect of indometacin in endotoxic shock in vivo*

Species	Dose of endotoxin	Indometacin dose	Effect	Rate of survival	Reference
Dog	0.4 mg/kg i.v. bolus	20 mg/kg (pre)	↑ MABP; attenuated initial phase response	not assessed	Erdös et al (1967)
Cat	2 mg/kg i.v. bolus	10 mg/kg (pre)	↓ pulmonary vascular resistance	improved at 6 hrs	Parratt and Sturgess (1974)
	2 mg/kg i.v. bolus	10 mg/kg (post)	↑ MABP	no effect at 6 hrs	Parratt and Sturgess (1975a)
Dog	1 mg/kg i.v. bolus	2–4 mg/kg (pre) plus 90 min infusion (post)	no effect on initial hemodynamic response, attenuated 2nd-phase response	not assessed	Anderson et al (1975a)
Dog	2.5 mg/kg i.v. bolus	2.5 mg/kg × 3 doses (post)	little hemodynamic effect on 2nd-phase hemodynamic response	increased survival at 5 hrs	Hilton and Wells (1976)
Dog	1.5 mg/kg i.v. bolus	10 mg/kg (pre and post)	attenuated initial phase hemodynamic response	increased survival at 72 hrs	Fletcher and Ramwell (1977b, 1978a)
Baboon	20 mg/kg i.v. (LD_{100}) bolus	1.5 mg/kg (pre and 3 hrs post)	attenuated initial phase hemodynamic response	no effect	Fletcher et al (1976)
Baboon	3 mg/kg i.v. (LD_{50}) bolus	1.5 mg/kg (pre and 3 hrs post)	attenuated initial phase hemodynamic response	increased survival at 72 hrs	Fletcher and Ramwell (1978b)

TABLE 8   *(continued)*

Species	Dose of endotoxin	Indometacin dose	Effect	Rate of survival	Reference
Rat	0.075 mg/kg plus lead acetate (LD_{50}) i.v. bolus	10 mg/kg (pre)	not assessed	no effect	Reichgott and Engelman (1975)
Rat	2 mg/kg i.v. bolus (LD_{80})	1 mg/kg (pre)	↑ MABP	no effect at 72 hrs	Goto et al (1980)
Dog	sepsis 4×10^9 organism i.v. (LD_{100})	20 mg/kg (pre and post)	↑ MABP ↓ portal venous pressure	increased survival	Culp et al (1971)
Rat	sepsis 9×10^9 organism i.v. (LD_{50})	3 mg/kg i.p. (1, 24, 48 hrs post)		enhanced survival at 24 hrs, but no enhanced survival at 72 hrs	Short et al (1981)

MABP = mean arterial blood pressure.

xin was improved. Posttreatment did not prove to be beneficial. Anderson et al (1975a), who described enhanced prostaglandin biosynthesis following endotoxin administration in their original paper, also reported on the effects of indometacin in their model. In contrast to the above cited data in the cat, indometacin did not modify the hemodynamic response of the initial phase in the dog to the bolus injection of endotoxin. PGE and PGF production was inhibited in this study. Posttreatment of endotoxic dogs with indometacin was carried out by Hilton and Wells (1976). They could not demonstrate any significant beneficial effect of multiple-dose administration of the drug on the delayed-phase hemodynamic response to endotoxin, but survival as assessed at 5 hours was significantly increased. Several studies were carried out by Fletcher and Ramwell using indometacin in both canine and primate models of lethal and sublethal endotoxemia. In the dog (1977b, 1978a), pretreatment with indometacin in LD_{50} to LD_{60} models attenuated the initial fall in mean arterial blood pressure after the bolus endotoxin administration. Acetylsalicylic acid, tested concurrently, did not have this effect. Both indometacin and acetylsalicylic acid were able to attenuate endotoxin-mediated increases in pulmonary vascular resistance to the same extent. Prostanoid production was abolished by both drugs. These drugs also increased the rate of 72-hour survival. Since the arachidonic acid metabolism was inhibited by both NSAIDs, the data suggest that the attenuation of early endotoxin-mediated hemodynamic changes in the bolus model by certain NSAIDs is not solely prostanoid related. In an LD_{100} baboon model, pretreatment with indometacin attenuated the initial endotoxin-mediated hemodynamic changes, but did not increase survival (Fletcher et al 1976). In a less severe baboon model (LD_{50}) (Fletcher and Ramwell 1978b), the same dose of indometacin significantly increased 72-hour survival. The rat seems to respond differently than other species to indometacin treatment. In this species, pretreatment of rats with indometacin in either an LD_{50} (Reichgott and Engelman 1975) or an LD_{80} model (Goto et al 1980) did not enhance long-term survival. Prostanoid concentrations were not assessed. The effect of indometacin has also been examined in the septic model of shock. An LD_{100} dose of live *E. coli* bacteria to dogs produced only the delayed hemodynamic response. Indometacin was able to significantly maintain mean arterial pressure, to attenuate increases in portal venous pressure, and to enhance survival (Culp et al 1971). Short et al (1981) demonstrated a similar salutary survival effect of indometacin in an LD_{50} rat sepsis model.

7.1.3. *Meclofenamate, flurbiprofen, ibuprofen*

Several other NSAIDs have been tested in various endotoxic shock systems (Table 9). Sodium meclofenamate, when administered prior to the bolus administration of endotoxin in the cat, was reported to attenuate the first-phase hemodynamic effects of the endotoxin and to delay the onset of the secondary-phase changes (Parratt and Sturgess 1975c, 1977). A multiple, post-treatment regime in this same model was shown to maintain mean arterial blood pressure, cardiac output and arterial pH, and to enhance the rate of animal survival at 6 hours after endotoxin injection (Parratt and Sturgess 1975b).

The same investigators also used flurbiprofen in the cat endotoxic shock model (Parratt and Sturgess 1976). Pretreatment with flurbiprofen almost completely abolished the initial, endotoxin-mediated increase in pulmonary vascular resistance, the fall in mean arterial pressure, and the decrease in cardiac output. However, in contrast to the effect of sodium meclofenamate in this model, sodium flurbiprofen had no effect on the secondary hemodynamic changes which occurred following endotoxin. Survival at 6 hours was unaffected by flurbiprofen. Endogenous prostanoid production was not measured so it is difficult to interpret these data as they relate to arachidonic acid metabolism.

Once again, there is evidence that the apparent beneficial actions of certain NSAIDs during endotoxin shock may be unrelated to their ability to inhibit the production of endogenous prostaglandin-like materials. This fact is again reflected in a study reported by Wise et al (1980a). Ibuprofen was used in a dose-related study in an endotoxic rat model. Pretreatment of the rats with low doses of ibuprofen (0.1 to 3.75 mg/kg) had significant value for survival. At these concentrations of the drug, both thromboxane and prostacyclin production were abolished. A larger dose of ibuprofen (30 mg/kg) afforded significantly less survival protection although the prostanoid production was also abolished at this dose. Thirty mg/kg of ibuprofen is not in itself lethal. This type of biphasic, prostaglandin-independent response of the NSAIDs on animal survival following endotoxin has also been shown true for acetylsalicylic acid in the rat model (Halushka et al 1981). The dose–response characteristics aside, ibuprofen was able to significantly enhance survival as assessed at 24 hours after endotoxin.

Ibuprofen has also been studied in the chronic unanesthetized sheep endotoxic shock model (Adams and Traber 1982). A dose of 14 mg/kg intravenously was given both before and 1.75 hours after endotoxin administration. This dose of ibuprofen was able to significantly attenuate the first-phase pulmonary and systemic hemodynamic effects of endotoxin, but had little or no effect on the delayed changes in several cardiovascular and lung injury parameters. Neither prostanoid biosynthesis nor survival were quantified. Thus, ibuprofen in the sheep acted much as flurbiprofen acted in the endotoxic cat model of Parratt and Sturgess (1976).

Different results were observed in a canine shock model. Jacobs et al

TABLE 9 *Use of nonsteroidal antiinflammatory drugs in endotoxin shock in vivo*

Species	Dose of endotoxin	NSAID and dose	Effect	Rate of survival	Reference
Cat	2 mg/kg i.v. bolus	meclofenamate (pre) 0.5–5.0 mg/kg i.v.	attenuation of the first phase hemo-dynamic response	increased at 6 hrs	Parratt and Sturgess (1975c, 1977)
Cat	2 mg/kg i.v. bolus	meclofenamate (post) 2 mg/kg/hr for 6 hrs	attenuation of first-phase response; delay of 2nd-phase hemo-dynamic response	increased at 6 hrs	Parratt and Sturgess (1975b)
Cat	2 mg/kg i.v. bolus	flurbiprofen (pre) 0.1–1.0 mg/kg i.v.	abolished the first phase hemodynamic response; did not affect the 2nd-phase hemodynamic response	no effect at 6 hrs	Parratt and Sturgess (1976)
Rat	20 mg/kg i.v. bolus	ibuprofen 0.1–3.7 mg/kg (pre)	$\downarrow$ 6-keto-PGF$_{1\alpha}$ $\downarrow$ TXB$_2$	increased at 24 hrs	Wise et al (1980a)
		30 mg/kg (pre)	$\downarrow$ 6-keto-PGF$_{1\alpha}$ $\downarrow$ TXB$_2$	slight increase at 24 hrs	
Sheep	0.75 µg/kg i.v. bolus	ibuprofen 14 mg/kg (pre and 1.75 hrs post)	attenuated first-phase response to endotoxin; did not affect 2nd-phase hemodynamic response in lung	not assessed	Adams and Traber (1982)
Dog	2 mg/kg i.v. bolus	ibuprofen (post) 25 mg/kg	increased MABP and CO at 2 hrs	not assessed	Jacobs et al (1982)

CO = cardiac output; MABP = mean arterial blood pressure.

(1982) posttreated dogs receiving endotoxin at the peak of the first-phase response (1–5 min post-endotoxin) with a high-dose (25 mg/kg) ibuprofen regimen. While the effect of the drug on the first-phase response to bolus endotoxin injection cannot be assessed in this study, the mean arterial blood pressure and cardiac output were well maintained in the ibuprofen-treated animals at 2 hours after endotoxin. Survival was not assessed.

7.1.4. *Thromboxane synthetase inhibitors*

There is a consensus in the literature regarding the role of thromboxane-like materials during the shock state. It is assumed that these materials are detrimental factors. Since other cyclooxygenase products have demonstrated salutary effects upon exogenous administration, several groups of investigators have examined the effect of specific thromboxane synthetase inhibitors in several endotoxin shock models (Table 10). The rationale is to specifically remove the influence of a negative factor while allowing the endogenous production of potentially positive factors to proceed. This may be a better therapeutic maneuver than is the use of cyclooxygenase blockade.

Cook et al (1980) used this avenue of approach with both imidazole, a reported thromboxane synthetase inhibitor, and 13-azaprostanoic acid, a reported thromboxane antagonist. These drugs were used in a lethal model of endotoxic shock in the rat. Both imidazole and 13-azaprostanoic acid pretreatment significantly decreased mortality by approximately 50% in this model. Pretreatment with indometacin also reduced mortality by about 50%. Thus, similar effects were seen regardless of whether cyclooxygenase products other than thromboxane were formed or not. This does not suggest a beneficial action of prostacyclin, for example, in the LD_{100} rat model. These same investigators (Wise et al 1980b) also tested 7-(l-imidazolyl)-heptanoic acid (7-IHA) in the rat model. While 7-IHA did not affect endogenous PGE production, it did significantly inhibit TXB_2 biosynthesis. Serum GOT and GPT elevations were attenuated by 7-IHA and 24-hour survival was greatly improved. In the cat, imidazole maintained mean arterial blood pressure and attenuated the endotoxin-induced increase in plasma lysosomal enzyme activity (Smith et al 1980). Long-term survival was not assessed.

Another thromboxane synthetase inhibitor, dazoxiben (OKY 1581), was used to assess the role of thromboxane-like materials in the pulmonary vascular response to endotoxin of the baboon (Casey et al 1982). Pretreatment with the OKY compound abolished the initial, endotoxin-mediated increase in pulmonary vascular resistance while totally inhibiting thromboxane production. Concurrent measurements of plasma 6-keto-$PGF_{1\alpha}$ concentrations reflected a 26-fold increase as compared to endotoxic animals receiving vehicle. These data suggest that OKY 1581 is an effective thromboxane syn-

TABLE 10   *Effect of thromboxane synthetase inhibition on the physiologic responses to endotoxin administration in vivo*

Species	Dose of endotoxin	TXA synthetase inhibitor and dose	Effect	Rate of survival	Reference
Rat	20 mg/kg i.v. bolus 20 mg/kg i.v. bolus	imidazole 30 mg/kg (pre) 13-azaprostanoic acid 30 mg/kg (pre)	$\downarrow$ TXB$_2$	increased survival at 24 hrs increased survival at 24 hrs	Cook et al (1980)
Rat	20 mg/kg i.v. bolus	7-(1-imidazole)-heptanoic acid 30 mg/kg (pre)	$\downarrow$ TXB$_2$ $\downarrow$ plasma GOT; GPT; acid phosphatase	increased survival at 24 hrs	Wise et al (1980b)
Cat	5 mg/kg i.v. bolus	imidazole 25 mg/kg/hr (30 min post for 4.5 hrs)	$\downarrow$ TXB$_2$ $\downarrow$ cathepsin D $\downarrow$ MABP	not assessed	Smith et al (1980)
Baboon	6 mg/kg i.v. bolus	OKY 1581 2 mg/kg (pre)	$\downarrow$ pulmonary vascular resistance response $\downarrow$ TXB$_2$ $\uparrow$ 6-keto-PGF$_{1\alpha}$	not assessed	Casey et al (1982)
Sheep	1 µg/kg i.v. bolus	RO-22-3581 30 mg/kg (pre)	$\downarrow$ pulmonary vascular resistance response $\downarrow$ TXB$_2$ $\uparrow$ 6-keto-PGF$_{1\alpha}$	not assessed (1982)	Watkins et al
Rat	sepsis LD$_{70\text{-}80}$ *E. coli*	imidazole 30 mg/rat i.p. (post)	$\downarrow$ TXB$_2$ $\uparrow$ 6-keto-PGF$_{1\alpha}$	no effect at 24 hrs	Short et al (1983)

MABP = mean arterial blood pressure.

thetase inhibitor in vivo and that thromboxanes play a pressor role in the baboon lung after endotoxin. The data also suggest that the inhibition of thromboxane synthetase results in a shunting of available endoperoxide intermediates into the remaining arachidonic acid cascade pathways. The long-term physiologic effects of the 26-fold increase in prostacyclin production cannot be directly assessed from this study. However, this large, acute increase in prostacyclin synthesis had no apparent negative short-term effect which could be documented. These same types of effects resulting from the inhibition of thromboxane production with subsequent shunting of endoperoxides to the prostacyclin pathway have been observed by Watkins et al (1982) in the sheep lung-lymph preparation.

The effect of thromboxane synthetase inhibitors in septic shock is less well characterized. Short et al (1983) used imidazole in a rat sepsis model. By posttreating with 30 mg/rat of imidazole intraperitoneally 1 hour after the intraperitoneal injection of an LD_{70} dose of *E. coli* bacteria, survival was not significantly enhanced. Endogenous thromboxane synthesis was completely inhibited by this dose of imidazole, and endogenous prostacyclin production was elevated. It appears that the roles of thromboxane and prostacyclin-like materials differ significantly between the bolus model and the septic model of endotoxic shock.

7.2. Essential fatty acid deficiency

A final avenue of approach in investigating the effects of inhibiting prostanoid production during the shock state is to produce animals which are deficient in arachidonic acid, the precursor molecule for the bisenoic series of eicosanoids. Under these conditions, essential fatty acid-deficient rats are better able to withstand an endotoxin challenge than are normal rats (Cook et al 1979, 1981b). If arachidonic acid is supplied back into their diets, then the resistance to endotoxin is lost. These data suggest that the essential fatty acids or some material formed from these compounds are involved in the pathophysiologic changes related to the bolus model of endotoxemia.

7.3. Summary

The use of nonsteroidal antiinflammatory drugs to treat endotoxin shock has yielded quite varying results in regard to animal survival. It appears that pretreatment with NSAIDs can lead to improved survival in the bolus endotoxin model in several species, but that posttreatment alone with these agents does not have any survival value. In addition, there are several examples of studies where survival effects were independent of endogenous eicosanoid biosynthesis. It must be realized that each NSAID may have a profile of pharmacologic actions which is unique to that drug. For example, indometacin appears to attenuate both the initial and delayed-phase hemodynamic responses to endotoxin whereas flurbiprofen and ibuprofen do not. All of these drugs inhibit cyclooxygenase activity and improve survival. Each drug must be viewed separately, and their actions other than cyclooxygenase inhibition must be characterized before the data can be adequately interpreted.

The lethality of the model also appears to affect the efficacy of NSAID therapy. In general, the more lethal the model, the less protective is the NSAID administration in enhancing survival. It is also apparent that short-term hemodynamic effects of these types of drugs are not of prognostic value in endotoxin shock. The abolition of the first-phase cardiovascular and pulmonary responses to the bolus administration of endotoxin by several NSAIDs has no apparent significant effect on long-term survival rates. The most specific type of inhibition which increases survival rates in the bolus model of endotoxemia is that of thromboxane synthetase. The use of these types of inhibitors has generally shunted endoperoxide intermediates into the prostacyclin pathway while also enhancing survival. Whether the increased prostacyclin production is part of the beneficial mechanism of the thromboxane synthetase inhibitors remains to be determined.

A final observation to be noted is that the bolus and septic models of endotoxin shock appear to differ significantly in regard to NSAIDs. In two out of three septic studies reported, the use of indometacin or imidazole had no long-term effect on animal survival. Additional studies in the septic model are required. The available literature thus suggests that NSAIDs do affect the course of events associated with endotoxin shock and that eicosanoids are involved in several specific events. However, the extent of individual eicosanoid involvement in the pathophysiology of endotoxin shock remains to be fully elucidated.

8. CONCLUSIONS

The original hypothesis which has been tested states that the arachidonic acid cascade is activated within a cell when that cell sustains an injury. The hypothesis further states that once the endogenous prostanoids are formed, they may either participate in the physiologic compensatory activities following the injury or may participate in the pathophysiologic events of the injury process.

The available literature strongly supports the first part of this hypothesis in regard to endotoxin shock in vivo. In all species studied, the bolus administration of endotoxin has resulted in increased plasma or lymph concentrations of one or more arachidonic metabolites. This activation is usually rapid with 'spikes' of thromboxane-like materials being a common early occurrence in most species. Since different time courses for the release of various prostanoids from specific tissues or organs have been observed, the prostanoid response to endotoxin in vivo need not be global in nature. Rather, the temporal profiles may reflect the sequential involvement of different tissues or cell types in the course of the shock state. Each of these tissues may have different enzymatic machinery and produce their own profile of arachidonic acid metabolites. The degree of activation of the arachidonic acid cascade by endotoxin in vivo appears to be maximal at doses of endotoxin which are not lethal to healthy animals. Thus, the presence of high circulating concentrations of various prostanoids is probably not responsible for producing the shock state. However, if high concentrations of prostanoids are attained in animals in the decompensatory stages of endotoxin shock, then the physiologic effects can be very different. Although early prostanoid production appears to occur in the bolus model of endotoxin shock, the septic model of circulatory shock is characterized by a slower, progressive production of prostacyclin-like materials. The septic model does not display the early prostanoid spike characteristic of the bolus model. One must therefore question the validity of the bolus model of endotoxin shock in regard to the study of prostanoid involvement in those events which occur clinically.

The rapidity of the changes in circulating prostanoid concentration following endotoxin suggest that de novo synthesis has occurred. However, there are data which indicate that endotoxin in vivo may decrease the ability of specific tissues to catabolize the circulating prostanoids. This mechanism may explain some portion of the increase in circulating prostanoid concentrations during shock. However, this has not been a general finding and there is other evidence to indicate that endotoxin administered in vivo to certain tissues may actually inhibit prostanoid formation.

In contrast to the ability of endotoxin to enhance prostanoid production by most tissues in vivo, endotoxin has very little effect when applied in vitro. This suggests that there is a complex in vivo interaction between endotoxin and some other body components which can instruct most tissues to begin producing prostanoids. The phagocytic cells such as macrophages are very interesting in this regard. These phagocytic cells do respond to endotoxin in vitro and produce large amounts of arachidonic acid metabolites. Studies in endotoxin-susceptible and -resistant mice have demonstrated the possible relationship between the phagocytes, the arachidonic acid cascade, and the pathophysiologic events of endotoxin shock.

The question of whether enhancement or blockade of the arachidonic acid cascade is most beneficial in endotoxin shock is a most difficult one to answer. In the bolus model of endotoxemia, posttreatment with A-, E- and F-types of prostaglandins has no effect on survival while pretreatment with these materials produces only marginally better results. Prostacyclin appears to possess a significant salutary effect. Both pre- and posttreatment in the bolus endotoxin model improves survival. While beneficial in the bolus model, the use of prostacyclin in the septic model may prove deleterious since this model is characterized by an agonal increase in endogenous prostacyclin production. Thus, the question of whether exogenous prostanoids are beneficial or detrimental depends upon the species, the type of shock and whether the animal is in a compensatory or decompensatory stage. Generalizations should not be made. Each model must be well characterized and the time course and type of prostanoid involvement must be known before an intelligent decision can be made to use exogenous prostanoids as a therapeutic maneuver.

The same philosophy is true for the use of cyclooxygenase blockade as an intervention during endotoxin shock. Both time course and metabolic profile factors should be characterized before the arachidonic acid system should be manipulated. Several studies have shown that substrate can be shunted from one prostanoid pathway to another with varying physiologic results. The production of inappropriate prostanoids could conceivably be detrimental. These considerations aside, many nonsteroidal antiinflammatory drugs do have a beneficial effect in the bolus model of endotoxin shock. It is very difficult at this point to separate the arachidonate versus nonarachidonate related effects of the 'cyclooxygenase inhibitors'. Many of these drugs exert potent effects on systems other than prostanoid production. Several studies have shown that the survival effects of a number of these drugs are independent of their ability to inhibit prostanoid production. However, studies carried out in essential fatty acid-deficient animals do suggest that some products of arachidonic acid metabolism may be detrimental to animal survival following endotoxin. Additional studies are needed in both characterizing the eicosanoid profile, including the lipoxygenase products, following prostanoid blockade in the endotoxin shock model and in characterizing the noneicosanoid effects of the specific blocking agent of interest.

In conclusion, endogenous eicosanoids are involved in the sequence of events associated with both the bolus and septic models of endotoxic shock. Each model must be viewed independently as must the involvement of each individual prostanoid. Therapeutic interventions involving both prostanoid administration and specific prostanoid blockade may be advantageous if each is applied to a well-defined model.

9. FUTURE CONSIDERATIONS

There are three general areas of prostanoid research as it relates to circulatory shock which should be pursued. The first of these I have already mentioned. The various models of endotoxin shock must be better characterized as to which types of prostanoids are produced and the time course of their production. This is especially true of the septic models. The modest amount of information which does exist in regard to sepsis suggests that the eicosanoid involvement may be quite different from the bolus model of endotoxin shock. If our long-term goal is the clinical management of sepsis, then research efforts should be channeled in this direction. Continued emphasis on the involvement of prostanoids in the bolus models of endotoxemia may prove scientifically productive, but clinically irrelevant.

The second area related to eicosanoids in endotoxin shock which should be investigated is the multifaceted therapeutic approach to manipulation of the arachidonic acid system. This approach has worked well with other systems as exemplified by the antibiotic/steroid treatment of the shock state. Once the prostanoid involvement in a specific shock model has been established, specific pathway inhibition coupled with exogenous prostanoid administration or the administration of other therapeutic agents may provide a synergistic effect. Very little work has been done in this field.

The final area of interest is perhaps the most exciting. This is the mechanism of eicosanoid interaction with other biological systems and with the endotoxin molecule. The in vivo reactions wherein endotoxin stimulates arachidonic acid metabolism in nonphagocytic cells remains to be fully elucidated. The ability of endotoxin to stimulate phagocytic cells in vitro to produce prostanoids suggests the importance of macrophage modulation of eicosanoid production as playing a role in endotoxic shock. The work done in C3H/HeN and C3H/HeJ mice regarding the endotoxin–macrophage–eicosanoid relationships are of significant interest. The relationships between the eicosanoid system and other humoral systems such as the complement system, the coagulation system, the lymphokines, chemotactic factors, free radicals, and the endogenous opiate system are all areas where investigation should be pursued.

REFERENCES

Adams T, Traber DL (1982) The effects of a prostaglandin synthetase inhibitor, ibuprofen, on the cardiopulmonary response to endotoxin in sheep. *Circ. Shock* 9, 481-489.
Anderson FL, Jubiz W, Tsagaris TJ, Kuida H (1975a) Endotoxin-induced prostaglan-

din E and F release in dogs. *Am. J. Physiol. 228*, 410-414.

Anderson FL, Tsagaris TJ, Jubiz W, Kuida H (1975b) Prostaglandin F and E levels during endotoxin-induced pulmonary hypertension in calves. *Am. J. Physiol. 228*, 1479-1482.

Armstrong JM, Lattimer N, Moncada S, Vane JR (1978) Comparison of the vasodepressor effects of prostacyclin and 6-oxo-prostaglandin $F_{1\alpha}$ with those of prostaglandin E_2 in rat and rabbits. *Br. J. Pharmacol. 62*, 105-129.

Bergstrom S, Danielsson H, Samuelsson B (1964) The enzymatic formation of prostaglandin E_2 from arachidonic acid. *Biochim. Biophys. Acta 90*, 207-210.

Bhatnagar R, Schade U, Rietschel E, Decker K (1982) Involvement of prostaglandin E and adenosine 3′,5′-monophosphate in lipopolysaccharide-stimulated collagenase release by rat Kupffer cells. *Eur. J. Biochem. 125*, 125-130.

Blackwell GJ, Flower RJ, Herman AG (1976) Effect of endotoxin on 15-hydroxyprostaglandin dehydrogenase in the rabbit jejunum and lung. *Arch. Int. Pharmacodyn. Ther. 220*, 325-326.

Bottoms GD, Templeton CB, Fessler JF, Johnson MH, Roesel OF, Ewert KM, Adams SB (1982) Thromboxane, prostaglandin I_2 (epoprostenol), and the hemodynamic changes in equine endotoxin shock. *Am. J. Vet. Res. 43*, 999-1002.

Brown JB (1928) The highly unsaturated fatty acid of liver lipids. *J. Biol. Chem. 80*, 455-460.

Bult H, Herman AG (1978) In vitro inhibition by endotoxin of the anti-aggregatory action of prostacyclin produced by aortic rings. *Arch. Int. Pharmacodyn. Ther. 234*, 335-336.

Bult H, Herman AG (1979) The role of thromboxane A_2 in endotoxin-induced aggregation of guinea pig platelets in vitro. *Agents Actions 4*, 147-155.

Bult H, Beetens J, Vercruysse P, Herman AG (1978) Blood levels of 6-keto-$PGF_{1\alpha}$, the stable metabolite of prostacyclin during endotoxin-induced hypertension. *Arch. Int. Pharmacodyn. Ther. 236*, 285-286.

Bult H, Rampart M, Van Houe C, Herman AG (1979) Effects of endotoxin on biosynthesis of prostacyclin by isolated rabbit peritoneum. *Arch. Int. Pharmacodyn. Ther. 242*, 228-229.

Butler RR, Wise WC, Halushka PV, Cook JA (1982) Thromboxane and prostacyclin production during septic shock. *Adv. Shock Res. 7*, 133-145.

Casey LC, Fletcher JR, Zmudka MI, Ramwell PW (1982) Prevention of endotoxin-induced pulmonary hypertension in primates by the use of a selective thromboxane synthetase inhibitor, OKY-1581. *J. Pharmacol. Exp. Ther. 222*, 441-446.

Cefalo RC, Lewis PE, O'Brien WF, Fletcher JR, Ramwell PW (1980) The role of prostaglandins in endotoxemia: comparisons in response in the nonpregnant, maternal, and fetal models. *Am. J. Obstet. Gynecol. 137*, 53-57.

Collier JG, Herman AG, Vane JR (1973) Appearance of prostaglandins in the renal venous blood of dogs in response to acute systemic hypotension produced by bleeding or endotoxin. *J. Physiol. 230*, 19P.

Cook JA, Wise WC, Callihan CS (1979) Resistance of essential fatty acid-deficient rats to endotoxic shock. *Circ. Shock 6*, 333-342.

Cook JA, Wise WC, Halushka PV (1980) Elevated thromboxane levels in the rat during endotoxic shock. *J. Clin. Invest. 65*, 227-230.

Cook JA, Wise WC, Halushka PV (1981a) Thromboxane A_2 and prostacyclin production by lipopolysaccharide-stimulated peritoneal macrophages. *J. Reticuloendothelial Soc. 30*, 445-450.

Cook JA, Wise WC, Knapp DR, Halushka PV (1981b) Sensitization of essential fatty-acid deficient rats to endotoxin by arachidonate pretreatment: role of thromboxane A_2. *Circ. Shock 8*, 69-76.

Cook JA, Halushka PV, Wise WC (1982) Modulation of macrophage arachidonic acid metabolism: potential role in the susceptibility of rats to endotoxic shock. *Circ. Shock 9*, 605-617.

Culp JR, Erdös EG, Hinshaw LB, Holmes DD (1971) Effects of antiinflammatory drugs in shock caused by injection of live E. coli cells. *Proc. Soc. Exp. Biol. Med. 137*, 219-223.

Demling RH, Smith M, Gunther R, Flynn JT, Gee MH (1981a) Pulmonary injury and prostaglandin production during endotoxemia in conscious sheep. *Am. J. Physiol. 240*, H348-H353.

Demling RH, Smith M, Gunther R, Gee M, Flynn JT (1981b) The effect of prostacyclin infusion on endotoxin-induced lung injury. *Surgery 89*, 257-263.

Erdös EG, Hinshaw LB, Gill CC (1967) Effect of indomethacin in endotoxic shock in the dog. *Proc. Soc. Exp. Biol. Med. 125*, 916-919.

Ferreira SH, Vane JR (1967) Prostaglandins: their disappearance from and release into the circulation. *Nature 216*, 868-873.

Feuerstein N, Ramwell PW (1981) In vivo and in vitro effects of endotoxin on prostaglandin release from rat lung. *Br. J. Pharmacol. 73*, 511-516.

Feuerstein N, Bash JH, Woody JN, Ramwell PW (1981) 3-Deazaadenosine, a transmethylase inhibitor, suppresses the effect of lipopolysaccharide on release of prostacyclin and thromboxane. *J. Pharm. Pharmacol. 33*, 401-402.

Fletcher JR, Ramwell PW (1977a) Altered lung metabolism of prostaglandins during hemorrhagic and endotoxic shock. *Surg. Forum 28*, 184-186.

Fletcher JR, Ramwell PW (1977b) Modification, by aspirin and indomethacin, of the haemodynamic and prostaglandin-releasing effects of E. coli endotoxin in the dog. *Br. J. Pharmacol. 61*, 175-185.

Fletcher JR, Ramwell PW (1978a) E. coli endotoxin shock in the dog: treatment with lidocaine or indomethacin. *Br. J. Pharmacol. 64*, 185-191.

Fletcher JR, Ramwell PW (1978b) Lidocaine or indomethacin improves survival in baboon endotoxic shock. *J. Surg. Res. 24*, 154-160.

Fletcher JR, Ramwell PW (1980) The effects of prostacyclin (PGI_2) on endotoxin shock and endotoxin-induced platelet aggregation in dogs. *Circ. Shock 7*, 299-308.

Fletcher JR, Ramwell PW, Herman CM (1976) Prostaglandins and the hemodynamic course of endotoxic shock. *J. Surg. Res. 20*, 589-594.

Fletcher JR, Ramwell PW, Harris RM (1981) Thromboxane, prostacyclin and hemodynamic events in primate endotoxin shock. *Adv. Shock Res. 5*, 143-148.

Flynn JT (1978) Endotoxic shock in the rabbit: the effects of prostaglandin and arachidonic acid administration. *J. Pharmacol. Exp. Ther. 206*, 555-566.

Flynn JT (1982) Hepatic arachidonic acid metabolism following in vitro perfusion and endotoxin administration. *Prostaglandins, Leukotrienes Med. 9*, 363-371.

Flynn JT, Demling RH (1982) Inhibition of endogenous thromboxane synthesis by exogenous prostacyclin during endotoxemia in conscious sheep. *Adv. Shock Res.* *7*, 199-207.

Flynn JT, Lefer AM (1977) Prostaglandin metabolism during circulatory shock. *Biochim. Biophys. Acta 497*, 775-784.

Flynn JT, Bridenbaugh GA, Lefer AM (1975) Clearance of prostaglandin $F_{2\alpha}$ during circulatory shock. *Life Sci. 17*, 1699-1706.

Gilbert RL (1960) Mechanisms of the hemodynamic effects of endotoxin. *Physiological Rev. 40*, 245-279.

Goldblatt MW (1935) Properties of human seminal plasma. *J. Physiol. 84*, 208-218.

Goto F, Fujita T, Otani E, Yamamuro M (1980) The effect of indomethacin and adrenergic receptor blocking agents on rats and canine responses to endotoxin. *Circ. Shock 7*, 413-424.

Greenway CU, Murthy VS (1971) Mesenteric vasoconstriction after endotoxin administration in cats pretreated with aspirin. *Br. J. Pharmacol. 43*, 259-269.

Hales CA, Sonne L, Peterson N, Kong D, Miller M, Watkins WD (1981) Role of thromboxane and prostacyclin in pulmonary vasomotor changes after endotoxin in dogs. *J. Clin. Invest. 68*, 497-505.

Hall RC, Hodge RL, Irvine R, Katic F, Middleton JM (1972) The effects of aspirin on the response to endotoxin. *Aust. J. Exp. Med. Sci. 50*, 589-601.

Halushka PV, Wise WC, Cook JA (1981) Protective effects of aspirin in endotoxic shock. *J. Pharmacol. Exp. Ther. 218*, 464-469.

Hamberg M, Svensson J, Samuelsson B (1975) Thromboxanes: a new group of biologically active compounds derived from prostaglandin endoperoxides. *Proc. Nat. Acad. Sci. USA 72*, 2994-2998.

Harms BA, Bodai BI, Smith M, Gunther R, Flynn J, Demling RH (1981) Prostaglandin release and altered microvascular integrity after burn injury. *J. Surg. Res. 31*, 274-280.

Harper MJK, Bodkhe RR, Friedrichs WE (1980) Effect of endotoxin treatment on prostaglandin metabolism by rabbit uterus and oviduct. *J. Reprod. Fertil. 58*, 101-108.

Harris RH, Zmudka M, Maddox Y, Ramwell PW, Fletcher JR (1980) Relationships of TxB_2 and 6-keto $PGF_{1\alpha}$ to the hemodynamic changes during baboon endotoxic shock. In: Samuelsson B, Ramwell PW, Paoletti R (Eds), *Advances in Prostaglandin Thromboxane Research Vol 7*, pp 843-849. Raven Press, New York.

Harris RH, Schmeling JW, Fletcher JR, Ramwell PW (1982) Endotoxin interaction with canine platelets fails to stimulate thromboxane production. *Proc. Soc. Exp. Biol. Med. 162*, 397-400.

Hartley P (1909) On the nature of the fat contained in the liver, kidney and heart. *J. Physiol. 38*, 351-374.

Hedqvist P (1969) Modulating effect of prostaglandin E_2 on noradrenaline release from the isolated cat spleen. *Acta Physiol. Scand. 75*, 511-516.

Herman AG, Vane JR (1976a) Effect of indomethacin on endotoxin-induced production of prostaglandins in the isolated rabbit jejunum. In: Samuelsson B, Paoletti R (Eds), *Advances in Prostaglandin Thromboxane Research, Vol 2*, pp 557-560. Raven Press, New York.

Herman AG, Vane JR (1976b) Release of renal prostaglandins during endotoxin-induced hypotension. *Eur. J. Pharmacol. 39*, 79-90.

Hilton JG, Wells CH (1976) Effect of indomethacin and nicotinic acid on E. coli endotoxin shock in anesthetized dogs. *J. Trauma 16*, 968-972.

Hinshaw LB, Solomon LA, Erdös EG, Reins DA, Gunter BJ (1967) Effects of acetylsalicylic acid on the canine response to endotoxin. *J. Pharmacol. Exp. Ther. 157*, 665-671.

Hirose T, Ikeda T, Aoki E, Hara N (1978) The protective effect of PGI_2 on increased lung vascular permeability caused by endotoxin in dogs. *Nippon Kyobu Shikkan Gakkai 16*, 410-417.

Hüttemeier PC, Watkins WD, Peterson MB, Zapol WM (1982) Acute pulmonary hypertension and lung thromboxane release after endotoxin infusion in normal and leukopenic sheep. *Circ. Res. 50*, 688-694.

Isakson PC, Shofer F, McKnight RC, Feldhaus RA, Raz A, Needleman P (1977) Prostaglandins and the renin-angiotensin system in canine endotoxemia. *J. Pharmacol. Exp. Ther. 200*, 614-622.

Jacobs ER, Soulsby ME, Bone RC, Wilson FJ, Hiller FC (1982) Ibuprofen in canine endotoxin shock. *J. Clin. Invest. 70*, 536-541.

Kahn A, Brachet E (1981) Involvement of prostaglandins in the local action of endotoxin. *Prostaglandins Leukotrienes Med. 6*, 23-28.

Kessler E, Hughes RC, Bennett EN, Nadela SM (1973) Evidence for the presence of prostaglandin-like material in the plasma of dogs with endotoxin shock. *J. Lab. Clin. Med. 81*, 85-94.

Korbut R, Ocetkiewicz A, Gryglewski RJ (1975) Release of a prostaglandin E-like substance into mixed venous blood during endotoxin hypotension in cats. *Pol. J. Pharmacol. Pharm. 27*, 439-443.

Krausz MM, Utsunomiya T, Feuerstein G, Wolfe JHN, Shepro D, Hechtman HB (1981) Prostacyclin reversal of lethal endotoxemia in dogs. *J. Clin. Invest. 67*, 1118-1125.

Kurland JI, Bockman R (1978) Prostaglandin E production by human blood monocytes and mouse peritoneal macrophages. *J. Exp. Med. 147*, 952-957.

Kurzrok R, Lieb C (1930) Biochemical studies of human semen: II. The action of semen on the human uterus. *Proc. Soc. Exp. Biol. Med. 28*, 268-272.

Lefer AM, Tabas J, Smith EF (1980) Salutary effects of prostacyclin in endotoxic shock. *Pharmacology 21*, 206-212.

Malik KM, McGiff JC (1976) Cardiovascular actions of prostaglandins. In: Karim SMM (Ed), *Prostaglandins: Physiological, Pharmacological and Pathological Aspects*, pp 103-200. MTP Press, Ltd.

Moncada S, Gryglewski R, Bunting S, Vane JR (1976) An enzyme isolated from arteries transforms prostaglandin endoperoxides to an unstable substance that inhibits platelet aggregation. *Nature 263*, 663-665.

Mowry DT, Brode WR, Brown JB (1942) Studies on the chemistry of the fatty acids. X. The structure of arachidonic acid as evidenced by oxidative degradation and selective hydrogenation. *J. Biol. Chem. 142*, 679-691.

Nakano J, Prancan AV (1973) Metabolic degradation of prostaglandin E_1 in the lung and kidney of rats in endotoxic shock. *Proc. Soc. Exp. Biol. Med. 144*, 506-508.

Northover BJ, Subramanian G (1962) Analgesic-antipyretic drugs as antagonists of endotoxin shock in dogs. *J. Pathol. Bacteriol. 83*, 463-468.

Oettinger W, Heil K, Walter G, Jensen U, Zumtohel V (1980) Pulmonary metabolism of prostaglandin $F_{2\alpha}$ ($PGF_{2\alpha}$) in human septic shock. *Eur. Surg. Res. 12*, 128-129.

Parratt JR, Sturgess RM (1974) The effect of indomethacin on the cardiovascular and metabolic response to E. coli endotoxin in the cat. *Br. J. Pharmacol. 50*, 177-183.

Parratt JR, Sturgess RM (1975a) E. coli endotoxin shock in the cat: treatment with indomethacin. *Br. J. Pharmacol. 53*, 485-488.

Parratt JR, Sturgess RM (1975b) The effects of the repeated administration of sodium meclofenamate, an inhibitor of prostaglandin synthetase, in feline endotoxin shock. *Circ. Shock 2*, 301-310.

Parratt JR, Sturgess RM (1975c) The protective effect of sodium meclofenamate in experimental shock. *Br. J. Pharmacol. 53*, 466P.

Parratt JR, Sturgess RM (1976) The effect of a new anti-inflammatory drug, flurbiprofen, on the respiratory, haemodynamic and metabolic responses to E. coli endotoxin shock in the cat. *Br. J. Pharmacol. 58*, 547-551.

Parratt JR, Sturgess RM (1977) The possible roles of histamine, 5-hydroxytryptamine and prostaglandin $F_{2\alpha}$ as mediators of the acute pulmonary effects of endotoxin. *Br. J. Pharmacol. 60*, 209-219.

Peskar BM, Weiler H, Kroner EE, Peskar BA (1981) Release of prostaglandins by small intestinal tissues of man and rat in vitro and the effect of endotoxin in the rat in vivo. *Prostaglandins 21 (Suppl)*, 9-14.

Prancan A, Simon D, Pope L (1981) Platelet thromboxane production during endotoxic shock. *Agents Actions 11*, 648-650.

Quilley CP, McGiff JC, Lee WH, Sun FF, Wong PYK (1980) 6-Keto PGE_1: a possible metabolite of prostacyclin having platelet anti-aggregatory effects. *Hypertension 2*, 524-528.

Raflo GT, Wangensteen SL, Glenn TM, Lefer AM (1973) Mechanism of the protective effects of prostaglandins E_1 and $F_{2\alpha}$ in canine endotoxin shock. *Eur. J. Pharmacol. 24*, 86-95.

Ramwell PW, Leovey EMK, Sintetos AL (1977) Regulation of the arachidonic acid cascade. *Biol. Reprod. 16*, 70-88.

Reichgott MJ, Engelman K (1975) Indomethacin: lack of effect on lethality of endotoxin in rats. *Circ. Shock 2*, 215-219.

Reines HD, Cook JA, Halushka PV, Wise WC, Rambo W (1982) Plasma thromboxane concentrations are raised in patients dying with septic shock. *Lancet 2*, 174-175.

Shatney CH, Lillehei RC (1976) Effects of prostaglandins in canine endotoxic shock. *Acta Biol. Med. Ger. 35*, 1141-1149.

Short BL, Gardiner M, Walker RI, Jones SR, Fletcher JR (1981) Indomethacin improves survival in gram-negative sepsis. *Adv. Shock Res. 6*, 27-36.

Short BL, Gardiner WM, Mishik AN, Ramwell PW, Walker D, Fletcher JR (1983) Thromboxane synthetase inhibitors in septic shock. *Adv. Shock Res. 10*, 143-148.

Smith EF, Tabas JH, Lefer AM (1980) Beneficial actions of imidazole in endotoxin shock. *Prostaglandins Med. 4*, 215-225.

Smith ME, Gunther R, Gee M, Flynn J, Demling RH (1981) Leukocytes, platelets, and thromboxane A_2 in endotoxin-induced lung injury. *Surgery 90*, 102-107.

Sorrells K, Erdos EG, Massion WH (1972) Effect of prostaglandin E_1 on the pulmonary vascular response to endotoxin. *Proc. Soc. Exp. Biol. 140*, 310-313.

Splawinski JA, Wojtaszek B, Swies J (1979) Endotoxin fever in rats: is it triggered by a decrease in breakdown of prostaglandin E_2? *Neuropharmacology 18*, 111-115.

Stuart MJ (1981) Effect of endotoxin on arachidonic acid release and thromboxane B_2 production by human platelets. *Am. J. Hematol. 11*, 159-164.

Tanigawa T, Suzuki T, Takayama H, Takagi A (1982) Changes in prostaglandin levels in cultures of SV-40 transformed macrophage cell lines in relation to their phenotypic expression. *Microbiol. Immunol. 26*, 59-66.

Utsunomiya T, Krausz MM, Kobayashi M, Shepro D, Hechtman HB (1982) Myocardial protection with prostacyclin after lethal endotoxemia. *Surgery 92*, 101-108.

Van Dorp DA, Beerthuis RK, Nugteren DH, Vonkeman H (1964) Enzymatic conversion of all-cis polyunsaturated fatty acids into prostaglandins. *Nature (London) 203*, 839-841.

Vane JR (1971) Inhibition of prostaglandin synthesis as a mechanism of action for aspirin-like drugs. *Nature (London) New Biol. 231*, 232-235.

Villa S, De Gaetano G, Semerano N (1981) Increased vascular prostacyclin activity in rats after endotoxin administration. *Experientia 37*, 494-495.

Von Euler US (1934) Zur Kenntnis der pharmakologischen Wirkungen von Nativsekreten und Extrakten mannlicher accessorischer Geschlechtsdrusen. *Naunyn-Schmiedeberg's Arch. Exp. Pathol. Pharmakol. 175*, 78-84.

Von Euler US (1935) A depressor substance in the vesicular gland. *J. Physiol. 84*, 21P.

Wahl LM, Rosenstreich DL, Glode LM, Sandberg AL, Mergenhagen SE (1979) Defective prostaglandin synthesis by C3H/HeJ mouse macrophages stimulated with endotoxin preparations. *Infect. Immun. 23*, 8 13.

Watkins WD, Hüttemeier PC, Kong D, Peterson MB (1982) Thromboxane and pulmonary hypertension following E. coli endotoxin infusion in sheep: effects of an imidazole derivative. *Prostaglandins 23*, 273-285.

Webb PJ, Westwick J, Scully MF, Zahavi J, Kakkar VV (1981) Do prostacyclin and thromboxane play a role in endotoxic shock? *Br. J. Surgery 68*, 720-724.

Wise WC, Cook JA, Eller T, Halushka PV (1980a) Ibuprofen improves survival from endotoxic shock in the rat. *J. Pharmacol. Exp. Ther. 215*, 160-164.

Wise WC, Cook JA, Halushka PV, Knapp DR (1980b) Protective effects of thromboxane synthetase inhibitors in rats in endotoxic shock. *Circ. Res. 46*, 854-859.

Handbook of Endotoxin, Vol. 2: Pathophysiology of Endotoxin
L.B. Hinshaw, editor
© Elsevier Science Publishers B.V., 1985
ISBN 0 444 90385 2
$0.85 per article per page (transactional system)
$0.20 per article per page (licensing system)

CHAPTER 11

The contact system in septic shock

CHARLES G. COCHRANE

Inflammatory reactions, including those participating in septic shock, result from the activation of plasma protein systems, inflammatory cells and a combination of the two. In plasma, there are three major systems of proteins that are known to play a role in the inflammatory process: the complement, contact and clotting systems.

In this presentation a brief description will be given of the contact system of plasma and its activation on surfaces bearing negative charges and by cellular enzymes; some of the biologic activities of its components that relate to septic shock; and an analysis of data on its participation in septic shock.

1. THE CONTACT SYSTEM OF PLASMA

The components of the contact system are presented in Figure 1, and a hypothesis on their assembly and activation on a negatively charged surface in Figure 2 (below). The physical characteristics and biochemistry of activation of the components in solution and on a negatively charged surface are the subject of a recent review from this laboratory (Cochrane and Griffin 1982).

When plasma contacts a negatively charged surface, a group of proteins interact on the surface to produce a sequence of conversions of proenzymes to enzymes. This occurs as a burst of activity lasting for seconds. Hageman factor (HF), prekallikrein, high molecular weight (MW) kininogen and clotting Factor XI (Table 1, below) are the principal molecules that undergo initial proteolytic cleavage as activation occurs. This cleavage appears to be essential for the rapid activation of each component. HF binds rapidly to the surface (Table 1, below) in whole plasma, and the peptide chain is cleaved so as to produce chains of 28 000 and 52 000 mol.wt. which are held together by a small disulfide bridge. Cleavage then occurs on the N-terminal site of the disulfide bridge, allowing the smaller chain which bears the enzymatic site of HF to dissociate into the fluid phase (Revak and Cochrane 1976). The larger fragment remains surface-bound.

286

The enzyme in plasma responsible for the rapid cleavage of HF is kallikrein. Its precursor, prekallikrein, is itself cleaved into heavy and light chains and activated by HF. Thus a reciprocal activation by these two molecules was proposed as a mechanism of activation of the contact system (Cochrane et al 1973). The reciprocal enzymatic cleavage and activation of HF and prekallikrein is augmented by the fact that when bound to the surface, HF is more than 100 times more susceptible to enzymatic cleavage (Griffin 1978). In the absence of prekallikrein, HF is cleaved and activated slowly, which suggests that other enzymes may play a secondary role. In the absence of prekallikrein, however, the clotting and kinin-generating activities are markedly retarded (Wuepper 1973).

Prekallikrein and Factor XI exist in plasma as a complex with high MW kininogen (Mandle, Colman and Kaplan 1976; Thompson et al 1977). The high MW kininogen, acting stoichiometrically with HF (Griffin and Cochrane 1976), brings prekallikrein and Factor XI to the negatively charged surface where they are able to interact with HF (Wiggins et al 1977). The light chain of high MW kininogen bears a domain extremely rich in histidine residues which is apparently responsible for the adherence of the molecule to negatively charged surfaces (Han et al 1975). A portion of the light chain is also responsible for the complexing of high MW kininogen with prekallikrein and Factor XI (Thompson et al 1978; Waldman et al 1977). Thus, high MW kininogen acts as a cofactor, in a stoichiometric relationship with HF, to promote the reciprocal activation of prekallikrein and HF (Griffin and Cochrane 1976; Meier et al 1977) and the cleavage and activation of Factor XI by HF.

Kallikrein rapidly dissociates from high MW kininogen on the surface and cleaves and activates other surface-bound HF molecules (Cochrane and

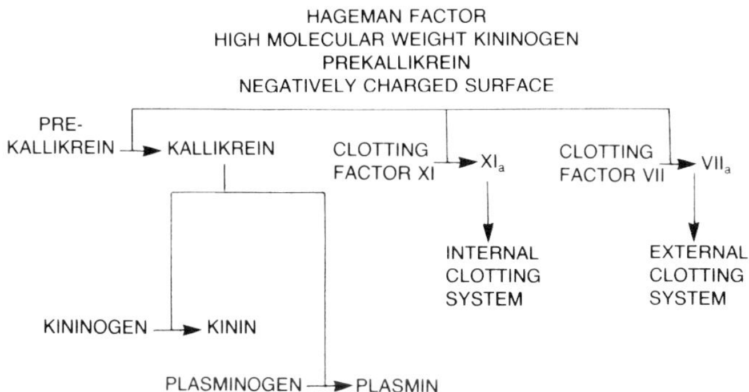

Fig. 1 *The sequences of activation of components of the Hageman factor system.*

TABLE 1 *Components of the contact activation system of human plasma*

	Molecular weight	Concentration in citrated plasma
Hageman factor	80 000	24 µg/ml
Prekallikrein	80 000	50 µg/ml
Factor XI	160 000 (dimer)	4 µg/ml
High molecular weight kininogen	110 000	70 µg/ml
Plasminogen	90 000	200–400 µg/ml

Revak 1980b). The dissociating kallikrein also rapidly cleaves high MW kininogen both on the surface and in fluid phase.

The initial event that triggers the activation of HF and prekallikrein is not precisely understood. While for years it has been thought that HF is activated on binding to a surface, this has lately been questioned. In the absence of prekallikrein or high MW kininogen, HF still becomes surface-bound, but does not activate and does not undergo cleavage for several minutes (well after the burst of activity that is seen in normal plasma) (Revak et al 1977). Single-chain, zymogen HF binds ^{3}H-diisopropylphosphofluoride (^{3}H-DFP) extremely slowly, and this is not influenced by contact with a surface (Fujikawa et al 1977; Griffin 1977). Data supporting surface activation have been presented in which zymogen and activated HF were found to activate prekallikrein at the same rate (Heimark et al 1980). The prekallikrein was present in great excess (more than 60 times the concentration of normal plasma, relative to the amount of HF used), and the possibility remained that rapid reciprocal cleavage and activation of the two molecules took place. This is especially true in view of the speed of reciprocal activation of kallikrein (Cochrane and Revak 1980a). It is possible that an undetected exogenous enzyme activates either prekallikrein or HF, although no evidence for this has been obtained. In addition, the possibility that HF and prekallikrein are 'active zymogens' has been raised (Cochrane and Griffin 1979). In support of this theory is the finding that both HF and prekallikrein take up ^{3}H-DFP slowly in a manner similar to that of trypsinogen. A final mechanism that may account for the initiation of activity involves autoactivation of HF (Wiggins and Cochrane 1979). When molecules of HF bind to a negatively charged surface in sufficient concentration, they activate one another through limited proteolytic cleavage. The possibility cannot be excluded that a few molecules of HFa exist that initiate the reaction (Silverberg et al 1980).

A summary of the molecular assembly leading to the activation of HF is given in Figure 2. In the upper left of the figure, HF and the complex of prekallikrein (PK) and high MW kininogen (HMWK) are depicted in the

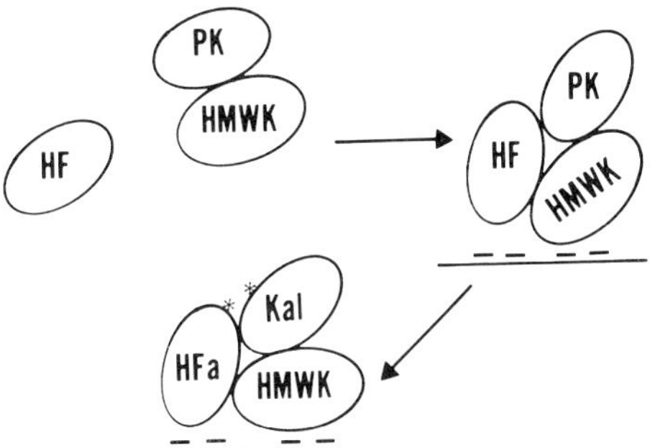

Fig. 2   *Proposed mechanism of activation of Hageman factor (HF) and prekallikrein (PK); the cofactor is high MW kininogen (IIMWK). The proteins are seen in solution in the upper left. Upon contact with a negatively charged surface (right), the proteins are assembled, leading to activation of HF and PK; * = active site. HMWK is cleaved in the process. See text for details.*

plasma. It should be noted that Factor XI and HMWK also exist in plasma as a complex. When presented with a negatively charged surface (right side of the figure), HF and the PK-HMWK complex become bound to the surface. This occurs by virtue of positively charged residues on the heavy chain of HF and a histidine-rich portion (termed Fragment 1–2) of the light chain of HMWK. HF and PK are thus brought into apposition and activation of each takes place. This is aided by apparent conformational changes occurring in the HF on surface contact, in that the surface-bound molecule is approximately 500 times more sensitive to enzymatic cleavage than unbound HF. The initial event in triggering activation could be from an as yet undetected extraneous enzyme which could cleave and activate either HF or PK. However, HF and PK slowly take up ^{3}H-DFP into the active site of their light chains over a period of 24 hours, thereby rendering the molecules nonactivatable, and zymogen HF, when brought onto a negatively charged surface in apposition with other zymogen HF molecules, leads to cleavage and 'autoactivation' of the population of surface-bound HF. This latter event fails to occur when the HF molecules are less densely dispersed on the surface. We thus propose that HF and prekallikrein are 'active zymogens' and when brought into contact with each other they can induce a triggering event leading to proteolytic cleavage and activation. The kallikrein binds to high MW kininogen with relatively low affinity and rapidly dissociates, leading to cleavage of HF molecules bound elsewhere on the surface. This movement of kallikrein in fluid phase probably accounts for most of the activation of HF. The kallikrein at the same time cleaves high MW kininogen both in fluid phase and on the surface.

1.1. Activation of proteins of the contact system by cellular enzymes

Several cells are now known to contain enzymes capable of cleaving and activating components of the HF system. Endothelial cells contain one or more enzymes capable of cleaving HF into fragments of 28 000 mol.wt. When Wiggins et al (1980) assessed the cleaved HF for activity, it was found, in turn, to cleave and activate both prekallikrein and Factor XI when bound to a negatively charged surface. The presence of the cofactor, high MW kininogen, greatly enhanced the reaction with Factor XI, presumably by bringing Factor XI to the surface where it encountered the cell enzyme-activated HF. The endothelial cell enzyme was inhibited by DFP (2mM) but not by soy-bean trypsin inhibitor (50 µg/ml), hirudin (8 mg/ml) or purified antibodies to prekallikrein or plasmin. Thus this cell, which is located at the focal point of the initial inflammatory damage, contains an activator of the HF system.

Recent studies have revealed another pathway by which cellular constituents react with components of the HF system (Newball et al 1980). Washed fragments of normal human lung and isolated human mast cells were submitted to anaphylactic challenge, and the supernatant fluid was examined for the presence of histamine and enzymes capable of cleaving and activating components of the HF system. Following such challenge it was possible to isolate enzymes that cleaved HF, prekallikrein, Factor XI and high MW kininogen, but not components of the complement system (C3, C5, C4, Factor B) or plasminogen. The high MW kininogen released kinin during its interaction with enzyme, as revealed by contraction of the estrus rat uterus. The HF was cleaved in disulfide-linked fragments but little activity of the cleaved HF was detected. Prekallikrein was cleaved by an enzyme separable from the above enzymes on SP-Sephadex and DEAE-Sephacel (Meier et al 1983). Kallikrein thus formed cleaved its peptide substrate Bz-pro-phe-arg-PNA. The data indicate that mast cells release enzymes that activate prekallikrein and release kinin.

2. PATHOPHYSIOLOGIC EFFECT OF THE CONTACT SYSTEM PERTAINING TO SEPTIC SHOCK

2.1. Hageman factor

2.1.1. Vascular permeability

The relationship of a permeability factor that is generated when plasma contacts glass in the presence of HF was established many years ago. The importance of glass in the generation of a permeability factor (Spector 1957) or a kinin-like activity (Armstrong et al 1954) pointed to the necessity of HF in the evolution of kinin activity (Margolis 1958, 1960). Webster and Ratnoff (1961) showed that permeability failed to develop in HF-deficient plasma upon contact with surfaces and that addition of HF to the mixture reconstituted the permeability-inducing factor. Ratnoff and Miles (1964) then employed partly activated HF to induce an increase in vascular permeability of guinea pigs. They found that a minimal dose of 2 μg in 0.1 ml volume induced increased permeability in the skin. Since the HF was only partly active and partly purified, it was uncertain whether this represented a true minimal dose.

When HF was added to dilute plasma, the permeability effect was enhanced, leaving the investigators to suspect that the HF had activated a second factor, thought to be the permeability factor of dilution (PF/dil). PF/dil, which was generated in plasma diluted in the presence of glass (Kagen et al 1963; Miles and Wilhelm 1955; Spector 1957; Wilhelm et al 1958), was thought to act as a substrate for the activated HF. PF/dil was later found to be removed from solution by insolubilized antibodies to HF (Johnston et al 1974). PF/dil and active HF were found to chromatograph similarly (Oh-Ishi and Webster 1975). It is probable, therefore, that PF/dil and activated HF are one and the same.

More recently, fully activated, purified guinea pig HF has been assessed for its permeability activity. Increased permeability was apparent using as little as 1 ng HFa in 0.1 ml (a concentration of 3×10^{-10}M), that is, about 10–100 times more active than bradykinin in this assay (Yamamoto and Cochrane 1981). The HFa was found to be identical to the permeability factor found in extracts of guinea pig skin (Kozono et al 1980; Yamamoto and Kambara 1978; Yamamoto et al 1980).

2.1.2. *Hypotension*

Following earlier reports of hypotensive responses in human beings to infusions of albumin (Bland et al 1973; Harrison et al 1971a, 1971b), a study was made of the nature of the hypotensive agent in commercial albumin preparations. These preparations were associated with a high incidence of hypotensive episodes, with a fall in arterial blood pressures ranging between 20 mmHg and circulatory collapse. β-HFa was found in the albumin preparations as determined by physical properties, its ability to activate prekallikrein, and by its inhibition with specific antibodies to HF (Alving et al 1978). The activity was reportedly not related to bradykinin or kallikrein in the preparation. Recent studies in this laboratory (Kozono and Cochrane 1983) have shown that β-HFa (5 μg) injected intravenously in guinea pigs caused a reduction of mean arterial pressure to approximately 50% and induced neutropenia. Both responses began in seconds and lasted about 30 minutes. These effects were associated with cleavage of [^{125}I] high MW kininogen present in the circulation. The hypotensive response was blocked by prior intravenous injection of soy-bean inhibitor that blocked any kallikrein formed and by removal of circulatory prekallikrein with specific antibody, and could also be evoked by intravenous injection of kallikrein and bradykinin. The neutropenic response was not inhibited by SBT1 and was not evoked by injection of kallikrein.

2.1.3. Leukocyte accumulation in blood vessels and tissues

When exposed to ellagic acid and placed in a vascular bed of rabbit ear chambers, human HF induced margination and infiltration of leukocytes (Graham et al 1965). The reaction reached a maximum after about 90 minutes as opposed to a more transient effect of bradykinin.

2.2. High molecular weight kininogen

Aside from being the parent molecule of bradykinin generated by the hydrolytic action of plasma kallikrein, high MW kininogen also contains the histidine-rich residue Fragment 1–2. In the case of bovine high MW kininogen, Fragment 1–2 and its cleavage products, Fragment 1 and Fragment 2, are released by the action of plasma kallikrein, a property not demonstrated with the human proteins. The bovine Fragment 1–2 and Fragment 2, but not Fragment 1, produce an increase in vascular permeability when administered intradermally. Matheson et al (1976), measuring the diameter of extruded blue dye in skin, observed a permeability reaction with 10^{-8} to 10^{-9} moles of Fragment 1–2, about 100 times less potent than the reaction to bradykinin. An additive effect between Fragment 1–2 and bradykinin was measured. Oh-Ishi et al (1977b) have noted in independent studies that Fragment 1–2 and Fragment 2, but not Fragment 1, induce vascular permeability to an extent of 1/100th that of bradykinin in rabbit skin. PGE$_2$, but not bradykinin, was found to enhance the permeability activity. It should be kept in mind that rabbits respond poorly to bradykinin or HFa in comparison to guinea pigs and primates.

Fragment 1–2 and Fragment 2 were also found to have 10 000 times less activity than bradykinin in the contraction of rat uterine and guinea pig ilial smooth muscle and 100–1000 times less activity than bradykinin in the hypotensive response in rats (Oh-Ishi et al 1977a).

2.2.1. Bradykinin

The physiologic and pathologic effects of bradykinin are of extreme diversity and have received intensive study over the past three to four decades. Both in vitro and in many species of animals in vivo, the pharmacologic effects of this nonapeptide are now well recognized. Increase in vascular permeability, smooth muscle contraction, diminished arterial resistance and hypotension, vascular margination and infiltration of leukocytes into tissues, pain, and stimulation of arachidonate formation are among the commonly studied responses to bradykinin. The actions have been subject to review and will not be specifically discussed here. Comparisons with active components of the contact system have been given above in this section. Bradykinin has been found in fluid such as plasma and joint fluid in numerous forms of experimental and clinical inflammatory disease. The reader is referred to recent compilations of studies of kinins (Erdös 1979; Fujii et al 1979a, 1979b; Pisano and Austen 1976).

3. PARTICIPATION OF THE CONTACT SYSTEM IN SEPTIC SHOCK

Activation of the contact system has been frequently implicated in bacterial sepsis and shock. Investigators have approached the question of involvement of the contact system in various ways: examining plasma during sepsis for lowered concentrations of components, searching for active proteins and peptides of the system, changes in inhibitors of the activated components and complexes between activated components and their inhibitors, generally C1 inh.

During sepsis a fall in concentration of functional prekallikrein in patients has been observed by several groups (Aasen et al 1979; Colman et al 1978; Hirsch et al 1974; Mason et al 1970; O'Donnell et al 1976; Robinson et al 1975), along with a fall in kallikrein inhibitory activity (Aasen et al 1979; Mason et al 1970; O'Donnell et al 1976). In patients developing typhoid fever, evidence was obtained that kallikrein was generated. Immunoelectrophoresis of the plasma revealed an altered migration of prekallikrein-kallikrein to a position where complexes of C1 inh and kallikrein migrate. This correlated with the presence of free arginine esterase in the plasma (Colman et al 1978). In this study, levels of prekallikrein, measured by immunologic means, and HF, measured functionally, were unchanged. Functional levels of high MW kininogen rose, as if in response to the infection. In other studies, HF (Aasen et al 1979; Mason et al 1970) and kininogen (Aasen et al 1979; Hirsch et al 1974) levels of the plasma were reportedly diminished in patients with endotoxin shock.

Bradykinin activity was found in the blood after intravenous administration of bacterial endotoxin (Kimball et al 1972). Levels rose rapidly, reaching a peak by 60 minutes.

Studies using experimental animals have not substantiated unequivocally the participation of the contact system in the development of septic shock.

In experimental hypotensive shock, a fall in kininogen level (detected as trypsin-releasable kinin) and the appearance of free kinin have been correlated with a fall in peripheral arterial resistance in rhesus monkeys (Nies et al 1968) and in monkeys infected with *Salmonella typhimurium* (Wing et al 1978). Similar changes have been detected in dogs (Aasen et al 1978; Gallimore et al 1978). In rabbits, initial studies supported a fall in kininogen (Erdös and Miwa 1968) in endotoxin shock, but subsequent studies of the turnover of human ^{125}I-labeled high MW kininogen failed to record a significant change in its disappearance rate and the circulating ^{125}I-labeled high MW kininogen failed to undergo cleavage despite severe shock (Cochrane and Revak 1980a). Similar studies were conducted in Rhesus monkeys given intravenous infusion of whole *Escherichia coli* organisms (which induced shock with a fall in mean arterial pressures to approximately 50% of normal). A small increase in rate of elimination of [^{125}I] high MW kininogen was observed over a 4–6 hour period. Cleavage of the kininogen was not apparent by analysis of the plasma in acrylamide electrophoresis. However, [^{125}I] high MW kininogen instilled intrabronchially and retrieved by lavage did undergo cleavage in bacteremic, but not in normal animals (Cochrane, Revak and Rice, unpublished observations). This suggested that local activation of the contact system could take place without significant reflection in the peripheral blood space.

In vitro, mixtures of purified HF and prekallikrein are readily activated after interaction with purified bacterial lipopolysaccharides (Morrison and Cochrane 1974).

These various studies suggest the participation of the contact system in septic shock, although the data are not unequivocal. Variation in data on plasma may result from activation of components occurring locally, without sufficient involvement to be reflected in the systemic circulation. Since edema, hypotension and coagulation are features common to the activated contact system and septic shock, there is reason to suspect a relationship between the two.

Nevertheless, the data to date are weak: in almost all studies of levels of components during shock, only functional activities were obtained, and almost without exception, the levels of components were not related to levels of other, unrelated proteins. This is especially critical in shock states in which fluid balance changes rapidly and where therapeutic administration of fluids is the rule. In addition, no account is given to the possible changes (or lack thereof) in circulation time of components upon activation. Such data are essential for the determination of heightened consumption during shock. With few exceptions, physicochemical evidence of activation of components has not been obtained, such as specific cleavage of the component of binding to inhibitor proteins. In addition, while studies are directed at participation of protein components of the contact system in septic shock, the causal relationship of the contact system and the pathophysiologic effects of shock can be examined only by specific inhibition of activated components of the contact system.

REFERENCES

Aasen AO, Frolish W, Saugstad OD, Amundsen E (1978) Plasma kallikrein activity and prekallikrein levels during endotoxin shock in dogs. *Eur. Surg. Res. 10*, 50.

Aasen AO, Gallimore MJ, Lyngass K, Larsbraaten M, Amundsen E, Smith N (1979) In: *Proceedings of the 7th International Congress on Thrombosis and Haemostasis*, p 253.

Alving BM, Hojima Y, Pisano JJ, Mason BM, Buckingham RE, Mozen MM, Finlayson JS (1978) Hypotension associated with prekallikrein activator (Hageman-factor fragments) in plasma protein fraction. *N. Engl. J. Med. 229*, 66.

Armstrong D, Kech CA, Jipson JB, Stewart JW (1954) Development of pain-producing substance in human plasma. *Nature (London) 174*, 791.

Bland JHL, Laver MB, Lowenstein E (1973) Vasodilator effect of commercial 5 percent plasma protein fraction solutions. *J. Am. Med. Assoc. 224*, 1721.

Cochrane CG, Griffin JH (1979) Molecular assembly in the contact phase of the Hageman factor system. *Am. J. Med. 67*, 657.

Cochrane CG, Griffin JH (1982) The biochemistry and pathophysiology of the contact system of plasma. In: Dixon FJ, Kunkel HG (Eds), *Advances in Immunology, Vol 33*, pp 241-306. Academic Press, New York.

Cochrane CG, Revak SD (1980a) The participation of high molecular weight kininogen in hypotensive shock and intravascular coagulation. *Clin. Immunol. Immunopathol. 15*, 367.

Cochrane CG, Revak SD (1980b) Dissemination of contact activation in plasma by plasma kallikrein. *J. Exp. Med. 152*, 608.

Cochrane CG, Revak SD, Wuepper KD (1973) The activation of Hageman factor in solid and fluid phases. A critical role. *J. Exp. Med. 138*, 1564.

Colman RW, Edelman R, Scott CF, Gilman RM (1978) Plasma kallikrein activation and inhibition during typhoid fever. *J. Clin. Invest. 61*, 287.

Erdös E (Ed) (1979) *Handbook of Experimental Pharmacology, Vol 25 (Suppl)*. Springer-Verlag, Berlin-New York.

Erdös E, Miwa I (1968) Effect of endotoxin shock on the plasma kallikrein-kinin system of the rabbit. *Fed. Proc. 27*, 92.

Fujii S, Moriya H, Suzuki T (Eds) (1979a) *Adv. Exp. Med. Biol. 120A*.

Fujii S, Moriya H, Suzuki T (Eds) (1979b) *Adv. Exp. Med. Biol. 120B*.

Fujikawa K, Walsh KA, Davie EW (1977) Isolation and characterization of bovine Factor XII (Hageman factor). *Biochemistry 16*, 2270.

Gallimore MJ, Aasen AO, Lyngass KHM, Larsbratten M, Amundsen E (1978) Falls in plasma levels of prekallikrein, high molecular weight kininogen, and kallikrein inhibitors during lethal endotoxin shock in dogs. *Thromb. Res. 12*, 307.

Graham R, Ebert RH, Ratnoff OD, Moses JM (1965) Pathogenesis of inflammation. II. *In vivo* observations of the inflammatory effects of activated Hageman factor and bradykinin. *J. Exp. Med. 121*, 807.

Griffin JH (1977) Molecular mechanism of surface-dependent activation of Hageman factor (coagulation Factor XII). *Fed. Proc. 36*, 324.

Griffin JH (1978) The role of surface in the surface-dependent activation of Hageman factor (Factor XII). *Proc. Natl Acad. Sci. USA 75*, 1998.

Griffin JH, Cochrane CG (1976) Mechanisms for the involvement of high molecular weight kininogen in surface-dependent reactions of Hageman factor. *Proc. Natl Acad. Sci. USA 73*, 2554.

Han YN, Komiya M, Iwanaga S, Suzuki T (1975) Studies on the primary structure of bovine high molecular weight kininogen. *J. Biochem. 77*, 55.

Harrison GA, Robinson M, Stacey RV (1971a) Hypotensive effects of stable plasma protein solution (SPPS): a preliminary communication. *Med. J. Aust. 2*, 1040.

Harrison GA, Torda JA, Schiff P (1971b) Hypotensive effects of stable plasma protein solution (SPPS): a preliminary communication. *Med. J. Aust. 2*, 1308.

Heimark RL, Kurachi K, Fujikawa K, Davie EW (1980) Surface activation of blood coagulation, fibrinolysis and kinin formation. *Nature (London) 208*, 456.

Hirsch EF, Nakayima T, Oshima G, Erdös EG, Herman CM (1974) Kinin system responses in sepsis after trauma in man. *J. Surg. Res. 17*, 147.

Johnston AR, Cochrane CG, Revak SD (1974) The relationship between PF/dil and activated human Hageman factor. *J. Immunol. 113*, 103.

Kagen JL, Liddy JP, Becker EL (1963) The presence of two permeability globulins in human serum. *J. Clin. Invest. 42*, 1353.

Kimball HR, Melman KL, Wolff SM (1972) Endotoxin-induced kinin production in man. *Proc. Soc. Exp. Biol. Med. 139*, 1078.

Kozono K, Cochrane CG (1983) Failure of human plasma kallikrein to activate human neutrophil functions. *J. Clin. Invest.*, submitted.

Kozono K, Yamamoto T, Kambara T (1980) A protease-like permeability factor in guinea pig skin: contact activation of the latent permeability factor and its nature as a prekallikrein activator. *Am. J. Pathol. 100*, 619.

Mandle R, Colman RW, Kaplan AP (1976) Identification of prekallikrein and high molecular weight kininogen as a complex in human plasma. *Proc. Natl Acad. Sci. USA 11*, 4179.

Margolis J (1958) Activation of a permeability factor in plasma by contact with glass. *Nature (London) 181*, 635.

Margolis J (1960) Hageman factor and capillary permeability. *Aust. J. Biol. Med. Sci. 37*, 239.

Mason JM, Kleeberg V, Dolan P, Colman RW (1970) Plasma kallikrein and Hageman factor in gram-negative bacteremia. *Ann. Intern. Med. 73*, 545.

Matheson RT, Miller DR, Lacombe MJ, Han YN, Iwanaga S, Kato H, Wuepper

KD (1976) Flaujeac factor deficiency. Reconstitution with highly purified bovine high molecular weight-kininogen and delineation of a new permeability-enhancing peptide released by plasma kallikrein from bovine high molecular weight-kininogen. *J. Clin. Invest. 58*, 1395.

Meier HL, Webster ME, Mandle R, Colman RW, Kaplan AP (1977) Enhancement of surface dependent Hageman factor activation by high molecular weight kininogen. *J. Clin. Invest. 60*, 18.

Meier HL, Kaplan AP, Lichtenstein LM, Revak SD, Cochrane CG, Newball HH (1983) Anaphylactic release of a prekallikrein activator from human lung *in vitro*. *J. Clin. Invest. 72*, 574.

Miles AA, Wilhelm DL (1955) Enzyme-like globulins from serum reproducing vascular phenomena of inflammation; activable permeability factor and its inhibitor in guinea pig serum. *Br. J. Exp. Pathol. 36*, 71.

Morrison DC, Cochrane CG (1974) Direct evidence for Hageman factor (Factor XII) activation by bacterial lipopolysaccharides (endotoxins). *J. Exp. Med. 140*, 797.

Newball HH, Meier HL, Kaplan AP, Revak SD, Cochrane CG, Lichtenstein LM (1980) Anaphylactic release of human lung kinin-generating activities. *Fed. Proc. 39*, 906.

Nies AS, Forsyth RP, Williams HE, Melmon KL (1968) Contribution of kinins to endotoxin shock in unanesthetized Rhesus monkeys. *Circ. Res. 22*, 155.

O'Donnell TF, Clowes GH Jr, Talamo RC (1976) Kinin activation in the blood of patients with sepsis. *Surg. Gynecol. Obstet. 143*, 539.

Oh-Ishi S, Webster ME (1975) Vascular permeability factors (PF/Nat and PF/Dil) – their relationship to Hageman factor and the kallikrein-kinin system. *Biochem. Pharmacol. 24*, 591.

Oh-Ishi S, Katori M, Han YM, Kanagawa S, Kato H, Suzuki T (1977a) Possible physiological role of new peptide fragments released from bovine high molecular weight kininogen by plasma kallikrein. *Biochem. Pharmacol. 26*, 115.

Oh-Ishi S, Tanaka K, Katori M, Han YM, Kato H, Iwanaga S (1977b). Further studies on biological activities of new peptide fragments derived from high molecular weight kininogen: an enhancement of the vascular permeability increase of the fragments by prostaglandin E2. *Life Sci. 20*, 695.

Pisano JJ, Austen KF (Eds) (1976) *Chemistry and Biology of the Kallikrein-Kinin System in Health and Disease.* No. 76–791, Department of Health, Education and Welfare/National Institutes of Health, Bethesda, MD.

Ratnoff OD, Miles AA (1964) The induction of permeability-increasing activity in human plasma by activated Hageman factor. *Br. J. Exp. Pathol. 45*, 328.

Revak SD, Cochrane CG (1976) The relationship of structure and function in human Hageman factor. The association of enzymatic and binding activities with separate regions of the molecule. *J. Clin. Invest. 57*, 852.

Revak SD, Cochrane CG, Griffin JH (1977) The binding and cleavage characteristics of human Hageman factor during contact activation. A comparison of normal plasma with plasmas deficient in Factor XI, prekallikrein, or high molecular weight kininogen. *J. Clin. Invest. 59*, 1167.

Robinson JA, Kloduycky ML, Lock HH, Racic MR, Gunner RM (1975) Endotoxin, prekallikrein, complement and systemic vascular resistance. Sequential measurements in man. *Am. J. Med. 59*, 61.

Silverberg M, Dunn JT, Kaplan AP (1980) Autoactivation of human Hageman factor. Demonstration utilizing a synthetic substrate. *J. Biol. Chem. 255*, 7281.

Spector WG (1957) Activation of a globulin system controlling capillary permeability in inflammation. *J. Pathol. Bacteriol. 74*, 67.

Thompson E, Mandle R, Kaplan AP (1977) Association of Factor XI and high molecular weight kininogen in human plasma. *J. Clin. Invest. 60*, 1376.

Thompson RE, Mandle R, Kaplan AP (1978) Characterization of human HMW kininogen: procoagulant activity associated with the light chain of kinin-free HMW-kininogen. *J. Exp. Med. 147*, 488.

Waldman R, Scicli AG, Scicli GM, Guimaraes JA, Carretero OA, Kato H, Han YN, Iwanaga S (1977) Significant role of fragment 1–2 plus light chain of bovine high molecular weight kininogen in contact mediated coagulation. *Thromb. Haemostasis 38*, 14.

Webster M, Ratnoff OD (1961) Role of Hageman factor in the activation of vasodilator activity in human plasma. *Nature (London) 192*, 622.

Wiggins RC, Cochrane CG (1979) The autoactivation of Hageman factor. *J. Exp. Med. 150*, 1122.

Wiggins RC, Bouma BM, Cochrane CG, Griffin JH (1977) Role of high molecular weight kininogen in surface-binding and activation of coagulation Factor XI and prekallikrein (Hageman factor, contact activation, fibrinolysis). *Proc. Natl Acad. Sci. USA 74*, 4636.

Wiggins RC, Loskutoff DJ, Cochrane CG, Griffin JH, Edgington TS (1980) Activation of rabbit Hageman factor by homogenates of cultured rabbit endothelial cells. *J. Clin. Invest. 65*, 197.

Wilhelm DL, Mill PJ, Sparrow EM, Mackey ME, Miles AA (1958) Enzyme-like globulins from serum reproducing the vascular phenomena of inflammation. IV. Activable permeability factor and its inhibitor in the serum of the rat and rabbit. *Br. J. Exp. Pathol. 39*, 228.

Wing DA, Yamada T, Hayley HB, Pettit GW (1978) Model for disseminated intravascular coagulation: bacterial sepsis in Rhesus monkeys. *J. Lab. Clin. Med. 92*, 239.

Wuepper KD (1973) Prekallikrein deficiency in man. *J. Exp. Med. 138*, 1345.

Yamamoto T, Cochrane CG (1981) Guinea pig Hageman factor as a vascular permeability enhancement factor. *Am. J. Pathol. 105*, 164.

Yamamoto T, Kambara T (1978) A protease-like permeability factor in the guinea pig skin. I. Partial purification and characterization. *Biochim. Biophys. Acta 540*, 55.

Yamamoto T, Kozono K, Okamoto T, Kato H, Kambara T (1980) Purification of guinea-pig plasma prekallikrein. Activation by prekallikrein activator derived from guinea-pig skin. *Biochim. Biophys. Acta 614*, 511.

Handbook of Endotoxin, Vol. 2: Pathophysiology of Endotoxin
L.B. Hinshaw, editor
© Elsevier Science Publishers B.V., 1985
ISBN 0 444 90385 2
$0.85 per article per page (transactional system)
$0.20 per article per page (licensing system)

CHAPTER 12

Endorphins in endotoxin shock*

NELSON J. GURLL

1. INTRODUCTION

The discovery of the existence of endogenous opioid substances with in vivo and in vitro activity has led to intensive and extensive investigation of their role in health and disease. While much of this scientific inquiry has been about their effects on pain and behavior, there is a burgeoning data base which supports a role for endogenous morphine-like substances (endorphins for short) in cardiovascular function under normal and pathological conditions.

Background information about endorphins and opiate receptors will be given in this chapter. I will describe the clinical, anatomical, and pharmacological clues for the involvement of endorphins in cardiovascular function and shock. Then the experimental data indicating a role for endorphins in the cardiovascular pathophysiology of endotoxin shock will be discussed largely on the basis of information obtained using the 'pure opiate antagonist' naloxone. Mechanisms and sites of action will be discussed incorporating new information gained by the use of specific ligands for the multiple opiate receptors. Finally, I will try to indicate future directions for research in this area and possible clinical application of this knowledge.

2. OPIATE RECEPTORS AND ENDORPHINS

2.1. Discovery

A decade ago three groups reported the existence of stereospecific binding sites in the brain (Pert and Snyder 1973; Simon et al 1973; Terenius 1973) which met the criteria proposed (Goldstein et al 1971) as critical for the

---

* Work referred to in this chapter was supported by funds from the Medical Research Service of the Veterans Administration and by Department of Defense (U.S. Army) contracts DAMD 17-80-C-0094 and DAMD 17-81-C-1177.

demonstration of specific opiate receptors in neuronal membranes. Theoretically these receptors ought to bind endogenous substances, and the search was on to find and identify these ligands. An endogenous opiate substance was identified, isolated from the brain (Hughes 1975) and later synthesized as a pentapeptide, called enkephalin, with striking opiate-receptor binding and opioid activity (Hughes et al 1975). The enkephalin molecule was of two forms with the C-terminal end being either leucine or methionine, thus giving rise to the designations leu-enkephalin and met-enkephalin. Opiate-like peptides were similarly identified in human cerebrospinal fluid (Terenius and Wahlström 1975) and mammalian brain (Pasternak et al 1975). A longer peptide called β-endorphin (β-END), with 31 amino acid residues and potent opiate-receptor binding and opioid effects, was discovered soon after in the pituitary gland (Cox et al 1975; Li and Chung 1976; Teschemacher et al 1975).

Some fascinating structural similarities were noted in these first-discovered endorphins. The N-terminal pentapeptide of β-endorphin is met-enkephalin. The entire structures of met-enkephalin and β-endorphin are contained in the C-terminal portion of a larger peptide called β-lipotropin (β-LPH). β-LPH had been discovered and its 91 amino acid sequence determined previously (Li et al 1965). This is an interesting sidelight in our story: β-LPH is a molecule that was fully characterized in terms of its structure before any useful or specific biological function was found.

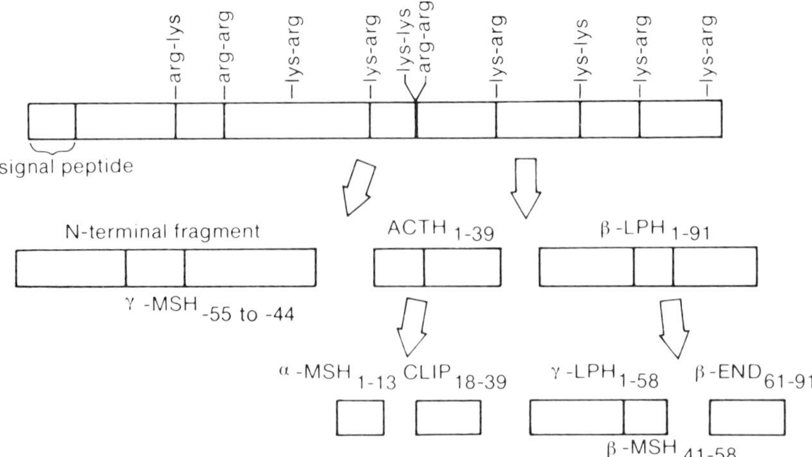

Fig. 1 *Processing of proopiomelanocortin. The peptide fragments are bracketed by a pair of basic amino acid residues which are potential sites for proteolysis yielding peptides shown.*

2.2. **Localization**

β-LPH was localized to the corticotropes of the anterior lobe and the cells of the intermediate lobe of the rat pituitary gland (Moon et al 1973). Antisera to both β-LPH and adrenocorticotropin (ACTH) stain all of the granules in both corticotropes and cells of the intermediate lobe on electron microscopy (Pelletier et al 1977).

There is an important biochemical relationship between β-endorphin, β-LPH, and ACTH (Figure 1). They are all derived from a common precursor molecule (called proopiomelanocortin), as revealed by studies of the biosynthetic product of an ACTH-producing mouse pituitary cell line (Mains et al 1977) and of the cell-free translation product directed by messenger RNA of the pituitary gland (Nakanishi et al 1977; Roberts and Herbert 1977). Later β-endorphin-like immunoreactivity was found in cells containing ACTH and β-LPH (Bloom et al 1977), thus completing the association.

The brain also contains β-endorphin (Bloom et al 1978) and β-LPH (Watson et al 1977) as determined by immunocytochemistry. ACTH immunoreactivity was also shown in brain extracts by Krieger et al (1977). Anatomical studies using immunocytochemical techniques have located ACTH, β-endorphin, and β-LPH in one major neuronal cell group in the basal arcuate nucleus of the hypothalamus (Watson et al 1978). From these cells long axons are distributed throughout the limbic system to the thalamus, periaqueductal central gray, locus ceruleus, and medullary nuclei. These results lend support to the commonality of cellular origin and biochemical precursor for ACTH, β-endorphin, and β-LPH in both pituitary and brain.

Enkephalins are found in the brain in distributions which are similar whether measured biochemically (Simantov et al 1976a) or immunohistochemically (Elde et al 1976). As revealed by subcellular fractionation studies, enkephalin is localized to the synaptosomal fractions that contain nerve terminals (Simantov et al 1976b). In contrast to β-endorphin, enkephalin-containing neurons have short axons and are widely distributed in brainstem nuclei and limbic forebrain, which are areas enriched with opiate receptors (Elde et al 1976). One finds enkephalin-positive material in the nucleus tractus solitarius, nucleus ambiguous, and dorsal motor nucleus of the vagus (Elde et al 1976) and in the locus ceruleus and floor of the fourth ventricle (Simantov et al 1977). A great amount of enkephalin-immunoreactive fibers are found in the sympathetic lateral column where large numbers of epinephrine and norepinephrine terminals exist (Fuxe et al 1979).

Opioid peptides are also found in peripheral tissues. Enkephalins have been found in sympathetic ganglia (DiGiulio et al 1978) as well as the vagus nerve (Lundberg et al 1979). The chromaffin cells of the medulla contain both catecholamines and enkephalins; when stimulated, these cells release both products simultaneously and in equimolar amounts (Viveros et al 1979). Some enkephalin content has been found in the gut as part of the myenteric plexus of nerves (Hughes et al 1977).

Localization of these peptides by immunohistochemistry and radioimmunoassay corresponds in general to the distribution of radioligand binding. Opiate receptors are distributed in the brainstem especially near the terminals of enkephalin-rich neurons. This general theme of receptor distribution corresponding to localization of enkephalins and β-endorphin finds an exception in the pituitary where opiate receptors are found predominantly in the posterior lobe and endorphins (β-endorphin) in anterior and intermediate lobes.

Localization of opiate receptors in the periphery is less well resolved. The vagus nerves contain opiate receptors which are transported along axons from cell bodies in the nodose ganglion; this distribution of receptors flows to the brain as well as the periphery (Young et al 1980). Opiates bind to membrane receptors of the chromaffin cells of the adrenal medulla (Kumakura et al 1980). Pulmonary J-receptors may contain opiate-receptor binding sites which are sensitive to enkephalins (Sapru et al 1981). Opiate binding has been found in nonneural membranes of the kidney and liver (Simantov et al 1978). Vascular beds may also contain opiate receptors, a hypothesis that is based on microcirculatory dilation with opiates which is blocked with naloxone (Altura et al 1980). Some opiate-receptor binding has been shown for the heart which may (Burnie 1981) or may not (Simantov et al 1978) be stereospecific.

The enkephalins have a very short half-life, are inactivated by widely distributed enzyme systems, and are thought to function locally in the central nervous system and gut as neurotransmitters or neuromodulators. The synthesis and storage of enkephalin-containing nerve cells and fibers in the brain are characteristic of neuromodulators. Furthermore, they are ideally located for expression or modulation of the many pharmacological effects of opiates.

Anatomical studies of ACTH, β-LPH and β-endorphin show that they could also serve a neuromodulator or neurotransmitter role. β-Endorphin, however, has a longer biological half-life and may function as a neurohormone. Adrenal enkephalins might serve as neurohormones too if they could resist inactivation or be released close to a site of action.

2.3. Location–function relationships

How might these endogenous opioid systems work? Obviously the endorphin neuronal systems could interact as neuromodulators or neurotransmitters in the central nervous system control areas for the autonomic nervous system. On the other hand, peripherally released β-endorphin and enkephalins might also function as hormones to alter peripheral or central cardiovascular activity. The endorphins might act peripherally on the heart and the vascular tree; they might also function centrally. β-Endorphin released from the pituitary can have access to central cardiovascular control sites directly (Bergland et al 1980). β-Endorphin and even the enkephalins might gain access to autonomic centers within the brain by way of the circumventricular organs (which lack a blood–brain barrier) like the area postrema and the subfornical region of the hypothalamus. The former is densely populated with opiate receptors and is critically located to influence brainstem autonomic sites which modulate cardiovascular function, since all afferent baroreceptor fibers pass through it. The subfornical region above the hypothalamus might give access for these substances to the hypothalamus which has a profound effect on cardiovascular function, especially in emotion and affective states.

Adrenal enkephalins are less likely to exert their effects at the central nervous system since they have a shorter half-life than β-endorphin. However, adrenal venous enkephalin content is high and might reach pulmonary J-receptors in a concentration high enough to have an effect, as suggested by Holaday (1983). Enkephalins from the adrenal may also play a neuromodulator role in adrenal medullary secretion and release (vide infra).

Although proopiomelanocortin contains the amino acid sequence for both met-enkephalin and β-endorphin, it is the precursor only for the latter. Despite structural homologies, their different anatomical distributions make it unlikely that enkephalins are breakdown products of β-endorphin. Recent gene-cloning techniques have revealed other precursor molecules for the enkephalins. Proenkephalin A is the precursor for leu-enkephalin and met-enkephalin (Gubler et al 1982; Figure 2); proenkephalin B is the precursor for α-neo-endorphin, leu-enkephalin, and the dynorphins (Kakidani et al 1982; Figure 3).

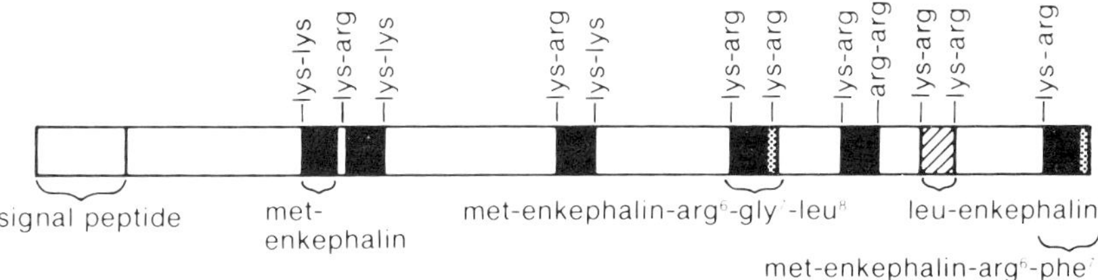

Fig. 2 *Processing of proenkephalin A (modified from Gubler et al 1982). Solid bars indicate met-enkephalin and hatched bars leu-enkephalin sequences.*

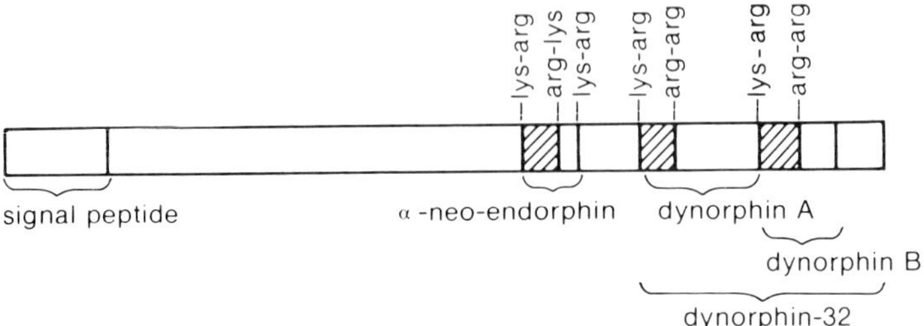

Fig. 3 *Processing of proenkephalin B (modified from Kakidani et al 1982). Solid bars indicate met-enkephalin and hatched bars leu-enkephalin sequences.*

3. THE RESPONSE TO STRESS

3.1. Neuroanatomical basis for central cardiovascular control

The response of the autonomic nervous system to stress is important in shock states so some brief review of its anatomy is indicated. Changes in peripheral hemodynamics are detected by stretch receptors in arteries and the heart and are relayed to the central nervous system by the vagus and glossopharyngeal nerves. Other inputs to the central nervous system via the vagus and glossopharyngeal nerves are from chemoreceptors such as pulmonary J-receptors which are in the alveoli adjacent to pulmonary capillaries (Galosy et al 1981).

Several nuclei in the brainstem provide for maintenance of cardiovascular homeostasis via the autonomic nervous system. The major parasympathetic components that modulate baroreceptor and chemoreceptor reflexes are in the nucleus tractus solitarius (NTS), nucleus ambiguous (NA), and dorsal vagal nucleus (DVN). The NTS and areas within the reticular formation are the primary central nervous system relays for the baroreceptor and chemoreceptor arcs. The NTS innervates the NA and DVN. The primary parasympathetic vagal efferents for baroreceptor output originate from the NA, which is also probably the primary site for integration of reflex bradycardia. The DVN also gives off vagal efferents which have both inotropic and chronotropic effects on the heart.

The NTS also relays information to the sympathetic portion of the autonomic nervous system. The primary common pathway for preganglionic sympathetic outflow to heart, vessels, and adrenal medulla is the intermediolateral nucleus (ILN) of the spinal cord. The ILN receives projections directly from the NTS, reticular system, locus ceruleus, and hypothalamus.

The hypothalamus is central to the orchestration of somatic, endocrine, autonomic, and emotional states which are manifested in the cardiovascular system. The hypothalamus has reciprocal innervation with brainstem parasympathetic nuclei. It not only receives neuronal information from the NTS, but projects to the NTS, NA, and DVN. Efferent sympathetic output from the hypothalamus extends to the ILN.

This picture is not so simple as portrayed, however. There are multiple hypothalamic nuclei which might be involved in cardiovascular regulation as shown by electrical stimulation studies. Virtually any combination of effects can be produced by stimulating one hypothalamic area or another. Because of its anatomical position and widespread connections it probably functions in an integrative fashion rather than in a simple reflex way. Other neural systems have also been implicated in cardiovascular control including cortical, limbic, and cerebellar nuclei. Stimulation of these areas can have potent cardiovascular effects and may be sensitive to opiates.

3.2. Baroreceptor response to shock

The cardiovascular response to a disturbance is the net result of a variety of influences including direct local effects, hormonal release (especially adrenal catecholamines), and autonomic neural input. Perhaps the best understood are the baroreceptor responses to hemorrhage. The autonomic effects of stress include increased adrenal catecholamine secretion, tachycardia due to increased cardiac sympathetic and reduced vagal efferent activity, and peripheral vasoconstriction (in certain vascular beds, with certain types of shock and in defined species). The reflex effects are produced mainly from reduced inputs from arterial baroreceptors and cardiac stretch receptors. The central nervous system can greatly influence the responses for arterial baroreceptor reflexes to arterial pressure changes (Korner 1971). This 'resetting' of reflexes can be produced by activation of other afferents, central command factors, and drugs acting at specific sites in the central nervous system. Thus these interactions add a degree of variability to the responses seen, out of proportion to those that might be predicted from a purely reflex response.

3.3. Sympathomedullary responses

A variety of stressful stimuli including endotoxin injection result in an increased production and release of catecholamines from the adrenal medulla. This release is neurally mediated (Egdahl 1959) and is part of the sympathetic autonomic response to the perturbation. Adrenal enkephalins released by stimulation of the splanchnic nerve have been shown to depress heart rate and blood pressure in anesthetized dogs depleted of adrenal catecholamines by pretreatment with reserpine (Hanbauer et al 1982). This effect was reversible by naloxone but required catechol depletion to unmask it.

The sympathetic nervous system response has effects on the heart and vasculature. In addition to the effects of circulating epinephrine and norepinephrine on the cardiovascular system, there is also local release of norepinephrine as a neurotransmitter.

3.4. Hormonal responses

There are a variety of hormonal responses to stress. Hemorrhage or endotoxin elicits an increased secretion and release of vasopressin from the hypothalamic-posterior pituitary system. Serum ACTH and cortisol rise after endotoxin injection (Egdahl 1959). Endotoxin may induce this rise in serum ACTH by a direct action at the hypothalamic region since damage to this area or pharmacological blockade of hypothalamic releasing factors prevents the rise in ACTH (Moberg 1971). Since β-endorphin and ACTH respond to a variety of perturbations, including stress, synchronously (Guillemin et al 1977), one might assume that β-endorphin would be stimulated and released in endotoxemic shock like ACTH. The renin-angiotensin system is activated by decreased renal perfusion with resultant increases in serum aldosterone.

There are a variety of effects of opioids on these neuroendocrine systems which might be important in stress and have been summarized by Holaday and Loh (1981) as follows. In general, acute opiate administration increases serum ACTH and chronic administration decreases ACTH, as measured by changes in plasma corticosteroids. High doses of morphine (50 mg/kg) increase β-endorphin levels in plasma, and chronic treatment decreases them. Opiate withdrawal in tolerant/dependent rats causes large increases in ACTH and β-endorphin immunoreactivity.

Since opiate-receptor activation stimulates the release of β-endorphin and since β-endorphin mimics the effects of morphine, these results suggest the possibility of a positive feedback system whereby β-endorphin could potentiate its own release. There are two other systems which can also result in biological amplification of β-endorphin and ACTH release which might be important in stress, shock, and trauma. β-Endorphin stimulates the release of vasopressin, and vasopressin is a corticotropin-releasing factor (CRF). As a CRF vasopressin thus stimulates release of β-endorphin and ACTH. A second system involves amines. Opiate peptides inhibit the release of tuberoinfundibular dopamine and dopamine inhibits the release of intermediate lobe β-endorphin.

If β-endorphin is tonically regulating its own release by such a mechanism, one would expect that naloxone would decrease ACTH, corticosteroids, and β-endorphin. However, 10 mg/kg naloxone does not change β-endorphin immunoreactivity in unstressed rats. It is doubtful whether tonic endorphinergic systems regulate β-endorphin release since a positive feedback system would be continuously turned on. Superanalgesic doses of morphine are required to increase plasma β-endorphin. The physiologically obtainable levels of β-endorphin required to activate its own release may only be achieved under extreme stress.

Naloxone blocks other stress-induced increases in plasma corticosterone in rats. High concentrations of naloxone increase plasma corticosterone in opiate-naive rats followed by significant decreases in corticosterone. Therefore, the very high doses used in animal studies and in human trials could reflect a nonspecific effect of naloxone on central nervous system arousal. Alternatively, since some investigators consistently find an increase in plasma cortisol in humans after naloxone, an inhibitory influence of endorphins on basal ACTH secretion can exist.

High doses of morphine required for ACTH activation could uncover antagonist properties of morphine since morphine, like naloxone, decreases the stress-induced rise in corticosteroids.

β-Endorphin may also play a functional role in the adrenal gland. Synthetic β-endorphin induces corticosteroid synthesis in isolated adrenal cells of rats. The steroidogenic effect of β-endorphin appears to be mediated by activation of an ACTH receptor.

Morphine is antidiuretic in animals and man. It is generally accepted that opiates produce antidiuresis by liberating ADH from the posterior pituitary, but morphine has been shown to produce diuresis in man and rats. In addition, morphine causes antidiuresis in Brattleboro rats which lack ADH. Oxytocin systems become hyperfunctional in this strain of rats and may be the explanation for this effect since oxytocin is released by morphine and has some antidiuretic properties.

Parenteral or intracerebroventricular injection of β-endorphin produces oliguria in rats, and cross-tolerance to this is obtained in the chronically morphinized rat. Other workers have shown that β-endorphin stimulates the release of ADH, but opiate receptors may not be involved since the effect was not reversible by naloxone. In contrast, naloxone reduces resting levels of ADH and inhibits the rise in ADH due to orthostatic stimuli in man; therefore, endogenous opiate peptides may be important physiological regulators of ADH in man. Opiate antagonists inhibit ADH release in man during ad lib fluid intake and in response to an osmotic stimulus by elevating the osmotic threshold for ADH release. At variance with these results is the finding that β-endorphin decreases ADH levels in rat plasma, an effect that was not blocked by naloxone.

Although the most clear-cut hormonal effect of opiates is stimulation of ADH release, the evidence cited by Holaday and Loh (1981) suggests that this effect is not easily reproducible. Moreover, although the presence of opiate receptors in the posterior pituitary and their absence in the anterior pituitary suggest a direct pituitary site for endorphin-induced ADH release, these variable effects of opiates in increasing and/or decreasing ADH are not always reversible by naloxone and may therefore not involve classical opiate receptors. In fact, the kidney has opiate receptors which demonstrate reverse stereospecificity, and this may allow for parenteral opiates to directly affect kidney function independent of ADH release.

More recent work may clarify the apparent discrepancies about the effects of opiates on diuresis. β-Endorphin-induced oliguria in hydrated rats is associated with a decrease in excretion of sodium and potassium. These effects were not cross-tolerant with morphine, were centrally mediated, and the urinary electrolyte changes were not comparable to those found with exogenous ADH. The opiates apparently produced their renal effects via a central mechanism independent of ADH release. Thus diuresis or antidiuresis per se is inadequate as the sole indication of an ADH-mediated effect of opiates.

3.5. Clinical observations and clues to endorphins in shock

Walter B. Cannon observed that soldiers injured in World War I exhibited a shock state seemingly out of proportion to the injuries suffered (Cannon 1918). From experiences in World War II Beecher commented that soldiers were often pain-free despite severe injuries (Beecher 1946). In civilian trauma, various invasive techniques (venous cutdown, abdominal incision for placement of peritoneal dialysis catheters, emergency room thoracotomy) can be done without anesthesia and yet with little or no reaction of pain (personal observations).

When confronted with this, the clinical scientist might presume that such refractoriness to pain might be due to a decrease in global central nervous system function due to hypoperfusion. To some extent this was supported by studies of shock in dogs. Such a bias, however, found no support when primate models were studied. Cerebral blood flow in the nonhuman primate is well maintained in hemorrhage and endotoxemia (Rutherford et al 1976).

The rise of plasma β-endorphin due to stress is a ready explanation for these observations since it is a potent analgesic agent. Dr. David Livingstone in his scholarly journals (1872) reported the sensations he noted when suddenly attacked by a lion:

> I heard a shout. Starting and looking half round, I saw the lion just in the act of springing upon me. I was on a little height; he caught my shoulder as he sprang and we both came to the ground below together. Growling horribly close to my ear, he shook me as a terrier does a rat. The shock produced a stupor similar to that which seems to be felt by a mouse after the first shake of a cat. It caused a sort of dreaminess in

> which there was no sense of pain or feeling of terror, although quite
> conscious of all that was happening. It was like what patients partially
> under the influence of chloroform describe, who see all the operation
> but feel not the knife. This singular condition was not the result of any
> mental process. The shake annihilated fear, and allowed no sense of
> horror in looking round the beast. The peculiar state is probably pro-
> duced in all animals killed by carnivora; and if so, is a merciful provision
> by our benevolent Creator for lessening the pain of death.

Such observations conjure up the image of an endorphin effect. Furthermore, endogenous and exogenous opiates depress the cardiovascular system
and may account for the cardiovascular dysfunction found in shock.

There are also a number of less documented reasons for suspecting that
opiates depress cardiovascular function and may prove deleterious in shock.
For many years a common clinical teaching was to avoid the use of morphine
in a patient in shock; it was not required and might exacerbate the shock
state.

4. PHARMACOLOGICAL STUDIES

Alkaloid opiates and endogenous neuropeptides have important cardiovascular effects when administered in pharmacological doses. Although one must
be cautious about inferring physiological effects from pharmacological actions, at least the pharmacological studies give some direction to the investigation of mechanisms and sites of physiological relevance. The field of opiate
research has been burdened by yet another problem. Different and apparently contradictory results have been reported which depend on the type
and dosage of opioid, site of injection and action, species studied, and state
of consciousness. Opiate effects, moreover, can be biphasic, either temporar-
ily or dose-dependently (Holaday and Loh 1981). Anesthesia also makes an
important difference in the responses, turning pressor into depressor responses in general (Sander et al 1982). The opiates may work centrally to
obtain these cardiovascular effects but require intact sympathetics and
parasympathetics for full expression. Peripheral vasodilatory effects of
opiates may involve histamine release, inhibition of normal vasomotor tone,
and/or attenuation (by histamine or opioid) of vasoconstriction due to
neurohumoral substances (Altura et al 1980). Opiates can gain access to the
central nervous system sites of action without having to cross the blood–brain
barrier (vide supra). In addition, opioid peptides exhibit a moderate cere-
brovascular permeability (Rapoport et al 1980) which might allow for feedback to regulate their own release and explain how they can have central
effects even when given systemically.

4.1. Alkaloid opiates

4.1.1. Intravenous injection

Much of our understanding about opiate effects has come from studies using morphine; endorphin effects are often inferred from data obtained with morphine, and the conclusions arrived at in this way are certainly correct when their cardiovascular actions are concerned. Intravenous injection of morphine produces hypotension (Schmidt and Livingston 1933). Although this might be due to the histamine-releasing properties of morphine (Eckenhoff and Oech 1960; McIntosh and Paton 1949), it is thought to be mediated primarily at autonomic cardiovascular control centers in the central nervous system since it is abolished by vagotomy in the rat, but not in the cat (Evans et al 1952). In cats and rats antihistamines have little or no effect on morphine-induced hypotension, yet antihistamines acting at H_1 and H_2 receptors do attenuate the depression in blood pressure, cardiac output and myocardial contractility due to intravenous morphine in dogs (Lind et al 1981). Bradycardia is also produced by morphine (Gokhale et al 1963).

Morphine apparently depresses cardiovascular function in anesthetized rats by vagally mediated bradycardia and decreased vasomotor sympathetic tone, as well as reflexly via vagal afferents (Fennessy and Rattray 1971). Depression by morphine of mean arterial pressure, cardiac output, and left ventricular contractility is attenuated by antihistamines (H_1 and H_2) and cardiac denervation, alone and in combination, and is restored by naloxone (Lind et al 1981).

4.1.2. Central injection

The alkaloid opiate fentanyl produces bradycardia, hypotension and attenuated baroreceptor reflexes when perfused through the fourth cerebral ventricle of conscious dogs, but has no effect in the third or lateral ventricles (Freye and Arndt 1979). These effects are thought to be mediated by opiate receptors in the floor of he fourth ventricle close to the nucleus ambiguous and vagal efferents travelling to the heart (Atweh and Kuhar 1977; Kuhar et al 1973). Both fentanyl and dextromoramide produce a dose-dependent decrease in blood pressure and heart rate in anesthetized dogs (Laubie et al 1974). These effects were thought to be due to central depression of sympathetic tone since both drugs reduced splanchnic nerve activity and were more potent when given intracisternally than intravenously.

Dual effects are found with central injection of morphine, depending upon the site of injection. Morphine given intracerebroventricularly increases heart rate and blood pressure, but given intracisternally it produces bradycar-

310

dia and hypotension (Feldberg and Wei 1977). The former may be due to sympathetic stimulation by actions near the third ventricle. The latter may be mediated at structures near the obex of the medulla with resultant inhibition of sympathetic output to the cardiovascular system and increased vagal tone to the heart (Feldberg and Wei 1978a).

4.2. Endorphins

4.2.1. Intravenous injection

Intravenous injection of β-endorphin produces a prolonged hypotension in rats anesthetized with urethane (Lemaire et al 1978). This effect was seen at doses as low as 30 μg/kg and involves a scrotonergic pathway since the hypotension was attenuated by 5-HT blockers and enhanced by pharmacological potentiation of 5-HT. On a molar basis, β-endorphin is more potent than morphine in causing bradycardia (Wei et al 1980). Endorphins may have their cardiovascular depressant effects as a result of actions on vagal central parasympathetic centers since vagotomy abolishes and atropine attenuates the bradycardia found with a potent enkephalin analog. One way to reconcile these findings is to suggest that these cardiovascular responses to endorphins involve both central serotonergic and peripheral cholinergic mechanisms.

Dual effects have been found with β-endorphin when given intravenously at doses of 0.7 to 2.8 mg/kg; it then produces tachycardia transiently followed by a prolonged and sustained hypotension (Holaday 1983). In these experiments morphine produced bradycardia and hypotension and was only half as potent, on a molar basis, as β-endorphin in its hypotension.

Leu-enkephalin produces hypertension followed by hypotension, and met-enkephalin caused only hypotension in anesthetized cats (Moore and Dowling 1980). Naloxone blocked the hypotension due to either enkephalin, but diphenhydramine blocked the hypotension due to leu-enkephalin but not that due to met-enkephalin. Intravenous injection of leu-enkephalin produced short-lived tachycardia and hypertension when injected in nanomolar amounts in conscious rats (Schaz et al 1980).

4.2.2. Central injection

These differences might be better understood when central injection data are examined since the effects concerned are thought to come about because of an action in the central nervous system. Met-enkephalin applied to the ventral surface of the brainstem results in hypotension and bradycardia (Flórez and Mediavilla 1977). These effects are opiate-mediated since they are naloxone-sensitive. Cisternal injection of β-endorphin in chloralose-anesthetized dogs increases heart rate and blood pressure initially and transiently, followed after some delay by naloxone-reversible decreased rate and pressure (Laubie et al 1977). Synthetic enkephalins but not met-enkephalin inhibited sympathetic tone resulting in hypotension and bradycardia. Similar dual responses were found with β-endorphin and morphine when given intracisternally (Bolme et al 1978). There was initial tachycardia and hypertension followed by intense and prolonged bradycardia and hypotension sensitive to naloxone. D-ala^2-met-enkephalinamide (DAME), a synthetic peptide which is protected from the enzymatic degradation that limits the activity of natural enkephalins, produced hypotension and naloxone-sensitive bradycardia. Leu-enkephalin, met-enkephalin, and α-endorphin had predominantly pressor effects which were naloxone-sensitive. Similar differential effects of endorphins were found as in the studies with morphine: intraventricular administration producing tachycardia and hypertension, cisternal injection just the opposite (Feldberg and Wei 1978b). Owen et al (1981) found that DAME injected into the third cerebral ventricle of the conscious monkey produced bradycardia and hypotension which were reversed with naloxone.

4.3. Effects on baroreceptor reflexes

Inhibition of baroreceptor reflexes is well documented for opioids. Opiates induce orthostatic hypotension in dogs (Hill 1895), and morphine blunts the baroreceptor reflexes associated with tilting in man (Drew et al 1946). Endogenous opioids share this feature. Enkephalin analogs (with μ opiate-receptor specificity) attenuate and naloxone augments the reflex bradycardia associated with phenylephrine injection (Petty and Reid 1981). DAME intracerebroventricularly attenuates the baroreceptor reflex (vagal bradycardia, prolonged pulse interval) to intravenous injection of the potent pressor substance angiotensin (Yukimura et al 1981). Naloxone perfused through the fourth cerebral ventricle of dogs stereospecifically attenuates the halothane-induced bradycardia, hypotension, and baroreflex activity as determined by the response to clamping of the carotid artery (Arndt and Freye 1979). Naloxone potentiates and morphine attenuates the pressure responses to stimulation of afferent (vagus and superior laryngeal) nerves (Montastruc et al 1981). Hypotension and bradycardia, which occur as a response to the release of endorphins induced by electroconvulsion, are enhanced by naloxone (Belenky and Holaday 1979).

5. ENDOGENOUS OPIATE SYSTEMS IN ENDOTOXIN SHOCK

Although the anatomical and pharmacological data presented suggest the involvement of endorphins and opiate receptors in cardiovascular function and regulation, more compelling support comes from the use of 'pure' opiate-receptor antagonists like naloxone to probe opiate involvement. Much of what follows is in reference to endotoxin shock, but liberal use is also made of data obtained in other shock paradigms.

5.1. Naloxone in endotoxemic shock

Opiate overdose and circulatory shock are similar in their manifestations of hypothermia, hypotension, and analgesia. In view of this and the fact that endogenous and exogenous opiates depress cardiovascular function, Holaday and Faden (1978) hypothesized that endorphins are activated in and contribute to the cardiovascular depression found in most shock states. Gram-negative endotoxins like *Escherichia coli* lipopolysaccharide have been widely used as models of septic shock (Gilbert 1960). To test the possible involvement of opiate systems in endotoxin shock, awake unrestrained rats were given a dose of *E. coli* endotoxin of 4 mg intravenously and then naloxone (10 mg/kg intravenously) when the mean arterial pressure (MAP) fell to 65–70 mmHg (Holaday and Faden 1978). In response to naloxone MAP rose by 20 mmHg and pulse pressure by 15 mmHg; pretreatment with naloxone at a dose of 10 mg/kg every 30 minutes prevented the fall in MAP due to endotoxin; moreover, naloxone reversed the fall in MAP in rats given 8 mg of endotoxin. Although the authors did not show that survival improved, they concluded that endorphins may be some of the endogenous substances released in and contributing to the cardiovascular derangements in endotoxin shock since naloxone was devoid of cardiovascular effects in nonshocked animals.

The hypotension in the conscious rat after endotoxin injection was reversed in a dose-dependent manner with naloxone at doses of 0.1–10 mg/kg intravenously (Faden and Holaday 1980). Maximum effectiveness in reversing hypotension due to *E. coli* endotoxin at a dose of 60 mg/kg intravenously (LD_{70}) was found with naloxone at 1 mg/kg. MAP increased by 20 mmHg when naloxone was given at 10 mg/kg bolus and 5 mg/kg/hr infusion for 24 hours in animals shocked by a dose of *E. coli* endotoxin of 40 mg/kg (LD_{25}). Although more naloxone-treated animals survived than saline-treated ones, this difference was not statistically significant.

5.1.1. *Hemodynamics of naloxone response*

These observations were extended to the pentobarbital-anesthetized dog to study the hemodynamics of this response. Naloxone at 2 mg/kg bolus plus an intravenous infusion of 2 mg/kg/hr attenuated the fall in MAP and maximum rate of increase of left ventricular pressure (LV dP/dt$_{max}$) and reversed the fall in cardiac output due to a dose of 0.1 mg/kg *E. coli* endotoxin (Reynolds et al 1980). These results, expressed as changes from baseline, are shown in Figures 4–6. Naloxone had little or no effect on heart rate, pulmonary arterial pressures, portal pressure, or superior mesenteric arterial blood flow. LV dP/dt$_{max}$ is a function of preload, after-load, and left ventricular myocardial contractility. Since preload (pulmonary arterial pressure) and after-load (calculated peripheral vascular resistance) were unchanged, it was suggested that the rise in LV dP/dt$_{max}$ was due to increased myocardial contractility. Naloxone was given 15 minutes after the endotoxin and improved survival from 20% to 80% in this model (LD$_{80}$). These results suggest that the myocardial depression of endotoxemic shock is, at least in part, mediated by endorphins or opiate-receptors. Naloxone had little or no cardiovascular effect in unstressed dogs, emphasizing the point that endogenous substances must be released before naloxone blockade can be effective. Such an endogenous substance stimulated by shock or stress and sensitive to naloxone would, by definition, most likely be an endorphin.

5.1.2. *Metabolic effects*

Similar beneficial results with naloxone obtain in other models of endotoxic shock with different anesthetics as well as in other types of shock. Anesthesia, its presence and type, is an important determinant in opiate responses (vide supra) and in metabolic responses. Some studies have documented

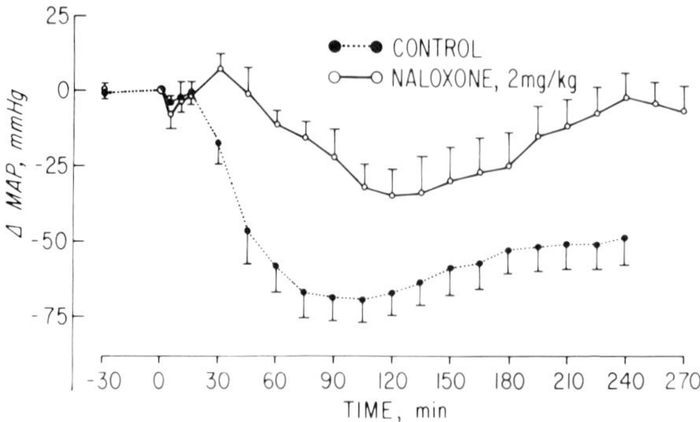

Fig. 4 *Naloxone in canine endotoxin shock attenuates the fall in mean arterial pressure from preshock values (Δ MAP, mmHg). Naloxone treatment (2 mg/kg bolus plus 2 mg/kg/hr infusion intravenously) began at t = 15 min after endotoxin injection at t = 0. Control animals received 0.9% NaCl in equivalent volumes.*

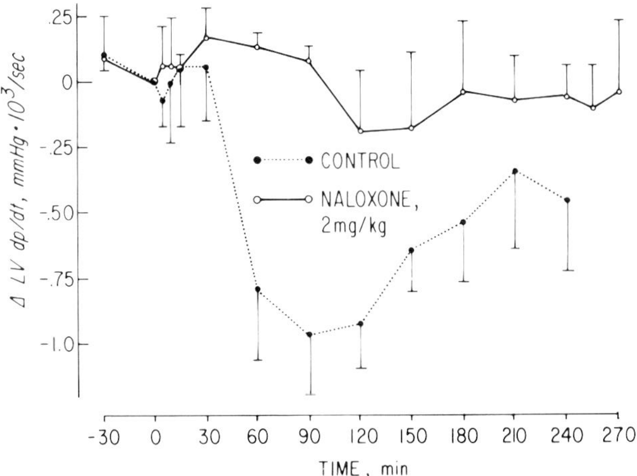

Fig. 5   *Naloxone attenuates the fall in left ventricular contractility (Δ LV dP/dt$_{max}$, mmHg $\times$ 10^3/sec) in canine endotoxin shock. Reprinted from Reynolds et al (1980), with permission. Legend as in Figure 4.*

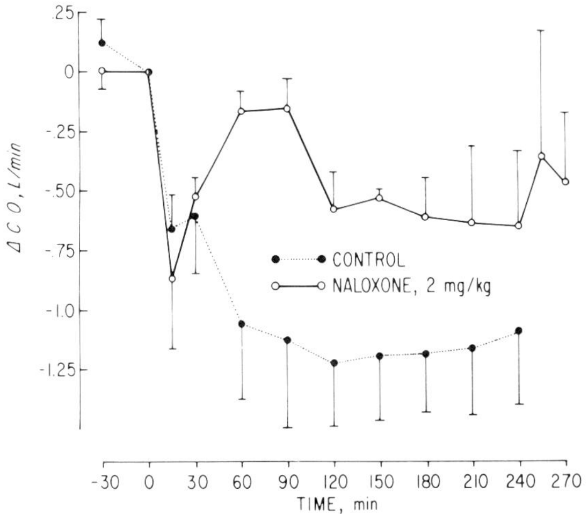

Fig. 6   *Naloxone reverses the fall in cardiac output (Δ CO, l/min) in canine endotoxin shock. Legend as in Figure 4.*

metabolic as well as cardiovascular responses. Pretreatment with naloxone prevented or attenuated the endotoxin-induced decreases in left ventricular work, coronary blood flow, myocardial oxygen consumption, and efficiency of left ventricular work in anesthetized dogs (Brückner et al 1981). Naloxone increases blood pressure and cardiac output as well as blood cyclic-AMP and glucagon levels in the pig subjected to live *E. coli* sepsis (Gahhos et al 1982). Acidosis, hypoglycemia, and hemoconcentration due to *E. coli* endotoxin (2 mg/kg intravenously) was attenuated by naloxone given simultaneously with endotoxin at 10 mg/kg followed by 5 mg/kg/hr for 6 hours in pentobarbital-anesthetized dogs (Raymond et al 1981). Even naloxone treatment at 2 mg/kg intravenously delayed for 90 minutes after an intravenous dose of *E. coli* endotoxin of 1.5 mg/kg increased MAP, cardiac output, and left ventricular stroke work in dogs anesthetized with N_2O/O_2; there was no improvement, however, in oxygen consumption and serum lactate (Thijs et al 1983). These metabolic effects might be merely due to improved tissue perfusion, so no conclusions can be drawn about the specificity of the finding. Naloxone increases survival rate and prolongs survival time in unanesthetized dogs given *E. coli* endotoxin (Raymond et al 1981).

5.1.3. Other models of shock

Naloxone improves MAP and cardiac output in live *E. coli* sepsis (Gahhos et al 1982; Weissglas et al 1982), a model perhaps more reminiscent of human septic shock than endotoxemic paradigms. Naloxone improves MAP, pulse pressure, and survival rate in hypovolemic rats (Faden and Holaday 1979). In hemorrhaged dogs, naloxone increases MAP, cardiac output and left ventricular contractility, and converts a 100% lethal model to 100% survival (Vargish et al 1980). These effects are dose-dependent (Gurll et al 1981a), obtain even when the shed blood is not reinfused (Gurll et al 1982a), and may be relevant to endotoxin-induced plasma loss and sequestration. In a model of neurogenic shock, naloxone reverses the hypotension due to cervical cord transection in cats and rats, which implicates parasympathetic mechanisms in this response (Faden et al 1980; Holaday and Faden 1980). Naloxone also improves hemodynamics in splanchnic occlusion shock (Lind et al 1980).

5.1.4. Criteria for opiate involvement

Naloxone blockade is a necessary but insufficient criterion for implication of endogenous opiate mechanisms in the system being studied. Although touted as a pure opiate-receptor antagonist, naloxone has apparent opiate-agonist activity in some in vitro and behavioral test systems; it is able to modify the effects of some nonopiate depressants and nonopiate antinociceptive agents; naloxone may interact with drugs that affect systems involving γ-aminobutyric acid (GABA) and dopamine as transmitters. Rather strict criteria for implication of opiate mechanisms have been suggested including: (a) inactivity of stereoisomers devoid of opioid antagonist activity; (b) similar effects with other opioid antagonists; (c) direct release of opioid peptides; (d) cross-tolerance with morphine; and (e) potentiation of effects by inhibiting breakdown of endogenous opioid peptides (Sawynok et al 1979). Most of these criteria have been met at least in one or more models of circulatory shock.

5.1.5. Stereoisomeric effects

The active (−)isomer of naloxone improved blood pressure in conscious rats given *E. coli* endotoxin; the (+)isomer, which does not bind to opiate receptors in vitro or in vivo, had no effect (Faden and Holaday 1980). The active (−)isomer of a new opiate antagonist (Win 44,441-3) was beneficial in raising blood pressure in cat hemorrhage while the inactive (+)isomer (Win 44,441-2) was ineffective (Curtis and Lefer 1982a). This is strong evidence for a stereospecific action of naloxone, probably by competitive antagonism with endogenous opiate peptides at the receptor level.

5.1.6. Other opioid antagonists are effective

Another 'pure' opiate receptor antagonist, naltrexone, is effective in increasing MAP, cardiac output and LV dP/dt$_{max}$ in canine hemorrhagic shock (Gurll et al 1982b). Meptazinol, a new analgesic with opiate receptor antagonist properties, has led to improved cardiovascular function in hemorrhagic (Chance et al 1981) and endotoxemic shock (Paciorek and Todd 1982) in rats. The opiate antagonists Win 44,441 (Curtis and Lefer 1982a) and nalorphine (Hock et al 1983) have also proved to be useful in the treatment of cat hemorrhagic shock. The responses to these agents resemble those found with naloxone by these investigators using their particular models.

5.2. Endorphin levels in endotoxemic shock

A dose of *E. coli* endotoxin of 1.5 mg/kg intravenously in dogs anesthetized with N$_2$O and O$_2$ resulted in increases in plasma β-endorphin from less than 100 pg/ml to values of 97–770 pg/ml (Thijs et al 1983). Small doses of *E. coli* endotoxin (450 ng/kg) in unanesthetized, partly restrained sheep increases plasma β-endorphin by a factor of 12, and this increase precedes the development of hypotension (Carr et al 1982). There is, in these sheep experiments, a second peak of β-endorphin immunoreactivity (at about 22-fold above baseline) which is associated with a fall in the ratio of β-endorphin to β-LPH suggesting early depletion of a pool rich in β-endorphin and subsequent release of a pool rich in unprocessed precursor. Naloxone augmented the endotoxin-induced opioid secretion, both early and late, in the sheep (Carr et al 1982) as well as in the anesthetized dog (Thijs et al 1983). This result suggests that endorphins exert a short-loop feedback inhibition of their own secretion. Live *E. coli* infusion increases plasma β-endorphin immunoreactivity by 55±38 pg/ml to 225±34 pg/ml, but this effect, in contrast, is not influenced by naloxone treatment (Rees et al 1983). Some caution must be exercised, at any rate, in the interpretation of plasma concentrations of substances in shock states where there is a loss of plasma volume (due to endotoxin-induced splanchnic pooling and histamine release causing increased vascular permeability) and hence increase in plasma concentrations of substances. Indeed, the decreases in plasma β-endorphin due to naloxone might merely be due to reversal of hemoconcentration already shown with naloxone treatment (Raymond et al 1981).

Such a precaution might also be given for the demonstrated increase in venous β-endorphin, met-enkephalin, and leu-enkephalin concentration in hemorrhagic shock (Lang et al 1982). These data about concentration do not convey actual secretion rates, but since adrenal vein leu-enkephalin concentration was 10 times the femoral vein concentration, the authors argued that the adrenal gland was the main source of this peptide in plasma.

Another caveat needs to be mentioned regarding elevated concentrations of endorphins in shock: coexistence of the shock state and elevated plasma levels of endorphins does not mean a cause-and-effect relationship. However, such an inference is possible when one considers other features: endorphins do depress cardiovascular function and the shock state is reversed by naloxone (vide supra).

5.3. Cross-tolerance

The fourth criterion for opiate involvement has been met in endotoxemic shock. Morphine-tolerant dogs are unusually sensitive to a dose of 1 mg/kg *E. coli* endotoxin with larger falls in MAP and cardiac index and early death compared to naive animals given endotoxin without morphine exposure (Hess et al 1981). Morphine-tolerant animals had blunted responses to bilateral carotid occlusion and peripheral injection of epinephrine; these findings suggested a defect in sympathetic response similar to that observed with opiates acutely (vide supra). The morphine-tolerant animals were unable to compensate hemodynamically, and their cardiovascular responses to endotoxemic shock resembled those of a rigid tube hydraulically.

5.4. Noncardiovascular effects

Naloxone may have noncardiovascular effects which may be important in determining survival. Naloxone at 1 mg/kg bolus plus 1 mg/kg/hr in pentobarbital-anesthetized dogs increased all cardiovascular parameters to a similar degree as 2 mg/kg plus 2 mg/kg/hr, but survival at the lower dose was not improved (Reynolds et al 1980).

Naloxone, as mentioned before, attenuated acidosis, hypoglycemia, and hemoconcentration effects which might be mediated by the cardiovascular system; on the other hand, they might not be (Raymond et al 1981). Some local effects of naloxone have been found seemingly independent of hemodynamics. Naloxone at 0.4 mg/kg plus 0.6 mg/kg/2 hrs reversed the fall in gastric mucosal oxygen tension and transmural electrical potential difference found with live *E. coli* sepsis in dogs without affecting systemic or local blood pressure or flow (Rees et al 1982). The characteristic fall in potential difference and mucosal oxygen tension was mimicked by β-endorphin infused into the arterial supply of the stomach studied at 20 μg/kg over 1 hour. These results suggest that opiates may have important local effects in sepsis and shock.

Naloxone has effects on the formed elements in blood. Naloxone reverses the platelet trapping in the lung due to endotoxemia in the dog (Almqvist et al 1981). Endotoxin in the rat exposed to *E. coli* endotoxin or ampicillin-treatment of *Borellia duttonii* produces a fall in body temperature, leukocyte count, and platelet count, all of which are prevented by naloxone in a dose-related fashion (Wright and Weller 1980). As these authors indicate and as we have suggested as well (vide supra), naloxone sensitivity does not necessarily implicate endorphins. However, the fact that naloxone preserves these blood cells may be an important reason for its beneficial effect in endotoxemia and sepsis.

6. OTHER AGENTS WITH ANTI-OPIATE EFFECTS IN SHOCK

The potential clinical application of opiate-receptor antagonists in human shock might be limited by their effect on pain perception. By blocking opiate-mediated analgesia, naloxone might enhance pain in shock states. One way out of this impasse is to use opiate-receptor antagonists which are specific for the type of opiate receptor mediating the adverse effects on cardiovascular function but not the analgesic effects of endorphins; such receptor-specific antagonists are being developed and investigated (vide infra). There are a number of mixed opiate agonists/antagonists with analgesic activity which might have a therapeutic advantage over naloxone for this reason and have been evaluated. Meptazinol reverses the cardiovascular depression in rat hemorrhage (Chance et al 1981) and endotoxemia (Paciorcek and Todd 1982). Nalbuphine improves cardiovascular function and survival in canine hemorrhage shock (Hunt et al 1983) and nalorphine improves blood pressure in feline hemorrhage (Hock et al 1983).

Thyrotropin-releasing hormone (TRH) is a tripeptide which reverses selected opiate actions without affecting analgesic latencies through its own nonopiate receptors and effectors (Holaday et al 1978; Tache et al 1977). Unlike naloxone, it has intrinsic cardiovascular activity and increases MAP by about 10 mmHg in conscious unstressed rats when given at a dose of 2 mg/kg (Holaday et al 1981a). TRH in rats given endotoxin increased MAP twice as much as in unstressed rats. Furthermore, TRH increased pulse pressure, heart rate, respiratory rate, and survival (Holaday et al 1981b). These effects were dose-dependent and were more pronounced than those found with naloxone. TRH has also been found to reverse the cardiovascular depression noted with endotoxemia in monkeys but not that in dogs (Gurll et al 1983). Moreover, TRH failed to improve survival significantly in endotoxemic monkeys (Reynolds et al 1982) but did in hypovolemic monkeys (Gurll et al 1982c). The therapeutic effects of TRH seem to be central in origin and mediated by the sympathomedullary outflow from the central nervous system since the effects were found with intracerebroventricular injection and were abolished with adrenal demedullectomy (Holaday et al 1982a). Peripheral sites of action are also probably involved in the cardiovascular responses to intravenous TRH since they were not abolished by adrenal demedullectomy.

7. MECHANISMS AND SITES OF ACTION

7.1. Opiate receptors

In view of the ubiquitous distribution of opiate receptors, the effects of naloxone might be at any one of many potential sites. First it would be important to know whether naloxone works centrally or peripherally. The involvement of the central nervous system opiate receptors has been investigated. Infusion of naloxone at 1 µg/kg/min into the ventriculocisternal system of pentobarbital-anesthetized dogs attenuated the fall in MAP and pulse pressure due to a dose of 0.5 mg/kg of *E. coli* endotoxin (Janssen and Lutherer 1980). The infusion was started 30 minutes before and was ended 180 minutes after the endotoxin was given. Subsequent MAP and pulse pressure were higher. Moreover, the naloxone pretreatment increased MAP and pulse pressure slightly compared with controls, which suggests the possibility of a tonically active central cardiovascular depressor system involving opiate systems. Whether tonically active or not, this cardiovascular depressor opiate system appears to be activated by endotoxin shock. Since the dose of naloxone used is ineffective when given systemically, the authors conclude that the beneficial effects of naloxone are centrally mediated. Naloxone similarly has been shown to work centrally in spinal shock (Holaday and Faden 1980) and hemorrhagic shock (Holaday et al 1981c). The iodomethylate of naloxone, which does not cross the blood–brain barrier easily, was effective in reversing endotoxemic shock in rats when given centrally but not peripherally (Rios and Jacob 1981). Janssen (1983) has reviewed the evidence for a central site of naloxone action in endotoxin shock. He concludes that the effects of naloxone are mediated by nuclei that (a) contain opiate receptors, (b) are found in rostral and caudal regions of the brainstem, and (c) interact with the fastigial nuclei of the cerebellum.

Even so, central mediation still requires peripheral hormonal or neural pathways to explain cardiac and peripheral vascular end results. The above results also do not exclude peripheral actions of naloxone. Indeed, tritiated naloxone is taken up after intravenous injection or infusion in peripheral tissues importantly involved in cardiovascular function, like adrenal, lung, and heart, as well as in those involved in drug metabolism, like liver and kidney (Curtis and Lefer 1982b). The tissue:plasma ratios of ^{3}H-naloxone were lowest in brain tissues and were similar in normotensive and hypotensive cats with a notable exception: in hypotensive cats an increase in the tissue:plasma ratios of ^{3}H-naloxone was found only in the adrenal. Whether this is a function of change in blood flow or reflects some other important homeostatic adjustment remains unresolved.

Obvious peripheral sites to investigate are the autonomic nervous system, pituitary, adrenal medulla, adrenal cortex, and blood-borne substances acting on the heart and vasculature. Total adrenalectomy or adrenal demedullation completely blocks the pressor response to naloxone in conscious rats given endotoxin (Holaday et al 1982a). These results were found with both central and peripheral injections of naloxone, in contrast to the discrepancies noted with TRH (vide supra). Moreover, these procedures enhanced susceptibility to endotoxemic shock while preserving adrenal cortical function. The authors concluded that the beneficial effects of naloxone involve central nervous system actions that are mediated peripherally by sympathomedullary discharge. In support of this is the observation that adrenalectomy blocks and splanchnicectomy attenuates the pressor response to naloxone in surgical stress on cats (Dashwood and Feldberg 1979). Naloxone given intravenously or intracisternally to anesthetized cats increased arterial pressure in association with an increase in preganglionic splanchnic nerve activity (Manugian et al 1981). Intravenous naloxone increased blood pressure and sympathetic nerve activity in endotoxic shock in anesthetized cats (Koyama et al 1983).

In contrast, we have shown that the effects of naloxone in hemorrhagic shock (increased MAP, cardiac output, and LV dP/dt$_{max}$) abolished by adrenalectomy are restored by adrenal corticosteroid in physiological doses (Patton et al 1983). Of course the differences between the results of this study and those mentioned above may merely be due to differences in shock models and species studied, but obviously further work needs to be done along these lines. Some resolution of these differences is possible by the realization that corticosteroids increase catecholamine effects: corticosteroids are required for epinephrine biosynthesis (Wurtman et al 1972), potentiate peripheral vascular effects of catecholamines (Besse and Bass 1966) and inhibit the catabolism of catecholamines (Kalsner 1969).

Opiate receptors have been demonstrated in cardiac tissue but may not be specific. The effects of naloxone on isolated heart muscle contraction have been investigated with contradictory results. Curtis and Lefer (1980) found a moderate positive inotropic effect on cat papillary muscles with naloxone at concentrations only slightly higher than estimated plasma values during shock. Weissglas et al (1981), however, were unable to find any inotropic effect of naloxone on guinea pig papillary muscles even at supratherapeutic doses.

Other peripheral opiate receptors may be important. For example, the adrenal medullary cells contain opiate receptors which probably modulate adrenomedullary secretion (Kumakura et al 1980). Opiate-receptor vascular binding may modulate peripheral effects of catecholamines and other vasoactive substances (Altura et al 1980).

7.2. Endorphins

Given that naloxone works by blocking opiate receptors, which endogenous opiate ligand is being blocked? There are a variety of candidate depressor substances, perhaps some not yet discovered, but we should look specifically at several possibilities.

If adrenal enkephalins were the important cardiovascular depressants among the endorphins, one might have expected that adrenalectomy would protect the animals against the cardiovascular effects of shock. Instead, after adrenalectomy, susceptibility to shock becomes many times greater in hemorrhage and endotoxemia (vide supra). Perhaps this is due to lifting of the negative feedback inhibition of pituitary endorphin production.

What about β-endorphin? The pituitary release of β-endorphin is under negative feedback inhibition by corticosteroid, and corticosteroids regulate endorphin production (Sabol 1978). If pituitary β-endorphin is the culprit in the endorphinergic cardiovascular depression of shock, then hypophysectomy ought to prevent the cardiovascular depression in shock and protect animals from shock as well as blunt the naloxone response (since there is nothing to block). Hypotension after *E. coli* endotoxin and pretreatment with naloxone developed only in hypophysectomized rats and not in intact rats (Davis et al 1983). Only the rats with intact pituitaries and circulating levels of β-endorphin could demonstrate the protective effect of naloxone; on the other hand, pituitary β-endorphin was not required to generate endotoxic shock in the hypophysectomized animals. Hypophysectomy also has been shown to attenuate the pressor response to naloxone in conscious hemorrhaged rats (Holaday et al 1981c). These hypophysectomy studies must be reviewed cautiously since hypophysectomy also results in adrenal atrophy. This atrophy might be responsible for the absence of naloxone response in hypophysectomized rats since adrenalectomy abolishes the response to naloxone in canine hemorrhagic shock and murine endotoxic shock (vide supra).

Opiate effects are additive. Intravenous injection of either β-endorphin or morphine worsened the hypotension found with *E. coli* endotoxin in conscious rats (Holaday et al 1982b). Moreover, naloxone reversed the additive hypotension. Morphine injections worsen the hypotension produced by *E. coli* injections in pigs (Gahhos et al 1982).

7.3. Nonopiate effects

Some concern has been raised, because the doses of naloxone used in shock are higher than those customarily required to block opiate receptors, that the effects of naloxone are not opiate specific. Indeed, naloxone has a number of pharmacological effects which, nonetheless, may be important in shock. Naloxone is antiproteolytic and stabilizes lysosomal membranes in hemorrhagic shock in cats, with resultant decreases in plasma amino nitrogen, cathepsin D, and myocardial depressant factor concentrations (Curtis and Lefer 1980). On the other hand, these particular actions may not be crucial in shock since the opiate antagonist Win 44,441-3, which neither stabilizes lysosomal membranes nor inhibits proteolysis, improves blood pressure similarly (to naloxone) in this same model (Curtis and Lefer 1982a). Naloxone also demonstrates lipid antioxidant properties in vitro and thus may be a general protector of membrane function which may be important in shock (Koreh et al 1981).

8. IMPORTANT INTERACTIONS

In all of this work there are a number of important variables to consider. Because of the ubiquity of opiate receptors and endogenous opioid peptides and in view of the panoply of opioid effects, it is not surprising that opiates interact with a number of systems. Some of these interactions may be quite important in the multifaceted picture we call shock.

8.1.　pH and temperature

Blood pH is important in shock states. On the one hand, naloxone prevents the development of systemic acidosis in dogs subjected to endotoxemia (Raymond et al 1981) or live *E. coli* sepsis (Rees et al 1982). On the other hand, acidosis severely attenuates the effects of opiate agonists and antagonists (Kaufman et al 1975). Indeed, we have shown that the effect of naloxone in monkeys subjected to hemorrhagic shock is pH-sensitive, the effect being less and less marked as the acidosis is more and more severe (Gurll et al 1982d). Blood pH must be restored to normal for naloxone to work.

Temperature is another important consideration. Reduced ambient temperature blunts the response to naloxone in endotoxin shock (Janssen et al 1982). Perhaps this operates by means of body temperature since a cool core temperature also interferes with the naloxone response in hemorrhagic shock in monkeys (Gurll et al 1982d). Endorphin systems are temperature-sensitive and activated by a rise in temperature (Deeter and Mueller 1981). Moreover, dogs with elevated rectal temperatures respond to hemorrhage with greater decreases in blood pressure and greater increases in ACTH (and presumably β-endorphin), renin, and vasopressin than normothermic dogs (Wood et al 1981).

8.2.　Adrenal corticosteroids

We have suggested that corticosteroids might have their beneficial effects chiefly by negative feedback inhibition of β-endorphin release. Indeed, high doses of dexamethasone or methylprednisolone attenuated the beneficial effects of naloxone in hemorrhagic shock in dogs (Vargish et al 1983). Since there would be less β-endorphin to block, there should be less reason for naloxone to be effective. In contrast, steroids have been shown to have additive effects to naloxone in pigs with septic shock caused by *E. coli* (Gahhos et al 1982). Perhaps these differences are due to differences in doses and animal models. Hydrocortisone at a dose of 50 mg intravenously restores the effects of naloxone in canine hemorrhage lost after adrenalectomy (Patton et al 1983). Perhaps low doses of corticosteroid potentiate, while higher doses inhibit, the naloxone response.

8.3. Catecholamines

Endorphin and catecholamine systems interact. Opiate receptors and peptides have been found in the locus ceruleus which is a major locus of noradrenergic content in the brain. Similarly, catecholamines and opiates and their receptors share some common loci like the intermediolateral nucleus of the spinal cord and the adrenal medulla. Therefore, interactions between opiates and catecholamines may be at a variety of locations from the central nervous system to the periphery.

Evidence for central interaction derives from work using the α_2-adrenergic agonist clonidine. As stated above, morphine and endogenous opiates can produce hypotension and bradycardia resulting partly from reduced sympathetic tone after activation of opiate receptors in the medulla oblongata. Clonidine activates α_2-adrenergic receptors in the same area and, like morphine, decreases pulse rate and blood pressure, produces analgesia, and induces symptoms (after withdrawal) which resemble those of opiate withdrawal. Drugs which interact to inhibit the binding of clonidine to its receptors elicit symptoms which look like opiate withdrawal, and clonidine reverses the symptoms of opiate withdrawal (Gold et al 1978). Furthermore, naloxone reverses the antihypertensive effects of clonidine (Farsang and Kunos 1979). These similarities and effects suggest an interaction of opiates and central α-adrenergic systems.

Interactions can also occur in the spinal cord and adrenal medulla, and enkephalins are thought to function here as neuromodulators. Opiate agonists, probably acting on opiate receptors on medullary cells, inhibit the nicotine-stimulated release of catecholamines (Yang et al 1980). Opioid peptides can interact with and modulate effects of circulating vasoactive humoral substances (Altura et al 1980). Circulating enkephalins may modulate the effects of catecholamines on the heart as well. Opioid peptides had no effect on rat atria to a concentration of 10^{-5} M, but 10^{-7} M enkephalin antagonized the chronotropic effects of norepinephrine (Eiden and Ruth 1982).

8.4. Interactions with other mediators

There are a variety of interactions that may be important and will be difficult to sort out because they involve activation of other mediators of shock responses. Morphine releases histamine. Histamine and catecholamines can release each other. β-Endorphin and arginine vasopressin stimulate each other's release. There are important interactions between opiates and adrenal corticosteroids. Resolution of whether observed effects are due to opiates or a secondarily released mediator will require careful pharmacological and physiological investigation.

9. FUTURE

9.1. Use in man

With all the available studies, it was only a matter of time before the data were applied to humans. Several letters to the editor have demonstrated the effects of naloxone in increasing cardiovascular function in human septic, hemorrhagic, and cardiogenic shock which have been summarized (Gurll 1983). In one nonrandomized trial, 13 patients were given naloxone when they went into shock (systolic blood pressure less than 90 mmHg for at least 45 minutes with urine output less than 15 ml/hr and/or impaired mental status). Sepsis was the major factor in 11 of the 13. Eight of 9 patients not receiving steroids responded to naloxone (0.4 mg intravenous bolus) by an increase in systolic blood pressure of 30 ± 5 mmHg (Peters et al 1981). Three of the 9 patients survived including the nonresponder. Four patients were 'hypoadrenocorticotropic' (3 on steroids, 1 after pituitary irradiation) and did not respond to naloxone although one survived. Five of 10 patients with sepsis responded to naloxone 0.3 mg/kg intravenously with an increase in MAP of 30 mmHg but no increase in heart rate, cardiac index, central venous pressure or pulmonary capillary wedge pressure (Groeger et al 1982). In this study, in contrast, steroids made no difference: 4 patients were receiving steroids, 2 each in the responder and the nonresponder groups. Time did make a difference: responders had been in shock for 8 hours or less, nonresponders more than 24 hours. The importance of naloxone treatment early in shock corroborates findings in animal models.

We regard these early adventures in use of naloxone in human shock ill-advised. A well-designed clinical trial needs to be done. Animal studies have, however, identified important factors that must be incorporated in the design: time, pH, temperature, steroids, to name just a few.

9.2. Primates

Before starting human studies, we investigated the effects of naloxone in anesthetized monkeys subjected to hemorrhagic or endotoxemic shock. In hemorrhage, naloxone at 2 mg/kg plus 2 mg/kg/hr increased MAP and LVdP/dt_{max} and also improved survival (Gurll et al 1982e). In endotoxemia, naloxone at the same dose improved cardiovascular hemodynamics and survival (Gurll et al 1981b). We also showed in both models a fivefold elevation in β-endorphin and β-LPH. Acidosis and cold were shown to have adverse effects on the naloxone response. Furthermore, β-endorphin release appears to be related to acid–base balance and temperature, being more pronounced when pH and temperature were higher. Primate studies are useful and should be pursued because of their relevance to human physiology.

9.3. Mechanisms

Much work needs to be done to look at the central and peripheral mechanisms for the action of β-endorphin and the enkephalins. Particularly promising has been the hypothalamo-pituitary-adrenal axis. Central nervous system injections, into specific nuclei, may better define central opiate-sensitive cardiovascular depression in shock and its reversal by naloxone. Pharmacological dissection, using agonists and antagonists, will be required to sort out autonomic effects, but this task has been aided by the development of specific receptor ligands.

9.4. Multiple opiate receptors

Multiple opiate receptors have been described, classically designed morphine-like (μ) and nalorphine-like (ϰ and σ), which mediate specific and separable effects (Martin et al 1976). To this were added the δ (Lord et al 1977) and ε (Schulz et al 1979) receptors, sites for binding of specific ligands (enkephalins and β-endorphin respectively). These receptors may be distributed differently and probably subserve different functions (Chang and Cuatrecasas 1981).

The development of opiate antagonists which specifically bind to one receptor and not the others will help in sorting out which effect is due to which receptor. To some extent this has been achieved with the mixed opiate-receptor agonists/antagonists like nalbuphine and nalorphine which antagonize μ receptors and are agonists at the ϰ and σ receptors. The fact that these agents work in hemorrhagic shock (vide supra) suggest that μ receptors are the ones involved in the cardiovascular depression of shock (at least hemorrhagic), although no statement can be made about δ and ε receptors.

Several experimental agents with various receptor affinities have been used in hemorrhagic shock; only those which were δ and μ antagonists were effective in increasing blood pressure in hemorrhage in cats; specifically excluded from effectiveness were agents that were ϰ antagonists (Curtis and Lefer 1983). These authors conclude that agents which are agonists at ϰ receptors and antagonists at δ and μ receptors might be used in the therapy of shock. These agents have the receptor-binding characteristics of the mixed agonist/antagonist analgesic agents which have been found to be useful in shock (vide supra).

The chief disadvantage of naloxone use is its potential for interfering with pain relief. This may be overcome using TRH, nalbuphine, or nalorphine. All along we have argued that naloxone is merely a tool to probe the involvement of endorphins in shock and should not be indiscriminantly applied to the human condition (Gurll 1983). Naloxone is like a broad-edged sword, and pharmacological dissection with it can lead to confusion results. Fortunately more selective, more rapier- and epee-like, agents are being developed which will, we hope, allow for a more precise elucidation of the role of endorphins and opiate receptors in shock. In the meantime, cautious extrapolation of basic research data to the clinical treatment of patients should be the watchword.

The usefulness of the selective approach can be seen in what follows. Injection of δ agonists into the third ventricle of anesthetized rats produced hypotension similar to that seen in endotoxemia (Holaday and Faden 1982). Therefore, the δ antagonist peptide M 154,129 was injected at 30–60 mg/kg intravenously in endotoxemic rats with reversal of hypotension without altering analgesic effects of simultaneously administered morphine (Holaday et al 1982c); the μ_1 antagonist β-FNA (fumarate derivative of naltrexone) had no effect on the blood pressure response to endotoxin in conscious rats (Holaday 1983). They implicate the involvement of the δ receptor in the cardiovascular pathophysiology of endotoxemic shock. Resolution of this conflict (μ or δ receptor as site of action of endorphins in shock) will await the development and use of specific opiate-receptor agonists and antagonists.

10. SUMMARY

Opioid peptides and receptors are widely distributed in sites important for cardiovascular function and homeostasis. Endogenous and exogenous opioid peptides have potent cardiovascular effects. Endorphins elevated in shock and acting on opiate receptors have been implicated in the pathophysiology of various shock states, including endotoxemia, because opiate receptor antagonists improved hemodynamics, metabolism, and survival. The development of specific opiate-receptor agonists and antagonists may help to explore the mechanisms and sites involved and may lead to the development of a clinically useful treatment.

ACKNOWLEDGMENTS

The author would like to thank Mrs. Pat Piper for her accurate and timely preparation of the manuscript, Dr. John Holaday for many thought-provoking and exciting discussions about endorphins, and Dr. David G. Reynolds for his encouragement, advice, and scholarly example.

REFERENCES

Almqvist P, Kuenzig M, Schwartz SI (1981) Effect of naloxone on endotoxin-induced pulmonary platelet sequestration. *Surg. Forum 32*, 304-306.

Altura BM, Altura BT, Carella A, Turlapaty PDMV, Weinberg J (1980) Vascular smooth muscle and general anesthetics. *Fed. Proc. 39*, 1584-1591.

Arendt JO, Freye E (1979) Opiate antagonist reverses the cardiovascular effects of inhalation anesthesia. *Nature (London) 277*, 399-400.

Atweh SF, Kuhar MJ (1977) Autoradiographic localization of opiate receptors in rat brain. I. Spinal cord and lower medulla. *Brain Res. 124*, 53-67.

Beecher MK (1946) Pain in wounded men in battle. *Ann. Surg. 123*, 96-105.

Belenky GL, Holaday JW (1979) The opiate antagonist naloxone modifies the effects of electroconvulsive shock (ECS) on respiration, blood pressure, and heart rate. *Brain Res. 177*, 414-417.

Bergland R, Blume H, Hamilton A, Monica P, Paterson R (1980) Adrenocortico-tropic hormone may be transported directly from the pituitary to the brain. *Science 210*, 541-543.

Besse JC, Bass AD (1966) Potentiation by hydrocortisone of responses to catecholamines in vascular smooth muscle. *J. Pharmacol. Exp. Ther. 154*, 224-238.

Bloom F, Battenberg E, Rossier J, Ling N, Leppaluoto J, Vargo TM, Guillemin R (1977) Endorphins are located in the intermediate and anterior lobes of the pituitary gland, not in the neurohypophysis. *Life Sci. 20*, 43-48.

Bloom F, Battenberg E, Rossier J, Ling N, Guillemin R (1978) Neurons containing β-endorphin in rat brain exist separately from those containing enkephalin: immunocytochemical studies. *Proc. Natl Acad. Sci. USA 75*, 1591-1595.

Bolme P, Fuxe K, Agnati LF, Bradley R, Smythies J (1978) Cardiovascular effects of morphine and opioid peptides following intracisternal administration in chloralose anesthetized rats. *Eur. J. Pharmacol. 59*, 287-291.

Brückner JB, Faber du Faur Jv, Danner R (1981) Naloxone in the treatment of endotoxin shock. *Br. J. Anaesth. 53*, 111P-112P.

Burnie J (1981) Naloxone in shock. *Lancet 1*, 942.

Cannon WB (1918) A consideration of the nature of wound shock. *J. Am. Med. Assoc. 70*, 611-621.

Carr DB, Bergland R, Hamilton A, Blume H, Kasting N, Arnold M, Martin JB, Rosenblatt M (1982) Endotoxin-stimulated opioid peptide secretion: two secretory pools and feedback control in vivo. *Science 217*, 845-848.

Chance E, Todd MH, Waterfall JF (1981) A comparison of the cardiovascular effects of meptazinol, morphine and naloxone in haemorrhagic shock in rats. *Br. J. Pharmacol. 74*, 930P.

Chang K-J, Cuatrecasas P (1981) Heterogeneity and properties of opiate receptors. *Fed. Proc. 40*, 2729-2734.

Cox BM, Opheim KE, Teschemacher H, Goldstein A (1975) A peptide-like substance from pituitary that acts like morphine. 2. Purification. *Life Sci. 16*, 1777-1782.

Curtis MT, Lefer AM (1980) Protective actions of naloxone in hemorrhagic shock. *Am. J. Physiol. 239* H416-H421.

Curtis MT, Lefer AM (1982a) Beneficial action of a new opiate antagonist (Win 44,441-3) in hemorrhagic shock. *Eur. J. Pharmacol. 78*, 307-313.

Curtis MT, Lefer AM (1982b) Tissue uptake of ^{3}H-naloxone in control and in hemorrhagic shock cats. *Arch. Int. Pharmacodyn. Ther. 255*, 48-58.

Curtis MT, Lefer AM (1983) Actions of opiate antagonists with selective receptor interactions in hemorrhagic shock. *Circ. Shock 10*, 131-145.

Dashwood MR, Feldberg W (1979) Central inhibitory effect of released opiate peptides on adrenal medulla, revealed by naloxone in the cat. *J. Physiol. 290*, 22P-23P.

Davis SD, McDonald WJ, Kendall JW, Potter DM (1983) Endotoxin hypotension: prevented by naloxone in intact but not in hypophysectomized rats. *Clin. Res. 31*, 23A.

Deeter WT, Mueller GP (1981) Differential effects of warm- and cold-ambient temperature on blood levels of β-endorphin and prolactin in the rat. *Proc. Soc. Exp. Biol. Med. 168*, 369-372.

DiGiulio AM, Yang H-YT, Lutold B, Fratta W, Hong JS, Costa E (1978) Characterization of enkephalin-like material extracted from sympathetic ganglia. *Neuropharmacology 17*, 989-992.

Drew JH, Dripps RD, Comroe J (1946) Clinical studies on morphine. II. The effect of morphine upon the circulation of man and upon the circulatory and respiratory responses to tilting. *Anesthesiology 7*, 44-61.

Eckenhoff JE, Oech SR (1960) The effects of narcotics and antagonists upon respiration and circulation in man: a review. *Clin. Pharmacol. Ther. 1*, 483-524.

Egdahl RH (1959) The differential response of the adrenal cortex and medulla to bacterial endotoxin. *J. Clin. Invest. 38*, 1120-1125.

Eiden LE, Ruth JA (1982) Enkephalins modulate responsiveness of rat atria in vitro to norepinephrine. *Peptides 3*, 475-478.

Elde R, Hökfelt T, Johansson O, Terenius L (1976) Immunohistochemical studies using antibodies to leucine-enkephalin: initial observations on the nervous system of the rat. *Neuroscience 1*, 349-351.

Evans AGJ, Nasmyth PA, Stewart HC (1952) The fall of blood pressure caused by intravenous morphine in the rat and the cat. *Br. J. Pharmacol. 7*, 542-552.

Faden AI, Holaday JW (1979) Opiate antagonists: a role in the treatment of hypovolemic shock. *Science 205*, 317-318.

Faden AI, Holaday JW (1980) Naloxone treatment of endotoxic shock: stereospecificity of physiologic and pharmacologic effects in the rat. *J. Pharmacol. Exp. Ther. 212*, 441-447.

Faden AI, Jacobs TP, Holaday JW (1980) Endorphin-parasympathetic interaction in spinal shock. *J. Auton. Nerv. Syst. 2*, 295-304.

Farsang C, Kunos G (1979) Naloxone reverses the antihypertensive effect of clonidine. *Br. J. Pharmacol. 67*, 161-164.

Feldberg W, Wei E (1977) The central origin and mechanism of cardiovascular effects of morphine as revealed by naloxone in cats. *J. Physiol. 272*, 99P-100P.

Feldberg W, Wei E (1978a) Central sites at which morphine acts when producing cardiovascular effects. *J. Physiol. 272*, 57P.

Feldberg W, Wei E (1978b) Central cardiovascular effects of enkephalins and C-fragment of lipotropin. *J. Physiol. 280*, 18P.

Fennessy MR, Rattray JF (1971) Cardiovascular effects of intravenous morphine in the anesthetized rat. *Eur. J. Pharmacol. 14*, 1-8.

Flórez J, Mediavilla A (1977) Respiratory and cardiovascular effects of met-enkephalin applied to the ventral surface of the brain stem. *Brain Res. 138*, 585-590.

Freye E, Arndt JO (1979) Perfusion of the fourth cerebral ventricle with fentanyl induces naloxone-reversible bradycardia, hypotension, and EEG synchronisation in conscious dogs. *Naunyn-Schmiedeberg's Arch. Pharmacol. 307*, 123-128.

Fuxe K, Bolme P, Jonsson G, Agnati LF, Goldstein M, Hökfelt T, Schwarcz R, Engel J (1979) On the cardiovascular role of noradrenaline, adrenaline and peptide containing neuron systems in the brain. In: Meyer P, Schmitt H (Eds), *Nervous System and Hypertension. Perspectives in Nephrology and Hypertension*, pp 1-17. Wiley-Flammarion, New York.

Gahhos FN, Chiu RCJ, Hinchey EJ, Richards GK (1982) Endorphins in septic shock: hemodynamic and endocrine effects of an opiate receptor antagonist and agonist. *Arch. Surg. 117*, 1053-1057.

Galosy RA, Clarke LK, Vasko MR, Crawford IL (1981) Neurophysiology and neuropharmacology of cardiovascular regulation and stress. *Neurosci. Biobehav. Rev. 5*, 137-175.

Gilbert RP (1960) Mechanisms of the hemodynamic effects of endotoxin. *Physiol. Rev. 40*, 245-279.

Gokhale SD, Gulati OD, Joshi NY (1963) Effect of some blocking drugs on the pressor response to physostigmine in the rat. *Br. J. Pharmacol. 21*, 273-284.

Gold MS, Redmond DE, Kleber HD (1978) Clonidine blocks acute opiate-withdrawal symptoms. *Lancet 2*, 599-602.

Goldstein A, Lowney LL, Pal BK (1971) Stereospecific and nonspecific interactions of the morphine cogener levorphanol in subcellular fractions of mouse brain. *Proc. Natl Acad. Sci. USA 68*, 1742-1747.

Groeger JS, Carolon GC, Howland WS (1982) High-dose naloxone in septic shock. *Circ. Shock 9*, 169.

Gubler U, Seeburg P, Hoffman BJ, Gage LP, Udenfriend S (1982) Molecular cloning establishes proenkephalin as precursor of enkephalin-containing peptides. *Nature (London) 295*, 206-208.

Guillemin R, Vargo T, Rossier J, Minick S, Ling N, Rivier C, Vale W, Bloom F (1977) β-Endorphin and adrenocorticotropin are secreted concomitantly by the pituitary gland. *Science 197*, 1367-1369.

Gurll NJ (1983) Naloxone in endotoxic shock: experimental models and clinical perspective. *Adv. Shock Res. 10*, 63-71.

Gurll NJ, Vargish T, Reynolds DG, Lechner RB (1981a) Opiate receptors and endorphins in the pathophysiology of hemorrhagic shock. *Surgery 89*, 364-369.

Gurll NJ, Reynolds DG, Vargish T, Lutz SA, Ganes E (1981b) Primate endotoxemic shock reversed by opiate receptor blockade with naloxone. *Physiologist 24*, 118.

Gurll NJ, Reynolds DG, Vargish T, Lechner RB (1982a) Naloxone without transfusion prolongs survival and enhances cardiovascular function in hypovolemic shock. *J. Pharmacol. Exp. Ther. 220*, 621-624.

Gurll NJ, Reynolds DG, Vargish T, Lechner RB (1982b) Naltrexone improves survival rate and cardiovascular function in canine hemorrhagic shock. *J. Pharmacol. Exp. Ther. 202*, 625-628.

Gurll NJ, Reynolds DG, Holaday J, Ganes E (1982c) Improved cardiovascular function and survival using thyrotropin-releasing hormone (TRH) in primate hemorrhagic shock. *Physiologist 25*, 342.

Gurll NJ, Reynolds DG, Vargish T, Ganes E (1982d) Body temperature and acid-base balance determine cardiovascular responses to naloxone in primate hemorrhagic shock. *Fed. Proc. 41*, 1135.

Gurll NJ, Reynolds DG, Vargish T, Ganes E (1982e) Primate hemorrhagic shock responds to naloxone. *Circ. Shock 9*, 189.

Gurll NJ, Reynolds DG, Vargish T, Ganes E, Holaday J (1983) Endorphins mediate cardiovascular (CV) depression in primate shock which is reversible with naloxone (NAL) or thyrotropin releasing hormone (TRH). *Proc. Int. Union Physiol. Sci. 15*, 209.

Hanbauer I, Govoni S, Majane EA, Yang H-YT, Costa E (1982) In vivo regulation of the release of met-enkephalin-like peptides from dog adrenal medulla. *Adv. Biochem. Psychopharmacol. 33*, 209-215.

Hess ML, Smith JM, Eaton LR, Kleinman W, Okabe E (1981) Chronic opiate receptor occupation and increased lethality in endotoxemia. *Circ. Shock 8*, 313-322.

Hill L (1895) The influences of the force of gravity on the circulation of the blood. *J. Physiol. 18*, 15-23.

Hock CE, Curtis MT, Jaffe JS, Lefer AM (1983) Beneficial actions of nalorphine during hemorrhagic shock in cats. *Proc. Soc. Exp. Biol. Med. 173*, 76-81.

Holaday JW (1983) Cardiovascular effects of endogenous opiate systems. *Ann. Rev. Pharmacol. Toxicol. 23*, 541-594.

Holaday JW, Faden AI (1978) Naloxone reversal of endotoxin hypotension suggests role of endorphins in shock. *Nature (London) 275*, 450-451.

Holaday JW, Faden AI (1980) Naloxone acts as central opiate receptors to reverse hypotension, hypothermia, and hypoventilation in spinal shock. *Brain Res. 189*, 295-299.

Holaday JW, Faden AI (1982) Selective cardiorespiratory differences between third and fourth ventricular injections of 'mu' and 'delta' opiate agonists. *Fed. Proc. 41*, 1468.

Holaday JW, Loh HH (1981) Neurobiology of β-endorphin and related peptides. In: Li CH (Ed), *Hormonal Proteins and Peptides, Vol 10*, pp 203-291. Academic Press, New York.

Holaday JW, Tseng L-F, Loh H, Li CH (1978) Thyrotropin-releasing hormone antagonizes beta-endorphin hypothermia and catalepsy. *Life Sci. 22*, 1537-1544.

Holaday JW, D'Amato RJ, Faden AI (1981a) Thyrotropin-releasing hormone improves cardiovascular function in experimental endotoxic and hemorrhagic shock. *Science 213*, 216-218.

Holaday JW, Ruvio BA, Faden AI (1981b) Thyrotropin-releasing hormone improves blood pressure and survival in endotoxic shock. *Eur. J. Pharmacol. 74*, 101-105.

Holaday JW, O'Hara M, Faden AI (1981c) Hypophysectomy alters cardiorespiratory variables: central effects of pituitary endorphins in shock. *Am. J. Physiol. 241*, H479-H485.

Holaday JW, D'Amato RJ, Ruvio BA, Faden AI (1982a) Action of naloxone and TRH on the autonomic regulation of circulation. *Adv. Biochem. Psychopharm. 33*, 353-361.

Holaday JW, Ruvio BA, Sickel J (1982b) Morphine exacerbates the cardiovascular pathophysiology of endotoxic shock in rats. *Circ. Shock 9*, 169.

Holaday JW, Ruvio BA, Robles LE, Johnson CE, D'Amato RJ (1982c) M 154,129, a putative delta antagonist, reverses endotoxic shock without altering morphine analgesia. *Life Sci. 31*, 2209-2212.

Hughes J (1975) Isolation of an endogenous compound from the brain with pharmacological properties similar to morphine. *Brain Res. 88*, 295-308.

Hughes J, Smith TW, Kosterlitz HW, Fothergill LA, Morgan BA, Morris HR (1975) Identification of two related pentapeptides from the brain with potent opiate agonist activity. *Nature (London) 258*, 577-579.

Hughes J, Kosterlitz HW, Smith TW (1977) The distribution of methionine-enkephalin and leucine-enkephalin in the brain and peripheral tissues. *Br. J. Pharmacol. 61*, 639-647.

Hunt LB, Reynolds DG, Gurll NJ, Ganes EM (1983) The efficacy of an opiate antagonist/agonist drug in canine hemorrhagic shock. *Circ. Shock 10*, 235.

Janssen HF (1983) Studies on a central site of action for naloxone in endotoxin shock. In: Reichard SM, Reynolds DG, Adams HR (Eds), *Advances in Shock Research, Vol 10*, pp 73-81. Alan R. Liss, New York.

Janssen HF, Lutherer LO (1980) Ventriculocisternal administration of naloxone protects against several hypotension during endotoxin shock. *Brain Res. 194*, 608-612.

Janssen HF, Pugh JL, Lutherer LO (1982) Reduced ambient temperature blocks the ability of naloxone to prevent endotoxin-induced hypotension. *Adv. Shock Res. 7*, 117-124.

Kakidani H, Furutani Y, Takahashi H, Noda M, Morimoto Y, Hirose T, Asai M, Inayama S, Nakanishi S, Numa S (1982) Cloning and sequence analysis of cDNA for porcine β-neo-endorphin/dynorphin precursor. *Nature (London) 298*, 245-249.

Kalsner S (1969) Mechanisms of hydrocortisone potentiation of responses to epinephrine and norepinephrine in rabbit aorta. *Circ. Res. 24*, 383-395.

Kaufman JJ, Semo NM, Koski WS (1975) Microelectrometric titration measurement of the pK_a's and partition and drug distribution coefficients of narcotics and narcotic antagonists and their pH and temperature dependence. *J. Med. Chem. 18*, 647-655.

Korner PI (1971) Integrative neural cardiovascular control. *Physiol. Rev. 51*, 312-367.

Koreh K, Seligman ML, Flamm ES, Demopoulos HB (1981) Lipid antioxidant properties of naloxone in vitro. *Biochem. Biophys. Res. Commun. 102*, 1317-1322.

Koyama S, Santiesban HL, Ammons WS, Manning JW (1983) The effects of naloxone on the peripheral sympathetics in cat endotoxin shock. *Circ. Shock 10*, 7-13.

Krieger DT, Liotta A, Brownstein MJ (1977) Presence of corticotropin in brain of normal and hypophysectomized rats. *Proc. Natl Acad. Sci. USA 74*, 648-652.

Kuhar MJ, Pert CB, Snyder SH (1973) Regional distribution of opiate receptor binding in monkey and human brain. *Nature (London) 245*, 447-450.

Kumakura K, Karoum F, Guidotti A, Costa E (1980) Modulation of nicotinic receptors by opiate receptor agonists in cultured adrenal chromaffin cells. *Nature (London) 283*, 489-492.

Lang RE, Bruckner UB, Hermann K, Kempf B, Rascher W, Sturm V, Unger T, Ganten D (1982) Effect of hemorrhagic shock on the concomitant release of endorphin and enkephalin-like peptides from the pituitary and adrenal gland in the dog. In: Costa E, Trabucci M (Eds), *Advances in Biochemical Psychopharmacology*, pp 363-369. Raven Press, New York.

Laubie M, Schmitt H, Canellas J, Roquebert J, Demichel P (1974) Centrally mediated bradycardia and hypotension induced by narcotic analgesics: dextromoramide and fentanyl. *Eur. J. Pharmacol. 28*, 66-75.

Laubie M, Schmitt H, Vincent M, Redmond G (1977) Central cardiovascular effects of morphinomimetic peptides in dogs. *Eur. J. Pharmacol. 46*, 67-71.

Lemaire I, Tseng R, Lemaire S (1978) Systemic administration of β-endorphin: potent hypotensive effect involving a serotonergic pathway. *Proc. Natl Acad. Sci. USA 75*, 6240-6242.

Li CH, Chung D (1976) Isolation and structure of an untriakontapeptide with opiate-like activity from camel pituitary glands. *Proc. Natl Acad. Sci. USA 73*, 1145-1148.

Li CH, Barnafi L, Chretien M, Chung D (1965) Isolation and amino acid sequence of β-LPH from sheep pituitary glands. *Nature 208*, 1093-1094.

Lind RE, Gurll NK, Jenkins JT, Reynolds DG (1980) Possible role of endorphins and opiate receptors in superior mesenteric artery occlusion shock. *Surg. Forum 31*, 48-51.

Lind RE, Reynolds DG, Ganes EM, Jenkins JT (1981) Morphine effects on cardiovascular performance. *Am Surg. 47*, 107-111.

Livingstone D (1872) *Adventures and discovering in the interior of Africa*, p 15. Hubbard Bros., Philadelphia.

Lord JAH, Waterfield AA, Hughes J, Kosterlitz HW (1977) Endogenous opioid peptides: multiple agonists and receptors. *Nature (London) 267*, 495-499.

Lundberg JM, Hökfelt T, Kewenter J, Pettersson G, Ahlman H, Edin R, Dahlström A, Nilsson G, Terenius L, Uvnäs-Wallenstein K, Said S (1979) Substance P-, VIP- and enkephalin-like immunoreactivity in the human vagus nerve. *Gastroenterology 77*, 468-471.

Mains RE, Eipper BA, Ling N (1977) Common precursor to corticotropins and endorphins. *Proc. Natl Acad. Sci. USA 74*, 3014-3018.

Manugian V, Koyama S, Santiesteban HL, Ammons WS, Manning JW (1981) Possible role of naloxone on sympathetic activity and blood pressure in cat. *Fed. Proc. 40*, 522.

Martin WR, Eades CG, Thompson JA, Huppler RE, Gilbert PE (1976) The effects of morphine- and nalorphine-like drugs in the nondependent and morphine-dependent chronic spinal dog. *J. Pharmacol. Exp. Ther. 197*, 515-532.

McIntosh FC, Paton WDM (1949) The liberation of histamine by certain organic bases. *J. Physiol. (London) 109*, 190-219.

Moberg GP (1971) Site of action of endotoxins on hypothalamic-pituitary-adrenal axis. *Am. J. Physiol. 220*, 397-400.

Montastruc J-L, Montastruc P, Morales-Olivas F (1981) Potentiation by naloxone of pressor reflexes. *Br. J. Pharmacol. 74*, 105-109.

Moon HD, Li CH, Jennings BM (1973) Immunohistochemical and histochemical studies of pituitary β-lipotrophs. *Anat. Rec. 175*, 529-538.

Moore RH, Dowling DA (1980) Effects of intravenously administered leu- or met-enkephalin on arterial blood pressure. *Reg. Pept. 1*, 77-87.

Nakanishi S, Inoue A, Taii S, Numa S (1977) Cell-free translation product containing corticotropin and β-endorphin encoded by messenger RNA from anterior lobe of bovine pituitary. *FEBS Lett. 84*, 105-109.

Owen MD, Gisolfi CV, Reynolds DG, Gurll NJ (1981) Effect of d-Ala^2-met^5-enkephalinamide (DAME) administered intracerebroventricularly on selected cardiovascular responses in the non-human primate. *Circ. Shock 8*, 189.

Paciorek PM, Todd MH (1982) Comparison of the cardiovascular effects of meptazinol and naloxone following endotoxic shock in anesthetized rats. *Br. J. Pharmacol. 75*, 128P.

Pasternak GW, Goodman R, Snyder SH (1975) An endogenous morphine-like factor in mammalian brain. *Life Sci. 16*, 1766-1769.

Patton ML, Gurll NJ, Reynolds DG, Vargish T (1983) Adrenalectomy abolishes and cortisol restores naloxone's beneficial effect on cardiovascular function and survival in canine hemorrhagic shock. *Circ. Shock 10*, 317-327.

Pelletier G, Leclerc R, Labrie F, Cote J, Chretien M, Les M (1977) Immunohistochemical localization of β-lipotropic hormone in the pituitary gland. *Endocrinology 100*, 770-776.

Pert CB, Snyder SH (1973) Opiate receptor: demonstration in nervous tissue. *Science 179*, 1011-1014.

Peters WP, Johnson MW, Friedman PA, Mitch WE (1981) Pressor effect of naloxone in septic shock. *Lancet 1*, 529-532.

Petty MA, Reid JL (1981) Opiate analogs, substance P, and baroreceptor reflexes in the rabbit. *Hypertension 3 (Suppl 1)*, 1142-1147.

Rapoport SI, Klee WA, Pettigrew KD, Ohno K (1980) Entry of opioid peptides into the central nervous system. *Science 207*, 84-86.

Raymond RM, Harkema JM, Stoffs WV, Emerson TE (1981) Effects of naloxone therapy on hemodynamics and metabolism following a superlethal dosage of Escherichia coli endotoxin in dogs. *Surg. Gynecol. Obstet. 152*, 159-162.

Rees M, Payne JG, Bowen JC (1982) Naloxone reverses tissue effects of live Escherichia coli sepsis. *Surgery 91*, 81-86.

Rees M, Bowen JC, Payne JG, MacPhee AA (1983) Plasma β-endorphin immunoreactivity in dogs during anesthesia, surgery, Escherichia coli sepsis, and naloxone therapy. *Surgery 93*, 386-390.

Reynolds DG, Gurll NJ, Vargish T, Lechner RB, Faden AI, Holaday JW (1980) Blockade of opiate receptors with naloxone improves survival and cardiac performance in canine endotoxic shock. *Circ. Shock 7*, 39-48.

Reynolds DG, Gurll NJ, Holaday J, Ganes E (1982) Thyrotropin-releasing hormone (TRH) in primate endotoxic shock. *Physiologist 25*, 309.

Rios L, Jacob J (1981) Comparisons des effets du chlorhydrate et de l'iodométhylate de naloxone sur le choc endotoxinique chez le rat anesthésie. *Arch. Inst. Pasteur Tunis 58*, 313-327.

Roberts JL, Herbert E (1977) Characterization of a common precursor to corticotropin and β-lipotropin: cell-free synthesis of the precursor and identification of corticotropin peptides in the molecule. *Proc. Natl Acad. Sci. USA 74*, 4826-4830.

Rutherford RB, Balis JV, Trow RS, Graves GM (1976) Comparison of hemodynamic and regional blood flow changes at equivalent stages of endotoxin and hemorrhagic shock. *J. Trauma 16*, 886-897.

Sabol SL (1978) Regulation of endorphin production by glucocorticosteroids in cultured pituitary tumor cells. *Biochem. Biophys. Res. Commun. 82*, 560-567.

Sander G, Giles T, Kastin A, Kaneish A, Coy D (1982) Leucine-enkephalin: reversal of intrinsic cardiovascular stimulation by pentobarbital. *Eur. J. Pharmacol. 78*, 467-470.

Sapru HN, Willette RN, Krieger AJ (1981) Stimulation of pulmonary J receptors by an enkephalin-analog. *J. Pharmacol. Exp. Ther. 217*, 228-234.

Sawynok J, Pinsky C, LaBella FS (1979) Minireview of the specificity of naloxone as an opiate antagonist. *Life Sci. 25*, 1621-1632.

Schaz K, Stock G, Simon W, Schlör KH, Unger T, Rockhold R, Ganten D (1980) Enkephalin effects on blood pressure, heart rate, and baroreceptor reflex. *Hypertension 2*, 395-407.

Schmidt CF, Livingston AE (1933) The action of morphine on the mammalian circulation. *J. Pharmacol. Exp. Ther. 47*, 411-441.

Schulz R, Faase E, Wüster M, Herz A (1979) Selective receptors for β-endorphin on the rat vas deferens. *Life Sci. 24*, 843-849.

Simantov R, Kuhar MJ, Pasternak GW, Snyder SH (1976a) The regional distribution of a morphine-like factor enkephalin in monkey brain. *Brain Res. 106*, 189-197.

Simantov R, Snowman AM, Snyder SH (1976b) A morphine like factor 'enkephalin' in rat brain: subcellular location. *Brain Res. 107*, 650-657.

Simantov R, Kuhar MJ, Uhl GR, Snyder SH (1977) Opioid peptide enkephalin: immunohistochemical mapping in rat central nervous system. *Proc. Natl Acad. Sci. USA 74*, 2167-2171.

Simantov R, Childers SR, Snyder SH (1978) [³H] opiate binding: anomalous properties in kidney and liver membranes. *Molec. Pharmacol. 14*, 69-76.

Simon EJ, Hiller JM, Edelman I (1973) Stereospecific binding of the narcotic analgesic ³H-etorphine to rat brain homogenate. *Proc. Natl Acad. Sci. USA 70*, 1947-1949.

Tache Y, Lis M, Collu R (1977) Effects of thyrotropin-releasing hormone on behavioral and hormonal changes induced by β-endorphin. *Life Sci. 21*, 841-846.

Terenius L (1973) Characteristics of the 'receptor' for narcotic analgesics in synaptic membrane fraction from rat brain. *Acta Pharmacol. Toxicol. 33*, 377-384.

Terenius L, Wahlström A (1975) Morphine-like ligand for opiate receptors in human CSF. *Life Sci. 16*, 1759-1764.

Teschemacher H, Opheim KE, Cox BM, Goldstein A (1975) A peptide-like substance from pituitary that acts like morphine. 1. Isolation. *Life Sci. 16*, 1771-1776.

Thijs LG, Balk E, Tuynman HARE, Koopman PAR, Bezemer PD, Mulder GH (1983) Effects of naloxone on hemodynamics, oxygen transport, and metabolic variables in canine endotoxin shock. *Circ. Shock 10*, 147-160.

Vargish T, Reynolds DG, Gurll NJ, Lechner RB, Holaday JW, Faden AI (1980) Naloxone reversal of hypovolemic shock in dogs. *Circ. Shock 7*, 31-38.

Vargish T, Gurll NJ, Reynolds DG, Lutz SA, Ganes EM (1983) Hemodynamic changes following corticosteroid and naloxone infusion in dogs subjected to hypovolemic shock without resuscitation. *Life Sci. 33*, 489-493.

Viveros OH, Diliberto EJ, Hazum E, Chang K-J (1979) Opiate-like materials in the adrenal medulla: evidence for storage and secretion with catecholamines. *Molec. Pharmacol. 16*, 1101-1108.

Watson SJ, Barchas JD, Li CH (1977) β-lipotropin: localization of cells and axons in rat brain by immunocytochemistry. *Proc. Natl Acad. Sci. USA 74*, 5155-5158.

Watson SJ, Akil H, Richard CW, Barchas JD (1978) Evidence for two separate opiate peptide neuronal systems and the coexistence of β-lipotropin, β-endorphin and ACTH immunoreactivities in the same hypothalamic neurons. *Nature (London) 275*, 226-228.

Wei ET, Lee A, Chang JK (1980) Cardiovascular effects of peptides related to the enkephalins and β-casomorphin. *Life Sci. 26*, 1517-1522.

Weissglas IS, Chiu RCJ, Shrier A, Hinchey EJ (1981) Endogenous opiates in shock: the direct effect of beta-endorphin on the myocardium. *Surg. Forum 32*, 236-239.

Weissglas IS, Hinchey EJ, Chiu RCJ (1982) Naloxone and methylprednisolone in the treatment of experimental septic shock. *J. Surg. Res. 33*, 131-135.

Wood CE, Shinsako J, Keil LC, Ramsay DJ, Dallman MF (1981) Hormonal and hemodynamic responses to 15 ml/kg hemorrhage in conscious dogs: responses correlate to body temperature. *Proc. Soc. Exp. Biol. Med. 167*, 15-19.

Wright DJM, Weller MPI (1980) Inhibition by naloxone of endotoxin-induced reactions in mice. *Br. J. Pharmacol. 70*, 99P-100P.

Wurtman RJ, Pohorecky LA, Baliga BS (1972) Adrenocorticol control of the biosynthesis of epinephrine and proteins in the adrenal medulla. *Pharmacol. Rev. 24*, 411-426.

Yang H-YT, Hexam T, Costa E (1980) Minireview: opioid peptides in adrenal gland. *Life Sci. 27*, 1119-1125.

Young WS, Wamsley JK, Zarbin MA, Kuhar MJ (1980) Opioid receptors undergo axonal flow. *Science 210*, 76-78.

Yukimura T, Stock G, Stumpf H, Unger T, Ganten D (1981) Effects of [D-Ala^2]-methionine-enkephalin on blood pressure, heart rate, and baroreceptor reflex sensitivity in conscious cats. *Hypertension 3*, 528-533.

Handbook of Endotoxin, Vol. 2: Pathophysiology of Endotoxin
L.B. Hinshaw, editor
© Elsevier Science Publishers B.V., 1985
ISBN 0 444 90385 2
$0.85 per article per page (transactional system)
$0.20 per article per page (licensing system)

CHAPTER 13

Endotoxin effects on germfree animals

EDWARD BALISH

1. INTRODUCTION

Many characteristics of germfree animals clearly demonstrate the impact that intestinal microorganisms have on their host. Germfree animals manifest better growth rates (Gordon and Pesti 1971), longer life spans (Gordon et al 1966; Gordon and Pesti 1971; Pollard 1971), lower oxygen consumption (Levenson 1978), underdeveloped lymphoid tissues (Gordon and Wostmann 1960; Kim et al 1966; Kim and Watson 1970; Miyakawa 1959; Pollard 1967; Thorbecke et al 1957; Thorbecke and Benacerraf 1959), and a relatively inactive immune system (Balish et al 1972; Benner et al 1982; Kim 1979; Sell 1964; Wostmann et al 1970). When the germfree animals' immune system encounters antigens, the response is complete and often several times more vigorous than that of their conventional counterparts (Bosma et al 1967; Kim et al 1979; Ebersole and Molinari 1977; Kiyono et al 1980). Myelopoiesis appears to proceed slower (Chang and Pollard 1973; Metcalf and Foster 1967, 1969; Morley et al 1972; Quesenberry et al 1978) in germfree animals, and they have fewer circulating leukocytes than their conventional counterparts (Gordon and Pesti 1971; Joshi et al 1979). The digestive tract manifests the most obvious anomalies of germfree life because germfree rodents have enlarged ceca which can be 10 to 20% of their body weight (Gordon and Pesti 1971; Wostmann 1975). Compared to their conventional counterparts, germfree animals have a slower intestinal epithelial cell renewal rate, decreased villus height and absorption area, hypocellularity of the lamina propria (Crabbe et al 1965, 1968; Gordon and Bruckner-Kardos 1961; Gordon and Pesti 1971; Wostmann 1975), and there are increased amounts of mucin, water, and secreted enzymes present in their alimentary tract (Gordon and Pesti 1971; Loesch 1968; Wostmann 1975). Recent international symposia call attention to many other facets of germfree life (Fliedner et al 1979; Sasaki et al 1981).

1.1. Germfree traits

Relatively few studies have been carried out to elucidate how specific species of intestinal bacteria can restore 'conventional traits' to germfree animals. It

338

is known that some bacteria in pure culture can reduce cecal size (Czuprynski et al 1983; Gustafsson et al 1970; Skelly et al 1962); however, for the most part, several hundred species of bacteria seem to be required to restore germfree animals to a 'conventional state' (Freter and Abrams 1972; Koopman and Kennis 1980; Syed et al 1970).

1.2. Specific pathogen-free and flora-defined animals

It is worth calling attention to the fact that most rodents purchased from commercial sources for research purposes are designated as either specific pathogen-free (SPF) or flora-defined (FD). These SPF or FD animals are not germfree. It is very likely that the original animals used for starting the SPF or FD colony were derived by cesarean section, raised by germfree foster mothers, and at weaning age they were associated (colonized) with an 'innocuous cocktail' of intestinal microbes. Very likely most commercial FD or SPF mice are associated with 8–10 different, but mostly unspeciated, microbes known as the 'Schaedler flora' (Schaedler et al 1965). Commercial animals are usually maintained and bred in this FD or SPF state by being housed in flexible film isolators or 'clean' animal care facilities and fed sterile (or acidified) water and autoclaved diet. Flora-defined or SPF animals often do not have gram-negative enteric bacteria in their alimentary tract while they are maintained in the 'clean' or 'barrier' type facilities of the commercial suppliers. Commercial rodents acquire a more complex intestinal flora from the environment of the animal care facilities that they are brought into for experimentation. Very often the environmental microbes they acquire are resistant species of gram-negative bacteria that have survived antimicrobial agents and disinfectants used to clean the animal facilities. Thus, the source of their intestinal bacteria (i.e., from humans or animals) is unknown. The latter situation is not unlike the neonatal intensive care units where colonization of cesarean-delivered 'germfree' neonates with resistant microbes from the hospital environment is a frequent problem for clinicians (Blakey et al 1982; Stark and Lee 1982).

I mention this information on FD and SPF animals here because we shall see that the acquisition of gram-negative bacteria has profound effects on them. The innovative studies of Dubos and Schaedler, and their colleagues (Dubos and Schaedler 1960; Dubos et al 1969; Mushin and Dubos 1965, 1966; Schaedler and Dubos 1961, 1962), point out dramatically that colonization of SPF mice with *Escherichia coli* has a very marked influence on the hosts' growth rate, reproductive performance, susceptibility to endotoxin and capacity to resist infectious agents. Thus, the growth and persistence of gram-negative bacteria in the alimentary tract can affect the health and well-being of the host in many ways (Dubos and Schaedler 1960; Dubos et al 1963).

How many of these intestinal-flora 'effects' on the host are caused by endotoxin alone is not known at this time, and we also know very little about the symbiotic effects of different components of the microflora on the host. It is well known that gram-positive and gram-negative species can combine to produce a lethal or toxic effect that exceeds the effect produced by each species separately (Kim and Watson 1965; Schlievert 1982, 1983). Since gram-negative bacteria and their endotoxins (Berg 1983; Chedid and Parant 1971; Elin and Wolff 1976; Magliolo et al 1975; Walker et al 1975) can gain access to the systemic circulation from the intestinal tract, it is very likely that endotoxin plays a key role in many aspects of the morphology, physiology, biochemistry, hematopoiesis, immunology, and endocrinology of the host.

1.3. Endotoxin effects – natural toxicity or hypersensitivity

Interest in the effects of endotoxin on germfree animals eminated primarily from debates about whether endotoxin has a primary toxicity or whether the myriad of biological effects of endotoxin are always and only the consequence of immediate or delayed hypersensitivity (Chedid and Parant 1971). Since most mammals are exposed to gram-negative bacteria (aerobic and anaerobic) throughout their life, it is very likely that they are continuously exposed to endotoxin (Berg 1983; Chedid and Parant 1971). If reactivity to endotoxin is dependent on 'natural' past exposure to microbial antigens, then germfree animals should be less sensitive to the effects of endotoxins because they have not been in contact with viable intestinal bacteria of any type. We know that germfree animals are not entirely free from antigens or endotoxin. Endotoxins are notoriously heat-stable and are present in the autoclaved diets usually fed to germfree animals (Ebersole and Molinari 1977). Since the question of endotoxin toxicity is of prime importance to understanding the many biological effects that endotoxin is capable of evoking, one would think that research on the effects of endotoxin on germfree animals would have been pursued with greater vigor. As it turns out, no research has yet been carried out to study the effect of endotoxin on 'endotoxin-free' research animals, and relatively few studies have compared and contrasted the effects of endotoxin on germfree and conventional animals.

2. LETHALITY

2.1. Specific pathogen-free mice

Dubos and Schaedler (1960) and Schaedler and Dubos (1961, 1962) made the observation that the absence of coliform bacteria from mice was associated with enhanced resistance to the lethal effects of endotoxin from *Escherichia coli*. Their NCS mice (not germfree), which were free of coliforms, were resistant to the intravenous injection of 600 μg of *E. coli* endotoxin whereas the standard Swiss (SS) strain (from which the NCS strain was derived by cesarean section), which had coliforms in their intestinal tract, were killed by an intravenous injection of 100 μg of endotoxin. Dubos and Schaedler (1960) and Schaedler and Dubos (1961, 1962) went on to show that they could restore endotoxin sensitivity to NCS mice by colonizing their intestinal tracts with an *E. coli* strain that was isolated from the endotoxin-sensitive SS strain. The experiments of Dubos and Schaedler (1960) and Schaedler and Dubos (1961, 1962) were not carried out with germfree animals. Their endotoxin-resistant NCS mouse strain very likely had gram-negative bacteria in their alimentary tract, for example, species of *Bacteroides*, a slow lactose fermenter which acquired the characteristics of *E. coli* when it was used to colonize the alimentary tract of germfree mice (Schaedler et al 1965), and possibly a species of *Pseudomonas* (Jensen et al 1963).

2.2. Germfree animals

Several investigators have compared the relative lethality of endotoxin for germfree and conventional rodents and pigs. Landy et al (1961) could not show any differences in the lethal effects of endotoxin for germfree and conventional mice, and Abernathy and Landy (1967) observed no differences in the lethal effects of endotoxin for germfree and conventional guinea pigs. Conversely, Jensen et al (1963) and Sacquet and Charlier (1965) demonstrated that germfree mice were more resistant to the lethal effects of endotoxin than conventional mice. In a more recent study, Kiyono et al (1980) found that germfree BALB/c mice were more resistant to the lethal effects of phenol-extracted endotoxin from *E. coli* (LD$_{50}$ 918 μg, intravenously) than conventional BALB/c mice (LD$_{50}$ 188 μg, intravenously).

Kim and Watson (1966) studied the lethal effects of endotoxin for piglets and demonstrated that *Salmonella abortus equi* endotoxin, phenol-extracted, was more lethal for germfree colostrum-deprived piglets than colostrum-fed piglets. Kim and Watson (1966) stated that the enhanced susceptibility of 'immunologically virgin' piglets to the lethal effects of endotoxin was good evidence that endotoxin has intrinsic toxicity, and the toxicity of endotoxin is not completely dependent on antigen–antibody reactions.

Since endotoxins are present in the sterilized diets fed to germfree animals (Ebersole and Molinari 1977), it is very likely that germfree animals are continuously exposed to dietary lipopolysaccharide. Pleasants et al (1981) and Hashimoto et al (1978) have raised mice on a chemically-defined, antigen-free diet, but no data are available on the susceptibility of these 'antigen-free' mice to endotoxins.

2.3. Antibiotic decontaminated animals

When antibiotics are used to decontaminate their alimentary tracts, animals acquire germfree characteristics (e.g., distended cecum).

Several investigators have tested the susceptibility of antibiotic decontaminated rodents to endotoxin and they demonstrated that decreased populations of bacteria in the intestinal tract enhanced the animals' resistance to the lethal effects of endotoxin (Galley et al 1975; Walker and Sheil 1976).

The innate toxicity of endotoxin for germfree and conventional animals is far from clear. The study of Kim and Watson (1966) on germfree colostrum-deprived pigs shows an enhanced sensitivity of germfree piglets to the lethal effects of endotoxin. Jensen et al (1963), Sacquet and Charlier (1965), and especially the recent study of Kiyono et al (1980) demonstrate a remarkable resistance of germfree rodents to the lethal effects of endotoxin. The investigations cited above used different species and strains of animals and endotoxin preparations from different species of bacteria. Although it is obvious that endotoxin was lethal for the germfree animals studied – that is, none were absolutely resistant to the toxic effect – germfree mice and antibiotic decontaminated mice can display a remarkable resistance to the lethal effects of endotoxin. Thus, decreased exposure to viable gram-negative bacteria does appear to be associated with enhanced resistance to endotoxin. The fact that no one has yet assessed endotoxin lethality in germfree rodents fed an 'endotoxin-free' diet still leaves the question about the innate toxicity of endotoxin open.

One other point is worth making on the lethality of endotoxin for gnotobiotic or antibiotic decontaminated animals and this relates to endotoxin tolerance. Repeated parenteral exposure to endotoxin actually results in a state of 'tolerance' to endotoxin (Chedid and Parant 1971). One would assume that animals associated with gram-negative bacteria would experience constant or intermittent exposure to endotoxin and would perhaps manifest greater resistance to endotoxin than germfree animals which were exposed to minimal amounts of endotoxin. Clearly, this does not seem to be the case, since the conventional animals that are exposed to gram-negative bacteria manifest greater susceptibility to the lethal effects of endotoxin than germfree mice or mice monoassociated with a pure culture of *E. coli* (Kiyono et al 1980).

Lethality, however, is only one aspect of wide variety of biological effects that endotoxin has on animals. Other investigators have used germfree animals to study the following responses to endotoxin: hematopoiesis, leukopenia, polyclonal activation of B-cells, lysosomal enzymes of macrophages, tumoricidal activity of macrophages, body temperature, and resistance to infectious diseases.

When assessing the lethality of endotoxin, it is important to know the health status of the research animals being used. For example, Suter (1962, 1964) demonstrated that conventional or gnotobiotic mice that are infected with *Mycobacterium bovis* (BCG) are much more sensitive to the lethal effects of endotoxin than animals not infected with BCG. It has also been demonstrated recently that endotoxins can enhance the toxic effects of *Staphylococcus aureus* toxins (Schlievert 1982, 1983). Therefore, colonization with gram-negative bacteria or intercurrent infections can alter the response of research animals to endotoxin challenge.

3. SHOCK

Hemorrhagic and septic shock are of prime clinical importance (Nagler and Levenson 1971). Very few studies have been carried out to compare the response of germfree and conventional animals to hemorrhagic or endotoxic shock. Germfree animals are more resistant to burn shock (Markley et al 1969) and tourniquet shock (Markley et al 1967) than their conventional counterparts. Other investigators failed to show any differences in the susceptibility of germfree and conventional rodents to hemorrhagic shock (Heneghan 1968; McNulty and Linares 1960; Zweifach et al 1958). Actually, very little work has been carried out on endotoxin-induced shock in gnotobiotic animals. This is surprising since shock is such an important area of research, and knowledge of the vasoactive, pyrogenic, and metabolic effects that endotoxin evokes in germfree animals is basic to our understanding of endotoxin effects in conventional animals. I was also unable to find any printed information on the induction of local or generalized Shwartzman reactions in germfree animals.

4. METABOLISM AND BODY TEMPERATURE

Typically, hyperthermia develops within a short time after endotoxin is injected into the circulatory system. The fever generally persists for 6 to 12 hours, then it subsides (Cluff 1971). Abernathy and Landy (1967) demonstrated that germfree guinea pigs manifested a hyperthermic response to endotoxin (intraperitoneal injection of *Salmonella typhosa* endotoxin, 2 or 4 mg/kg); however, conventional guinea pigs showed a hypothermic response to 2, 4, or 6 mg/kg of endotoxin. Germfree guinea pigs did manifest a hypothermic response when they were injected with 6 mg/kg of *S. typhosa* endotoxin. Levenson (1978) reported that the body temperature of conventional rats was about 1 °C higher than that of germfree rats. Levenson (1978) also indicated that conventional rats have a metabolic rate that is 15–20% higher than that of their germfree counterparts.

Once again, the intestinal flora apparently has an effect on metabolism and body temperature. Since so few studies have been carried out to assess the effect of endotoxin on body temperature, metabolism, etc., of germfree animals, it is difficult at this time to draw conclusions about different effects of endotoxin on the metabolism and body temperature of germfree and conventional animals.

5. HEMATOPOIESIS

Endotoxin affects hematopoiesis in several ways (Walker 1983). Endotoxin increases substances in serum which stimulate the red cell line (erythropoietin), myelopoiesis (colony stimulating factor, CSF), and thrombopoiesis (thrombopoietin).

5.1. Granulopoiesis and thrombopoiesis

Hashimoto and Hashimoto (1968) compared bone marrow from germfree and conventional rats and observed less hematopoietic activity in germfree than in conventional rats. Stimulation of hematopoiesis takes place when germfree rats and mice are colonized with intestinal flora (Gordon and Pesti 1971). Boggs et al (1967) observed an increase in the rate of release of neutrophils from the marrow of germfree animals following the intraperitoneal administration of 10 µg of *Salmonella typhosa* endotoxin. Fliedner et al (1966) demonstrated that myelocyte maturation took longer (82 hours) in germfree animals than conventional animals (10 hours).

Baseline white blood cell counts are usually lower in germfree animals than in their conventional counterparts (Gordon and Pesti 1971; Joshi et al 1979; Yale and Balish 1976). Injection of endotoxin into conventional animals usually causes a granulocytopenia within the first few hours after injection and then a leukocytosis occurs (Cluff 1971). Abernathy and Landy (1967) demonstrated that conventional guinea pigs manifest a persistent leukopenia for 6 hours after a peritoneal injection with either 2, 4, or 6 mg/kg of *S. typhosa* endotoxin; however, germfree guinea pigs underwent a leukocytosis during the 6-hour period following injection with 2 or 4 mg/kg, and leukocytosis was evident even after some were injected with 6 mg/kg of endotoxin. Thus, germfree guinea pigs had a higher threshold of resistance to the development of endotoxin-induced leukopenia than their conventional counterparts. McLeod and Balish (1978b) demonstrated that the intraperitoneal injection of 4 mg/kg of *S. typhosa* endotoxin into germfree and conventional rats caused immediate (1–3 hours) decreases in the number of platelets, lymphocytes, and polymorphonuclear leukocytes in both groups of animals. The only difference observed between the responses of germfree and those of conventional rats to endotoxin was that germfree rats cleared the endotoxin from blood sooner than conventional rats, and the germfree rats corrected the endotoxin-induced thrombocytopenia sooner than the conventional rats (McLeod and Balish 1978b).

Tlaskalová-Hogenová et al (1980) observed that germfree piglets showed a more severe neutropenia in response to an injection of *E. coli* endotoxin (200 µg intravenously) than their conventional counterparts. The latter investigators also reported that germfree animals, in comparison to conventional controls, manifested a heightened leukocytosis by 24 hours after the injection of endotoxin.

5.2. Colony stimulating factor

Sera from conventional mice contains colony stimulating factor (CSF), which is probably a protein in nature and has a molecular weight of approximately 50 000 (Stanley et al 1968). CSF stimulates the proliferation of certain bone marrow cells in vitro to form granulocytes and mononuclear cells (Bradley and Metcalf 1966).

Joshi et al (1979) assessed changes in serum colony forming units (CFU), bone marrow colony forming cells (CFU-C) and granulocyte levels in germfree mice and in mice that were removed from their environment and colonized with intestinal bacteria. Joshi et al (1979) observed lower levels of serum CSF in germfree mice than in conventionalized controls, and colonization of germfree mice with bacteria increased the granulocyte counts in the ex-germfree mice.

Metcalf et al (1967, Metcalf and Forster 1969) reported that serum CSF levels were lower in germfree mice than in their conventional counterparts. Morley et al (1972) reported that there was no increase in serum CSF in irradiated germfree mice and postulated that intestinal bacteria play a role in increasing CSF in irradiated conventional mice. Chang and Pollard (1973) demonstrated that irradiated gnotobiotic mice (colonized with *E. coli*) had serum levels of CSF that were comparable to the levels observed in conventional mice following irradiation. Mice colonized with *Streptococcus faecalis* showed no increase in CSF following irradiation (Chang and Pollard 1973).

Staber et al (1978) have shown that chemically pure preparations of gram-negative bacteria can modulate myelopoiesis in rodents. Quesenberry et al (1972, 1974) also suggested that the hosts' microflora regulates myelopoiesis. It is very likely that endotoxin from gram-negative intestinal bacteria stimulates hematopoiesis, but the mechanism(s) is far from clear.

Thus, the germfree state appears to be associated with decreased hematopoiesis and lower levels of circulating leukocytes. Obviously, gram-negative bacteria can affect hematopoiesis in many ways; however, studies that have assessed the primary effects of endotoxin on hematopoiesis in germfree animals are too few in number and need further clarification.

Recent clinical research has placed emphasis on removing gram-negative pathogens from the alimentary tract of patients to prevent opportunistic gram-negative infection (Wade et al 1983). The intestinal flora is important in the regulation of serum levels of CSF activity and of marrow granulopoiesis. Walker (1983) calls attention to the fact that antibiotic decontamination of patients may also decrease hematopoiesis. Such decreased hematopoietic activity could further predispose the patient to opportunistic infections or make them less responsive to certain chemotherapeutic regimens.

6. MACROPHAGE FUNCTION

6.1. Number and enzyme activity

Compared with conventional animals' macrophages, those from gnotobiotic animals have been shown to be fewer in number (Gordon and Pesti 1971; Johnson and Balish 1980; Morland et al 1979) and defective in several functions: phagocytosis via C3b-receptors, chemotactic responsiveness, and lysosomal enzyme activity (Heise and Myrvik, 1966; Jungi and McGregor 1978; Morland et al 1979; Starling and Balish 1981). Starling and Balish (1981) showed that associating germfree rats with gram-negative bacteria increased the number of pulmonary alveolar macrophages and enhanced their lysosomal enzyme activity. Other investigators have reported that Kupffer cells from germfree rats also have decreased lysosomal activity in comparison to their conventional counterparts (Berg and Midtvedt 1976). Although no studies used endotoxin injections to enhance the macrophage activity of germfree animals in vivo, it is evident that gram-negative bacteria in the alimentary tract can increase the number of macrophages and increase their lysosomal enzyme activity (Heise and Myrvik 1966; Starling and Balish 1981).

6.2. Tumoricidal activity

A number of effector cells can kill tumor cells in vitro. Among these are natural killer (NK) cells, cytotoxic T-cells and activated macrophages (Moore 1978). It is generally believed that macrophages must be nonspecifically activated in order to be able to exert significant cytotoxic activity (Adams and Snyderman 1979). Recently, the ability of nonactivated macrophages to kill tumor cells has been described by several investigators (Johnson and Balish 1980; Koller 1978; Meltzer 1976; Pels and Otter 1979; Tagliabue et al 1979).

Generally, investigators have reported that macrophages from germfree mice have impaired direct tumor cytotoxic activity compared to macrophages from conventional animals (Meltzer 1976). Johnson and Balish (1981a) used resident macrophages from germfree and conventional rats to study their in vitro tumoricidal (direct and antibody-mediated) activity. They (Johnson and Balish 1981b) observed that resident peritoneal macrophages from germfree, monoassociated, or conventional rats had comparable cytotoxic activity against several allogeneic tumor cells. When germfree rats were conventionalized (i.e. allowed to acquire a complex microflora), their peritoneal macrophages manifested a transient decrease in cytotoxicity (Johnson and Balish 1981b). The latter investigators also reported that peritoneal macrophages from germfree rats have impaired endotoxin-induced tumoricidal activity when compared with macrophages from conventional rats. While endotoxin was not able to activate alveolar macrophages from germfree or monoassociated (with *Staphylococcus epidermidis*) rats, it did enhance tumoricidal activity of macrophages from conventional rats (Johnson and Balish 1981b).

Thus, the gram-negative microbial flora, and very likely the endotoxin from these bacteria, have noticeable effects on the number, enzyme activity and tumoricidal capacity of macrophages. Far too few research efforts have delved into these interesting observations, which could have far-reaching effects on resistance to infectious and neoplastic diseases.

7. POLYCLONAL B-CELL ACTIVATION

The capacity of endotoxin to stimulate B-cells in vitro is a well known phenomenon. Very few studies have been carried out to elucidate the effects of endotoxin on lymphocytes from germfree animals.

Kiyono et al (1980) and McGhee et al (1980) quantitated the mitogenic response of spleen cells from germfree, *Escherichia coli*-monoassociated, and conventional BALB/c mice to phytohemagglutinin, concanavalin A, pokeweed mitogen and lipopolysaccharide. They observed that spleen cells from gnotobiotic and conventional mice gave a similar blastogenic response to concanavalin A, phytohemagglutinin and pokeweed mitogen; however, in spite of the increased resistance of germfree mice to the lethal effects of injected endotoxin (Kiyono et al 1980), their spleen cells gave a better blastogenic response to lipopolysaccharide (phenol-extracted) in vitro than spleen cells from conventional animals. The germfree animals' spleen cells also responded to lower doses of endotoxin than spleen cells from conventional mice (Kiyono et al 1980).

Wells and Balish (1979) used spleen cells from germfree and conventional rats and showed that they responded to concanavalin A, phytohemagglutinin and pokeweed mitogen; however, they were unable to show any mitogenic response of spleen cells from germfree or conventional rats to four lipopolysaccharide preparations from *E. coli*, glycolipid from *Salmonella minnesota* R595 and with *S. minnesota* lipid A.

Kim (1979) observed that polyclonal B-cell activators (PBA) such as lipopolysaccharides, lipid A, and dextran sulfate were not able to activate B-cells from the peripheral blood of immunologically virgin piglets, but they did so for B-cells from conventional piglets. Kim (1979) suggested that the response to PBA may be due to nonspecific activation of already committed B-cells and that uncommitted 'virgin' B-cells may not be able to respond to B-cell activation with endotoxin. Conversely, Tlaskalová-Hogenová et al (1980) reported that lymphocytes from germfree piglets do respond to lipopolysaccharide stimulation.

To date, mouse and human lymphocytes have been used in most endotoxin mitogenicity studies. Lymphocytes from germfree mice appear to be more sensitive to endotoxin stimulation in vitro than lymphocytes from their conventional counterparts. It is interesting to note that germfree mice are more resistant to the lethal effects of endotoxin but they show enhanced immunogenic and blastogenic responses to endotoxin. C3H/HeJ mice, which are also resistant to the lethal effects of endotoxin, are resistant to mitogenic and immunogenic effects of endotoxin as well (Borek 1982). The mechanism responsible for the germfree animals' resistance to the lethal effects of endotoxin should be elucidated.

8. ADJUVANT ACTIVITY AND RESISTANCE TO INFECTIOUS DISEASES

Endotoxin injections can enhance the resistance of conventional animals to infection (Cluff 1971). The adjuvant activity is thought to be associated with the overall reticuloendothelial system stimulation that is evoked by endotoxin treatments. Surprisingly few experiments have been carried out to assess the adjuvant effects of endotoxin on germfree animals.

Landy et al (1961) demonstrated that endotoxin injection (5 µg of *Salmonella enteritidis* endotoxin intraperitoneally) increased the level of bactericidal antibody, to a variety of gram-negative bacteria, to a similar degree in both germfree and conventional mice. MacDonald and Carter (1979) showed that germfree mice could be induced to generate a 'normal' delayed-type hypersensitivity response to sheep red blood cells if they were either colonized with *E. coli* or injected with endotoxin. Kim (1979) and Tlaskalová-Hogenová et al (1980) demonstrated that endotoxin injections enhanced the resistance of germfree piglets to gram-negative pathogens.

Thus, endotoxin appears to have adjuvant activity in germfree animals and is apparently able to enhance their resistance to pathogenic microbes; however, far too little work has been done to make critical evaluations of the adjuvant effects of endotoxin on germfree animals.

Clearly, endotoxin plays an important role in the ontogeny of the immune response (Cebra et al 1983; Kim 1979). It is also evident that colonization of the gastrointestinal tract with pure cultures of bacteria or a complex microflora enhances innate and acquired immunity to homologous and heterologous microorganisms (Balish et al 1972; Gordon and Pesti 1971; McLeod and Balish 1978a; Minden et al 1972; Robbins et al 1972; Tlaskalová-Hogenová et al 1983).

Understanding the basis of endotoxin effects on the germfree animals' immune system will make vital contributions to our knowledge of the mechanisms that underlie the ontogeny of the innate, cellular and humoral immunity to infectious and neoplastic diseases.

9. SYSTEMIC AND ORAL TOLERANCE

It is well known that the repeated injection of endotoxins into healthy humans or animals induces a state of tolerance to its toxic and pyrogenic effects (Chedid and Parant 1971; Cluff 1971). The tolerance that develops appears to involve an early refractoriness on the part of cells that synthesize and release endogenous pyrogens and a later phase that appears to be related to the production of antiendotoxin antibodies against the O-antigen polysaccharide chain, the core oligosaccharide and lipid A, which may be the most important antigen (Lüderitz 1966).

Because conventional animals are exposed to viable gram-negative bacteria and their endotoxins throughout life, it could be speculated that they should be more tolerant (resistant) to endotoxin than their germfree counterparts who at most are exposed to endotoxin present in their autoclaved diets (Ebersole and Molinari 1977). In terms of lethality this does not appear to be the case since germfree mice manifest a 5-fold (or more) increase in resistance to the lethal effects of endotoxin in comparison to their conventional counterparts (Kiyono et al 1980). Kiyono et al (1980) and McGhee et al (1980) have also shown that germfree mice manifest greater mitogenic and immunogenic responses to endotoxin than their conventional counterparts and the germfree animals formed antibody to lower doses of the trinitrophenylated lipopolysaccharide than their conventional counterparts. It may very well be that some tolerance to endotoxin is manifest in all conventional animals that have viable gram-negative bacteria in their alimentary tract.

Immune tolerance is readily induced to a variety of antigens administered by the oral route (Chase 1946; Thomas and Parott 1974; Tomasi 1980). Recently Michalek et al (1983) reported that immune tolerance to sheep red blood cells could not be induced in conventional, LPS-unresponsive, C3H/ HeJ mice. Immune tolerance to sheep red blood cells could not be induced in germfree mice either, unless they were administered lipopolysaccharide prior to oral treatment with sheep red blood cells (Michalek et al 1983; Wannenmuehler et al 1982). The latter investigators concluded that responsiveness to lipopolysaccharide is a prerequisite for the induction of oral tolerance. Conversely, Saklayen et al (1983) recently demonstrated that immune tolerance to hen albumin could be induced in C3H/HeJ mice by antigen feeding prior to parenteral immunization and suggested that ability to respond to lipopolysaccharide is not a prerequisite for induction of oral tolerance to a soluble protein.

Although more research on the induction of immune tolerance to endotoxin and other antigens has to be carried out, perhaps more attention should be paid to the mechanisms of immune tolerance induction in germfree animals and especially in antigen- or endotoxin-free germfree animals. Endotoxin appears to be involved in many important aspects of the ontogeny of the immune system and far too little is known about the primary effects of endotoxin on these systems.

10. CONCLUSIONS

It is obvious that intestinal bacteria and endotoxin can modulate an animal's response to endotoxin. Perhaps the different responses evoked by endotoxin in different mammalian species are in part due to differences in the composition of their intestinal microflora. It is clear that far too little effort has been put into studies on the primary endotoxin effects on germfree animals.

Germfree mice are clearly more resistant to the lethal effects of endotoxin than their conventional counterparts. It is also evident that the germfree animals' resistance to the lethal effects of endotoxin is not brought about by an unresponsive state since Kiyono et al (1980) showed that germfree animals' splenocytes were more reactive to endotoxin stimulation in vitro and manifested a better blastogenic response to phenol-extracted endotoxin than spleen cells from their conventional counterparts. Conversely, spleen cells from genetically resistant C3H/HeJ mice do not respond to lipid A in vitro or in vivo. The mechanism responsible for the resistance of germfree mice to lethal effects of endotoxin has not yet been explained but could be of great importance to our understanding of hemorrhagic or endotoxin shock.

The effects of endotoxin on hematopoiesis are well documented. It is well known that gram-negative microbes very often occur in large numbers in the alimentary tract of newborn infants. We have pointed out in this chapter how gram-negative bacteria and their endotoxins can influence many different aspects of hematopoiesis and the ontogeny of the immune response. It would be well worth-while to delineate how pure cultures of gram-negative bacteria in the intestinal tract can influence the ontogenetic development and immunocompetence of gnotobiotic neonates. Very little effort has been put forth to elucidate such effects of intestinal microorganisms on the host.

Clearly, there has been far too little research on the effects of endotoxins on neonatal and adult germfree animals. When we compare germfree animals with conventional animals, we see differences in their morphology, physiology, hematology, biochemistry, immunology, and endocrinology. How much of a role endotoxin plays in bestowing 'conventional traits' on germfree animals is not known at this time; however, the work that has been done so far indicates that endotoxin from intestinal bacteria may play a vital role in stimulating or suppressing many important biological processes of 'conventional' life.

If nothing else, studies that have been carried out on the effects of endotoxin and gram-negative bacteria on research animals should point out how important it is to know the microbial profile of research animals. Differences in intestinal microflora could account for a good deal of the variability observed in studies on the effects of endotoxin in different species and strains of research animals.

REFERENCES

Abernathy RS, Landy JJ (1967) Increased resistance to endotoxin in germfree guinea pigs. *Proc. Soc. Exp. Biol. Med. 124*, 1279-1283.

Adams DO, Snyderman R (1979) Do macrophages destroy nascent tumors? *J. Natl Cancer Inst. 62*, 1341-1346.

Balish E, Yale CE, Hong R (1972) Serum proteins of gnotobiotic rats. *Infect. Immun. 6*, 112-118.

Benner R, van Oudenaren A, Björklund M, Ivars E, Holmberg P (1982) 'Background' immunoglobulin production: measurement, biological significance and regulation. *Immunol. Today 3*, 243-249.

Berg RD (1983) Translocation of indigenous bacteria from the intestinal tract. In: Hentgas D (Ed), *Human Intestinal Microflora in Health and Disease*, pp 333-352, Academic Press, New York.

Berg T, Midtvedt T (1976) The influence of infection on the content of lysosomal enzymes in rat Kupffer cells and hepatocytes. *Acta Pathol. Microbiol. Scand. (Sect. A) 84*, 415-421.

Blakey JL, Lubitz L, Barnes GL, Bishop RF, Campbell NT, Gillam GL (1982) Development of gut colonization in pre-term neonates. *J. Med. Microbiol. 15*, 519-529.

Boggs DP, Chervenick PA, Marsh JC (1967) Granulopoiesis in germfree mice. *Proc. Soc. Exp. Biol. Med. 125*, 325-330.

Borek F (1982) The murine LPS gene and the endotoxin responsiveness. *Afr. J. Clin. Exp. Immunol. 3*, 321-326.

Bosma MJ, Makinodan T, Walburg HE Jr (1967) Development of immunologic competence in germfree and conventional mice. *J. Immunol. 99*, 420-430.

Bradley TR, Metcalf D (1966) The growth of mouse bone marrow cells in vitro. *Aust. J. Exp. Biol. Med. Sci. 44*, 287-300.

Cebra JJ, Cebra ER, Clough ER, Fuhrman JA, Komisar JL, Schweitzer PA, Shahin RD (1983) IgA commitment: models for B-cell differentiation and possible roles for T-cells in regulating B-cell development. *Ann. N.Y. Acad. Sci. 409*, 25-38.

Chang CF, Pollard M (1973) Effects of microbial flora on levels of colony stimulating factor in serums of irradiated CFW mice. *Proc. Soc. Exp. Biol. Med. 144*, 177-180.

Chase MW (1946) Inhibition of experimental drug allergy by prior feeding of antigen. *Proc. Soc. Exp. Biol. Med. 41*, 257-259.

Chedid L, Parant M (1971) Role of hypersensitivity and tolerance in reduction to endotoxins. In: Kadis S, Weinbaum G, Ajl SJ (Eds), *Microbial Toxins Vol 5: Bacterial Endotoxins*, pp 415-457. Academic Press, New York.

Cluff LE (1971) Effects of lipopolysaccharides (endotoxins) on susceptibility to infections. In: Kadis S, Weinbaum G, Ajl SJ (Eds), *Microbial Toxins Vol 5: Bacterial Endotoxins*, pp 399-414. Academic Press, New York.

Crabbe PA, Carbonaro AO, Heremans JF (1965) The normal intestinal mucosa as a major source of plasma cells containing A-immunoglobulins. *Lab. Invest. 14*, 235-248.

Crabbe PA, Bazin H, Eyssen H, Heremans JF (1968) The normal microbial flora and major stimulation for proliferation of plasma cell synthesizing IgA in the gut. *Int. Arch. Allergy Appl. Immunol. 34*, 362-375.

Czuprynski CJ, Johnson WJ, Balish E, Wilkins T (1983) Pseudo membranous colitis in *Clostridium difficile*-monoassociated rats. *Infect. Immun. 39*, 1368-1376.

Dubos RJ, Schaedler RW (1960) The effects of intestinal flora on the growth rate of mice, and on their susceptibility to experimental infection. *J. Exp. Med. 111*, 407-417.

Dubos RJ, Schaedler RW, Costello R (1963) Composition, alteration and effects of the intestinal flora. *Fed. Proc. 22*, 1325-1329.

Dubos RJ, Lee C, Costello R (1969) Lasting biological effects of early environmental influences. *J. Exp. Med. 130*, 963-977.

Ebersole JL, Molinari JA (1977) Gastrointestinal antibody responses in axenic mice to topically administered *Escherichia coli. Infect. Immun. 16*, 938-946.

Elin RJ, Wolff S (1976) Biology of endotoxin. *Annu. Rev. Med. 27*, 127-141.

Fliedner TM, Fache I, Adolphi C (1966) Über die Umsatzkinitik der Leukocyten bei keimfreien Mäusen. *Schweiz. Med. Wochenschr. 37*, 1236-1238.

Fliedner TM, Heit H, Niethammer D, Pflieger H (Eds) (1979) *Clinical and Experimental Gnotobiotics. Proceedings on the VIth International Symposium on Gnotobiology.* Gustav Fischer Verlag, Stuttgart-New York.

Freter R, Abrams GD (1972) Function of various intestinal bacteria in converting germfree mice to the normal state. *Infect. Immun. 6*, 119-126.

Galley CB, Walker RI, Ledney GD, Gambrill MR (1975) Evaluation of biological activity of attenuated endotoxin in mice. *Exp. Hematol. 3*, 197-204.

Gordon HA, Bruckner-Kardos E (1961) Effects of normal microbial flora on various tissue elements in the small intestine. *Acta Anat. 44*, 210-225.

Gordon HA, Pesti L (1971) The gnotobiotic animal as a tool in the study of host-microbial relationships. *Bacteriol. Rev. 35*, 390-429.

Gordon HA, Wostmann BS (1960) Morphological studies on the germfree albino rat. *Anat. Rec. 137*, 65-70.

Gordon HA, Bruckner-Kardos E, Wostmann BS (1966) Aging in germfree mice: life tables and lesions observed at natural death. *J. Gerontol. 21*, 380-387.

Gustafsson BE, Midtvedt T, Strandberg K (1970) Effects of microbial contamination on the cecum enlargement of germfree rats. *Scand. J. Gastroenterol. 5*, 309-314.

Hashimoto F, Handra H, Umehara K, Sasaki S (1978) Germfree mice reared on an 'antigen-free' diet. *Lab. Animal Sci. 28*, 38-45.

Hashimoto M, Hashimoto N (1968) Histological study of bone marrow in breeding rats. In: Miyakawa M, Luckey T (Eds), *Advances in Germfree Research and Technology*, pp 149-161. Chem. Rubber Co. Press, Cleveland, Ohio.

Heise ER, Myrvik QW (1966) Levels of lysosomal hydrolases in alveolar and peritoneal macrophages from conventional and germfree rats. *Fed. Proc. 25*, 439-440.

Heneghan JB (1968) Hemorrhagic shock in unanesthetized gnotobiotic rats. In: Miyakawa M, Luckey T (Eds), *Advances in Gnotobiotic Research and Gnotobiology*, pp 166-170. Chem. Rubber Co. Press, Cleveland, Ohio.

Jensen SB, Mergenhagen SE, Fitzgerald RJ, Jordan HV (1963) Susceptibility of conventional and germfree mice to lethal effects of endotoxin. *Proc. Soc. Exp. Biol. Med. 113*, 710-714.

Johnson WJ, Balish EB (1980) Macrophage function in germfree, athymic (nu/nu) and conventional flora (nu/+) mice. *J. Reticuloendothel. Soc. 28*, 55-66.

Johnson WJ, Balish E (1981a) Director tumor cell and antibody-dependent cell-mediated cytotoxicity by macrophages from germfree and conventional rats. *J. Reticuloendothel. Soc. 29*, 205-213.

Johnson WJ, Balish E (1981b) Tumor cytotoxic activity of resident rat macrophages. *J. Reticuloendothel. Soc. 29*, 369-379.

Joshi JH, Entringer M, Robinson WA (1979) Bacterial stimulation of serum colony-stimulating activity and neutrophil production in germfree mice. *Proc. Soc. Exp. Biol. Med. 162*, 44-47.

Jungi TW, McGregor DD (1978) Impaired chemotactic responsiveness of macrophages from gnotobiotic rats. *Infect. Immun. 19*, 553-558.

Kim YB (1979) Role of antigen in the ontogeny of the immune response. In: Schlessinger D (Ed), *Microbiology 1979*, pp 343-348. ASM Press, Washington, D.C.

Kim YB, Watson DW (1965) Enhancement of susceptibility to endotoxin shock by purified streptococcal pyrogenic toxin. *Bacteriol. Proc. 65*, 45.

Kim YB, Watson DW (1966) Role of antibodies in reactions to gram-negative bacterial endotoxins. *Ann. N.Y. Acad. Sci. 133*, 727-745.

Kim YB, Watson DW (1970) A purified group A streptococcal pyrogenic exotoxin. *J. Exp. Med. 131*, 611-628.

Kim YB, Bradley SG, Watson DW (1966) Ontogeny of the immune response. 1. Development of immunoglobulins in germfree and conventional, caesarian-deprived piglets. *J. Immunol. 97*, 52-63.

Kim YB, Setcavage TM, Kim DJ, Chun HG, Scheffel JW (1979) Ontogenic development and differentiation of the immune system in the gnotobiotic miniature swine. In: Fliedner T, Heit H, Niethammer D, Pflieger H (Eds), *Clinical and Experimental Gnotobiotics. Proceedings of the VIth International Symposium on Gnotobiology*, pp 203-213. Gustav Fischer Verlag, Stuttgart-New York.

Kiyono H, McGhee JR, Michalek SM (1980) Lipopolysaccharide regulation of the immune response: comparison of responses to LPS in germfree, *Escherichia coli*-monoassociated and conventional mice. *J. Immunol. 124*, 36-41.

Koller R (1978) Macrophage-mediated natural cytotoxicity against various target cells in vitro. I. Macrophages from diverse anatomical sites and different strains of rats and mice. *Br. J. Cancer 37*, 732-736.

Koopman JP, Kennis HM (1980) Influence of normal mouse intestinal bacteria on caecal weight in mice. *Z. Versuchstierkd 22*, 224-229.

Landy M, Whitby JI, Michael JG, Woods MW, Newton WL (1961) Effect of bacterial endotoxin in germfree mice. *Proc. Soc. Exp. Biol. Med. 109*, 353-356.

Levenson SM (1978) The influence of indigenous microflora on mammalian metabolism and nutrition. *J. Parenterol. Enterol. Nutr. 2*, 75-107.

Loesche WJ (1968) Accumulation of endogenous protein in the cecum of the germfree rat. *Proc. Soc. Exp. Biol. Med. 129*, 380-384.

Lüderitz O, Staub AM, Westphal O (1966) Immunochemistry of O and R antigens of *Salmonella* and related *Enterobacteriaceae*. *Bacteriol. Rev. 30*, 192-255.

Mac Donald TT, Carter PB (1979) Requirement for a bacterial flora before mice generate cells capable of mediating the delayed hypersensitivity reaction to sheep red blood cells. *J. Immunol. 122*, 2624-2629.

Magliolo E, Sceuola D, Fumarola D (1975) Limulus test in *Escherichia coli* enteritis. *Lancet 1*, 407-409.

Markley K, Smallman E, Evans G (1967) Mortality due to endotoxin in germfree conventional mice after tourniquet trauma. *Am. J. Physiol. 212*, 541-548.

Markley K, Smallman E, Thornton S (1969) *Escherichia coli* endotoxemia after thermal trauma in conventional and germfree mice. *Fed. Proc. 28*, 271.

McGhee JR, Kiyono H, Michalek S, Babb JL, Rosenstreich DL, Mergenhagen S (1980) Lipopolysaccharide (LPS) regulation of the immune response: T-lymphocytes from normal mice suppress mitogenic and immunogenic responses to LPS. *J. Immunol. 4*, 1603-1611.

McLeod JC, Balish E (1978a) Homologous and cross-reacting antibodies in the sera of gnotobiotic rats. *Can. J. Microbiol. 24*, 365-371.

McLeod JC, Balish E (1978b) Endotoxin in germfree, gnotobiotic or conventional flora Sprague-Dawley rats. *Can. J. Microbiol. 24*, 1602-1606.

McNulty WP Jr, Linares R (1960) Hemorrhagic shock of germfree rats. *Am. J. Physiol. 198*, 141-144.

Meltzer MS (1976) Tumoricidal responses *in vitro* of peritoneal macrophages from conventionally housed and germfree nude mice. *Cell Immunol. 22*, 176-181.

Metcalf D, Foster RS (1967) Bone marrow colony stimulating activity of serum from mice with virus-induced leukemia. *J. Natl Cancer Inst. 39*, 1235-1245.

Metcalf D, Foster RS (1969) *In vitro* colony forming cells in the bone marrow of germfree mice. In: Mirand EA, Bach N (Eds), *Germfree Biology: Experimental and Clinical Aspects*, pp 383-388. Plenum Press, New York.

Metcalf D, Foster RS, Pollard M (1967) Colony stimulating activity of serum from germfree normal and leukemia mice. *J. Cell. Physiol. 70*, 131-132.

Michalek SM, McGhee JR, Kiyono H, Colwell DE, Eldridge JH, Wannenmuehler MJ, Koopman WJ (1983) The IgA response: induction aspects, regulatory cells and effector functions. *Ann. N.Y. Acad. Sci. 409*, 48-71.

Minden P, McClatchy JK, Farr RS (1972) Shared antigens between heterologous · bacterial species. *Infect. Immun. 6*, 574-582.

Miyakawa M (1959) The lymphatic system of germfree guinea pigs. *Ann. N.Y. Acad. Sci. 78*, 221-236.

Moore MA (1978) Macrophage heterogeneity: Role in hemopoiesis and cytotoxicity for normal and neoplastic targets. In: Cudkowicz G, Landy M, Shearer GM (Eds), *Natural Resistance Systems Against Foreign Cells, Tumors and Microbes*, pp 93-106. Academic Press, New York.

Morland B, Senievoll AI, Midtvedt T (1979) Comparison of peritoneal macrophages from germfree and conventional mice. *Infect. Immun. 26*, 1129-1133.

Morley AP, Quesenberry P, Bealmear P, Stohlman F Jr, Wilson R (1972) Serum colony stimulating factor levels in irradiated germfree and conventional CFW mice. *Proc. Soc. Exp. Biol. Med. 140*, 478-480.

Mushin R, Dubos RJ (1965) Colonization of the mouse intestine with *Escherichia coli. J. Exp. Med. 122*, 745-757.

Mushin R, Dubos RJ (1966) Coliform bacteria in the intestines of mice. *J. Exp. Med. 123*, 657-672.

Nagler AL, Levenson S (1971) Experimental hemorrhagic and endotoxic shock. In: Kadis S, Weinbaum G, Ajl SJ (Eds), *Microbial Toxins Vol 5: Bacterial Endotoxins*, pp 341-397. Academic Press, New York.

Pels E, Otter WD (1979) Natural cytotoxic macrophages in the peritoneal cavity of mice. *Br. J. Cancer 40*, 856-860.

Pleasants JR, Bruckner-Kardos E, Bartizal KF, Beaver MH, Wostmann BS (1981) Reproductive and physiological parameters of germfree C3H mice fed chemically-defined diet. In: Sasaki S, Ozawa A, Hashimoto K (Eds), *Recent Advances in Germfree Research*, pp 333-338. Tokai University Press, Tokyo.

Pollard M (1967) Germinal centers in germfree animals. In: Cottier CH, Odortchenko R, Congdon C (Eds), *Germinal Centers in Immune Responses*, pp 343-346. Springer-Verlag, New York.

Pollard M (1971) Senescence in germfree rats. *Gastroenterologia 17*, 333-338.

Quesenberry P, Morley A, Stohlman F, Richard K, Howard D, Smith M (1972)

Effect of endotoxin on granulopoiesis and colony stimulating factor. *N. Engl. J. Med. 286*, 227-231.

Quesenberry P, Bealmcar P, Ryan M, Stohlman F Jr (1974) Serum colony stimulating and inhibiting factors in irradiated germfree and conventional mice. *Br. J. Haematol. 28*, 531-539.

Quesenberry P, Cohen H, Levin J, Sullivan R, Bealmear P, Ryan M (1978) The effects of bacterial infection and irradiation on serum colony stimulating factor levels in tolerant and nontolerant CF_1 mice. *Blood 51*, 229-244.

Robbins JB, Meyerowitz RL, Whisant JK, Argman K, Schneuson R, Handsel ZT, Gotschlich EC (1972) Enteric bacteria cross-reactive with *Neisseria meningitidis* groups A and C and *Peptococcus pneumoniae* types I and III. *Infect. Immun. 6*, 651-656.

Sacquet E, Charlier H (1965) Sensibilité de diverses races de souris à l'action létale de l'endotoxine de *Salmonella typhosa*. Influence de la flore microbienne associée. *Ann. Inst. Pasteur 108*, 353-363.

Saklayen M, Pesce AM, Pollak VE, Michael JG (1983) Induction of oral tolerance in mice unresponsive to bacterial lipopolysaccharide. *Invest. Immun. 41*, 1383-1385.

Sasaki S, Ozawa A, Hashimoto K (Eds) (1981) *Recent Advances in Germfree Research. Proceedings of the VIIth International Symposium on Gnotobiology*. Tokai University Press, Tokyo, Japan.

Schaedler RW, Dubos RJ (1961) The susceptibility of mice to bacterial endotoxins. *J. Exp. Med. 113*, 559-570.

Schaedler RW, Dubos RJ (1962) The fecal flora of various strains of mice. Its bearing on their susceptibility to endotoxin. *J. Exp. Med. 115*, 1149-1161.

Schaedler RW, Dubos RJ, Costello R (1965) Association of germfree mice with bacteria from normal mice. *J. Exp. Med. 122*, 77-82.

Schlievert PM (1982) Enhancement of host susceptibility to lethal endotoxin shock by staphylococcal pyrogenic exotoxin type C. *Infect. Immun. 36*, 123-128.

Schlievert PM (1983) Alteration of immune function by staphylococcal pyrogenic exotoxin type C: possible role in toxic shock syndrome. *J. Infect. Dis. 147*, 391-398.

Sell S (1964) Immunoglobulins of the germfree guinea pig. *J. Immunol. 93*, 122-131.

Skelly JJ, Trexler PC, Tanami J (1962) Effect of a *Clostridium* species upon cecal size of gnotobiotic mice. *Proc. Soc. Exp. Biol. Med. 110*, 455-458.

Staber FG, Taresay L, Dukor P (1978) Modulation of myelopoiesis *in vivo* by chemically pure preparations of cell wall components from gram-negative bacteria. *Infect. Immun. 20*, 40-49.

Stanley ER, Robinson WA, Ada GL (1968) Properties of the colony stimulating factor in leukemic and normal mouse serum. *Aust. J. Biol. Med. 46*, 715-726.

Stark PL, Lee A (1982) The bacterial colonization of the large bowel of pre-term low birth weight neonates. *J. Hyg. 89*, 59-67.

Starling JR, Balish E (1981) Lysosomal enzyme activity in pulmonary alveolar macrophages from conventional, germfree, monoassociated and conventional rats. *J. Reticuloendothel. Soc. 30*, 497-505.

Suter E (1962) Hyperreactivity to endotoxin in mice infected with BCG. *J. Immunol.*

89, 377-381.

Suter E (1964) Hyperreactivity to endotoxin after infection with BCG. *J. Immunol.* *92*, 49-54.

Syed SA, Abrams GD, Freter R (1970) Efficiency of various intestinal bacteria in assuming normal functions of enteric flora after association with germfree mice. *Infect. Immun. 2*, 376-386.

Tagliabue A, Montovani A, Kilgallen M, Herberman RB, McCoy JL (1979) Natural cytotoxicity of mouse monocytes and macrophages. *J. Immunol. 122*, 2363-2368.

Thomas HC, Parott D (1974) The induction of tolerance to a soluble protein antigen by oral administration. *Immunology 27*, 631-639.

Thorbecke GJ, Benacerraf B (1959) Some histological and functional aspects of lymphoid tissue in germfree animals. II. Studies on phagocytosis *in vivo. Ann. N.Y. Acad. Sci. 78*, 247-253.

Thorbecke GJ, Gordon HA, Wostmann BS, Wagner M, Reyniers JA (1957) Lymphoid tissue and serum gamma globulin in young germfree chickens. *J. Infect. Dis. 101*, 237-251.

Tlaskalová-Hogenová H, Mandel L, Dlabac U, Prokesová L, Hoffman J, Kovaru F (1980) Pathophysiological and immunological effects of lipopolysaccharide in germfree and conventional piglets. In: Agarwal MK (Ed), *Bacterial Endotoxins and Host Response*. pp 143-166. Elsevier Biomedical Press, New York.

Tlaskalová-Hogenová H, Zterzl J, Stepánková R, Dlabac U, Vetuicka V, Rossman P, Mandel L, Rejnrk K (1983) Development of immunological capacity under germfree and conventional conditions. *Ann. N.Y. Acad. Sci. 409*, 96-113.

Tomasi TB Jr (1980) Oral tolerance. *Transplantation. 29*, 353-356.

Wade JC, de Jongh CA, Newman KA, Crowley J, Wrernik PH, Schimpf SC (1983) Selective antimicrobial modulation as prophylaxis against infection during granulocytopenia: trimethoprim-sulfamethoxazole vs. nalidixic acid. *J. Infect. Dis. 147*, 624-634.

Walker RI (1983) The regulation of bone marrow activity by the endogenous microflora and the contribution of cellular and humoral factors in the control of endotoxin. In: Van der Waaij D (Ed), *Antibiotic Choice: the Importance of Colonization Resistance*, pp 95-107. Research Studies Press, John Wiley and Sons, New York.

Walker RI, Sheil JM (1976) Contribution of granulocytopenia to endotoxin sensitivity of mice irradiated or undergoing graft-vs-host reaction. *Exp. Hematol. 4*, 329-338.

Walker RI, Ledney GD, Galley CB (1975) Aseptic endotoxemia in radiation injury and graft vs. host disease. *Radiol. Res. 62*, 242-244.

Wannenmuehler MJ, Kiyono H, Babb J, Michalek S, McGhee J (1982) Lipopolysaccharide (LPS) regulation of the immune response: LPS converts germfree mice to sensitivity to oral tolerance induction. *J. Immunol. 129*, 959-965.

Wells C, Balish E (1979) The mitogenic activity of lipopolysaccharides for spleen cells from germfree, conventional and gnotobiotic rats. *Can. J. Microbiol. 25*, 1087-1093.

Wostmann BS (1975) Nutrition and metabolism of the germfree mammal. *World Rev. Nutr. Diet. 22*, 40-92.

Wostmann BS, Pleasants JR, Bealmear P, Kincade PW (1970) Serum proteins and lymphoid tissue in germfree mice fed a chemically-defined, water-soluble, low molecular weight diet. *Immunology 19*, 443-448.

Yale CE, Balish E (1976) Blood and serum chemistry values of gnotobiotic beagles. *Lab. Animal Sci. 26*, 633-639.

Zweifach BW, Gordon HA, Wagner MA, Reyniers JA (1958) Irreversible hemorrhagic shock in germfree rats. *J. Exp. Med. 107*, 437-449.

Handbook of Endotoxin, Vol. 2: Pathophysiology of Endotoxin
L.B. Hinshaw, editor
© Elsevier Science Publishers B.V., 1985
ISBN 0 444 90385 2
$0.85 per article per page (transactional system)
$0.20 per article per page (licensing system)

CHAPTER 14

Mechanisms of endotoxin tolerance

CURTIS A. JOHNSTON AND SHELDON E. GREISMAN

1. INTRODUCTION

Man and most experimental animals are highly sensitive to the toxic effects of the lipopolysaccharide-protein complex found universally in gram-negative bacterial cell walls. These toxic effects are wide ranging and include fever, hypotension, hypoglycemia, coagulopathy, and death. Tolerance to these effects has been investigated in the hope that an understanding of the underlying mechanisms would allow more effective prophylaxis and therapy of gram-negative bacterial infections. Current concepts of the mechanisms of endotoxin tolerance have developed slowly, and many areas remain controversial. This chapter will consider the evolution of, and our current understanding of, the mechanisms of tolerance to the febrile and lethal activities of endotoxin.

1.1. Historical background

The acquisition of resistance to endotoxin proved an annoying problem to physicians who encountered this phenomenon almost 100 years ago, when using whole bacterial vaccines to induce fever therapeutically. Thus, it was widely recognized by the early 1900s that whenever fever in man was evoked with killed gram-negative bacilli, increasing quantities had to be infused intravenously to maintain elevated temperatures. Some patients eventually required as much as 250 ml of typhoid vaccine in a single day (Heyman 1945). Various methods were used to circumvent this resistance. Nelson (1931) found that resistance could be partly overcome by giving two doses of typhoid vaccine with each treatment; the second dose, given 2 hours later, was described as having the effect of 'exploding the charge' supplied by the first. Continuous intravenous infusions of typhoid vaccine were later recommended by Heyman (1945), the infusion rate to be increased as the fever began to decline. The difficulty in maintaining therapeutic fever with continuous infusions was described by Bennett (1956): 'When I was a house officer, we gave a fair amount of typhoid vaccine by constant intravenous drip in treatment of patients with neurosyphilis, and in those patients, on the ascending limb of the temperature curve, it was obvious that a tiny dose would cause the patient to have hyperpyrexia. On the descending limb of the curve, though, practically no amount would make the patient's temperature rise again.'

Acquired resistance to a purified pyrogenic preparation derived by alcoholic and other fractionation procedures from bacterial culture filtrates was initially demonstrated at the end of the last century by Centanni (1894). By repeated injections of animals with his 'pyrotoxina bacterica', progressive reduction of its fever-inducing activity was observed. This pyrotoxina, which appeared to be common to a wide variety of bacteria, was found to be stable to heat and not protein in nature. Centanni postulated that the progressive reduction in fever was due to a cellular mechanism and not based on a serological immune response (Centanni 1942).

Favorite and Morgan (1942) demonstrated that resistance to the pyrogenic fraction prepared from *Salmonella typhosa* by alcohol precipitation developed readily in man. Because circulating antibody titers to this pyrogen did not correlate with resistance, they too suggested that nonimmunologic mechanisms were responsible and the term tolerance was employed.

Investigators interested in the potential antitumor activity of endotoxin described the development of lethal tolerance in experimental animals. Using a preparation purified from *Serratia marcescens*, Shear and Perrault (1944) found that the administration of progressively greater daily doses for 9 days allowed the survival of mice challenged with 16 times the LD_{50} for untreated mice. The duration of this lethal tolerance was studied by Wharton and Creech (1949), who administered a single endotoxin injection to mice and assessed tolerance at intervals of 1 to 21 days afterwards. Lethal tolerance was found to develop within one day, reaching a peak by the third day, and waning over the subsequent two weeks. Although rising serum titers of agglutinating antibodies were detected, these did not appear to be correlated with the tolerance observed.

In a classic series of studies on tolerance performed in rabbits by Beeson (1946, 1947a, 1947b), immunologic mechanisms were again excluded. This conclusion was based upon apparent lack of either O-specificity or serum transferability of pyrogenic tolerance, as well as the rapid waning of such tolerance after discontinuing the daily endotoxin injections. In addition, tolerance was found to be accompanied by enhanced blood clearance of endotoxin. 'Blockade' of the reticuloendothelial system (RES) with thorotrast (thorium dioxide) retarded this enhanced clearance and 'abolished' tolerance. Thus it was proposed that pyrogenic tolerance was based upon enhanced RES uptake of circulating endotoxin, thereby protecting more susceptible tissues from injury.

This hypothesis was supported by Morgan (1948b), who also concluded that pyrogenic tolerance in rabbits was independent of circulating antibody because it extended to heterologous endotoxins, 'almost' disappeared while circulating antibody to the toxin still remained elevated, and could be evoked with poorly antigenic, that is, phenol-extracted, endotoxins. In further support of the nonimmunologic hypothesis, Freedman (1959, 1960a, 1960b) reported that protection against the pyrogenic and lethal effects of endotoxin induced by passively transferred sera from tolerant rabbits could apparently by related to enhanced RES phagocytic activity, as assessed by colloidal carbon clearance.

Developing in parallel with the above concepts was the hypothesis that endotoxin did not act directly to cause fever, but rather induced host cells to release an endogenous pyrogen (EP), which entered the circulation and acted upon the thermoregulatory centers to evoke fever. For almost two decades after Beeson (1948) reported the extraction of EP from rabbit polymorphonuclear leukocytes, these cells were regarded as the major source of EP in response to endotoxin. Indeed, Herion et al (1961), using nitrogen mustard in rabbits to produce agranulocytosis, concluded that polymorphonuclear leukocytes were the sole source of EP after intravenous endotoxin administration.

Atkins and Wood (1955) had furthermore reported that tolerant rabbits did not develop circulating EP when given endotoxin intravenously, but did so after pyrogenic tolerance was 'reversed' with thorotrast. Repeated intravenous administration of EP itself did not lead to tolerance (Atkins 1960). On the basis of these observations, by the early 1960s the hypothesis appeared secure that pyrogenic tolerance was mediated by a nonspecific enhancement of RES phagocytic activity, which diverted circulating endotoxin away from those cells (i.e., neutrophils) responsible for EP production (Atkins 1960).

A number of observations had been reported, however, that did not fit the above hypothesis. Origin of EP from neutrophils was challenged by the studies of pyrogenic tolerance in patients with agranulocytosis, conducted by Page and Good (1957). These investigators reported that such patients developed a febrile response to typhoid vaccine 'at least as high as that produced when normal numbers of neutrophils were present in the blood'. It was concluded that 'injury to neutrophils with liberation of endogenous pyrogen plays no important role in development of fever following the intravenous injection of pyrogens in man'. Moreover, when repeated intravenous injections of typhoid vaccine were given, it was found that 'refractoriness to endotoxin develops in the complete absence of neutrophils in the circulating blood or in the blood forming tissues and consequently demonstrates that these cells are not essential to this defense reaction'.

That EP is a macrophage secretory product was reported in 1967 by Atkins et al and Hahn et al, who studied rabbit peritoneal macrophages, and by Bodel and Atkins who studied human blood monocytes. Thus the very cells hypothesized to be nonspecifically removing endotoxin from the circulation, thereby protecting endotoxin-sensitive cells, were in fact those cells responsible for the production of EP.

1.2. Recognition of early and late phases of tolerance

The work of many earlier investigators (Abernathy 1957; Beeson 1947a, 1947b, Braude et al 1958; Shear and Perrault 1944) had shown that lethal and pyrogenic tolerance developed within one or a few days following single or multiple daily injections of endotoxin. In the course of studies on pyrogenic tolerance conducted by Greisman et al (1965) in the mid-sixties, it became clear that a distinct refractory state rapidly developed during a continuous intravenous infusion of endotoxin, in volunteers not previously injected with the toxin. This refractory state occurred within hours, in both the rabbit and in man, and was found to be specific for endotoxins as a class. In the rabbit, a single intravenous injection of endotoxin induced a similar state of tolerance, when tested at 24 hours. When tested at 48 hours, however, such tolerance had waned markedly (Fig. 1).

In both the rabbit and man, it was found that if 5 days were allowed to elapse between endotoxin injection and assessment of pyrogenic tolerance, a form of tolerance with completely different characteristics occurred. In contrast to the early refractory state, this late tolerance was transferrable with serum and exhibited marked O-specificity. Thus the concept evolved that two distinct mechanisms of tolerance to endotoxin exist, designated early-phase and late-phase tolerance (Fig. 1).

Early-phase tolerance appears within hours of a single injection of endotoxin and wanes over the subsequent days. The onset of late-phase tolerance, on the other hand, is correlated with the appearance of circulating antiendotoxin antibodies. Repetitive daily injections of endotoxin over the course of several days or more activate both early- and late phase mechanisms.

The early-phase pyrogenic tolerance appears to reflect a refractoriness of macrophages of the RES to the further release of endogenous pyrogen in response to circulating endotoxin. The late-phase tolerance, on the other hand, appears to be mediated by specific humoral antibodies. The characteristics and our current understanding of the mechanisms of these phases are reviewed in the following pages.

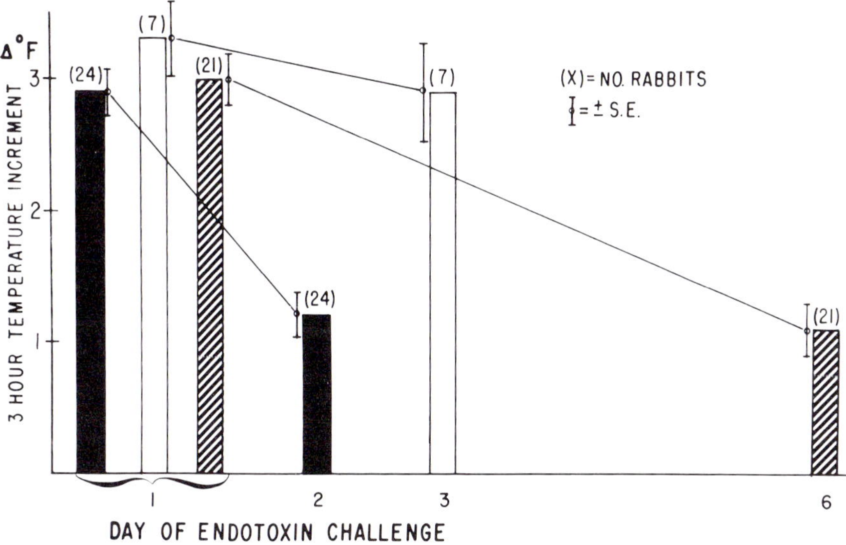

Fig. 1 *Temporal development of pyrogenic tolerance following a single intravenous injection of 0.5 μg/kg E. coli endotoxin. The initial responses of three groups of normal rabbits are comparable (day 1). Significant tolerance, designated early-phase tolerance, is seen at 24 hours (day 2). Animals tested at 48 hours (day 3) are no longer tolerant. By day 6 marked tolerance, designated late-phase tolerance, is again evident.*

2. CELLULAR MECHANISMS

2.1. The early refractory state

The mechanisms underlying the early-phase pyrogenic tolerance have been intensively studied with the use of continuous intravenous infusions of endotoxin (Greisman and Woodward 1965; Greisman et al 1966). In man, the initial fever and accompanying subjective discomfort, that is, headache, chills, myalgia, nausea, begin to wane after several hours despite the continuing or even increasing rates of infusion (Fig. 2). Volunteers express disbelief that the endotoxin is still being infused. This early-phase tolerance can be completely overlooked in man if the toxin is given in the usual fashion as a single intravenous bolus at daily intervals (Fig. 3).

Early-phase pyrogenic tolerance is transient and requires continuing endotoxin infusions (or closely spaced injections) for maximum maintenance. In man, if the initial infusion is discontinued after tolerance first appears and is resumed the following day, volunteers will have regained full responsive-

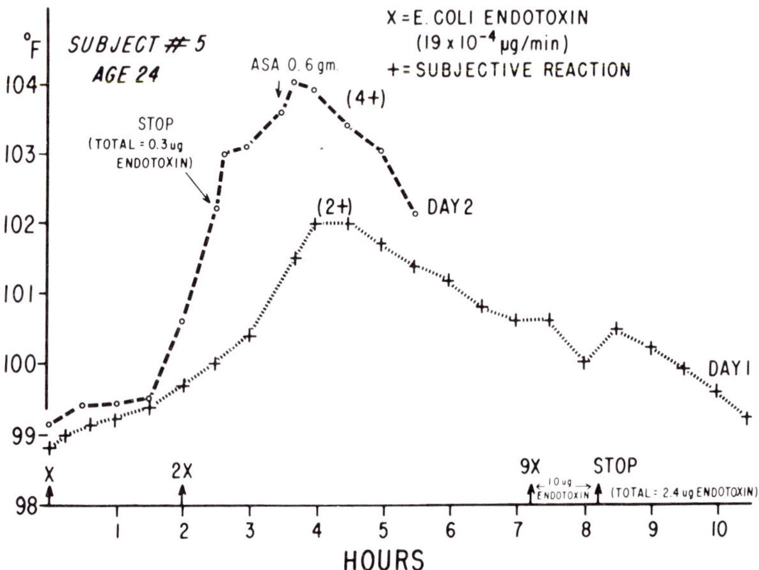

Fig. 2   *Typical pyrogenic response to an initial continuous intravenous infusion of endotoxin in a healthy volunteer. ASA = acetylsalicylic acid. Note the loss of the early tolerant state when the infusion is resumed on day 2. Reproduced from Greisman and Woodward (1965), with permission.*

ness – in fact, they usually hyperreact at this time (Fig. 2). In the rabbit, by contrast, pyrogenic tolerance persists for 24 hours, and then wanes over the course of several days (Fig. 1).

Within the sensitive dose–response range, the level of early-phase tolerance is directly proportional to the magnitude of the initial pyrogenic response.

The early-phase pyrogenic tolerance is specific for endotoxins as a class. No tolerance occurs to other pyrogens, such as staphylococcal enterotoxin, influenza virus, or old tuberculin in specifically sensitized animals (provided massive doses of endotoxin are not given). Thus early-phase tolerance does not simply reflect exhaustion of cellular stores of EP or its precursors. This is consistent with the finding that this phase of tolerance is relative and can be overcome by increasing the rate of endotoxin infusion. The normal response to other pyrogens also indicates that early tolerance is not based upon refractoriness to released EP. This was confirmed by infusions of sera containing preformed EP obtained from endotoxin-treated nontolerant animals.

Early tolerance is also not mediated simply by clearance of endotoxin from the circulation into those tissues that selectively detoxify the molecule, thereby diverting the endotoxin from 'target' cells. Pyrogenic quantities of endotoxin can be demonstrated in the circulation by passive transfer to normal animals at the time the donor animals, receiving the endotoxin infusion, have become afebrile and no longer have circulating EP. Early tolerance, moreover, is not based upon depletion of humoral substrates or leukocytes required for expression of fever, since infusion of large quantities of fresh

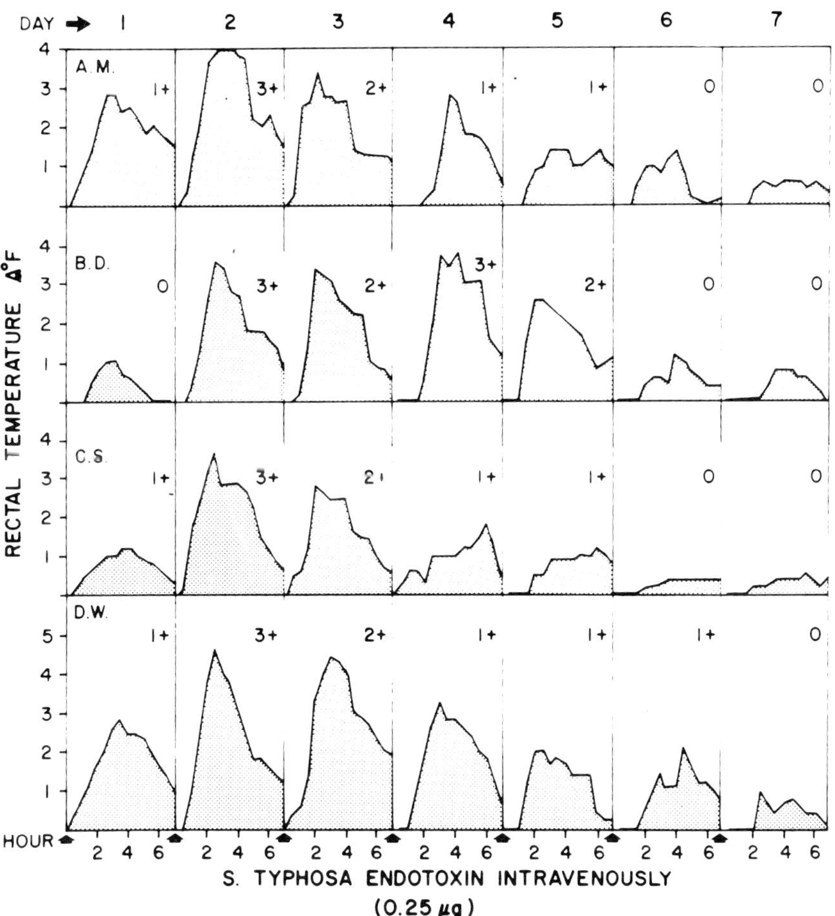

Fig. 3 *Development of tolerance to daily intravenous injections of endotoxin in healthy volunteers. Subjective toxic responses (headache, myalgia, chills, nausea) are graded from 0 to 4+. Note that tolerance does not appear to develop for several days by this method of administration and that responses are actually enhanced before tolerance develops. Reproduced from Greisman (1983), with permission.*

heparinized normal whole blood or plasma does not reverse the unresponsive state.

That early-phase pyrogenic tolerance is not mediated by circulating antibody is indicated by the facts that it cannot be transferred with plasma, that increments in antiendotoxin antibodies cannot be detected, that splenectomized animals and man readily demonstrate early tolerance, that 6-mer-

TABLE 1　*Studies on the specificity of early-phase tolerance actively induced by smooth LPS**

Feature studied	O-specificity of tolerance			Presumed role of antibody		Reference
	none	partial	complete	important	unimportant	
Pyrogenicity	+				+	Beeson 1947a, b
Pyrogenicity	+				+	Morgan 1948a, b
Lethality	+				deferred	Creech et al 1949
Lethality	+				+	Wharton and Creech 1949
Shwartzman reactivity	+				+	Cluff and Bennett 1951
Pyrogenicity *and* leuko-cyte changes		+			+	Cluff 1953
Pyrogenicity	+	+			+	Watson and Kim 1963
Pyrogenicity	+				deferred	Greisman et al 1964b
Pyrogenicity *and* lethality	+				+	Kim and Watson 1966
Pyrogenicity	+				+	Kim and Watson 1967
Shwartzman reactivity**		+	+		deferred	Kovats and Vegh 1967
Pyrogenicity	+				+	Greisman et al 1969b
Pyrogenicity	+				+	Milner 1973

*　In these studies, tolerance was induced by single or repetitive injections of endotoxin, and tolerance testing was performed within 1–2 days of the last injection.
**　In this study tolerance, induced by daily injections for 5 days, was assessed for the preparative, not the provocative, dose in the Shwartzman reaction.

captopurine-treated rabbits readily develop early tolerance, and finally, that no specificity of early tolerance exists between endotoxins from diverse bacterial species (Greisman et al 1969b; Milner 1973; Wolff et al 1965b; Table 1).

An early phase of lethal tolerance also occurs, with characteristics very similar to those of early pyrogenic tolerance. This early lethal tolerance has been demonstrated in mice 24 hours after a single intraperitoneal or intravenous injection by several investigators (Abernathy 1957; Urbaschek and Nowotny 1968; Wharton and Creech 1949). The presence of an early lethal tolerance was appreciated in early studies by Creech et al (1948), who reported: 'Extensive work is in progress to establish the nature of the pronounced tolerance which is established within 24 hours after the initial injection of polysaccharide. This problem is particularly interesting because of the nonspecific nature of this tolerance as developed in mice.'

Early-phase lethal tolerance appears to be dependent upon the physical characteristics of the test preparation of endotoxin. Using 'soluble' endotoxin preparations, Greer and Rietschel (1978b) have described a hyperreactivity to endotoxin lethality 24 hours after a single endotoxin injection in mice, followed by the appearance of marked tolerance by the fourth day. These 'soluble' preparations were treated with 'extensive heating and sonification'. By contrast, when mice were challenged with more conventional preparations of endotoxin (simply suspended in phosphate-buffered saline), designated 'insoluble', marked lethal tolerance was seen 24 hours after a single endotoxin injection.

As with pyrogenicity, early lethal tolerance extends to endotoxins of differing gram-negative species, including both smooth and rough strains (Abernathy 1957; Greer and Rietschel 1978b). Thus, no O-specificity is seen during early-phase tolerance (Table 1). Similarly, when searched for during the first or second days of lethal tolerance in guinea pigs (Urbaschek and Nowotny 1968) or mice (Wharton and Creech 1949), antiendotoxin antibodies have not been detectable. A sensitive enzyme-linked immunosorbent assay (ELISA) has also failed to show appreciable increments in specific antiendotoxin antibodies in rabbits before 72 hours of immunization (Walker and Beasley 1980).

2.2. Central role of the macrophage

In the case of the early phase of pyrogenic tolerance, it is clear that cells of the RES become refractory to the synthesis and/or release of EP in response to endotoxin. As mentioned above, the fever evoked by endotoxin has been shown to result from the secondary release of EP from host cells. These cells have been convincingly shown to be macrophages and not neutrophils (Hanson et al 1980; Windle et al 1983). Since the hepatic Kupffer cells are the primary site of localization of intravenously injected endotoxin (Braude et al 1955; Cremer and Watson 1957), these macrophages could comprise the major source of EP, and refractoriness of these cells could be of central importance. Several studies suggest that this indeed is the case.

Fukuda and Murata (1964, 1965) examined the pyrogenic activity of liver extracts from dogs given intravenous endotoxin and concluded: 'Thus, it is certain that the liver forms an easily extractable pyrogenic factor after taking up endotoxin in its reticuloendothelial cells ... The tolerance in fever response might result from the absence of the development of the hepatic pyrogen.' Elegant in vitro studies by Dinarello et al (1968) have confirmed these findings. Endotoxin was shown to stimulate isolated rabbit hepatic Kupffer cells to generate EP. Moreover, isolated Kupffer cells (but not blood leukocytes or lung macrophages) from rabbits made tolerant to intravenously injected endotoxin were found to be refractory to EP release by endotoxin. It was suggested that hepatic Kupffer cells may comprise the major source of EP after intravenous endotoxin administration, and that pyrogenic tolerance may be mediated by increased uptake of endotoxin by Kupffer cells which have become refractory to further EP release. Recent studies by Haeseler et al (1977) have confirmed the ability of isolated rabbit Kupffer cells to produce EP.

In vivo studies performed by Greisman and Woodward (1970) support the hepatic hypothesis of pyrogenic tolerance. By means of chronic indwelling portal vein cannulae, endotoxin was perfused directly into the hepatic circulation of rabbits, and pyrogenic responses compared with those after ear vein injections. If pyrogenic tolerance were based upon increased RES uptake of endotoxin, thereby protecting other cells, then less fever would be expected after direct hepatic perfusion with the toxin. This did not occur; indeed, higher fevers were elicited after intrahepatic infusion (Fig. 4). By contrast, after early-phase tolerance was induced by daily ear vein injections, direct hepatic perfusion with endotoxin resulted in marked reductions in fever.

Macrophages exposed to endotoxin display an early refractoriness to the secretion of a number of other mediators beside EP. An early acquired refractoriness develops in macrophages to the release of granulocyte-monocyte colony-stimulating activity (CSA) (Sullivan et al 1983). Human blood monocytes release large amounts of CSA in response to an initial exposure to endotoxin. Marked reductions in CSA release in response to a second endotoxin exposure develop over the following 72 hours. This effect was shown to be based upon cellular refractoriness to endotoxin, since it could not be related to enhanced degradation of endotoxin by monocytes, the accumulation of inhibitory metabolites, or a direct negative feedback inhibition by CSA itself. Rietschel et al (1982) have likewise shown that peritoneal

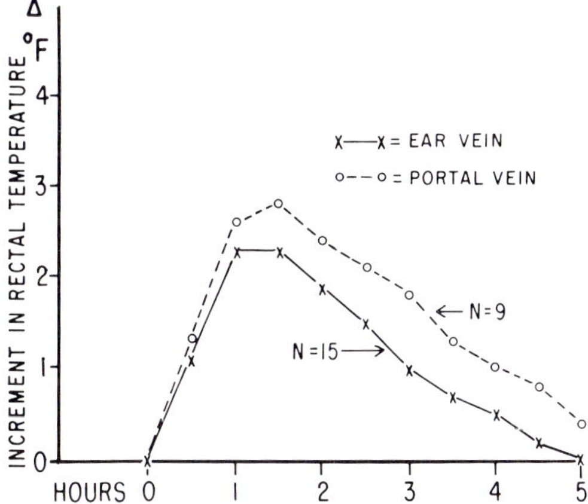

Fig. 4   *Comparative mean febrile response to 0.01 µg/kg E. coli endotoxin adminis-tered as an initial infusion via ear vein or via chronically implanted hepatic portal venous cannulae in healthy nontolerant rabbits. Reproduced from Greisman and Woodward (1970), with permission.*

macrophages from tolerant animals develop impaired secretion of prostaglan-dins (PGE_2 and $PGF_{2\alpha}$) in response to LPS. Tolerant macrophages secrete normal amounts, on the other hand, of prostaglandins in response to zymo-san.

Early refractoriness of macrophages to the release of the lysosomal enzyme β-glucuronidase also has been demonstrated during the tolerant state (Snyder et al 1979). Peritoneal macrophages from tolerant mice harvested 3 hours after challenge with a lethal dose of endotoxin released less β-glucuronidase in culture and retained more cellular enzyme, than macrophages from non-tolerant animals.

Thus, much evidence implicates an acquired refractoriness of macrophages to the release of biologically active mediators during early-phase tolerance. The cellular mechanisms underlying induced refractoriness to EP release have not been delineated as yet. The fact that this refractoriness occurs to endotoxins from diverse gram-negative organisms, but does not extend to pyrogens of other bacteria or viruses, suggests that specific recognition of a common grouping such as lipid A at the cellular level is involved. Greer and Rietschel (1978a) have shown that early tolerance to hypothermia in mice (this species shows a depression rather than elevation of core temperature in response to endotoxin injection) is due to the lipid A component.

Since early lethal tolerance can also be induced by free lipid A (Greer and Rietschel 1978b), it is likely that, as in early pyrogenic tolerance, cellular recognition of the lipid A moiety is of central importance. It is clear that the induction of early tolerance is not related to the degree of glycosylation of the endotoxin core. Thus, endotoxins from smooth organisms (Table 1) and rough mutants (Table 4, below), and purified lipid A, all readily elicit early tolerance.

For early-phase lethal tolerance, there is again evidence that RES macrophages play an important role. In 1965, Heilman reported that cultured macrophages from a variety of normal rabbit tissues were highly susceptible to the cytotoxic activity of endotoxins. Such cytotoxicity has been confirmed by Glode et al (1977) and Peavy and Brandon (1980), who in addition have found that macrophages from the genetically resistant C3H/HeJ mice exhibit marked resistance to the cytotoxic effects of endotoxin. Maier and Ulevitch (1981) have recently demonstrated high levels of endotoxin-induced cytotoxicity for hepatic macrophages of rabbits in vitro and in vivo.

A large number of secretory products of macrophages in response to endotoxin have been described, which may play important roles in lethality. Potentially important mediators include lysosomal enzymes, superoxides, prostaglandins, procoagulant factor, glucocorticoid-antagonizing factor (GAF) (Moore et al 1978), and macrophage insulin-like activity factor (MILA). Release of lysosomal enzymes has been postulated by Janoff et al (1962). Pabst and Johnston (1980) demonstrated that mouse macrophages exposed to endotoxins become primed to generate enhanced superoxide anion, and suggested that '... this same mechanism could also permit macrophages to mediate oxidative tissue injury'. Prostaglandin production has been demonstrated in response to endotoxin (Schade and Rietschel 1982). Procoagulant factor release may contribute to disseminated intravascular coagulopathy (DIC), as in typhoid fever (Hornick and Greisman 1978). This concept has been extended by Maier and Ulevitch (1981) to include hepatic macrophage involvement in the initiation of DIC during endotoxemic shock. Filkins (1980) has proposed that endotoxic hypoglycemia results from the combined effects of GAF impairment of gluconeogenesis and MILA augmentation of tissue glucose uptake.

The relative importance of direct cytotoxicity and the production of these diverse mediators in the pathogenesis of endotoxemic mortality is not yet clear. Peavy and Brandon (1980), correlating mortality and in vitro cytotoxicity during BCG (Bacille Calmette-Guérin) infection, reported: 'Our results demonstrate that increases or decreases in the susceptibility of mice to the lethal effects of LPS are paralleled by similar changes in susceptibility of their macrophages to LPS in vitro.' Maier and Ulevitch (1981) have concluded that 'the combined effects to cytotoxicity and selective stimulation and release of mediators in LPS-stimulated hepatic macrophages may play a central role in the endotoxemic shock syndrome'.

Further support for the importance of the RES in mediating endotoxic lethality is derived from studies of the effects of a variety of agents modulating RES functions. Stimulation of the phagocytic activity of the RES by such agents as zymosan, glucan, BCG, triolein, graft-versus-host reactions, or *Brucella* infection results in markedly enhanced susceptibility to endotoxin lethality. It has been proposed that the common denominator is the production of a 'systemically activated macrophage system' (Berendt et al 1980). Conversely, agents which depress RES phagocytic activity for inert colloidal particles, such as ethyl stearate or methyl palmitate, reduce endotoxin lethality (Benacerraf et al 1959; Cook et al 1980; Cooper and Stuart 1961; Di Luzio and Crafton 1970; Howard 1961; Lemperle 1966; Stuart and Cooper 1962; Suter et al 1958; Ueda et al 1966). Glucocorticoids, which reduce endotoxic cytotoxicity for macrophages in vitro, reduce host mortality following endotoxin challenge, whereas structurally related steroids without antiinflammatory activity fail to protect (Peavy and Brandon 1980).

· Two decades ago, Cooper and Stuart (1961) concluded: 'Certainly it seems that the more phagocytes present in the liver and spleen the more sensitive are animals to endotoxin; on the other hand, decrease in cell numbers appears to have, if anything, a slightly beneficial effect on the resistance of the animals.' This was subsequently supported by Agarwal's observations that splenectomized animals are more resistant to endotoxic lethality (Agarwal et al 1972).

Recent investigations employing the genetically endotoxin-resistant C3H/HeJ mouse indicate that sensitivity to endotoxin lethality is transferrable by radiosensitive cells present in bone marrow and spleen (Brown and Morrison 1982; Glode et al 1976; Michalek et al 1980; Rosenstreich and Vogel 1980). Macrophages are thought to be the important cells transferring sensitivity. Vogel and Mergenhagen (1982) have concluded that 'the susceptibility of a host to the toxic effects of LPS appears to be an accurate reflection of the state of activation exhibited by the host's macrophages'.

The relationship between macrophage responsiveness and susceptibility to endotoxin lethality has been further examined in the C3H/HeJ mouse through studies employing preinfection with BCG. Peavy and Brandon (1980) found that BCG infection of endotoxin-resistant C3H/HeJ mice, in contrast to normal mice, had no sensitizing effect on either endotoxin lethality or cytotoxicity for macrophages. Another group of investigators, however, has reported a marked increase in endotoxin lethality induced by preinfection of C3H/HeJ mice with BCG (Vogel et al 1980, 1982). Surprisingly, macrophages isolated from these BCG-treated C3H/HeJ mice did not demonstrate augmented release of EP or prostaglandins in vitro in response to endotoxin. T-lymphocytes from such BCG-treated mice, however, when cocultured with resistant C3H/HeJ macrophages, markedly increased their responsiveness to endotoxin (Vogel et al 1982).

Responses of the RES thus play a crucial role in the pathophysiology of endotoxin lethality. Tolerance to lethality presumably reflects reticuloendothelial cellular adaptations which result in diminished cytotoxicity and/or mediator release.

2.3. Phagocytic activity of the reticuloendothelial system

Beeson's classic studies of pyrogenic tolerance advanced the hypothesis that tolerance resulted from enhanced phagocytosis of circulating endotoxin by the RES. Central to the hypothesis was the finding that animals rendered tolerant by daily injections of endotoxin were resensitized by the administration of an RES 'blockading' agent, thorotrast (thorium dioxide). Pyrogenic tolerance thus appeared to be 'abolished' by this agent, which was also found to retard the clearance of circulating endotoxin as determined by passive transfer studies (Beeson 1946, 1947b).

Subsequent studies have shown that although thorotrast markedly increases the pyrogenic reactivity of both normal and tolerant animals, the striking differences in reactivity between normal and tolerant animals persist; that is, tolerance has not been 'abolished' (Greisman et al 1963; Wolff et al 1965a).

Dinarello et al (1968) observed that isolated lung macrophages and blood leukocytes from tolerant rabbits, in contrast to Kuppfer and spleen cells, had not acquired resistance to the generation of EP in response to endotoxin. Thus it was hypothesized that following 'blockade' of the RES with thorotrast, endotoxin is diverted to other EP-producing macrophages that have not become resistant. However, since normal rabbits also exhibit marked enhancement of pyrogenic responses after thorotrast, the concept of diversion of endotoxin from the RES to other EP-producing cells requires the additional assumption that in the nontolerant animal these other cells are intrinsically more responsive than those of the RES macrophages. Available data, both in vitro (Dinarello et al 1968) and in vivo (Greisman and Woodward 1970) do not support this presumption.

An alternative possibility is that thorotrast acts directly upon RES macrophages to enhance EP generation in response to endotoxin in both tolerant and normal animals, without abolishing the underlying cellular and humoral mechanisms responsible for tolerance. Evidence is available to support this possibility. Isolated liver cells from endotoxin-tolerant rabbits given thorotrast, although initially reported to remain refractory to EP release (Dinarello et al 1968), have been found upon more extensive study to regain their capacity to release EP in response to endotoxin (Dinarello 1969).

As in experimental animals, there is evidence that in man the general level of RES phagocytic activity per se is not correlated with tolerance to endotoxin. RES phagocytic activity has been studied in informed healthy volunteers given endotoxin daily for one week. During the development of pyrogenic tolerance, no enhancement of nonspecific RES phagocytic activity occurred, as measured by blood clearance of radioactively labeled aggregated human serum albumin (Greisman et al 1964b).

In another study, volunteers were infected with gram-negative bacteria (*Salmonella typhosa* or *Pasteurella tularensis*) inducing subacute febrile illnesses (Greisman et al 1964a, 1969a). During the latter part of the incubation period of these illnesses, progressively increasing pyrogenic reactivity developed to intravenous endotoxin. The hyperreactivity persisted throughout the illness and progressively declined during the early phase of convalescence. These changes in pyrogenic reactivity also showed no correlation with nonspecific RES phagocytic activity as measured by blood clearance of radioactively labeled aggregated human serum albumin (Fig. 5). In addition, the mechanisms of pyrogenic tolerance were found to remain functional within this hyperreactive framework. Thus, despite the enhanced initial responses, the early pyrogenic refractory state was readily inducible during illness, with both continuous intravenous infusions and daily single injections. Likewise, when both early- and late-phase tolerance mechanisms were activated prior to illness by repetitive daily intravenous injections of endotoxin, the tolerant volunteers remained significantly less responsive to the endotoxin during illness than did control nontolerant volunteers, although both groups became comparably ill and exhibited hyperreactivity to the endotoxin.

These findings simulate those in rabbits given thorotrast, in that the pyrogenic tolerance mechanisms remain functional within a hyperreactive framework. In typhoid fever in man, however, it can be more readily inferred that the hyperreactivity is based upon enhanced RES responsiveness to EP generation, rather than upon reduced RES clearance with diversion of endotoxin to more susceptible tissues, since the blood clearance rates of radiolabeled endotoxin are not retarded (Greisman et al 1969a).

Studies employing exchange transfusion also argue against a relationship between the rate of clearance of endotoxin from the circulation and lethal tolerance. Normal rabbits were injected intravenously with an LD_{80} dose of endotoxin and 20 minutes later, after the RES had taken up approximately 50% of the toxin, the remainder in the circulation was rapidly reduced by exchange transfusion to levels seen in tolerant animals (Greisman and DuBuy 1975). Although such artificial removal of endotoxin from the circulation of normal animals closely simulated the blood clearance seen in tolerant animals (Fig. 6), mortality rates were not significantly reduced. These findings have been confirmed by Gollan and McDermott (1979), who also were unable to prevent mortality in dogs given endotoxin intravenously and

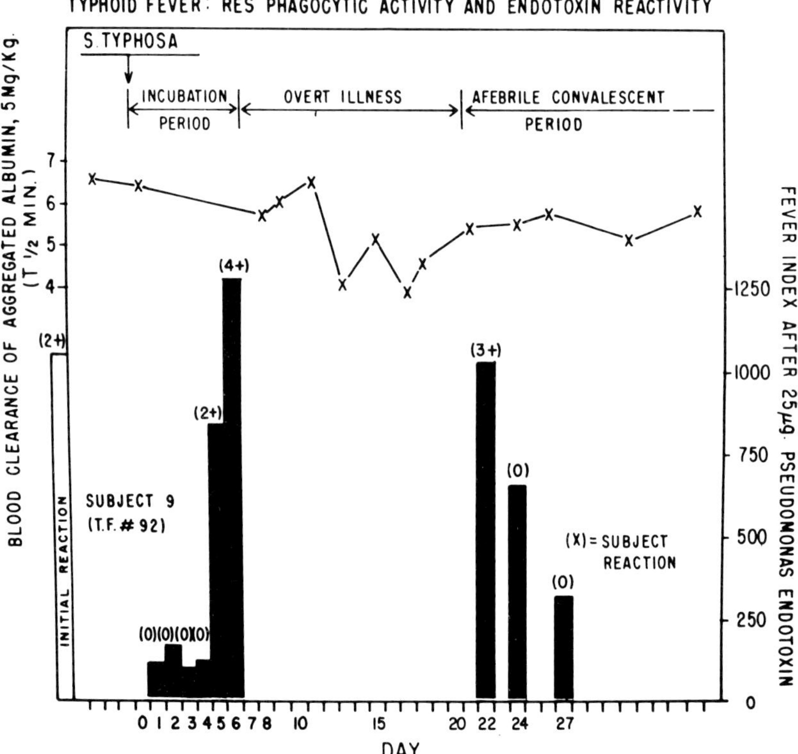

Fig. 5   *Typical pattern of response of informed volunteers to endotoxin during typhoid fever. Tolerance was induced before typhoid fever by daily intravenous injections of Pseudomonas endotoxin in increasing doses from 25 to 250 µg over a 30-day period. The initial control response is shown by the open bar on the left. Responses to 25 µg of the Pseudomonas endotoxin following oral infection with viable Salmonella typhosa are shown by the solid bars. Note the hyperreactivity during the latter portion of the incubation period and its subsidence during early convalescence. The nonspecific phagocytic activity of the RES as measured by the blood clearance of radiolabeled aggregated human serum albumin showed no correlation with changes in endotoxin reactivity. Numbers in parentheses represent the intensity of subjective toxic responses (headache, chills, myalgia, nausea) graded from 0 to 4+. Reproduced from Greisman (1983), with permission.*

exchange-transfused 5 minutes later. These investigators concluded: 'Since the endotoxin particles are rapidly phagocytosed by the reticuloendothelial system, even very early total blood exchange cannot dislodge them from their intracellular site.'

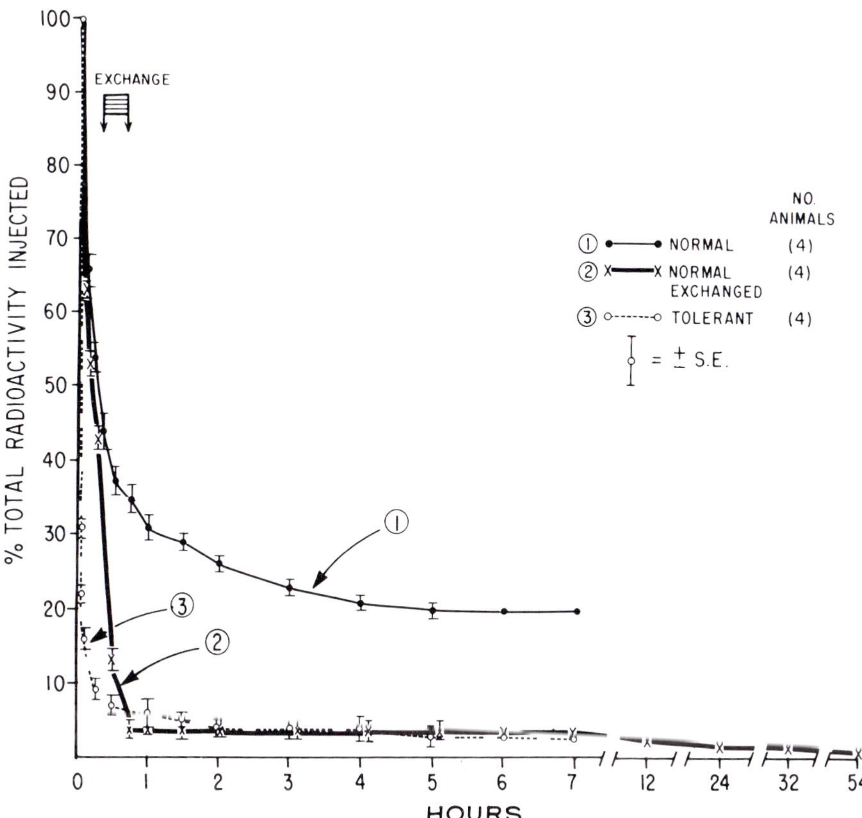

Fig. 6   *Blood clearance of 2500 μg of ^{51}Cr-tagged E. coli endotoxin in normal (curve
1), normal exchange-transfused (curve 2), and tolerant (curve 3) rabbits. Note the
prolonged circulation of high concentrations of endotoxin in normal animals and the
ability of exchange transfusions to simulate the endotoxin clearance curve of the tolerant
animal. Reproduced from Greisman and Dubuy (1975), with permission.*

Furthermore, in 1966 Chedid et al reported that, in contrast to endotoxins
from smooth gram-negative bacteria, endotoxins from rough mutants were
cleared equally rapidly from the blood of nontolerant and tolerant mice.
They concluded: 'It is, therefore, difficult to explain the increased resistance
of the tolerant mouse against R (rough) endotoxin solely on the basis of its
greater disappearance from the blood.'
Studies using RES-modifying agents in rats, similarly, have not shown any
correlation between RES clearance of endotoxin and endotoxin lethality.
Despite a significant reduction in susceptibility to endotoxin lethality induced

by methyl palmitate, no alteration in endotoxin clearance rates could be detected by Di Luzio and Crafton (1970). Hepatic and pulmonary localization was unaltered 30 minutes after injection. Furthermore, animals treated with both glucan and methyl palmitate exhibited normal phagocytic clearance of a lipid emulsion, but displayed enhanced susceptibility to endotoxin. Di Luzio and Crafton (1970) have concluded that '... alterations in endotoxin susceptibility are not readily explained on the rate of vascular clearance of endotoxin'.

Whereas alterations in RES phagocytic activity per se do not explain tolerance, there is in vitro evidence that the RES acquires the ability to 'detoxify' endotoxin when animals are given endotoxin repetitively as single daily intravenous injections. Rutenberg et al (1965) initially demonstrated that detoxification of endotoxin by rabbit splenic homogenates increased appreciably after repetitive injections, waned within 10 days after endotoxin was discontinued despite elevated antiendotoxin antibody titers, and increased again within several days upon resumption of endotoxin administration. These findings were confirmed and extended by Trejo and Di Luzio (1972), using homogenates of mouse liver and spleen. Thus, enhanced detoxification at the cellular level develops after several days of endotoxin administration.

The importance of enhanced detoxification of endotoxin is supported by the results of interfering with this mechanism. Using sublethal doses of carbon tetrachloride, Farrar et al (1968) demonstrated that normal rabbits became about 40 times more susceptible to the pyrogenic activity of endotoxin when tested two days afterwards. Five days after the carbon tetrachloride, febrile responses returned to baseline. Normal liver homogenates were capable of detoxifying endotoxin in vitro, while those from carbon tetrachloride-treated rabbits were not. Other than carbon tetrachloride, hepatotoxins that depressed RES phagocytic function but did not interfere with fatty acid oxidation did not impair the ability of liver homogenates to inactivate endotoxin (Wiznitzer et al 1960). The effects of thorotrast may be related to its reported ability to impair detoxification of endotoxin by RES (Wiznitzer et al 1960).

The importance of enhanced detoxification of endotoxin by RES in the first hours of early-phase tolerance has not been delineated. It was not discernable, however, in one series of studies on early pyrogenic tolerance (Greisman and Woodward 1965), which suggests either that other mechanisms for refractoriness of macrophages to the release of EP are operative at this time, or that the assay employed was not sufficiently sensitive to detect the onset of enhanced detoxification.

3. HUMORAL MECHANISMS

Whereas the early-phase tolerance is specific for endotoxins as a class, late-phase tolerance to smooth endotoxins is specific for the species and even the strain of endotoxin employed. Such specificity reflects the antigenic specificity of circulating antibodies. Thus, in contrast to early-phase tolerance, any consideration of late-phase antibody-mediated mechanisms must carefully consider the antigenic nature of the endotoxin preparations used for induction and testing of tolerance. Most investigations of the role of antibodies in endotoxin tolerance have made use of smooth gram-negative organisms or their purified endotoxins for tolerance induction. These elicit antibodies primarily directed against the O-polysaccharide side chains (Hiernaux et al 1983; Michael and Mallah 1981; Walker and Beasley 1980; Westphal et al 1983; Ziegler et al 1982).

Recent studies have employed rough bacterial mutants or purified and solubilized lipid A for the preparation of high-titered antisera with specificity for core polysaccharide or lipid A structures. These antibodies are far less readily elicited with endotoxins from smooth organisms (Galanos et al 1977; Hiernaux et al 1983; Michael and Mallah 1981; Walker and Beasley 1980; Ziegler et al 1982). As might be expected from these considerations, and as will be reviewed critically, antibodies directed against core polysaccharides and lipid A appear to play little role in tolerance induced by endotoxins from smooth organisms. However, studies of the potential of such antibodies, induced by hyperimmunization with rough mutants or purified lipid A, to protect against challenge with endotoxins from smooth strains merit careful consideration.

In the following discussion, endotoxins prepared from wild-type smooth organisms are termed smooth endotoxins. Likewise, endotoxins from rough mutant strains are termed rough endotoxins.

3.1. O-specific antibodies

It is now clear that when smooth endotoxins are administered, a late phase of pyrogenic and lethal tolerance can be produced that is mediated by circulating antibodies with specificity for the O-polysaccharide of endotoxin (Abernathy 1957; Greisman and Young 1969; Greisman et al 1964b, 1969b, 1973, 1975; Milner 1973; Ralovich et al 1974). Onset of this phase is delayed, requiring several days to appear after a single intravenous injection of a highly immunogenic endotoxin (Fig. 1). It does not occur after a single intravenous injection of comparably toxic but poorly immunogenic endotoxin preparations, such as phenol-water preparations.

The onset of late-phase tolerance correlates with the appearance of circulating antibodies with specificity for the O-polysaccharide of the administered endotoxin. Most importantly, late-phase tolerance exhibits high levels of O-specificity (Table 2). This tolerance is enduring, persisting for weeks to months.

Unlike early-phase tolerance, late-phase pyrogenic tolerance is not directly related to the magnitude of the initial pyrogenic response. Tolerance in informed healthy volunteers on day 7 was found to be comparable regardless of whether they were intravenously immunized with 0.01 or 0.001 µg/kg *Escherichia coli* endotoxin. The former dose elicited marked febrile and subjective toxic responses, whereas the latter elicited no reactions. The factor common to both doses of endotoxin was their ability to evoke rises in

TABLE 2   *Studies on the specificity of late-phase tolerance actively induced by smooth LPS**

Feature studied	O-specificity of tolerance			Presumed role of O-antibody		Reference
	none	partial	complete	important	unimportant	
Lethality	+				+	Zahl and Hutner 1944
Lethality	+	+			deferred	Abernathy 1957
Lethality	+				deferred	Schaedler and Dubos 1961
Lethality	+				deferred	Mergenhagen and Jensen 1962
Leukocyte changes	+				deferred	Ritts et al 1964
Pyrogenicity	+			+		Greisman et al 1969b
Pyrogenicity		+		+		Ralovich and Lang 1972
Pyrogenicity		+		+		Milner 1973
Pyrogenicity		+			deferred	Ralovich et al 1974
Hypotension**		+			deferred	Abdelnoor et al 1981

*   In these studies, tolerance was induced by single or repetitive endotoxin injections, and tolerance was tested at least 5 days after the last injection.

**   These conclusions are based upon studies employing lipid-deficient endotoxin or purified polysaccharide, administered on days 0 and 6 or 7, with tolerance tested the following day. Since these preparations lack lipid A, early tolerance was not induced.

O-specific antibody by the day of tolerance testing (Greisman et al 1975). Similar results have been observed in rabbits.

Experiments employing chemically modified endotoxins support the importance of O-specific antibodies. Noll and Braude (1961) demonstrated late-phase pyrogenic tolerance in rabbits by the use of a single intravenous injection of a nonpyrogenic *E. coli* endotoxin produced by treatment with lithium aluminum hydride. This contrasts with Westphal's (1963) study in which late-phase pyrogenic tolerance could not be induced with a preparation of *E. coli* endotoxin treated with hydrogen peroxide to destroy its sugar components but not its pyrogenicity.

Late-phase tolerance is transferable with serum, and with both IgG and IgM fractions (Table 3). It has long been known that antibodies with O-specificity appear in the serum of actively immunized animals and man (Boivin and Mesrobeanu 1938; Favorite and Morgan 1942; Perlman and Goebel 1946; Wharton and Creech 1949). In early studies, such antibodies were shown to convey tolerance to endotoxin lethality which was entirely O-specific when immune rabbit serum was passively transferred to mice (Boivin and Mesrobeanu 1938). Using gamma-globulin fractions from immune rabbit serum, Creech et al (1948) demonstrated strain specificity of the lethal tolerance conferred against purified *Serratia marcescens* polysaccharides by passive transfer. In early quantitative studies of precipitating antibody to Type III somatic antigen of *Shigella paradysenteriae*, Perlman and Goebel (1946) concluded that: 'Antibodies to intact dysentery bacilli apparently contain no toxin-neutralizing antibodies other than those which precipitate the purified somatic antigen.'

Absorption studies have shown that only the homologous O-polysaccharide reduces the protective activity of smooth antiendotoxic sera (Milner 1973; Rioux-Darrieulat et al 1978).

Tolerant responses, moreover, can be transferred with spleen cells from endotoxin-immunized donors. Such anamnestic pyrogenic tolerance partly extends to heterologous endotoxins, but is most marked against the O-specific endotoxin (Greisman and Young 1969).

In man, late-phase tolerance can be overcome in part by dividing the dose of endotoxin in two and administering the second portion 2 hours after the first. This has been attributed to the Danysz phenomenon, wherein the initial dose of toxin binds a high proportion of protective antibody, allowing the second half to act with less antibody inhibition (Greisman et al 1964b). This phenomenon has recently also been demonstrated by Ogata and Kanamori (1978) during in vitro neutralization of endotoxin by its O-specific antiserum. That the reversal of late-phase tolerance in man by dividing the dose of endotoxin is based, at least in part, upon binding of protective antibodies is suggested by studies employing ^{51}Cr-labeled *Pseudomonas* LPS (S.E. Greis-

TABLE 3 *Studies on the specificity of tolerance passively induced by antiserum to smooth LPS*

Feature studied	O-specificity of tolerance			Presumed role of O-antibody		Reference
	none	partial	complete	important	unimportant	
Lethality			+	+		Boivin and Mesrobeanu 1938
Lethality			+	+		Perlman and Goebel 1946
Pyrogenicity*					+	Beeson 1947a, 1947b
Lethality			+		deferred	Creech et al 1948
Lethality		+	+		deferred	Creech et al 1949
Lethality		+	+	+		Wharton and Creech 1949
Shwartzman reactivity*					+	Cluff and Bennett 1951
Pyrogenicity *and* leukocyte changes*					+	Cluff 1953
Lethality	+				+	Freedman 1959
Pyrogenicity	+				+	Kim and Watson 1965
Pyrogenicity, lethality *and* Shwartzmann reactivity			+	+		Nowotny et al 1965
Shwartzman reactivity			+	+		Radvany et al 1966
Lethality	+				+	Tate et al 1966
Lethality	+				+	Davis et al 1969
Pyrogenicity		+	+	+		Muller-Ruchholtz and Sprock-hoff 1969
Pyrogenicity			+	+		Ralovich and Lang 1971
Shwartzman reactivity		+			+	Braude and Douglas 1972
Shwartzman reactivity	+				+	Braude et al 1973
Pyrogenicity		+	+	+		Greisman et al 1973
Pyrogenicity			+	+		Milner 1973
Lethality *and* Shwartzman reactivity			+		deferred	Ziegler et al 1973b
Pyrogenicity			+		deferred	Ralovich et al 1974
Pyrogenicity			+	+		Ogata et al 1975
Lethality			+		deferred	McCabe et al 1977
Pyrogenicity			+	+		Ogata and Kanamori 1978
Abortion			+	+		Rioux-Darrieulat et al 1978
Lethality			+	+		Johns et al 1983
Lethality			+	+		Johnston and Greisman 1983

* No tolerance detected.

man and H.N. Wagner Jr., unpublished observations). In each of 8 tolerant volunteers, given daily intravenous injections, clearance from the circulation of the lipopolysaccharide (LPS) increased progressively. Reversal of this accelerated blood clearance of LPS, and of tolerance, was seen when the labeled LPS was given 2 hours, rather than 24 hours, after a prior half dose of unlabeled LPS. That anti-LPS antibodies with O-specificity mediate such clearance changes can be inferred from the observations of Goodman et al (1969). These investigators employed isolated perfused rabbit livers to examine the effect of humoral and cellular factors on the clearance of LPS from the circulation. Late-phase tolerant serum significantly enhanced the clearance of homologous ^{51}Cr-labeled *E. coli* LPS by the normal liver. Normal serum, on the other hand, did not enhance clearance in livers from either normal or tolerant animals. Cellular mechanisms appeared less important since livers from tolerant and normal animals exhibited comparable rates of LPS clearance when perfused with either normal or tolerant sera. In unpublished studies by J.S. Goodman and L.L. Treanor, both IgG and IgM fractions of tolerant serum were found to enhance hepatic uptake of LPS. Moreover, O-specificity was demonstrated by the finding that serum from animals made comparably tolerant to LPS from *Salmonella enteritidis* or *Shigella dysenteriae* failed to enhance hepatic clearance of the *E. coli* LPS.

The most definitive evidence for the capability of O-specific antibodies to confer late-phase tolerance has recently been obtained by immunizing rabbits with a pyrogen-free O-antigenic dodecasaccharide (Lindberg et al 1983). This material was prepared by cleaving the O-polysaccharide side chains from *Salmonella typhimurium* endotoxin by means of a bacteriophage glycanase and then coupling this haptene to human serum albumin. Despite the absence of lipid A and core polysaccharides in this immunogen, and the absence of any pyrogenic or Limulus lysate gelation activity, immunized animals developed significant pyrogenic tolerance to *S. typhimurium* endotoxin. Such tolerance did not extend to *Salmonella thompson.*

The general recognition of this late-phase tolerance and the role of O-specific antibodies has been slow in coming. The conclusions derived from studies conducted over the past 45 years with regard to the role of O-specific antibodies in tolerance are summarized in Tables 1–3. Although distinction between early- and late-phase tolerance was often not made in these studies, for critical analysis they are so distinguished in Tables 1 and 2. Table 3 lists the conclusions of passive transfer studies.

The early studies conducted by Boivin and Mesrobeanu (1938), employing passively transferred sera, clearly pointed to an important role for anti-O antibodies in lethal tolerance. Studies reported over the next 30 years, however, largely concluded that O-specific antibodies played no significant role in tolerance to endotoxin. Seven important reasons may be advanced for the neglect of the important role of O-specific antibodies:

a. Many investigators seeking to evaluate the role of antibody were unaware of the need to allow the early cellular mechanisms to subside before testing for specificity of tolerance. An apparent lack of specificity of tolerance will hence be observed. It is interesting to note that the tabulated data in the careful study by Mulholland et al (1965) documented clearly the occurrence, extent, and duration of early- and late-phase tolerances. Yet, these investigators questioned a role of antibody, since the two phases of tolerance were not recognized to be separate entities, and antibodies to endotoxin were not detectable during the early phase of tolerance. Zaldivar and Scher (1979) also questioned the role of antibody in lethal tolerance. They demonstrated marked lethal tolerance 8 days following a single intravenous endotoxin injection in B-lymphocyte-defective (CBA/N) mice, apparently incapable of antiendotoxin antibody synthesis. Persistence of early-phase tolerance for this length of time seems likely, probably related in part to the very large initial dose of endotoxin employed.

b. The degree of tolerance observed in the early phase can be markedly greater than that obtainable in the late phase. Thus, many studies have overlooked the presence of a significant late-phase tolerance when O-antibody titers were elevated, since the degree of such tolerance appeared relatively unimpressive.

c. 'Naturally occurring' antibodies with O-specificity have been identified in the sera of both experimental animals and man (Michael et al 1962). For example, due to the prevalence of such antibodies in man against *Pseudomonas aeruginosa*, commercial normal human gamma-globulin fractions have been found to confer distinct protection against *Pseudomonas* infection in animals (Rosenthal et al 1957). The need to examine pre-immune serum from the same donor, before drawing conclusions regarding the production of humoral factors with a broad spectrum (i.e., non-O-specific) of protective activity after immunization with gram-negative bacterial antigens has been recently emphasized (Greisman et al 1979; Peter et al 1982).

d. Endotoxin is a polyclonal B-cell activator. Its ability to induce increments in a variety of O-specific antibodies must be considered before alternative 'broad spectrum' tolerance mechanisms are postulated (Michael et al 1961).

e. Some investigators did not take precautions to ensure that serum or serum fractions employed in passive transfer studies were pyrogen-free. Contamination of such fractions with endotoxin will yield a tolerance that appears to be unrelated to O-antibodies, since it is based upon activation of the early-phase tolerance mechanisms. In one classic study, serum from endotoxin-tolerant rabbits immunized with smooth heat-killed *E. coli* O8 was fractionated on DEAE and Sephadex G-200 columns, and the fractions employed for passive transfer studies of endotoxin tolerance (Kim and Watson 1965). No precautions to free the columns of contaminating endotoxins were indicated, nor was the pyrogen content of the serum fractions specified. The findings were interpreted as evidence for a key role of antibody to common toxophore antigens, and the unimportance of O-specific antibody in the mediation of pyrogenic tolerance. However, normal rabbit serum fractions, prepared in similar fashion without strict precautions to remove contaminating endotoxins, are consistently pyrogenic and evoke classic nonspecific early phase tolerance (S.E. Greisman, unpublished observations).

f. Some investigators have employed poorly immunogenic preparations of smooth endotoxins, such as those extracted by the phenol-water method.

g. In studies of the late phases of pyrogenic and lethal tolerance, both active and passive, conclusions that anti-'O' antibodies or indeed any antiendotoxin antibodies were unimportant were generally based upon testing performed outside the sensitive dose–response range for fever or lethality. The use of excessively high test doses of endotoxin prevents detection of significant late-phase tolerance.

It must be noted that although late-phase pyrogenic tolerance conferred by sera or serum fractions is consistently most marked against the homologous O endotoxin, some degree of tolerance can be detected against heterologous O endotoxins. This was initially documented with the use of rabbits given plasma from *E. coli* endotoxin-immunized donors and tested with *S. typhosa* endotoxin (Greisman et al 1963), as well as human volunteers given plasma from *S. typhosa* endotoxin-immunized donors and tested with *Pseudomonas* endotoxin (Greisman et al 1964b). In quantitative studies of passively transferred antisera, however, such heterologous late-phase pyrogenic tolerance could be demonstrated only by using considerably greater quantities of plasma than required to provide homologous protection (Greisman et al 1973). Passive transfer studies by Milner (1973), Muller-Ruchholtz and Sprockhoff (1969), Ogata et al (1975), Ogata and Kanamori (1978), Ralovich and Lang (1971) and Ralovich et al (1974) confirm these findings (Table 3).

Lethal tolerance conferred by passively transferred antisera to smooth endotoxins also exhibits marked O-specificity (Boivin and Mesrobeanu 1938; Creech et al 1948; Johnston and Greisman 1983; McCabe et al 1977; Perlman and Goebel 1946). It must be noted that two early studies did not demonstrate O-specificity (Davis et al 1969; Tate et al 1966). These studies, from the same laboratory, concluded that the difference in degree of protection afforded mice by rabbit antisera to homologous and heterologous smooth endotoxins was not statistically significant. However, a subsequent investigation from this laboratory (Ziegler et al 1973b), examining the protection conferred by human antisera against endotoxic lethality in mice, reported that in contrast to homologous protection: 'Attempts to demonstrate heterologous protection by using *E. coli* antiserum against *S. typhimurium* endotoxin, and vice versa, were not successful.' The most recent publication from this laboratory (Ziegler et al 1982), referring to antiserum to smooth LPS, now concludes: 'Such antiserum protects against the biologic effects of lipopolysaccharide from the immunizing strain, but is much less effective against lipopolysaccharide from other strains.'

Thus, there is now general agreement that animals immunized with smooth endotoxins develop circulating O-specific antibodies protective against the pyrogenic and lethal activities of the homologous endotoxin (Tables 2 and 3). The mechanisms whereby antibodies with specificity for the O-polysaccharides protect against the toxic effects of the covalently linked lipid A remain a matter of speculation.

3.2. Core polysaccharide antibodies

The marked antigenic diversity between gram-negative bacteria of differing species, and even between strains of the same species, has impaired efforts to produce a 'universal' gram-negative vaccine. Whereas the O-polysaccharides differ markedly in structure, the differences in lipid A and core polysaccharide structures of various species are known to be far less (Westphal et al 1983). Many investigators have therefore explored the possibility that antibodies with specificity for these structures might confer broad spectrum protection against the effects of gram-negative endotoxins.

As mentioned above, antibodies to core polysaccharides and lipid A are difficult to elicit when O-polysaccharide antigens are attached to these structures. This has been postulated to result from 'masking' effects or from antigenic competition (Galanos et al 1977; Ziegler et al 1982). Since the rough mutants of Enterobacteriaceae do not attach O-polysaccharide chains to the LPS core, they have been useful for the production of high-titered antisera to core polysaccharide structures. The concept that antibodies to rough mutant bacteria might provide significant protection against a variety of smooth gram-negative bacteria was explored (and rejected) as early as the 1920s (Arkwright 1927; Ledingham 1926).

It is important to stress that core polysaccharide antibodies cannot be obtained in high titers by injection of conventional smooth endotoxins. This point is emphasized so as to avoid simple extrapolations of late-phase tolerance mechanisms activated by rough mutant antigens to those activated by smooth endotoxins. Studies of the degree of protection against parental or unrelated smooth endotoxins afforded by immunization with rough mutants are summarized in Table 4.

384

In experiments involving active immunization of rabbits, Milner (1973) found that whereas endotoxins from smooth and rough (Re chemotype) strains of *S. typhimurium* both conferred early-phase pyrogenic tolerance to smooth endotoxin, only animals immunized with the parental smooth endotoxin exhibited late-phase pyrogenic tolerance. Furthermore, despite passive transfer of large volumes of hyperimmune antisera prepared against a rough (Re) *S. minnesota* (strain 1167), Milner could not demonstrate pyrogenic tolerance to the endotoxin from the parental smooth organism.

Studies by Greisman et al (1973) also examined the ability of rough mutant antisera to reduce the fever-producing activity of smooth endotoxins. Rabbit antisera prepared against a rough mutant of *S. typhimurium* (Rd chemotype) and a UDP-galactose epimeraseless mutant (J5) of *E. coli* O111 (Rc chemotype) were found to confer slight, but statistically significant, tolerance in rabbits to the pyrogenic activities of the endotoxins from heterologous smooth strains. Such protection, however, was far less than that provided by O-specific antisera.

Two laboratories, in particular, have focused their studies on the therapeutic potential of rough mutant antibodies. Braude's laboratory has reported that rabbit antiserum prepared against a galactose-deficient rough mutant of *E. coli* O113 protected mice against lethal doses of endotoxin from the parent smooth *E. coli* strain as effectively as homologous O-specific antiserum (Davis et al 1969; Tate et al 1966). McCabe's laboratory has examined the passive protection against endotoxic lethality provided by antisera to the Re 595 mutant of *S. minnesota* and the J5 (Rc) mutant of *E. coli* (McCabe et al 1977). Both antisera conferred protection when mice were challenged with endotoxins from *S. typhi* or *S. minnesota*, although such protection was not as marked as that produced by O-specific antisera. Interpretation of this study is complicated by the fact that all animals were concomitantly treated with actinomycin D which markedly augments the susceptibility of mice to endotoxin, so that the test doses of endotoxin employed were markedly reduced. Actinomycin D also appears to alter the mechanisms of endotoxin lethality (Brown and Morrison 1982).

Recently, Johns et al (1983), of McCabe's laboratory, have extended this work both with and without actinomycin D treatment. They again observed protection by anti-Re 595 and anti-J5 rabbit antisera against lethality produced by both *S. typhosa* and *S. minnesota* smooth endotoxins in mice, although as previously this protection was not as marked as the protection seen with O-specific antisera. This appeared to be truly related to antibodies to the rough mutants, since absorption of the anti-Re antisera with red blood cells coated with smooth endotoxins did not remove the protective activity for these endotoxins, whereas red blood cells coated with Re LPS did. It is not immediately apparent, however, why antibodies unable to bind to smooth

TABLE 4   *Studies on tolerance to parental or unrelated smooth LPS induced by rough LPS*

Feature studied	Tolerance to parental smooth LPS			Tolerance to unrelated smooth LPS			Reference
	none	minimal	marked	none	minimal	marked	
A. *Early tolerance*							
Pyrogenicity			+			+	Kim and Watson 1967
Pyrogenicity			+				Milner 1973
Hypotension			+				Abdelnoor et al 1981
B. *Late tolerance*							
Pyrogenicity	+						Milner 1973
Pyrogenicity					+		Ralovich et al 1974
C. *Passive tolerance*							
Lethality			+				Tate et al 1966
Lethality			+				Kim and Watson 1967
Lethality			+				Davis et al 1969
Shwartzman reactivity			+			+	Braude and Douglas 1972
Shwartzman reactivity			+				Braude et al 1973
Pyrogenicity					+		Greisman et al 1973
Pyrogenicity	+						Milner 1973
Shwartzman reactivity		+					Ng et al 1974
Pyrogenicity				+			Ralovich et al 1974
Lethality			+			+	McCabe et al 1977
Shwartzman reactivity						+	Davis et al 1978
Lethality			+			+	Johns et al 1983
Lethality***	+			+			Johnston and Greisman 1984

* In these studies, tolerance was induced by single or repetitive injections of endotoxin, and tolerance testing was performed within 1–2 days of the last injection.

** In these studies, tolerance was induced by single or repetitive injections of endotoxin, and tolerance testing was performed at least 5 days after the last injection.

*** Antisera with elevated O-specific titers to the parental smooth LPS were excluded.

LPS in vitro would do so in vivo. It is not apparent either why antisera to the Re 595 and the J5 mutants protected almost equally effectively. There is serologic evidence that rough mutant antisera are not uniformly reactive with core antigens of differing species. Ruschmann found, as reported by Lüderitz et al (1968), that antisera to a range of rough *S. minnesota mutants* (Ra to Re) did not cross-react with a series of *E. coli* rough mutants, including a UDP-galactose epimeraseless mutant of *E. coli* O111, analogous to the J5 mutant. On the other hand, a study of a series of other *E. coli* rough endotoxins found that cross-reactivity did occur with antisera to *S. minnesota* rough mutants, but only with LPS of similar chemotype (Schmidt et al 1970). Hemagglutination inhibition studies of a variety of antisera to *Salmonella* rough mutants have confirmed high levels of rough specificities (Lindberg and Holme 1968). In vitro data would suggest, then, that the serological specificity of rough mutant antisera is sharply restricted to the chemotype of the immunizing LPS.

The present authors have attempted to confirm the ability of rough mutant antisera to confer protection against lethal quantities of heterologous smooth endotoxins (Johnston and Greisman 1984). Rabbit antisera prepared to the J5 (Rc) *E. coli* O111 mutant and the Re 595 *S. minnesota* mutant, possessing HA titers to the purified glycolipids of 1:640 to 1:2560, have been tested in mice challenged with smooth endotoxins from *E. coli* O111B4 and O127B8, *S. minnesota*, and *S. typhimurium*. Whereas homologous O-specific antisera were uniformly protective, protection was not observed with the rough mutant antisera. These findings are consistent with the minimal protection provided by the rough mutant antisera against the thousandfold lower test quantities of smooth endotoxins used in the pyrogenic tolerance studies, as discussed previously.

The explanation for the discrepancy in the ability of rough mutant antisera to transfer broad spectrum protection against endotoxin lethality is not readily apparent. It is possible that differences in methods of preparation of the bacterial vaccines or of the challenge endotoxins are important. Recent investigations by Karch et al (1983) have shown that the physical state of endotoxin significantly influences the isotype and even the subclass of specific antibody induced. Johns et al (1983) employed acetone-killed rough mutants for immunization and tested with endotoxins from phenol-water or phenol-chloroform-petroleum ether extracts, which were electrodialyzed, scraped from membranes, and solubilized in triethylamine salts. Johnston and Greisman (1984), on the other hand, employed boiled rough mutants for immunization and conventional Boivin-extracted endotoxins for tolerance testing.

One additional factor should be stressed. Johnston and Greisman (1984) excluded rough mutant antisera that exhibited increments in O-antibodies to the parental smooth endotoxins. Johns et al (1983), however, claimed that rough mutant antisera possessing significant titers of parental O-antibodies were no more protective against parental smooth LPS than those without these antibodies, and therefore such antisera were not excluded. This is inconsistent with their observations that O-specific antisera were consistently more protective than rough mutant antisera.

Several investigators have detected significant titers of O-specific antibodies to the parental smooth LPS in sera from animals immunized with rough organisms (Greisman et al 1973; Johns et al 1977; Kenny and Herzberg 1968; Schlecht and Westphal 1979; Tate et al 1966). Michael and Mallah (1981), studying smooth and rough (Rd_2 and Re) chemotypes of *S. minnesota*, have demonstrated that '... immunization with rough mutants produces bactericidal antibodies directed against the closely related (smooth) parental strain but not against smooth strains of unrelated bacterial species'. A biochemical basis can be found in the observations that *Salmonella* and *Shigella* rough mutants defective in the synthesis of core polysaccharide (*rfa* mutants) produce O-polysaccharide chains which cannot be attached to the defective LPS core (Beckman et al 1964; Johnston et al 1967; Lüderitz et al 1968). Presumably, then, in some manner the O-polysaccharide haptens, which are not expressed on the surface of the intact living *rfa* mutant, become immunogenic during the preparation of the killed bacterial vaccine. Therefore, in studies using such vaccines, the possibility must be considered that protection against smooth LPS from the parental strain may be related to the production of O-specific antibodies. Furthermore, even with use of rough mutant vaccines that do not contain O-specific haptens, rises in O-specific antibodies secondary to the polyclonal B-cell stimulating activity of the lipid A must be considered in analysis of protective activity. Impressive increments in antibodies to both parental and nonparental smooth LPS have been shown with the use of the J5 *E. coli* mutant, which, because of its lack of UDP-galactose epimerase, is unable to synthesize O-polysaccharide in the absence of exogenous galactose (Greisman et al 1973; Ziegler et al 1973a).

In conclusion, the ability of rough mutant antisera to confer pyrogenic tolerance to endotoxins from heterologous smooth gram-negative bacteria is detectable, though slight compared with O-specific antisera. Since minute quantities of endotoxins are used for pyrogenic testing, a role for haptenic and/or polyclonally induced O-specific antibodies cannot be excluded. The ability of rough mutants to induce core-specific antibodies that provide broad spectrum protection against lethal quantities of smooth endotoxins is presently controversial.

3.3. Lipid A antibodies

Since most of the toxic effects of endotoxin have been shown to derive from the lipid A portion of the molecule, one might predict that antibodies directed against this portion would be protective. The remarkable finding has been that by themselves, such antibodies did not protect against fever (Rietschel and Galanos 1977) or lethality (McCabe et al 1977). However, if the animal was given an injection of lipid A prior to the lipid A antiserum, marked pyrogenic tolerance was then observed to a subsequent injection of endotoxin (Rietschel and Galanos 1977). The interval between lipid A pre-treatment and subsequent endotoxin challenge was 48 hours, at which time the early-phase tolerance mechanisms had waned. Both the preparative and challenge injections were lipid A-specific. Thus, by this method, tolerance could be produced to endotoxins of diverse species.

The mechanism whereby a preparatory injection of lipid A allows expression of the protection by subsequent lipid A antiserum is not clear. Rietschel and Galanos conclude that '... besides specific humoral factors, other factors, perhaps cellular, are involved', and suggest that the mechanisms underlying the phenomenon they have observed may also mediate the early refractory state. Some problems with attributing a major role to lipid A antibodies are:

a.   The immunogenicity of lipid A as it exists in the native LPS molecule is very weak. This seems to be due to the presence of the polysaccharide moiety, which masks the lipid A and suppresses its immunogenicity (Galanos et al 1977). Significant rises in lipid A antibodies have not been detected in the sera of rabbits after hyperimmunization with smooth *S. typhi* (Walker and Beasley 1980) or the Re595 *S. minnesota* mutant (Johns et al 1977).

b.   Mice differ from many animal species in that they do not possess naturally occurring lipid A antibodies and do not respond readily to lipid A immunization, even using lipid A-coated bacterial antigens (Galanos et al 1977). Yet endotoxin tolerance to hypothermia (as mentioned, mice develop hypothermia rather than fever after endotoxin) and lethality can be readily induced in mice (Greer and Rietschel 1978a, 1978b).

c.   Pyrogenic tolerance to endotoxin can be induced by daily or continuous intravenous injections of endotoxin in immunosuppressed animals and man almost as readily as in normal antibody producers (Good and Varco 1955; Good and Zak 1956; Greisman et al 1975; Wolff et al 1965b; Zaldivar and Scher 1979).

3.4. Nonspecific serum factors

Early investigations of the effect of serum on the activities of endotoxin concluded that pyrogenicity was augmented by brief (30 min) incubation with serum (Cluff and Bennett 1957; Farr and Lequire 1950). Subsequent studies have found either no effect (Kimball and Wolff 1967), or a reduction in endotoxin pyrogenicity (Fig. 7). A number of serum factors have been hypothesized to abrogate the toxic effects of endotoxin, such as binding to

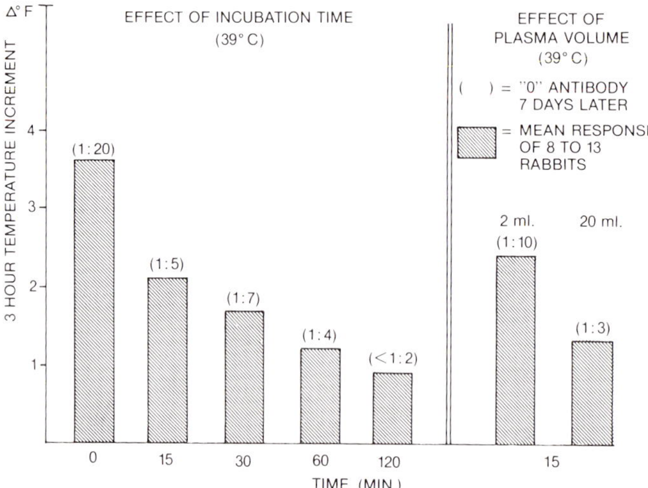

Fig. 7   *Reduction of the pyrogenicity and O-immunogenicity of E. coli 0127:B8 en-dotoxin by incubation with fresh normal, heparinized rabbit plasma. In the left panel, the endotoxin (0.5 μg/kg) was mixed with 10 ml aliquots of pooled rabbit plasma, incubated for the times shown at 39°C, and injected intravenously into normal rabbits. In the right panel, endotoxin (0.5 μg/kg) was mixed with either 2 or 20 ml aliquots of the same pooled rabbit plasma. Antibody titers were determined by a bentonite floccu-lation method.*

high-density lipoproteins (Ulevitch and Johnston 1978), degradation by a circulating esterase (Skarnes 1970) or lipase (Gupta and Reed 1968), interaction with the complement system (Johnson and Ward 1972), and inactivation by an alpha globulin (Johnson et al 1977).

Moreau and Skarnes (1973) have postulated a major role for intravascular detoxification of endotoxin by a heat-stable esterase in the serum of tolerant animals. These observations have not yet been confirmed, however, and it remains unclear what role, if any, such mechanisms play in vivo in tolerance to endotoxin.

4. SUMMARY

The characteristics of acquired tolerance to two of the toxic effects of bacterial endotoxins, fever and lethality, have been reviewed. At least two distinct mechanisms of tolerance to these effects can be delineated. An early phase of pyrogenic and lethal tolerance develops within hours of endotoxin injection which requires the presence only of the lipid A moiety for its induction.

The mechanism appears to involve an acquired refractoriness of macrophages of the RES to the release of biologically active mediators, in response to the lipid A component of endotoxin. Continued administration of endotoxin at closely spaced intervals augments this refractory state, presumably in part by induction of enhanced RES detoxification mechanisms. If the early phase of tolerance is allowed to subside by discontinuing the injections of endotoxin, a later phase of pyrogenic and lethal tolerance can be discerned, which exhibits specificity for the O-polysaccharide component of endotoxin. This late-phase tolerance, which is not as potent as that which can be achieved with the early-phase mechanisms, is mediated by O-specific antibodies. The mechanism whereby such O-specific antibodies confer tolerance is not yet clear.

Both early- and late-phase mechanisms are activated during tolerance induced by repetitive daily endotoxin injections. Such tolerance is therefore broad spectrum (i.e., lipid A-specific), yet demonstrably O-specific when studied by passive transfer.

There is currently no evidence that circulating antibodies to core polysaccharide or lipid A play any major role in endotoxin tolerance induced by smooth endotoxins. The potential of such antibodies, which can be induced in high titer only by immunization with rough mutants or purified lipid A, as well as the potential of humoral nonantibody mechanisms, to mediate broad spectrum pyrogenic tolerance to smooth endotoxins when passively transferred to normal animals, appears minimal. Their potential to mediate broad spectrum tolerance to lethality from smooth endotoxins currently remains controversial.

REFERENCES

Abernathy RS (1957) Homologous and heterologous resistance in mice given bacterial endotoxins. *J. Immunol. 78*, 387-394.

Abdelnoor AM, Johnson AG, Anderson-Imbert A, Nowotny A (1981) Immunization against bacteria- and endotoxin-induced hypotension. *Infect. Immun. 32*, 1093-1099.

Agarwal MK, Parant M, Parant F (1972) Role of spleen in endotoxin poisoning and reticuloendothelial function. *Br. J. Exp. Pathol. 53*, 485.

Arkwright AK (1927) The value of different kinds of antigen in prophylactic (enteric) vaccines. *J. Pathol. Bacteriol. 30*, 345-364.

Atkins E (1960) Pathogenesis of fever. *Physiol. Rev. 40*, 580-646.

Atkins E, Wood WB Jr (1955) Studies on the pathogenesis of fever. II. Identification of an endogenous pyrogen in the blood stream following the injection of typhoid vaccine. *J. Exp. Med. 102*, 499-516.

Atkins E, Bodel P, Francis L (1967) Release of an endogenous pyrogen *in vitro* from rabbit mononuclear cells. *J. Exp. Med. 126*, 357-384.

Beckmann I, Subbaiah TV, Stocker BAD (1964) Rough mutants of *Salmonella typhimurium*. (2) Serological and chemical investigations. *Nature (London) 201*, 1299-1301.

Beeson PB (1946) Development of tolerance to typhoid bacterial pyrogen and its abolition by reticuloendothelial blockade. *Proc. Soc. Exp. Biol. Med. 61*, 248-250.

Beeson PB (1947a) Tolerance to bacterial pyrogens. I. Factors influencing its development. *J. Exp. Med. 86*, 29-38.

Beeson PB (1947b) Tolerance to bacterial pyrogens. II. Role of the reticuloendothelial system. *J. Exp. Med. 86*, 39-44.

Beeson PB (1948) Temperature-elevating effect of a substance obtained from polymorphonuclear leucocytes (Abstract). *J. Clin. Invest. 27*, 524.

Benacerraf B, Thorbecke GJ, Jacoby D (1959) Effect of zymosan on endotoxin toxicity in mice. *Proc. Soc. Exp. Biol. Med. 100*, 796-799.

Bennett IL Jr (1956) Pyrogens. In: Springer GF (Ed), *Polysaccharides in Biology*, p 132, Josiah Macy Jr Foundation, New York.

Berendt MJ, Newborg MF, North RJ (1980) Increased toxicity of endotoxin for tumor bearing mice and mice responding to bacterial pathogens: macrophage activation as a common denominator. *Infect. Immun. 28*, 645-647.

Bodel P, Atkins E (1967) Release of endogenous pyrogen by human monocytes. *N. Engl. J. Med. 276*, 1002-1008.

Boivin A, Mesrobeanu L (1938) Recherches sur les antigènes somatiques et sur les endotoxines des bactéries. IV. Sur l'action anti-endotoxique de l'anticorps O. *Rev. Immunol. 4*, 40-52.

Braude AI, Douglas H (1972) Passive immunization against the local Shwartzman reaction. *J. Immunol. 108*, 505-512.

Braude AI, Carey FJ, Zalesky M (1955) Studies with radioactive endotoxin. II. Correlation of physiologic effects with distribution of radioactivity in rabbits injected with lethal doses of *E. coli* endotoxin labelled with radioactive sodium chromate. *J. Clin. Invest. 34*, 858-866.

Braude AI, Zalesky M, Douglas H (1958) The mechanism of tolerance to fever (Abstract). *J. Clin. Invest. 37*, 880-881.

Braude AI, Douglas H, Davis CE (1973) Treatment and prevention of intravascular coagulation with antiserum to endotoxin. *J. Infect. Dis. 128*, S157-S164.

Brown DE, Morrison DC (1982) Possible alteration of normal mechanisms of endotoxin toxicity *in vivo* by actinomycin D. *J. Infect. Dis. 146*, 746-750.

Centanni E (1894) Untersuchungen über das Infectionsfieber – das Fiebergift der Bacterien. *Dtsch. Med. Wochenschr. 20*, 148-150.

Centanni E (1942) Immunitätserscheinungen im experimentellen Fieber mit besonderer Berücksichtigung des pyrogenen Stoffes aus Typhusbazillen. *Klin. Wochenschr. 21*, 664.

Chedid L, Parant F, Parant M, Boyer F (1966) Localization and fate of ^{51}Cr-labeled somatic antigens of smooth and rough Salmonellae. *Ann. N.Y. Acad. Sci. 133*, 712-726.

Cluff LE, (1953) Studies of the effect of bacterial endotoxins on rabbit leukocytes. II. Development of acquired resistance. *J. Exp. Med. 98*, 349-364.

Cluff LE, Bennett IL Jr (1951) Acquired resistance to the Shwartzman phenomenon. *Proc. Soc. Exp. Biol. Med. 77*, 461-4.

Cluff LE, Bennett IL Jr (1957) Factors influencing the alteration of the pyrogenic action of endotoxin by serum. *Bull. Johns Hopkins Hosp. 101*, 281-291.

Cook JA, Dougherty WJ, Holt TM (1980) Enhanced sensitivity to endotoxin induced by the RE stimulant, glucan. *Circ. Shock 7*, 225-238.

Cooper GN, Stuart AE (1961) Sensitivity of mice to bacterial lipopolysaccharide following alteration of activity of the reticuloendothelial system. *Nature (London) 191*, 294-295.

Creech HJ, Hamilton MA, Nishimura ET, Hankwitz RF (1948) The influence of antibody-containing fractions on the lethal and tumor-necrotizing actions of polysaccharides from *Serratia marcescens* (*Bacillus prodigiosus*). *Cancer Res. 8*, 330-336.

Creech HJ, Hankwitz RF, Wharton DRA (1949) Further studies of the immunological properties of polysaccharides from *Serratia marcescens* (*Bacillus prodigiosus*) I. The effects of passive and active immunization on the lethal activity of the polysaccharides. *Cancer Res. 9*, 150-157.

Cremer N, Watson DW (1957) Influence of stress on distribution of endotoxin in RES determined by fluorescein antibody technic. *Proc. Soc. Exp. Biol. Med. 95*, 510-513.

Davis CE, Brown KR, Douglas H, Tate WJ III, Braude AI (1969) Prevention of death from endotoxin with antisera. I. The risk of fatal anaphylaxis to endotoxin. *J. Immunol. 102*, 563-572.

Davis CE, Ziegler EJ, Arnold K (1978) Neutralization of meningococcal endotoxin by antibody to core glycolipid. *J. Exp. Med. 47*, 1007-1017.

Di Luzio NR, Crafton CG (1970) A consideration of the role of the reticuloendothelial system (RES) in endotoxin shock. In: Bertelli A, Beck N (Eds), *Shock: Biochemical, Pharmacological, and Clinical Aspects*, pp 27-57. Plenum Press, New York.

Dinarello CA (1969) *The Role of the Liver in the Production of Fever.* Doctoral thesis, Yale University School of Medicine, New Haven, Connecticut.

Dinarello CA, Bodel PT, Atkins E (1968) The role of the liver in the production of fever and in pyrogenic tolerance. *Trans. Assoc. Am. Physicians 81*, 334-344.

Farr RS, LeQuire VS (1950) Leukocytic and pyrogenic effects of typhoid vaccine and augmentation by homologous plasma. *Proc. Soc. Exp. Biol. Med. 75*, 661-666.

Farrar WE Jr, Eidson M, Kent TH (1968) Susceptibility of rabbits to pyrogenic and lethal effects of endotoxin after acute liver injury. *Proc. Soc. Exp. Biol. Med. 128*, 711-715.

Favorite GO, Morgan HR (1942) Effects produced by the intravenous injection in man of a toxic antigenic material derived from *Eberthella typhosa*: clinical, hematological, chemical and serological studies. *J. Clin. Invest. 21*, 589-599.

Filkins JP (1980) Endotoxin-enhanced secretion of macrophage insulin-like activity. *J. Reticuloend. Soc. 27*, 507-511.

Freedman HH (1959) Passive transfer of protection against lethality of homologous and heterologous endotoxins. *Proc. Soc. Exp. Biol. Med. 102*, 504-506.

Freedman HH (1960a) Passive transfer of tolerance to pyrogenicity of bacterial endotoxin. *J. Exp. Med. 111*, 453-463.

Freedman HH (1960b) Reticuloendothelial system and passive transfer of endotoxin tolerance. *Ann. N.Y. Acad. Sci. 88*, 99-106.

Fukuda T, Murata K (1964) Hepatic pyrogen and cortisone-antipyresis in canine endotoxin fever. *Nature (London) 204*, 690-691.

Fukuda T, Murata K (1965) Species difference in the febrile and the leukocytic response to endotoxin: endogenous hepatic pyrogen in dogs. *Jpn. J. Physiol. 15*, 169-179.

Galanos C, Lüderitz O, Rietschel ET, Westphal O (1977) Newer aspects of the chemistry and biology of bacterial lipopolysaccharides, with special reference to their lipid A component. In: Goodwin TW (Ed) *Biochemistry of Lipids II, International Review of Biochemistry, Vol 14*, pp 239-335.

Glode LM, Mergenhagen SE, Rosenstreich DL (1976) Significant contribution of spleen cells in mediating the lethal effects of endotoxin *in vivo*. *Infect. Immun. 14*, 626-630.

Glode LM, Jacques A, Mergenhagen SE, Rosenstreich DL (1977) Resistance of macrophages from C3H/HeJ mice to the *in vitro* cytotoxic effects of endotoxin. *J. Immunol. 119*, 162-166.

Gollan LM, McDermott J (1979) Total hypothermic blood exchange in acute endotoxin shock. *Resuscitation 7*, 229-236.

Good RA, Varco RL (1955) A clinical and experimental study of agammaglobulinemia. *J. Lancet 75*, 245-271.

Good RA, Zak SJ (1956) Disturbances in gamma globulin synthesis as 'experiments of nature'. *Pediatrics 18*, 109-149.

Goodman JS, Rogers DE, Koenig MG (1969) The hepatic uptake of bacterial endotoxin. I. The influence of humoral and cellular factors. *Proc. Soc. Exp. Biol. Med. 132*, 372-375.

Greer GG, Rietschel ET (1978a) Lipid A-induced tolerance and hyperreactivity to hypothermia in mice. *Infect. Immun. 19*, 357-368.

Greer GG, Rietschel ET (1978b) Inverse relationship between the susceptibility of lipopolysaccharide (lipid A)-pretreated mice to the hypothermic and lethal effect of lipopolysaccharide. *Infect. Immun. 20*, 366-374.

Greisman SE (1983) Induction of endotoxin tolerance. In: Nowotny A (Ed), *Beneficial Effects of Endotoxins*, pp 149-178. Plenum Press, New York.

Greisman SE, DuBuy B (1975) Mechanisms of endotoxin tolerance. IX. Effect of exchange transfusion. *Proc. Soc. Exp. Biol. Med. 148*, 675-678.

Greisman SE, Hornick RB (1973) Mechanisms of endotoxin tolerance with special reference to man. *J. Infect. Dis. 128 (Suppl)*, 265-276.

Greisman SE, Woodward WE (1965) Mechanisms of endotoxin tolerance. III. The refractory state during continuous intravenous infusions of endotoxin. *J. Exp. Med. 121*, 911-933.

Greisman SE, Woodward CL (1970) Mechanisms of endotoxin tolerance. VII. The role of the liver. *J. Immunol. 105*, 1468-1476.

Greisman SE, Young EJ (1969) Mechanisms of endotoxin tolerance. VI. Transfer of the 'anamnestic' tolerant response with primed spleen cells. *J. Immunol. 103*, 1237-1241.

Greisman SE, Carozza FA Jr, Hills JD (1963) Mechanisms of endotoxin tolerance. I. Relationship between tolerance and reticuloendothelial system phagocytic activity in the rabbit. *J. Exp. Med. 117*, 663-674.

Greisman SE, Hornick RB, Woodward TE (1964a) The role of endotoxin during typhoid fever and tularemia in man. III. Hyperreactivity to endotoxin during infection. *J. Clin. Invest. 43*, 1747-1757.

Greisman SE, Wagner HN Jr, Iio M, Hornick RB (1964b) Mechanisms of endotoxin tolerance. II. Relationship between endotoxin tolerance and reticuloendothelial system phagocytic activity in man. *J. Exp. Med. 119*, 241-264.

Greisman SE, Young EJ, Woodward WE (1966) Mechanisms of endotoxin tolerance. IV. Specificity of the pyrogenic refractory state during continuous intravenous infusions of endotoxin. *J. Exp. Med. 124*, 983-1000.

Greisman SE, Hornick RB, Wagner HN Jr, Woodward WE, Woodward TE (1969a) The role of endotoxin during typhoid fever and tularemia in man. IV. The integrity of the endotoxin tolerance mechanisms during infection. *J. Clin. Invest. 48*, 613-629.

Greisman SE, Young EJ, Carozza FA Jr (1969b) Mechanisms of endotoxin tolerance. V. Specificity of the early and late phases of pyrogenic tolerance. *J. Immunol. 103*, 1223-1236.

Greisman SE, Young EJ, DuBuy B (1973) Mechanisms of endotoxin tolerance. VIII. Specificity of serum transfer. *J. Immunol. 111*, 1349-1360.

Greisman SE, Young EK, Workman JB, Ollodart RM, Hornick RB (1975) Mechanisms of endotoxin tolerance: the role of the spleen. *J. Clin. Invest. 56*, 1597-1607.

Greisman SE, DuBuy JB, Woodward CL (1979) Experimental gram-negative bacterial sepsis: prevention of mortality not preventable by antibiotics alone. *Infect. Immun. 25*, 538-557.

Gupta JD, Reed CE (1968) The role of serum lipase in the degradation of *Salmonella enteritidis* endotoxin. *J. Immunol. 101*, 308-316.

Haeseler F, Bodel P, Atkins E (1977) Characteristics of pyrogen production by isolated rabbit Kupffer cells *in vitro*. *J. Reticuloendothel. Soc. 22*, 569-581.

Hahn HH, Char DC, Postel WB, Wood WB Jr (1967) Studies on the pathogenesis of fever XV. The production of endogenous pyrogen by peritoneal macrophages. *J. Exp. Med. 126*, 385-394.

Hanson DF, Murphy PA, Windle BE (1980) Failure of rabbit neutrophils to secrete endogenous pyrogen when stimulated with staphylococci. *J. Exp. Med. 151*, 1360-1371.

Heilman DH (1965) The selective toxicity of endotoxin for phagocytic cells of the reticuloendothelial system. *Int. Arch. Allergy Appl. Immunol. 26*, 63-79.

Herion JC, Walker RI, Palmer JG (1961) Endotoxin fever in granulocytopenic animals. *J. Exp. Med. 113*, 1115-1125.

Heyman A (1945) The treatment of neurosyphilis by continuous infusion of typhoid vaccine. *Vener. Dis. Inform. 26*, 51-57.

Hiernaux JR, Jones JM, Rudbach JA, Rollwagen F, Baker PJ (1983) Antibody response of immunodeficient (xid) CBA/N mice to *Escherichia coli* 0113 lipopolysaccharide, a thymus-independent antigen. *J. Exp. Med. 157*, 1197-1207.

Hornick RB, Greisman SE (1978) On the pathogenesis of typhoid fever. *Arch. Intern. Med. 138*, 357-359.

Howard JG (1961) Increased sensitivity to bacterial endotoxin of F1 hybrid mice undergoing graft-versus-host reaction. *Nature (London) 190*, 1122.

Janoff A, Weissmann G, Zwiefach BW, Thomas L (1962) Pathogenesis of experimental shock. *J. Exp. Med. 116*, 451.

Johns MA, Bruins SC, McCabe WR (1977) Immunization with R mutants of *Salmonella minnesota*. II. Serological response to lipid A and the lipopolysaccharide of Re mutants. *Infect. Immun 17*, 9-15.

Johns M, Skehill A, McCabe WR (1983) Immunization with rough mutants of *Salmonella minnesota*. IV. Protection by antisera to O and rough antigens against endotoxin. *J. Infect. Dis. 147*, 57-67.

Johnson KJ, Ward PA (1972) The requirement for serum complement in the detoxification of bacterial endotoxin. *J. Immunol. 108*, 611.

Johnson KJ, Ward PA, Goralnick S, Osborn BS, Osborn MJ (1977) Isolation from human serum of an inactivator of bacterial lipopolysaccharide. *Am. J. Pathol. 88*, 559.

Johnston CA, Greisman SE (1984) Failure of rough mutant antisera (RMA) to provide broad spectrum protection against lethality from endotoxemia (Abstract). *Clin. Res. 32*, 513A.

Johnston JH, Johnston RJ, Simmons DAR (1967) The immunochemistry of *Shigella flexneri* O-antigens. The biochemical basis of smooth to rough mutation. *Biochem. J. 105*, 79-87.

Karch H, Gmeiner J, Nixdorff K (1983) Alteration of the immunoglobulin G subclass responses in mice to lipopolysaccharide: effects of nonbacterial proteins and bacterial membrane phospholipids or outer membrane proteins of *Proteus mirabilis*. *Infect. Immun. 40*, 157-165.

Kenny K, Herzberg M (1968) Antibody response and protection induced by immunization with smooth and rough strains in experimental Salmonellosis. *J. Bacteriol. 95*, 406-417.

Kim YB, Watson DW (1965) Modification of host responses to bacterial endotoxins. II. Passive transfer of immunity to bacterial endotoxin with fractions containing 19S antibodies. *J. Exp. Med. 121*, 751-759.

Kim YB, Watson DW (1966) Role of antibodies in reactions to gram-negative bacterial endotoxins. *Ann. N.Y. Acad. Sci. 133*, 727.

Kim YB, Watson DW (1967) Biologically active endotoxins from *Salmonella* mutants deficient in O- and R-polysaccharides and heptose. *J. Bacteriol. 94*, 1320-1326.

Kimball HR, Wolff SM (1967) Febrile responses of rabbits to bacterial endotoxin following incubation in homologous serum and plasma. *Proc. Soc. Exp. Biol. Med. 124*, 269-273.

Kovats TG, Vegh P (1967) Shwartzman reaction in endotoxin-resistant rabbits induced by heterologous endotoxin. *Immunology 12*, 445-453.

Ledingham JCG (1926) Some problems of natural immunity and prophylaxis. The Harben Lectures of 1925, Lecture 1. *J. State Med. 34*, 2-25.

Lemperle G (1966) Effect of RES stimulation on endotoxin shock in mice. *Proc. Soc. Exp. Biol. Med. 122*, 1012-1015.

Lindberg AA, Holme T (1968) Immunochemical studies on cell-wall polysaccharide of rough mutants of *Salmonella typhimurium*. *J. Gen. Microbiol. 52*, 55-65.

Lindberg AA, Greisman SE, Svenson SB (1983) Induction of endotoxin tolerance with nonpyrogenic O-antigenic oligosaccharide-protein conjugates. *Infect. Immun. 41*, 888-895.

Lüderitz O, Jann K, Wheat R (1968) Somatic and capsular antigens of gram-negative bacteria. In: Florkin M, Stotz EH (Eds), *Comprehensive Biochemistry Vol 26A: Extracellular and Supporting Structures*, pp 105-228. Elsevier Biomedical Press, Amsterdam-New York.

Maier RV, Ulevitch RJ (1981) The response of isolated rabbit hepatic macrophages (H-MO) to lipopolysaccharide (LPS). *Circ. Shock 8*, 165-181.

Maier RV, Mathison JC, Ulevitch RJ (1981) Interactions of bacterial lipopolysaccharides with tissue macrophages and plasma lipoproteins. In: Majde JA, Person RJ (Eds), *Pathophysiological Effects of Endotoxins at the Cellular Level*, pp 133-155, Alan R. Liss, New York.

McCabe WR, Bruins SC, Craven DE, Johns M (1977) Cross-reactive antigens: their potential for immunization-induced immunity to gram-negative bacteria. *J. Infect. Dis. 136 (Suppl)*, 161-166.

Mergenhagen SE, Jensen SB (1962) Specific induced resistance of mice to endotoxin. *Proc. Soc. Exp. Biol. Med. 110*, 139.

Michael JG, Mallah I (1981) Immune response to parental and rough mutant strains of *Salmonella minnesota*. *Infect. Immun. 33*, 784-787.

Michael JG, Whitby JL, Landy M (1961) Increase in specific bactericidal antibodies after administration of endotoxin. *Nature (London) 191*, 296-297.

Michael JG, Whitby JL, Landy M (1962) Studies on natural antibodies to gram-negative bacteria. *J. Exp. Med. 115*, 131-146.

Michalek SM, Moore RN, McGhee JR, Rosenstreich DL, Mergenhagen SE (1980) The primary role of lymphoreticular cells in the mediation of host responses to bacterial endotoxin. *J. Infect. Dis. 141*, 55-63.

Milner KC (1973) Patterns of tolerance to endotoxin. *J. Infect. Dis. 128* (Suppl), 237-245.

Moore RN, Goodrum KJ, Couch R Jr, Berry LJ (1978) Factors affecting macrophage function: glucocorticoid antagonizing factor. *J. Reticuloend. Soc. 23*, 321-332.

Moreau SC, Skarnes RC (1973) Host resistance to bacterial endotoxemia: mechanisms in endotoxin-tolerant animals. In: Kass H, Wolff SM (Eds), *Bacterial Lipopolysaccharides, the Chemistry, Biology, and Clinical Significance of Endotoxins*, pp 114-125. University of Chicago Press, Chicago.

Morgan HR (1948a) Resistance to the action of the endotoxins of enteric bacilli in man. *J. Clin. Invest. 27*, 706-709.

Morgan HR (1948b) Tolerance to the toxic action of somatic antigens of enteric bacteria. *J. Immunol. 59*, 129-134.

Mulholland JH, Wolff SM, Jackson AL, Landy M (1965) Quantitative studies of febrile tolerance and levels of specific antibody evoked by bacterial endotoxin. *J. Clin. Invest. 44*, 920-928.

Muller-Ruchholtz W, Sprockhoff H (1969) Spezifität der Toleranz bakterieller Pyrogene bei Kaninchen. *Z. Immunitaetsforsch. Allerg. Klin. Immunol. 136*, 50-61.

Nelson MO (1931) An improved method of protein fever treatment in neurosyphilis. *Am. J. Syph. 15*, 185-189.

Ng AK, Chang CM, Chen CH, Nowotny A (1974) Comparison of the chemical structure and biological activities of the glycolipids of *Salmonella minnesota* R595 and *Salmonella typhimurium* SL1102. *Infect. Immun. 10*, 938-947.

Noll HJ, Braude AI (1961) Preparation and biological properties of a chemically modified *Escherichia coli* endotoxin of high immunogenic potency and low toxicity. *J. Clin. Invest. 40*, 1935-1951.

Nowotny A, Radvany R, Neale NE (1965) Neutralization of toxic bacterial O-antigens with O-antibodies while maintaining their stimulus on non-specific resistance. *Life Sci. 4*, 1107-1114.

Ogata S, Kanamori M (1978) Effects of homologous O-antibody on host responses to lipopolysaccharide from *Yersinia enterocolitica*: neutralization of its pyrogenicity. *Microbiol. Immunol. 22*, 485-494.

Ogata S, Yamamoto A, Kanamori M (1975) Effect of antiserum against bacterial endotoxin on febrile response to lipopolysaccharide in host. *Jpn. J. Med. Sci. Biol. 28*, 312-315.

Pabst MJ, Johnston RB (1980) Increased production of superoxide anion by macrophages exposed *in vitro* to muramyl dipeptide or lipopolysaccharide. *J. Exp. Med. 151*, 101-114.

Page AR, Good RA (1957) Studies on cyclic neutropenia. *Am. J. Dis. Child. 94*, 623-661.

Peavy DL, Brandon CL (1980) Macrophages: primary targets for LPS activity. In: Agarwal MK (Ed), *Bacterial Endotoxins and Host Response*, pp 299-309, Elsevier/ North-Holland, Amsterdam-New York.

Perlman E, Goebel WF (1946) Studies on the Flexner group of dysentery bacilli. IV. The serological and toxic properties of the somatic antigens. *J. Exp. Med. 84*, 223-234.

Peter G, Chernow M, Keating MH, Ryff JC, Zinner SH (1982) Limited protective effect of rough mutant antisera in murine *Escherichia coli* bacteremia. *Infection 10*, 228-232.

Radvany R, Neale NL, Nowotny A (1966) Relation of structure to function in bacterial O-antigens. VI. Neutralization of endotoxic O-antigens by homologous O-antibody. *Ann. N.Y. Acad. Sci. 133*, 763-786.

Ralovich B, Lang C (1971) Antipyrogenic effect of immune serum. *Z. Immunitaetsforsch. 141S*, 251-257.

Ralovich B, Lang C (1972) Antipyrogenic effect in actively immunized rabbits. *Acta Phys. Acad. Sci. Hung. 42*, 75-78.

Ralovich B, Emody L, Lang C (1974) Problems of antiendotoxic immunity. *J. Hyg. Epidemiol. Microbiol. Immunol. 18*, 439-446.

Rietschel ET, Galanos C (1977) Lipid A antiserum-mediated protection against lipopolysaccharide and lipid A-induced fever and skin necrosis. *Infect. Immun. 15*, 34-49.

Rietschel ET, Schade U, Jensen H, Wollenweber HW, Lüderitz O, Greisman SE (1982) Bacterial endotoxins: chemical structure, biological activity, and role in septicemia. *Scand. J. Infect. Dis. (Suppl) 31*, 8-21.

Rioux-Darrieulat F, Parant M, Chedid L (1978) Prevention of endotoxin-induced abortion by treatment of mice with antisera. *J. Infect. Dis. 137*, 7-13.

Ritts RE, Young EJ, Arndt WF (1964) Observations on natural antibodies in endotoxin tolerance. In: Landy M, Braun W (Eds), *Bacterial Endotoxins*, pp 311-318. Rutgers University Press, New Brunswick, NJ.

Rosenstreich DL, Vogel SN (1980) Central role of macrophages in the host response to endotoxin. In: Schlessinger D (Ed), *Microbiology 1980*, pp 11-15. American Society for Microbiology, Washington, D.C.

Rosenthal SM, Millican RC, Rust J (1957) A factor in human gamma-globulin preparations active against *Pseudomonas Aeruginosa* infections. *Proc. Soc. Exp. Biol. Med. 94*, 214-217.

Rutenberg SH, Rutenburg AM, Smith EE, Fine J (1965) On the nature of tolerance to endotoxin. *Proc. Soc. Exp. Biol. Med. 118*, 620-623.

Schade U, Rietschel ET (1982) The role of prostaglandins in endotoxic activities. *Klin. Wochenschr. 60*, 743-745.

Schaedler RW, Dubos RJ (1961) The susceptibility of mice to bacterial endotoxins. *J. Exp. Med. 113*, 559-570.

Schlecht S, Westphal O (1979) Protective role of *Salmonella* R mutants in *Salmonella* infection in mice. *Zentralbl. Bakteriol. Parasitenkd. Infektionskr. Hyg. Abt. I Orig. A 245*, 71-88.

Schmidt G, Jann B, Jann K (1970) Immunochemistry of R lipopolysaccharides of *Escherichia coli*. Studies on R mutants with an incomplete core, derived from *E. coli* 08:K 27. *Eur. J. Biochem. 16*, 382-392.

Shear MJ, Perrault A (1944) Chemical treatment of tumors. IX. Reactions of mice with primary subcutaneous tumors to injection of a hemorrhage-producing bacterial polysaccharide. *J. Natl Cancer Inst. 4*, 461-476.

Skarnes RC (1970) Host defense against bacterial endotoxemia: mechanism in normal animals. *J. Exp. Med. 132*, 300-316.

Snyder SL, Eklund SK, Walker RI (1979) Release of beta-glucuronidase from peritoneal macrophages of normal and endotoxin-tolerant mice. *Can. J. Microbiol. 25*, 1245-1251.

Stuart AE, Cooper GN (1962) Susceptibility of mice to bacterial endotoxin after modification of reticuloendothelial function by simple lipids. *J. Pathol. Bacteriol. 83*, 245-254.

Sullivan R, Gans PJ, McCarroll LA (1983) The synthesis and secretion of granulocyte-monocyte colony-stimulating activity (CSA) by isolated monocytes: kinetics of the response to bacterial endotoxin. *J. Immunol. 130*, 800-807.

Suter E, Ullman GE, Hoffman RG (1958) Sensitivity of mice to endotoxin after vaccination with BCG (Bacillus Calmette-Guérin). *Proc. Soc. Exp. Biol. Med. 99*, 167-169.

Tate WJ, Douglas H, Braude AI, Wells WW (1966) Protection against lethality of *E. coli* endotoxin with 'O' antiserum. *Ann. N.Y. Acad. Sci. 133*, 746-762.

Trejo RA, Di Luzio NR (1972) Influence of endotoxin tolerance on detoxification of *Salmonella enteritidis* endotoxin by mouse liver and spleen. *Proc. Soc. Exp. Biol. Med. 141*, 501-505.

Ueda K, Ueda Y, Imaizumi K (1966) Studies on relation between endotoxin hyperreactivity and phagocytic activity of the reticuloendothelial system. *Jpn. J. Med. Sci. Biol. 19*, 127-135,

Ulevitch RJ, Johnston AR (1978) The modification of biophysical and endotoxic properties of bacterial lipopolysaccharides by serum. *J. Clin. Invest. 62*, 1313-1324.

Urbaschek B, Nowotny A (1968) Endotoxin tolerance induced by detoxified endotoxin (endotoxoid). *Proc. Soc. Exp. Biol. Med. 127*, 650-652.

Vogel SN, Mergenhagen SE (1982) Cellular basis of endotoxin susceptibility. In: Genco RJ, Mergenhagen SE (Eds), *Host-Parasite Interactions in Periodontal Diseases*, pp 160-168. American Society for Microbiology, Washington, D.C.

Vogel SN, Moore RN, Sipe JD, Rosenstreich DL (1980) BCG-induced enhancement of endotoxin sensitivity in C3H/HeJ mice. I. *In vivo* studies. *J. Immunol. 124*, 2004-2009.

Vogel SN, Weedon LL, Wahl LM, Rosenstreich DL (1982) BCG-induced enhancement of endotoxin sensitivity in C3H/HeJ mice. II. T cell modulation of macrophage sensitivity to LPS *in vitro. Immunobiol. 160*, 479-493.

Walker RI, Beasley WJ (1980) Evidence that antibodies to polysaccharides alter platelet responses to endotoxin in tolerant rabbits. *Can. J. Microbiol. 26*, 1241-1246.

Watson DW, Kim YB (1963) Modification of host responses to bacterial endotoxins. I. Specificity of pyrogenic tolerance and the role of hypersensitivity in pyrogenicity, lethality, and skin reactivity. *J. Exp. Med. 118*, 425-446.

Westphal O (1963) Discussion Section, Part VII. In: Landy M, Braun W (Eds), *Bacterial Endotoxins*, p 620. Rutgers University Press, New Brunswick, N.J.

Westphal O, Jann K, Himmelspach K (1983) Chemistry and immunochemistry of bacterial lipopolysaccharides as cell wall antigens and endotoxins. *Prog. Allergy 33*, 9-39.

Wharton DRA, Creech HJ (1949) Further studies of the immunological properties of polysaccharides from *Serratia marcescens* (*Bacillus prodigiosus*). II. Nature of the antigenic action and the antibody response in mice. *J. Immunol. 62*, 135-153.

Windle BE, Murphy PA, Cooperman S (1983) Rabbit polymorphonuclear leukocytes do not secrete endogenous pyrogens or interleukin 1 when stimulated by endotoxin, polyinosine:polycytosine, or muramyl dipeptide. *Infect. Immun. 39*, 1142-1146.

Wiznitzer T, Better N, Rachlin W, Atkins N, Frank ED, Fine J (1960) *In vivo* detoxification of endotoxin by the reticuloendothelial system. *J. Exp. Med. 112*, 1157-1166.

Wolff SM, Mulholland JH, Ward SB (1965a) Quantitative aspects of the pyrogenic response of rabbits to endotoxin. *J. Lab. Clin. Med. 65*, 268-275.

Wolff SM, Mulholland JH, Ward SB, Rubenstein M, Mott PD (1965b) Effect of 6-mercaptopurine on endotoxin tolerance. *J. Clin. Invest. 44*, 1402-1409.

Zahl PA, Hutner SH (1944) Protection against the endotoxins of some gram-negative bacteria conferred by immunization with heterologous organisms. *Am. J. Hyg. 39*, 189-196.

Zaldivar NM, Scher I (1979) Endotoxin lethality and tolerance in mice: analysis with the B-lymphocyte-defective CBA/N strain. *Infect. Immun. 24*, 127-131.

Ziegler EJ, Douglas H, Sherman JE, Davis CE, Braude AI (1973a) Treatment of *E. coli* and *Klebsiella* bacteremia in agranulocytic animals with antiserum to a UDP-gal epimerase-deficient mutant. *J. Immunol. 111*, 433-438.

Ziegler EJ, Douglas H, Braude AI (1973b) Human antiserum for prevention of the local Shwartzman reaction and death from bacterial lipopolysaccharides. *J. Clin. Invest. 52*, 3236-3238.

Ziegler EJ, McCutchan JA, Fierer J, Glauser MP, Sadoff JC, Douglas H, Braude AI (1982) Treatment of gram-negative bacteremia and shock with human antiserum to a mutant *Escherichia coli*. *N. Engl. J. Med. 307*, 1225-1230.

Subject index